Feedback Control Systems

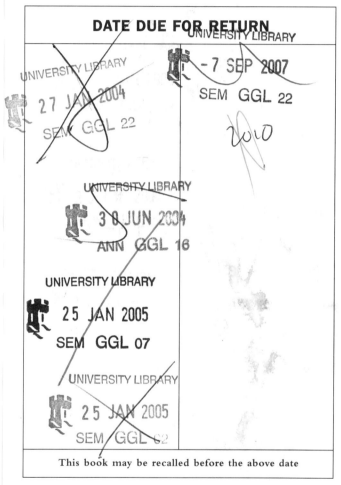
Prentice Hall International, Inc.

Publisher: **TOM ROBBINS**
Editor-in-chief: **MARCIA HORTON**
Acquisitions editor: **ALICE DWORKIN**
Production editor: **IRWIN ZUCKER**
Managing editor: **EILEEN CLARK**
Executive managing editor: **VINCE O'BRIEN**
Manufacturing buyer: **PAT BROWN**
Assistant vice president of production and manufacturing: **DAVID W. RICCARDI**
Cover director: **JAYNE CONTE**

©2000 by Prentice Hall
Prentice-Hall, Inc.
Upper Saddle River, New Jersey 07458

The author and publisher of this book have used their best efforts in preparing this book. These efforts include the development, research, and testing of the theories and programs to determine their effectiveness. The author and publisher make no warranty of any kind, expressed or implied, with regard to these programs or the documentation contained in this book. The author and publisher shall not be liable in any event for incidental or consequential damages in connection with, or arising out of, the furnishing, performance, or use of these programs.

Printed in the United States of America

10 9 8 7 6 5 4 3 2 1

ISBN 0-13-016124-1

Prentice-Hall International (UK) Limited, London
Prentice-Hall of Australia Pty. Limited, Sydney
Prentice-Hall Canada Inc., Toronto
Prentice-Hall Hispanoamericana, S.A., Mexico
Prentice-Hall of India Private Limited, New Delhi
Prentice-Hall of Japan, Inc., Tokyo
Prentice-Hall (Singapore) Pte. Ltd., Singapore
Editora Prentice-Hall do Brasil, Ltda., Rio de Janeiro
Prentice Hall, Upper Saddle River, New Jersey

1001795191

To
Bobby, Ellen, Pat, and Doeie (CLP)

Addigene (RDH)

Contents

Preface

The structure and philosophy of the previous editions of *Feedback Control Systems* remains unchanged in the fourth edition. However, the focus has been sharpened as a result of the experience using the first three editions and the reactions of colleagues who have taught from the book. Some explanations have been enhanced. Where appropriate, a number of examples have been improved. The majority of the end-of-chapter problems have been either altered or replaced.

The simulation program SIMULINK^{TM}, a block-diagram program to be used with MATLAB^{TM}, is introduced to illustrate the simulation of both continuous (analog) and discrete systems, and of nonlinear continuous systems. In addition, the symbolic mathematics of MATLAB is used in verifying the calculations of transforms, the solutions of differential equations, and in other appropriate applications. New capabilities of MATLAB have been added and most examples now contain short MATLAB programs. In this fourth edition, the MATLAB programs given in the examples may be downloaded from the internet site: ftp:// ftp.mathworks.com/pub/books/harbor. Students may then alter the data statements of these programs for the end-of-chapter problems, which eases the debugging problems.

This book is intended to be used primarily as a text for junior- and senior-level students in engineering curricula and for self-study by practicing engineers with little or no experience in control systems. For maximum benefits, the reader should have had some experience in linear-system analysis.

This book is based on material taught at Auburn University and at the University of West Florida and in intensive short courses offered in both the United States and Europe. The practicing engineers who attended these short courses have greatly influenced the material and the descriptions in this book, resulting in more emphasis placed in the practical aspects of analysis and design. In particular, much emphasis on understanding the difference between mathematical models and the physical systems that the models represent. While a true understanding of this difference can be acquired only through experience, the students should understand that a difference does exist.

The material of this book is organized into three principal areas: analog control systems, digital control systems, and nonlinear analog control systems. Chapter 1 in this book presents a brief introduction and an outline of the text. In addition, some control systems are described to introduce the reader to typical applications. Next a short history of feedback control systems is given. The mathematical models of some common components that appear in control systems are developed in Chapter 2.

Chapters 3 through 10 cover the analysis and design of linear analog systems; that is, control systems that contain no sampling. Chapter 2 develops the transfer function model of linear analog systems, and Chapter 3 develops the state-variable model.

Chapter 4 covers typical responses of linear analog systems, including the concept of frequency response. Since many of the characteristics of closed-loop systems cannot be adequately explained without reference to frequency response, this concept is developed early in the book. The authors believe that the frequency-response concept ranks in importance with the time-response concept.

Important control-system characteristics are developed in Chapter 5. Some of the applications of closed-loop systems are evident from these characteristics. The very important concept of system stability is developed in Chapter 6 along with the Routh-Hurwitz stability criterion. Chapter 7 presents analysis and design by root locus procedures, which are basically time-response procedures. The equally important frequency-response analysis and design procedures are presented in Chapters 8 and 9. Chapter 10 is devoted to modern control-system design. Pole-placement design is developed, and the design of state estimators is introduced.

The material of Chapters 3 through 10 applies directly to analog control systems; that of Chapters 11 through 13 applies to digital control systems. Essentially all the analog analysis and design techniques of Chapters 3 through 9 are developed again for digital control systems. These topics include typical responses, characteristics, stability, root-locus analysis and design, and frequency-response analysis and design.

Nonlinear system analysis is presented in Chapter 14. These methods include the describing-function analysis, linearization, and the state-plane analysis.

Three appendices are included. The first appendix reviews matrices. The second appendix reviews the Laplace transform, and the third appendix is a table of Laplace transforms and z-transforms.

Many examples are given, with an effort made to limit each example to illustrating only one concept. It is realized that in using this approach, many obvious and interesting characteristics of the systems of the examples are not mentioned; however, since this is a book for beginning students in feedback control, making the examples more complex would tend to add confusion.

Usually, nonlinear controls are not covered in introductory books in control. However, many of the important characteristics of physical systems cannot be explained on the basis of linear systems. For example, stability as a function of signal amplitude is one of the most common phenomena observed in closed-loop physical systems, and the describing function is included in Chapter 14 to offer an analysis procedure that explains this phenomenon. Lyapunov's first stability theorem is also presented to illustrate some of the pitfalls of linear-system stability analysis.

In general, the material of each chapter is organized such that the more advanced material is placed toward the end of the chapter. This placement is to allow the omission of this material by those instructors who wish to present a less intense course.

This book may be covered in its entirety as a three-hour one-semester course in analog control (Chapters 1 through 9), and a three-hour one-semester course in digital control and nonlinear control with an introduction to modern control (Chapters 10 through 14). The material may also be covered in two-quarter course sequence, with approximately five hours for each course. With the omission of appropriate material, the remaining material may be covered in courses with fewer credits. If a course in digital control is taught without the coverage of the first nine chapters, some of the material of the first nine chapters must be introduced; Chapters 11 through 13 rely on some of this material. A solutions manual containing the solutions to all problems at the ends of the chapters is available for teachers who have adopted the text for use in the classroom.

We wish to acknowledge the many colleagues, graduate students, and staff members of Auburn University and the University of West Florida who have contributed to the development of this book. We are especially indebted to Professor J. David Irwin, head of the department of Electrical Engineering of Auburn University, and to Professor T. F. Elbert, chairman of the Systems Science Department of the University of West Florida, for their aid and encouragement. Thanks to Peter Dorato, University of New Mexico and Ranjare Mukherjee, Michigan State University for their review of the manuscript. Finally, we express our gratitude and love for our families, without whom this undertaking would not have been possible.

The e-mail address for Charles L. Phillips is: phillcp@eng.auburn.edu

CHARLES L. PHILLIPS, *Auburn University*

ROYCE D. HARBOR, *University of West Florida*

1

Introduction

This book is concerned with the analysis and design of closed-loop control systems. In the *analysis* of closed-loop systems, we are given the system, and we wish to determine its characteristics or behavior. In the *design* of closed-loop systems, we specify the desired system characteristics or behavior, and we must configure or synthesize the closed-loop system so that it exhibits these desired qualities.

We define a closed-loop system as one in which certain of the system forcing signals (we call these *inputs*) are determined, at least in part, by certain of the responses of the system (we call these *outputs*). Hence, the system inputs are a function of the system outputs, and the system outputs are a function of the system inputs. A diagram representing the functional relationships in a closed-loop system is given in Figure 1.1.

An example of a closed-loop system is the temperature-control system in the home. For this system we wish to maintain, automatically, the temperature of the living space in the home at a desired value. To control any physical variable, which we usually call a *signal,* we must know the value of this variable; that is, we must measure this variable. We call the system for the measurement of a variable a *sensor,* as indicated in Figure 1.2. In a home temperature-control system, the sensor is a thermostat, which indicates a low temperature by closing an electrical switch and an acceptable temperature by opening the same switch.

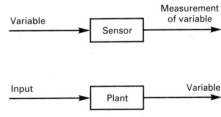

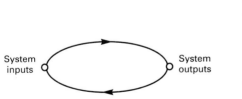

Figure 1.1   Closed-loop system.   **Figure 1.2**   Control-system components.

1

We define the *plant* of a control system as that part of the system to be controlled. It is assumed in this example that the temperature is increased by activating a gas furnace. Hence the plant input is the electrical signal that activates the furnace, and the plant output signal is the actual temperature of the living area. The plant is represented as shown in Figure 1.2. In the home-heating system, the output of each of the systems is connected to the input of the other, to form the closed loop. However, in most closed-loop control systems, it is necessary to connect a third system into the loop to obtain satisfactory characteristics for the total system. This additional system is called a *compensator,* a *controller,* or simply a *filter.*

The usual form of a single-loop closed-loop control system is given in Figure 1.3. The system input is a reference signal; usually we want the system output to be equal to this input. In the home temperature-control system, this input is the setting of the thermostat. If we want to change the temperature, we change the system input. The system output is measured by the sensor, and this measured value is compared with (subtracted from) the input. This difference signal is called the *error signal,* or simply the error. If the output is equal to the input, this error signal is zero, and the plant output remains at its current value. If the error is not zero, in a properly designed system the error signal causes a response in the plant such that the magnitude of the error is reduced. The compensator is a filter for the error signal, since usually satisfactory operation does not occur if the error signal is applied directly to the plant.

Control systems are sometimes divided into two classes. If the object of the control system is to maintain a physical variable at some constant value in the presence of disturbances, we call this system a *regulator.* One example of a regulator control system is the speed-control system on the ac power generators of power utility companies. The purpose of this control system is to maintain the speed of the generators at the constant value that results in the generated voltage having a frequency of 60 Hz in the presence of varying electrical power loads. Another example of a regulator control system is the biological system that maintains the temperature of the human body at approximately 98.6°F in an environment that usually has a different temperature.

The second class of control systems is the *servomechanism.* Although this term was originally applied to a system that controlled a mechanical position or motion, it is now often used to describe a control system in which a physical variable is required to follow, or track, some desired time function. An example of this type of system is an automatic aircraft landing system, in which the aircraft follows a ramp to the desired touchdown point. A second example is the control systems of a robot, in which the robot hand is made to follow some desired path in space.

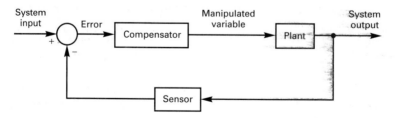

Figure 1.3   Closed-loop control system.

The preceding is a very simplified discussion of a closed-loop control system. The remainder of this book improves upon this description. In order to perform either mathematical analysis or design, it is necessary that we have a mathematical relationship between the input and the output for each of the blocks in the control system of Figure 1.3. The purpose of Chapter 2 is to develop these functional relationships for some common physical systems. Chapter 3 presents a different method of expressing these functional relationships.

We examine typical responses that occur in control systems in Chapter 4 and look at control system specifications in Chapter 5. Chapter 6 presents concepts and some analysis techniques for system stability. The root locus, one of the principal methods of analysis and design, is developed in Chapter 7. Chapters 8 and 9 present a second principal analysis and design method, the frequency response. Chapter 10 presents an introduction to a different method of design of control systems, which is classified as a modern control procedure.

In Chapters 2 through 9 it is assumed that no system signals appear in sampled form and in particular that no digital computers are used in the control of the system. The systems considered in these chapters are called *analog* systems, *continuous-data* systems, or *continuous-time* systems. Chapters 11 through 13 consider systems in which sampling does occur, and these systems are called *sampled-data* systems. If a digital computer is used in the control of these systems, the systems are then called *digital* control systems. *Discrete* systems is also used to refer to sampled-data systems and digital control systems. The terms *discrete-time* systems or simply *discrete* systems are also used to refer to sampled-data systems and digital control systems.

In the systems of Chapters 2 through 13, all systems are assumed to be linear (linearity is defined in Chapter 2). However, physical systems are not linear, and in general, nonlinear systems are difficult to analyze or design. Throughout this book we discuss the problems of the inaccurate representations in the functional relationships that we use to model physical systems. However, for some physical systems, the linear model is not sufficiently accurate, and nonlinearities must be added to the system model to improve the accuracy of these functional relationships. We consider some common nonlinearities and some properties and analysis methods for nonlinear systems in Chapter 14.

In the analysis of linear systems, we use the Laplace transform for analog systems and the z-transform for discrete systems. Appendix B presents the concepts and procedures of the Laplace transform, and the z-transform is covered in Chapter 11. Next, the control problem is presented, and then some control systems are discussed.

1.1 THE CONTROL PROBLEM

We may state the control problem as follows. A physical system or process is to be accurately controlled through closed-loop, or feedback, operation. An output variable, called the response, is adjusted as required by the error signal. This error signal is the difference between the system response, as measured by a sensor, and the reference signal, which represents the desired system response.

Generally, a controller, or compensator, is required to filter the error signal in order that certain control criteria, or specifications, be satisfied. These criteria may involve, but not be limited to

1. Disturbance rejection
2. Steady-state errors
3. Transient response characteristics
4. Sensitivity to parameter changes in the plant

Solving in control problem generally involves

1. Choosing sensors to measure the plant output
2. Choosing actuators to drive the plant
3. Developing the plant, actuator, and sensor equations (models)
4. Designing the controller based on the models developed and the control criteria
5. Evaluating the design analytically, by simulation, and finally, by testing the physical system
6. If the physical tests are unsatisfactory, iterating these steps

Because of inaccuracies in the models, the first tests on the physical control system are usually not satisfactory. The controls engineer must then iterate this design procedure, using all tools available, to improve the system. Generally, intuition, developed while experimenting with the physical system, plays an important part in the design process.

NASA space shuttle launch. (Courtesy of NASA.)

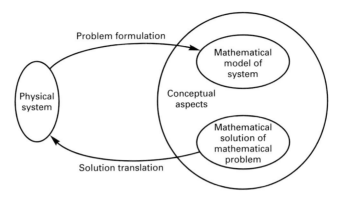

Figure 1.4 Mathematical solution of physical problems.

The relationship of mathematical analysis and design to physical-system design procedures is depicted in Figure 1.4 [1]. In this book, all phases shown in the figure are discussed, but the emphasis is necessarily on the conceptual part of the procedures—the application of mathematical concepts to mathematical models. In practical design situations, however, the major difficulties are in formulating the problem mathematically and in translating the mathematical solution back to the physical world. As stated earlier, many iterations of the procedures shown in Figure 1.4 are usually required in practical situations.

Depending on the system and the experience of the designer, some of the steps listed earlier may be omitted. In particular, many control systems are implemented by choosing standard forms of controllers and experimentally determining the parameters of the controller by following a specified step-by-step procedure with the physical system; no mathematical models are developed. This type of procedure works very well for certain control systems. For other systems, however, it does not. For example, a control system for a space vehicle obviously cannot be designed in this manner; the control system must respond in a satisfactory manner the first time that it is activated.

In this book mathematical procedures are developed for the analysis and design of control systems. The actual techniques may or may not be of value in the design of a particular control system. However, standard controllers are utilized in the developments in this book. Thus the analytical procedures develop the concepts of control system design and indicate the application of each of the standard controllers.

1.2 EXAMPLES OF CONTROL SYSTEMS

In this section some physical control systems are described, to acquaint the reader with these types of systems. In addition, the various physical components that make up control systems will become more evident.

1.2.1 Aircraft Landing Systems

The first example of a closed-loop control system is the case of a pilot landing an aircraft. The pilot has three basic tasks. First the aircraft must approach the airfield along the extended

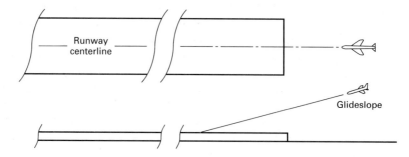

Figure 1.5   Landing an aircraft.

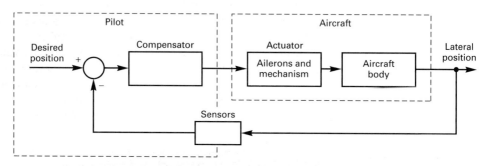

Figure 1.6   Aircraft landing system.

centerline of the runway. This approach is depicted in Figure 1.5. Also shown in this figure is the second task of maintaining the aircraft on the correct glideslope. The third task is that of maintaining the correct speed.

We will consider only the problem of keeping the aircraft on the extended centerline of the runway (controlling the aircraft's lateral position). A block diagram of the system, of the type of Figure 1.3, is given in Figure 1.6. It is assumed that the lateral position of the aircraft is controlled by the ailerons, the control surfaces near the tip of each wing. Actually, the rudder is also used to control the lateral position, but for simplicity we will consider only the ailerons. The ailerons, along with the mechanisms for varying their angle, form the actuator of the control system.

For this control system, a number of sensors are used by the pilot to determine the lateral position of the aircraft. The pilot has certain instrument readings available in the cockpit. In addition, usually the pilot can see the runway, which indicates the position of the aircraft relative to the runway. Hence the pilot has indications of the lateral position of the aircraft and also knows the desired lateral position. He/she manipulates the ailerons in an effort to bring the aircraft to the desired lateral position. For a control system, we usually call the difference between the desired position and the actual position the *system error*.

The remaining block in Figure 1.6 is the compensator. For this system the compensation function is performed by the pilot, who also performs part of the sensing function and the manipulation of the control surfaces. Given the position and the attitude of the aircraft, the pilot manipulates the ailerons in a certain manner to drive the perceived system error to

zero. The manner in which the pilot drives the error to zero is determined by intensive training of the pilot. For example, the ailerons would be manipulated in a certain manner to correct errors for a large passenger airliner. For a small single-engine plane, the ailerons would
be manipulated in a different manner. Hence we can say that the compensation function is
performed in the pilot's brain and is the result of intensive training, such that the same compensation is performed under the same circumstances for each landing.

Suppose now that we consider an automatic aircraft landing system, in which the
landing is independent of the pilot. For example, large passenger aircraft can be landed
automatically at many airports. A much more complex control system is the one that lands
aircraft automatically in U.S. Navy aircraft carriers [2].

The block diagram of Figure 1.6 also applies to automatic landing systems. However,
the ailerons are manipulated by the bank autopilot (which is also a closed-loop control system). The sensor is usually a radar that determines the position of the aircraft relative to the
extended centerline of the runway. The compensation function is accomplished in a digital
computer. As discussed in Chapter 11, the computer iteratively solves a set of equations to
determine the commands to be sent to the bank autopilot of the aircraft. For earlier versions
of the automatic landing of aircraft on U.S. Navy aircraft carriers, the compensator is of a
standard form (called a PID compensator) that will be studied in much detail in this book.
This is probably the most commonly used type of compensator in feedback control systems.

Two-crew member flight deck of the Boeing 757. The digital electronics include an
automatic flight control system. (Courtesy of Boeing Commercial Airplane Group.)

As an additional point, most activities of humans involve the human as a part of a closed-loop control systems. When we drive cars, we are constantly sensing the position and direction of travel of the car. If we decide that these are not satisfactory, we take action to correct them; that is, actions are taken to correct sensed error in the system. A second simple example is the task of sketching. We continuously observe the position of the pencil point, attempting to minimize the difference between the point's actual position and its desired position. [If you doubt that sketching is a form of feedback control, try sketching with your eyes closed (that is, open loop or no feedback).] Another example is that of catching a ball, which involves a closed-loop control of the position of the hands. We have all seen the effect of opening the loop (looking away) before the ball arrives at the hands.

As we have already mentioned, an example of a biological control system is the temperature-control system of the human body. This control system attempts to maintain the body temperature at a constant value. In general, the environment tends to vary the body temperature from the desired value. The body responds to a sensed error in temperature by perspiring, by increasing or decreasing blood flow, by shivering, and so on. This control system has one characteristic that control systems designed by humans do not usually have: it usually operates in a satisfactory manner for more than 70 years. Another characteristic of the human temperature control system *is* usually present in control systems that we design. If the magnitude of the error becomes too large, the system fails.

1.2.2 Chamber Temperature-Control System

The second control system that we describe in this section is the temperature-control system for an environmental test chamber [3]. This chamber is used by scientists to study the effects of temperature on the growth of plants. The system hardware is represented by the block of diagram of Figure 1.7. The temperature of the chamber is sensed by a thermistor, which is a temperature-sensitive resistor. Hence the temperature is sensed by measuring the resis-

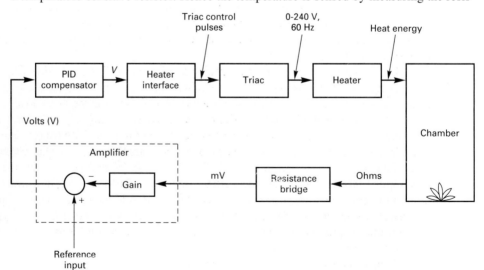

Figure 1.7 Chamber control system hardware diagram.

tance of the thermistor. For accuracy, this resistance is measured using a Wheatstone (resistance) bridge [4]. The signal output of the resistance bridge is of the order of millivolts; hence a voltage amplifier is required to amplify this voltage to a more usable level. Many transducers have outputs of very low voltage; thus sensor systems often include voltage amplifiers.

The desired temperature of the chamber is represented by a voltage also. The difference between the desired temperature (represented by a voltage) and the actual temperature (also represented by a voltage) is the error signal. An air conditioner attempts to maintain the temperature of the chamber at a value lower than the desired temperature. The voltage applied to the electric heater in Figure 1.7 is then used to increase the temperature to the desired value. Hence an increase in voltage increases the temperature, and a decrease in voltage decreases the temperature. The heater voltage is controlled by a triac, which is a solid-state electronic switch. The triac is controlled by the triac control hardware, which converts the error signal into timing pulses for the triac.

The compensator for this control system is a PID compensator, as is the case for the automatic aircraft landing system described earlier. In the physical system, the PID compensation is realized in a digital computer, as described in Chapter 13. A PID compensator can also be realized with a circuit of resistors, capacitors, and operational amplifiers, as described in Chapter 9.

1.3 SHORT HISTORY OF CONTROL

In this section a short history of automatic feedback control is presented [5–7]. The term *automatic* implies that the control is performed without human intervention. The term *feedback* seems to have been used first in 1920 by personnel of Bell Telephone Laboratories.

The earliest known feedback devices include a water clock from the second century B.C. Time was measured by measuring the water dropping at a constant rate from a reservoir through an orifice. To assure a constant rate of water drops, it was necessary to maintain a constant level of water in the reservoir; some type of automatic control was needed to accomplish this. It is interesting to note that we still design control systems for maintaining a constant liquid level in a vessel.

One version of the water clock is depicted in Figure 1.8. The float value in the upper vessel, which is both the sensor and actuator of the control system, maintains the water level of this vessel constant. The valve closes when the water level is at the desired value. If the water level is less than the desired value, the valve is open, causing the water level to rise. Note the similarity of this control to that currently used in most flush toilets. The water drops in the clock are measured by accumulating the water in a vessel that has a scale. The level of the water is an accurate indication of the time lapse since the last time that the vessel was emptied. Note that in fact the accumulation of water in the lower vessel is a form of integration (summation). This same procedure (summation) is sometimes used to perform numerical integration on a digital computer. This procedure was used 22 centuries ago and is still useful on present-day digital computers. It is very difficult for us to devise useful mathematical procedures that are new.

Cornelis Drebbel (1572–1633), a Dutch mechanic and chemist, invented a temperature regulator to maintain a constant temperature in a chamber, which is thought to have been used both in chicken incubators and in a general furnace for his chemical experimentation. The regulator was based on a device that allowed hot air to escape the chamber, once the desired temperature was reached. This regulator can be compared to the current home-heating system, in which hot air is circulated in the home once the temperature falls below the desired value.

An American, William Henry (1729–1786) of Lancaster, Pennsylvania, invented a temperature regulator that utilized an automatically manipulated flue damper to control combustion and hence temperature. The temperature sensor and actuator were based on pressure exerted by the expansion of heated air. The expanding air tended to close the damper, which decreased the rate of combustion, and contracting air tended to open the damper.

Advances relating to float-valve regulators for steam boilers were made by many, including James Brindley in 1758, Sutton Thomas Wood in 1784, and the Russian I. I. Polzunov in 1763. Advances in pressure regulators for steam engines were made by Denis Papin (1674–1712), by Robert Delap in 1799, and by Matthew Murray in 1799.

A major invention in the speed control of both windmills and steam engines was the centrifugal governor of Thomas Mead in 1787. Matthew Boulton and James Watt invented the centrifugal governor of the type of Figure 1.9 in 1788. An increase in the speed of the steam engine moved the balls out by centrifugal force, which, as a consequence, moved the mechanism up. This movement caused a valve in the steam line to reduce the flow of steam, which then decreased the speed of the steam engine. A decrease in the speed of the steam engine increased the flow of steam, which in turn increased the speed of the engine.

The devices just described worked approximately as described; however, in many cases the regulator would not maintain the system variable at a constant value, but instead would permit a small oscillation about the desired value. This effect was often called "hunt-

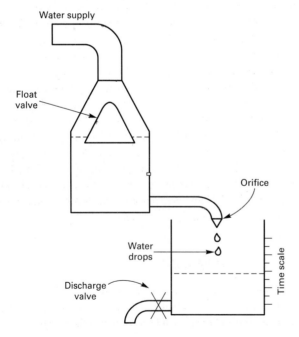

Figure 1.8   Simplified water clock.

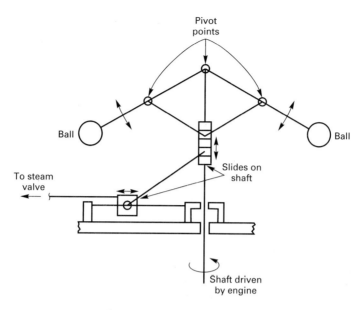

Figure 1.9   Steam engine governor.

ing." To eliminate this instability, it was necessary to mathematically model the physical devices; that is, mathematical equations were needed whose solutions described the operation of the devices. Such needs led to present-day mathematical modeling, analysis, and design.

The contributors to the mathematics used in modeling, analysis, and design of control systems are numerous. Pierre Simon Laplace (1749–1827) devised the Laplace transform, the basis of most analysis and design procedures for control systems. Others are Sir Isaac Newton (1642–1727), mathematical modeling and analysis; Brook Taylor, mathematical analysis (the Taylor's series); James Clerk Maxwell (1831–1879), mathematical modeling and analysis; Edward John Routh (1831–1907), the Routh criterion; Oliver Heaviside (1850–1925), mathematical analysis; Charles P. Steinmetz (1865–1923), frequency-response analysis using complex numbers; Harry Nyquist (1889–1976), the Nyquist criterion; Hendrik W. Bode (1905–1982), the Bode diagram; Harold S. Black (1898–1981), negative-feedback amplifiers; W. R. Evans, the root locus; and John von Neumann (1903–1957), basic operation of a digital computer. This list is by no means exhaustive. It does not necessarily include the most important contributors, nor the most important contributions of those listed. It is intended to give the reader an idea of the contributors in the area of automatic control and the time periods in which they worked. A list containing present-day contributors would be much too long to include, and any such list would certainly be debatable.

REFERENCES

1. W. A. Gardner. *Introduction to Random Processes.* New York: Macmillan, 1986.
2. R. F. Wigginton. "Evaluation of OPS-II Operational Program for the Automatic Carrier Landing System." Naval Electronic Systems Test and Evaluation Facility, Saint Inigoes, MD, 1971.

3. R. E. Wheeler. "A Digital Control System for Plant Growth Chambers," M.S. thesis, Auburn University, Auburn, AL, 1980.

4. J. D. Irwin. *Basic Engineering Circuit Analysis,* 5th ed. Upper Saddle River, NJ: Prentice Hall, 1996.

5. M. E. El-Hawary. *Control System Engineering.* Reston, VA: Reston Publishing Company, 1984.

6. W. A. Blackwell and L. L. Grigsby. *Introductory Network Theory.* Boston: PWS Engineering, 1985.

7. "Centennial Hall of Fame," *IEEE Spectrum,* April 1984.

2

Models of Physical Systems

In this chapter the topic of obtaining the mathematical models of physical systems is considered. By the term *mathematical model* we mean the mathematical relationships that relate the output of a system to its input. Perhaps one of the simplest models of a physical system is that known as Ohm's law (which more accurately might be called Ohm's model), which applies to the electrical phenomenon resistance. The model is given as

$$v(t) = i(t)R \tag{2-1}$$

In this equation, $v(t)$ is the voltage in volts, $i(t)$ is the current in amperes, and R is the resistance in ohms. If the resistance were connected across a known voltage source, the voltage would be the system input and the current, the system output (or response).

In this chapter we develop models for some common physical systems. For the remainder of this book, these models will be used to illustrate the analysis and design of control systems. We use models to *predict* how the system will respond under certain specified conditions, without actually testing the physical system.

We used the term *linear system* in Chapter 1. We now define this term. *A system is linear if superposition applies.* For example, suppose that the response of a system for an input $r_1(t)$ is $c_1(t)$ and for an input $r_2(t)$ is $c_2(t)$. If the system is linear, the response of the system for the input $[k_1 r_1(t) + k_2 r_2(t)]$ is $[k_1 c_1(t) + k_2 c_2(t)]$, where k_1 and k_2 are arbitrary constants.

2.1 SYSTEM MODELING

In the IEEE *Dictionary* [1] the *mathematical model* of a system is defined as a set of equations used to represent a physical system. *It should be understood that no mathematical model of a physical system is exact.* We may be able to increase the accuracy of a model by increasing the complexity of the equations, but we never achieve exactness. We generally strive to develop a model that is adequate for the problem at hand without making the model overly complex. It has been stated that the development of the models of the physical systems involved is from 80 to 90 percent of the effort required in control system analysis and design.

In this chapter the laws of physics are used to develop the models of some simple physical systems. These models are used later to illustrate analysis and design techniques for control systems. We consider only systems that are described by *linear time-invariant differential equations.* As is shown in Appendix B, these equations may be solved using the Laplace transform; but even more important, a system described by linear time-invariant differential equations may be represented by a *transfer function,* and all the input–output characteristics of the equations may be determined from this transfer function. Hence, to the extent that the equations accurately model the physical system, the transfer function yields the input–output characteristics of the physical system.

A very important point must be made concerning the preceding discussion. In analysis and design as covered in this book, we always work with the *mathematical model* of the physical system involved. We do not want to confuse the model with the actual physical system. The model may or may not accurately represent the actual characteristics of the physical system. The model may accurately represent the physical system for a certain specified input, but may be quite inaccurate for a different specified input. This point is illustrated by an example.

Example 2.1

Suppose that we have a common 1-Ω 2-W carbon composition resistor. Hence we are considering a physical system. If we apply a constant (dc) voltage of 1 V to the resistor, the mathematical model tells us that a current of 1 A will flow, from (2-1). If we physically connect the resistor across a 1-V dc power supply, approximately 1 A will flow through the resistor, depending on the

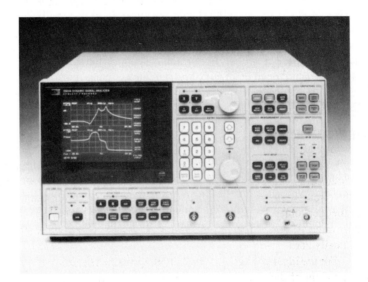

A dynamic signal analyzer which, as one of many functions, calculates the transfer function of a system from input–output measurements. (Courtesy of Hewlett-Packard Company.)

true resistance of the resistor, the characteristics of the power supply, and so forth. Since the power dissipated in a resistance is given by

$$p(t) = \frac{v^2(t)}{R} \qquad (2\text{-}2)$$

approximately 1 W will be dissipated in the resistor. Now suppose we go through the same experiment with a 10-V source. The mathematical model tells us that 10 A will flow through the resistance, and that 100 W will be dissipated in the resistance. However, since the physical resistor can dissipate only 2 W, the resistor would fail if connected to a 10-V power supply, resulting in no current. Or, depending on the characteristics of the physical power supply, a fuse might be blown. In any event, the current will not be approximately 10 A as predicted by the model. Thus the characteristics of the 1-Ω resistor can change, depending on the input signal (voltage) applied to the device.

This simple example illustrates that the model of a physical system is dependent on the input signal, among other things. The model can also be a function of temperature, humidity, speed (for instance, aircraft), and so on. A model of a particular physical system may be adequate under a given set of circumstances but be inadequate under other circumstances. Generally, only testing the physical system will answer the questions of adequacy. This problem leads to some very anxious moments in the initial flights of aircraft, space vehicles, and the like. Unfortunately, it also leads to some catastrophes.

The preceding discussion is intended to emphasize the difference between a physical system and *a* model of that system. Since a physical system does not have a unique model, we cannot talk about *the* model of a particular physical system. Recall that we can always increase the accuracy of a model by increasing its complexity, but we can never achieve exactness.

Now that we have emphasized the difference between the physical system and a model of that system, we will use the common approach to system analysis and design and be very careless in our language in discussing systems. In general, in this book we use the term *system* when we mean *mathematical model*. When we discuss a physical system, we refer to it by the term *physical system*.

2.2 ELECTRICAL CIRCUITS

In this section we develop models for some simple, useful electrical circuits [2]. The model of resistance was given in (2-1). This model is also given in Figure 2.1, along with the models of inductance and capacitance. For the short circuit, $v(t)=0$, and $i(t)$ is determined by the circuit connected across the terminals. For the open circuit, $i(t)=0$, and $v(t)$ is determined by the circuit across the terminals. Also defined in Figure 2.1 are the ideal voltage source and the ideal current source. The dashed lines indicate that the elements shown are parts of larger circuits. For example, the resistance must be a part of a larger circuit or else $v(t)$ is identically zero.

For the ideal voltage source, the voltage at the terminals of the source is as specified, independent of the circuit connected across these terminals. The current, $i(t)$, that flows through the source is determined by the circuit connected across the terminals. For the ideal current source, the current that flows through the current source is as specified, independent of the circuit connected across the source. The voltage, $v(t)$, that appears across

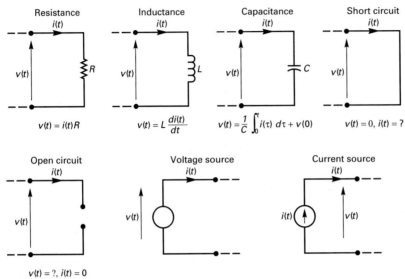

Figure 2.1   Electrical circuit elements.

the terminals of the current source is determined by the circuit connected across these terminals.

Consider a circuit that is an interconnection of the elements shown in Figure 2.1. The circuit equations are written using the models given in the figure along with Kirchhoff's voltage and current laws. Kirchhoff's voltage law may be stated as:

The algebraic sum of voltages around any closed loop in an electrical circuit is zero.

Kirchhoff's current law may be stated as:

The algebraic sum of currents into any junction in an electrical circuit is zero.

Simple examples of writing circuit equations are given next.

Example 2.2

As a first example, consider the circuit of Figure 2.2. In this circuit we consider $v_1(t)$ to be the circuit input and $v_2(t)$ to be the circuit outputHence we wish to write a set of equations such that the solution of these equations will yield $v_2(t)$ as a function of $v_1(t)$ or, equivalently in the Laplace transform variable, $V_2(s)$ as a function of $V_1(s)$. By Kirchhoff's voltage law, assuming zero initial condition on the capacitor,

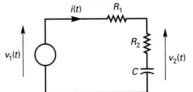

Figure 2.2   Circuit for Example 2.2.

$$R_1 i(t) + R_2 i(t) + \frac{1}{C} \int_0^t i(\tau)\, d\tau = v_1(t)$$

$$R_2 i(t) + \frac{1}{C} \int_0^t i(\tau)\, d\tau = v_2(t)$$

Hence these two equations form the mathematical model of the circuit in Figure 2.2.

Also, we can represent the model in terms of the Laplace transform variable s. Taking the Laplace transform of the last two equations yields

$$R_1 I(s) + R_2 I(s) + \frac{1}{sC} I(s) = V_1(s)$$

$$R_2 I(s) + \frac{1}{sC} I(s) = V_2(s)$$

Thus the circuit may be modeled by the two differential equations or, equivalently, by the two equations in the Laplace transform variable.

Consider a system with input $e(t)$ and output $c(t)$. If $C(s)$ is the Laplace transform of the system output and $E(s)$ is the Laplace transform of the system input, the transfer function $G(s)$ is *defined* by

$$G(s) = \frac{C(s)}{E(s)} \tag{2-3}$$

A transfer function can be written only for the case in which the system model is a linear time-invariant differential equation and the system initial conditions are ignored. Example 2.3 derives the transfer function of the circuit of Example 2.2.

Example 2.3

From the first Laplace-transformed equation,

$$R_1 I(s) + R_2 I(s) + \frac{1}{sC} I(s) = V_1(s)$$

we solve for $I(s)$:

$$I(s) = \frac{V_1(s)}{R_1 + R_2 + (1/sC)}$$

From the second Laplace-transformed equation,

$$R_2 I(s) + \frac{1}{sC} I(s) = V_2(s)$$

we substitute the value of $I(s)$ found earlier:

$$V_2(s) = \frac{R_2 + (1/sC)}{R_1 + R_2 + (1/sC)} V_1(s)$$

Rearranging this equation yields the transfer function $G(s)$.

$$G(s) = \frac{V_2(s)}{V_1(s)} = \frac{R_2 Cs + 1}{(R_1 + R_2)\, Cs + 1}$$

Hence for the circuit of Figure 2.2, the differential equations of Example 2.2 (or the Laplace transform of these equations) or the transfer function of this example are equally valid descriptions. As just seen, we can easily go from one form of the equations to a different form.

The preceding examples illustrate the development of the mathematical model of an electrical circuit. In physical control systems one additional circuit element is usually present—the *operational amplifier* (usually called the *op amp*) [3]. The op amp is normally used in sensor circuits to amplify weak signals and is also used in compensation circuits, as discussed in Section 7.13. The symbol used in circuit diagrams for an op amp is shown in Figure 2.3(a). The input labeled with the minus sign is called the *inverting input,* and the one labeled with the plus sign is called the *noninverting input.* The power supply connections are labeled V^+ for the positive voltage and V^- for the negative voltage. The op amp is normally shown as in Figure 2.3(b), without the power supply connections. In this figure v_d is the input voltage and v_o is the output voltage. The amplifier is designed and constructed such that the input impedance is very high, resulting in i^- and i^+ being very small. Additionally the amplifier gain is very large (to the order of 10^5 and larger), resulting in a very small allowable input voltage if the amplifier is to operate in its linear range. As an additional point, Kirchhoff's current law does not apply to the circuit of Figure 2.3(b), since two connections have been omitted.

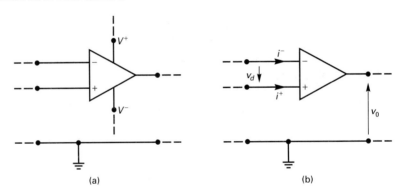

Figure 2.3 Operational amplifier.

For analysis purposes, we assume that the amplifier is ideal. The ideal op amp has zero input currents; that is, in Figure 2.3b), i^- and i^+ are each equal to zero. Additionally, the ideal amplifier operates in its linear range with infinite gain, resulting in the input voltage v_d being equal to zero. For most applications the ideal amplifier model is adequate.

Since the op amp is a very high gain device, feedback must be added to the amplifier in order to stabilize it. The feedback circuit is connected from the output terminal to the inverting input terminal; that is, the minus terminal. This connection results in negative, or stabilizing, feedback.

An example of a practical op-amp circuit is shown in Figure 2.4. This circuit is used to convert a current i_1 to a voltage v_o. Since i^- is zero, the input current i_1 must flow through R, resulting in

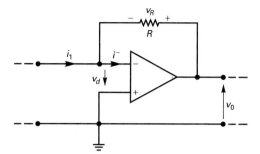

Figure 2.4 Current-to-voltage converter.

$$v_R = -i_1 R$$

Since v_d is also equal to zero, summing voltages around the outer loop through the ground connections yields

$$v_o = v_R = -i_1 R$$

Hence it is seen that the output voltage of the circuit is a constant (R) times the input current, with sign inversion.

Another very useful op-amp circuit is analyzed in the following example.

Example 2.4

Consider the circuit of Figure 2.5 As before, we assume v_d and i^- to be zero. Then the current i_1 flows through the $R_2 C$ circuit and, assuming zero initial conditions, the following equations may be written:

$$v_i = i_1 R_1$$

$$v_0 = -i_1 R_2 - \frac{1}{C} \int_0^t i_1(\tau)\, d\tau$$

The Laplace transform of these equations yields

$$V_i(s) = I_1(s) R_1$$

$$V_0(s) = -\left(R_2 + \frac{1}{sC}\right) I_1(s)$$

Solving the first equation for $I_1(s)$ and substituting the result into the second equation yields

$$\frac{V_0(s)}{V_i(s)} = -\left(\frac{R_2}{R_1} + \frac{1}{sR_1 C}\right) = -\frac{R_2 Cs + 1}{R_1 Cs} = G(s)$$

This relationship is, of course, the circuit transfer function if $V_i(s)$ is defined as the input and $V_o(s)$ is the output. Note that if R_2 is set to zero, the transfer function is that of an integrator with a gain of $-1/R_1 C$; that is.

$$G(s) = \left(\frac{-1}{R_1 C}\right) \frac{1}{s}$$

If, instead, the capacitor is replaced with a short circuit ($C \rightarrow \infty$), the circuit becomes an amplifier with a gain of $-R_2/R_1$. Both of these circuits are basic building blocks of an analog computer, which is described in Section 3.8.

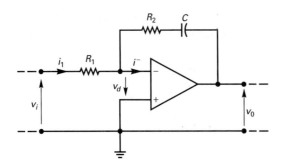

Figure 2.5   Circuit for Example 2.4.

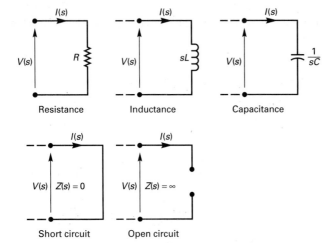

Figure 2.6   Impedance approach.

As a final topic in this section, we cover the circuit analysis technique that is called the *impedance approach.* It should be noted that this approach is valid only for the case that initial conditions are ignored. Recall that ignoring initial conditions is a requirement for developing a transfer function; in the developments and examples that follow, initial conditions are ignored unless otherwise stated.

Consider the basic circuit elements, shown again in Figure 2.6 The impedance approach is illustrated using the inductance, which is described by

$$v(t) = L\frac{di(t)}{dt} \tag{2-4}$$

The Laplace transform of this equation yields

$$V(s) = LsI(s) \tag{2-5}$$

which is of the form

$$V(s) = Z(s)I(s) \tag{2-6}$$

where, by definition, $Z(s)$ is the impedance of the element. In general, the impedance of a passive circuit (no sources in the circuit) is the ratio of the Laplace transform of the voltage across the circuit to the Laplace transform of the current through the circuit. The impedance is defined with respect to two specified terminals. The impedance of a resistance, an inductance, a capacitance, a short circuit, and an open circuit are shown in Figure 2.6. For a resistance,

$$Z(s) = R \tag{2-7}$$

and for a capacitance,

$$Z(s) = \frac{1}{sC} \tag{2-8}$$

The short-circuit impedance $Z(s)$ is zero, and the open-circuit impedance $Z(s)$ is infinite. For a circuit, impedances in series add. For impedances in parallel, the reciprocal of the equivalent impedance is equal to the sum of the reciprocals of the individual impedances. An example is given next.

Example 2.5

For the circuit of Figure 2.7, the equivalent impedance of the parallel RC circuit, $Z_p(s)$, is

$$\frac{1}{Z_p(s)} = \frac{1}{R} + \frac{1}{1/sC}$$

Hence,

$$Z_p(s) = \frac{R}{RCs + 1}$$

and the total impedance is given by

$$Z(s) = \frac{V(s)}{I(s)} = sL + Z_p(s) = \frac{RLCs^2 + Ls + R}{RCs + 1}$$

Note that the impedance approach simplifies the derivation of transfer functions for electrical circuits.

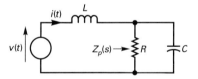

Figure 2.7 Circuit for Example 2.5.

The impedance approach is particularly useful in deriving the transfer function of operational amplifier circuits used as controllers. For example, consider the circuit of Figure 2.8. Note that this is the same circuit given in Example 2.4. We can write the circuit equations

$$V_i(s) = R_1 I_1(s)$$

and

$$V_o(s) = -\left(R_2 + \frac{1}{sC}\right)I_1(s)$$

Taking the ratio of the two equations yields

$$G(s) = \frac{V_o(s)}{V_i(s)} = -\frac{R_2Cs + 1}{R_1Cs}$$

which is the same transfer function derived by a much longer procedure in Example 2.4.

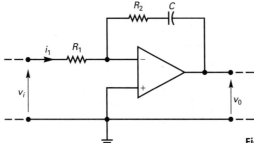

Figure 2.8   Example for impedance approach.

It is convenient at this point to derive a transfer function for a circuit that contains a single op amp, as shown in Figure 2.9. In this figure, $Z_i(s)$ is the input impedance and $Z_f(s)$ is the feedback impedance. The equations of the circuit are

$$V_i(s) = Z_i(s)I(s)$$

and

$$V_o(s) = -Z_f(s)I(s)$$

The ratio of the two equations yields the transfer function:

$$G(s) = \frac{V_o(s)}{V_i(s)} = -\frac{Z_f(s)}{Z_i(s)} \qquad (2\text{-}9)$$

This circuit is very useful in the implementation of compensators in control systems, as is shown in Section 7.13.

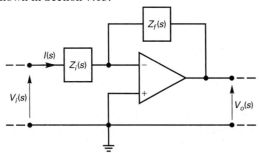

Figure 2.9   Inverting amplifier with generalized impedances.

2.3 BLOCK DIAGRAMS AND SIGNAL FLOW GRAPHS

In the preceding section the concept of a transfer function for a linear time-invariant system was presented. By definition the transfer function is the ratio of the Laplace transform of the

E(s) $\rightarrow$ [G(s)] $\rightarrow$ C(s)

(a)

E(s) •—— G(s) ——• C(s)

(b)

Figure 2.10 Block diagram and signal flow graph elements.

output variable to the Laplace transform of the input variable. Let $E(s)$ be the (Laplace transform of the) input variable, $C(s)$ be the output variable, and $G(s)$ be the transfer function. One method of graphically denoting the transfer function relationship

$$C(s) = G(s)E(s) \qquad (2\text{-}10)$$

is through a block diagram, as shown in Figure 2.10(a). For the block shown, the output is equal by definition to the transfer function given in the block multiplied by the input. The input and the output are defined by the directions of the arrowheads, as shown.

A signal flow graph is also used to denote graphically the transfer function relationship. The signal flow graph that represents (2-10) is given in Figure 2.10(b). Each signal is represented by a *node* in the signal flow graph, as shown by $E(s)$ and $C(s)$ in the figure. Each transfer function is represented by a *branch,* shown in the figure by the line and arrowhead, with the transfer function written near the arrowhead. Be definition, the signal out of a branch is equal to the transfer function of the branch multiplied by the signal into the branch.

Two very important points are to be made. First, a block diagram and a signal flow graph contain exactly the same information. There is no advantage to one over the other; there is only personal preference. Next, a block diagram (or signal flow graph) is the graphical representation of an equation or a set of equations, since the block diagram is constructed from the equations.

One additional element is required to represent equations by a block diagram. This element is the summing junction, which is illustrated in Figure 2.11 for this equation

$$C(s) = G_1(s)E_1(s) + G_2(s)E_2(s) - G_3(s)E_3(s) \qquad (2\text{-}11)$$

For the block diagram, the summing junction is represented by a circle, as in Figure 2.11(a). By definition, the signal out of the summing junction is equal to the sum of the signals into the junction, with the sign of each component determined by the sign placed near the arrowhead of the component. Note that whereas a summing junction can have any number of inputs, we show only one output.

For the signal flow graph, the function of the summing junction is inherently implemented by a node. A summing junction is represented by branches into a node, as illustrated in Figure 2.11(b). By definition, the signal at a node is equal to the sum of the signals from the branches connected into that node. The block diagram and the signal flow graph are illustrated by an example.

Example 2.6

Suppose that we are given the equations

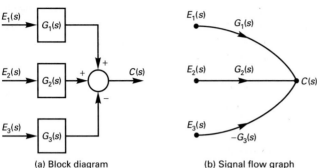

(a) Block diagram   (b) Signal flow graph   **Figure 2.11** Equivalent examples.

$$E(s) = R(s) - H(s)C(s)$$
$$C(s) = G(s)E(s)$$

It is specified that in these equations, $R(s)$ is the input signal, $C(s)$ is the output signal, $E(s)$ is an internal signal, and both $G(s)$ and $H(s)$ are transfer functions. The first equation defines a summing junction, as shown in Figure 2.12(a), and the second equation is used to complete the block diagram. Also shown in this figure is the signal flow graph for these equations. Note that the geometrical constructions of the block diagram and that of the signal flow graph are identical. This will always be the case if the describing equations are used in an identical manner for the construction of each one.

Note that for the preceding example, the equations may be solved for the output $C(s)$ as a function of the input $R(s)$. Substituting the first equation into the second yields

$$C(s) = G(s)R(s) - G(s)H(s)C(s) \qquad (2\text{-}12)$$

Solving this equation for $C(s)$ yields the system transfer function $T(s)$:

$$C(s) = \frac{G(s)}{1 + G(s)H(s)}R(s) = T(s)R(s) \qquad (2\text{-}13)$$

Hence we can represent this system with a single block, as shown in Figure 2.12(c).

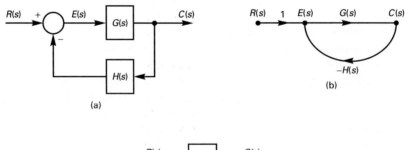

(a)

(b)

(c)

Figure 2.12 Equivalent examples.

The method given above is adequate in solving for the system transfer function provided that the system is described by only two or three equations. However, a better procedure is to use Cramer's rule [4]. This method first requires that the equations be expressed in vector-matrix form. For the last example, the vector-matrix equations may be written as

$$\begin{bmatrix} 1 & H(s) \\ G(s) & -1 \end{bmatrix} \begin{bmatrix} E(s) \\ C(s) \end{bmatrix} = \begin{bmatrix} R(s) \\ 0 \end{bmatrix} \tag{2-14}$$

These equations are of the form

$$\mathbf{Ax} = \mathbf{y} \tag{2-15}$$

where

$$\mathbf{A} = \begin{bmatrix} 1 & H(s) \\ G(s) & -1 \end{bmatrix} \quad \mathbf{x} = \begin{bmatrix} E(s) \\ C(s) \end{bmatrix} \quad \mathbf{y} = \begin{bmatrix} R(s) \\ 0 \end{bmatrix} \tag{2-16}$$

As shown in the review of matrices given in Appendix A, (2-15) can be solved for $\mathbf{x}$ as

$$\mathbf{x} = \mathbf{A}^{-1}\mathbf{y} \tag{2-17}$$

where $\mathbf{A}^{-1}$ denotes the inverse of matrix $\mathbf{A}$. Using this procedure to solve (2-14) results in the solution for both $E(s)$ and $C(s)$. Cramer's rule allows us to solve only for $C(s)$ in a somewhat simpler manner. For (2-14), Cramer's rule results in the solution as a ratio of the determinants

$$C(s) = \frac{\begin{vmatrix} 1 & R(s) \\ G(s) & 0 \end{vmatrix}}{\begin{vmatrix} 1 & H(s) \\ G(s) & -1 \end{vmatrix}} = \frac{G(s)}{1 + G(s)H(s)} R(s) = T(s)R(s) \tag{2-18}$$

Cramer's rule applies to the solution of a vector-matrix equation of the form of (2-15) and is expressed as the ratio of two determinants, as illustrated in (2-18). The denominator determinant is always the determinant of the matrix $\mathbf{A}$. Suppose that we want to solve for the variable in the ith row of the vector $\mathbf{x}$. Then the numerator determinant in the solution is obtained by replacing the ith column of the denominator with the vector $\mathbf{y}$. Note that in (2-18), the numerator determinant is simply the denominator determinant, with the second column replaced with the vector $\mathbf{y}$ of (2-14) and (2-16).

As shown in the preceding circuit examples, and as illustrated later, the models of linear time-invariant systems are linear simultaneous algebraic equations if the Laplace transform is used. As shown earlier, the equations may be solved by eliminating variables, by Cramer's rule, or by inverse matrix procedures. The latter two methods are generally preferred, since errors are generally easier to locate when using either of these methods. A different procedure is presented in the next section. This method, known as Mason's gain formula, is particularly suited to simple systems. However, as is demonstrated, it can be very difficult to find certain errors that may have been made in applying the procedure.

2.4 MASON'S GAIN FORMULA

This section presents a procedure that allows us to find the transfer function, by inspection, of either a block diagram or a signal flow graph. This procedure is called *Mason's gain formula* [5]. However, even though this procedure is relatively simple, it must be used with extreme care, since terms in either the numerator or the denominator of the transfer function can *easily* be overlooked. Furthermore, there is no method available that will give an indication in the case that terms have been overlooked. However, Mason's gain formula can be applied to simple systems with some confidence, after one gains experience.

In this section we give the rules for Mason's gain formula as applied to signal flow graphs. Exactly the same rules apply to block diagrams. First, definitions are given for two of the types of nodes that can appear in a flow graph. The nodes are the *source* (input) nodes and the *sink* (output) nodes.

Source node   A *source node* is a node for which signals flow *only away* from the node; that is, for the branches connected to a source node, the arrowheads are all directed away from the node.

Sink node   A *sink node* is a node for which signals flow *only toward* the node.

Figure 2.13(a) illustrates these definitions. In this figure the node identifying the signal $R(s)$ is a source node, and the right-hand node for $C(s)$ is a sink node. Note that an artifice of adding a branch of unity gain at the original node for $C(s)$ is used in order to create a sink node for $C(s)$. We can do this because the equation for $C(s)$, as portrayed by the signal flow graph, is not changed. In general, however, it is *not* possible to make an arbitrary node into a source node by adding a unity-gain branch, since this alters the mathematical expression for the original node. To illustrate this point, the expression for $E(s)$ in Figure 2.13(a) is

$$E(s) = R(s) - H(s)C(s)$$

If we add a unity-gain branch in an effort to make $E(s)$ into a source node, as shown in Figure 2.13(b), the expression for the node $E(s)$ becomes

$$E(s) = R(s) - H(s)C(s) + E(s)$$

which is clearly a different expression. This example illustrates a fundamental consideration. We may manipulate (change the configuration of) a flow graph or a block diagram, provided that the manipulations do not alter the equations upon which the flow graph or the block diagram is based.

Additional definitions required for the application of Mason's gain formula are as follows.

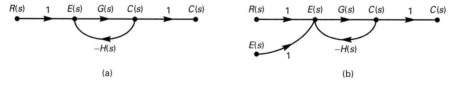

(a)   (b)

Figure 2.13   Signal flow graph.

Path   A *path* is a continuous connection of branches from one node to another with all arrowheads in the same direction; that is, all signals flow in the same direction from the first node to the second one.

Loop   A *loop* is a closed path (with all arrowheads in the same direction) in which no node is encountered more than once. Note that a source node cannot be a part of a loop, since each node in the loop must have at least one branch into the node and at least one branch out.

Forward path   A *forward path* is a path that connects a source node to a sink node, in which no node is encountered more than once.

Path gain   The *path gain* is the product of the transfer functions of all branches that form the path.

Loop gain   The *loop gain* is the product of the transfer functions of all branches that form the loop.

Nontouching   Two loops are *nontouching* if these loops have no nodes in common. A loop and a path are nontouching if they have no nodes in common.

All these definitions are illustrated using the signal flow graph of Figure 2.14. There are two loops in the flow graph, one with a gain of $-G_2 H_1$ and the other with a gain of $-G_4 H_2$. Note that these two loops do not touch. In addition, there are two forward paths connecting the input $R(s)$ and the output $C(s)$. One of these forward paths has a gain of $G_1 G_2 G_3 G_4 G_5$, and the other has a gain of $G_6 G_4 G_5$. Note that the forward path $G_6 G_4 G_5$ does not touch the loop $-G_2 H_1$ but does touch the other loop. The path $G_1 G_2 G_3 G_4 G_5$ touches both loops.

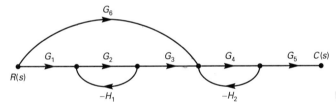

Figure 2.14   Signal flow graph.

Based on the preceding definitions, we may now state Mason's gain formula. The formula gives the transfer function from a source (input) node to a sink (output) node *only* and may be stated as

$$T = \frac{1}{\Delta} \sum_{k=1}^{p} M_k \Delta_k = \frac{1}{\Delta} (M_1 \Delta_1 + M_2 \Delta_2 + \cdots + M_p \Delta_p) \tag{2-19}$$

where T is the gain (transfer function) from the input node to the output node, p is the number of forward paths, and

$\Delta = 1 -$ (sum of all *individual* loop gains)

+ (sum of the products of the loop gains of all possible combinations of nontouching loops taken two at a time)

   − (sum of the products of the loop gains of all possible combinations of nontouching loops taken three at a time)

   + (sum of the products of the loop gains of all possible combinations of nontouching loops taken four at a time)

   − (...)

M_k = path gain of the kth forward path

Δ_k = value of Δ for that part of the flow graph not touching the kth forward path

Some examples are now given to illustrate the use of Mason's gain formula.

Example 2.7

As a first example, consider again the system shown in Figure 2.13(a). This system has a single loop, with a gain of

$$L_1 = -G(s)H(s)$$

and thus

$$\Delta = 1 - L_1 = 1 + G(s)H(s)$$

There is one forward path, with a gain of

$$M_1 = (1)G(s)(1) = G(s)$$

In addition, this forward path touches the only loop; that is, the forward path and the loop have at least one node in common. Hence, no loops remain in that part of the flow graph that does not touch the forward path, and

$$\Delta_1 = 1$$

The transfer function is then, from (2-19),

$$T(s) = \frac{M_1\Delta_1}{\Delta} = \frac{G(s)}{1 + G(s)H(s)}$$

This result checks that of (2-18). A MATLAB program that uses symbolic math to solve for $C(s)$ is given by

```
syms e c r g h
eq1=e-r+h*c;
eq2=c-g*e;
s=solve(eq1,eq2,e,c)
s.c
```

Suppose that in Example 2.7, we wish to express the output $C(s)$ as a function of $E(s)$. Mason's gain formula *cannot* be applied, since $E(s)$ is not a source node. Note also that the desired equation is $C(s) = G(s)E(s)$.

Example 2.8

As a second example of Mason's gain formula, consider again the system of Figure 2.14. Let L_i be the gain of the ith loop. Then the gains of the only two loops can be written as

$$L_1 = -G_2H_1 \qquad L_2 = -G_4H_2$$

Two forward paths are present in Figure 2.14; the forward path gains can be expressed as

$$M_1 = G_1G_2G_3G_4G_5 \qquad M_2 = G_6G_4G_5$$

The value of Δ can be written directly from Figure 2.14:

$$\Delta = 1 - (L_1 + L_2) + L_1L_2 = 1 + G_2H_1 + G_4H_2 + G_2G_4H_1H_2$$

The last term is present in this equation since the two loops do not touch; that is, the loops have no nodes in common.

　　The determination of the Δ_k of (2-19) is more difficult. As just stated, the value of Δ_1 is the value of Δ for that part of the flow graph not touching the first forward path. One method of evaluating Δ_1 is to redraw the flow graph with the first forward path removed. Of course, all nodes of the first forward path must also be removed. Figure 2.15(a) gives the result of removing the first forward path. The second part of the figure illustrates removing the second forward path. Hence Δ_1 is simply the Δ of the flow graph of Figure 2.15(a), and Δ_2 is that of Figure 2.15(b), and we can write

$$\Delta_1 = 1 \qquad \Delta_2 = 1 - (-G_2H_1)$$

since Figure 2.15(a) has no loops and Figure 2.15(b) has one loop. From (2-19) we write the transfer function of the system of Figure 2.14 as

$$T = \frac{M_1\Delta_1 + M_2\Delta_2}{\Delta} = \frac{G_1G_2G_3G_4G_5 + G_6G_4G_5(1 + G_2H_1)}{1 + G_2H_1 + G_4H_2 + G_2G_4H_1H_2}$$

Note that even for this relatively simple flow graph, many terms appear in the transfer function. Hence terms can easily be overlooked. The only methods available to verify the results of Mason's gain formula are algebraic methods such as Cramer's rule. This example is now checked using Cramer's rule.

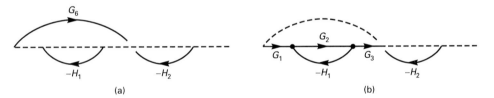

Figure 2.15　Flow graphs with forward path removed.

Example 2.9

In order to verify the transfer function obtained in Example 2.8, it is necessary to write a set of equations for the signal flow graph of Figure 2.14. First, each node must be denoted as a variable as shown in Figure 2.16. The system equations are then

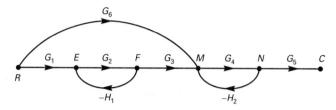

Figure 2.16 Flow graph with all signals labeled.

$$E = G_1R - H_1F \qquad \Rightarrow \qquad E + H_1F = G_1R$$
$$F = G_2E \qquad \Rightarrow \qquad -G_2E + F = 0$$
$$M = G_3F + G_6R - H_2N \Rightarrow -G_3F + M + H_2N = G_6R$$
$$N = G_4M \qquad \Rightarrow \qquad -G_4M + N = 0$$
$$C = G_5N \qquad \Rightarrow \qquad -G_5N + C = 0$$

The second set of equations have the unknown signals on the left side and the known signals on the right side.

These equations may be written in matrix form as

$$
\begin{bmatrix}
1 & H_1 & 0 & 0 & 0 \\
-G_2 & 1 & 0 & 0 & 0 \\
0 & -G_3 & 1 & H_2 & 0 \\
0 & 0 & -G_4 & 1 & 0 \\
0 & 0 & 0 & -G_5 & 1
\end{bmatrix}
\begin{bmatrix}
E \\ F \\ M \\ N \\ C
\end{bmatrix}
=
\begin{bmatrix}
G_1 \\ 0 \\ G_6 \\ 0 \\ 0
\end{bmatrix}
R
$$

which can be expressed as

$$\mathbf{Ax} = \mathbf{y}$$

To apply Cramer's rule we first find the determinant of the matrix $\mathbf{A}$, which is denoted as $\det \mathbf{A}$. The determinant is calculated using the method of minors given in Appendix A. Expanding the determinant of $\mathbf{A}$ about the first column,

$$
\det \mathbf{A} = (1)
\begin{vmatrix}
1 & 0 & 0 & 0 \\
-G_3 & 1 & H_2 & 0 \\
0 & -G_4 & 1 & 0 \\
0 & 0 & -G_5 & 1
\end{vmatrix}
+ G_2
\begin{vmatrix}
H_1 & 0 & 0 & 0 \\
-G_3 & 1 & H_2 & 0 \\
0 & -G_4 & 1 & 0 \\
0 & 0 & -G_5 & 1
\end{vmatrix}
$$

Now each of these determinants may be expanded about its first row, yielding

$$
\det \mathbf{A} =
\begin{vmatrix}
1 & H_2 & 0 \\
-G_4 & 1 & 0 \\
0 & -G_5 & 1
\end{vmatrix}
+ G_2H_1
\begin{vmatrix}
1 & H_2 & 0 \\
-G_4 & 1 & 0 \\
0 & -G_5 & 1
\end{vmatrix}
$$

Each of these determinants may be expanded about the last column.

$$\det \mathbf{A} = \begin{vmatrix} 1 & H_2 \\ -G_4 & 1 \end{vmatrix} + G_2 H_1 \begin{vmatrix} 1 & H_2 \\ -G_4 & 1 \end{vmatrix} = 1 + G_4 H_2 + G_2 H_1 (1 + G_4 H_2)$$

This expression is seen to be the same as that in the denominator of the transfer function of Example 2.8.

The numerator of the transfer function is the value of the determinant formed by replacing the fifth column of the matrix $\mathbf{A}$ by the excitation vector of the matrix equation. This replacement is necessary since C is the fifth variable in the vector of unknowns. Thus the required numerator determinant is

$$\begin{vmatrix} 1 & H_1 & 0 & 0 & G_1 R \\ -G_2 & 1 & 0 & 0 & 0 \\ 0 & -G_3 & 1 & H_2 & G_6 R \\ 0 & 0 & -G_4 & 1 & 0 \\ 0 & 0 & 0 & -G_5 & 0 \end{vmatrix}$$

This determinant is equal to the numerator of the transfer function derived in Example 2.8. It is left as an exercise for the reader to show this. A MATLAB program that solves for C(s) is given by

```
syms   c   e   f   r   m   n   g1   g2   g3   g4   g5   g6   h1   h2
eq1=e-g1*r+h1*f;
eq2=-g2*e+f;
eq3=m-g3*f-g6*r+h2*n;
eq4=n-g4*m;
eq5=c-g5*n;
s=solve (eq1, eq2, eq3, eq4, eq5, e, c, f, m, n)
s.c
```

We see from this example that Mason's gain formula can be considered to be a graphical procedure for evaluating Cramer's rule or for taking the inverse of a matrix.

As stated earlier, Mason's gain formula can be applied directly to block diagrams. In the definitions given above, the following replacements are made, with the term *signal* defined as any input or any output of a block or a summing junction

SIGNAL FLOW GRAPH		**BLOCK DIAGRAM**
input node	$\rightarrow$	input signal
output node	$\rightarrow$	output signal
branch	$\rightarrow$	block
node	$\rightarrow$	signal

Note again that we can identify any internal signal as the output signal.

2.5 MECHANICAL TRANSLATIONAL SYSTEMS

This section presents the models of the elements of linear translational mechanical systems and a method of writing equations for systems that are composed of an interconnection of these elements. The mechanical elements involved are *mass, damping (friction)*, and *elastance (spring)*. The symbols for these elements are given in Figure 2.17. Recall from our previous discussion of electrical circuits that the mathematical models may or may not accurately represent specific physical devices. This same statement applies to the mechanical elements in Figure 2.17. We define these elements by mathematical equations, as we did for electrical elements. Of course, these definitions must have reasonable application to physical systems. Note that motion can occur in only one dimension.

The mechanical linear translational elements are now defined. First consider *mass*. In Figure 2.17 $f(t)$ represents the applied force, $x(t)$ represents the displacement, and M represents the mass. It is assumed that all units are consistent. Then, in accordance with Newton's second law,

$$f(t) = Ma(t) = M\frac{dv(t)}{dt} = M\frac{d^2x(t)}{dt^2} \tag{2-20}$$

where $v(t)$ is velocity and $a(t)$ is acceleration. It is assumed that the mass is rigid; that is, the top connection point cannot move relative to the bottom connection point. Hence the position of the top point is also $x(t)$.

For the remaining two mechanical elements, the top connection point can move relative to the bottom connection point. Hence two displacement variables are required to describe the motion of these elements. First consider *friction*, as shown in Figure 2.17. A physical realization of this phenomenon is the viscous friction associated with oil, air, and so forth. A physical device that is modeled as friction is a shock absorber on an automobile. The mathematical model of friction is given by

$$f(t) = B\left[\frac{dx_1(t)}{dt} - \frac{dx_2(t)}{dt}\right] \tag{2-21}$$

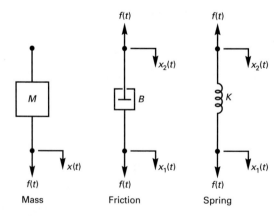

Figure 2.17   Mechanical linear translational elements.

where B is the damping coefficient. Note that the frictional force is directly proportional to the *difference* in the velocities across the element.

The final translational mechanical element to be defined is a *spring*. The defining equation, from Hooke's law, is given by

$$f(t) = K[x_1(t) - x_2(t)] \qquad (2\text{-}22)$$

Note here that the force developed is directly proportional to the *difference* in the displacement of one end of the spring relative to the other. These equations apply for the forces and the displacements in the directions shown by the arrowheads in Figure 2.17. If any of the directions are reversed, the sign on that term in the equations must be changed. In addition, the friction and spring elements have zero mass.

For these mechanical elements, friction dissipates energy but cannot store it. Both mass and a spring can store energy but cannot dissipate it.

We write equations for an interconnection of these mechanical elements using Newton's law, which may be stated as follows:

The sum of the forces on a body is equal to the mass of the body times its acceleration.

The linear translational elements and Newton's law are now illustrated with two examples.

Example 2.10

Consider the simple mechanical system of Figure 2.18. As in all the translational systems that we consider, it is assumed that motion can occur in only one dimension, and this dimension is indicated by $x(t)$ in the figure. No side-to-side motion is allowed. We sum forces on the mass, M. Three forces influence the motion of the mass, namely, the applied force, the frictional force, and the spring force. Hence we can write

$$M\frac{d^2x}{dt^2} = f(t) - B\frac{dx}{dt} - Kx$$

Note that each mechanical element connected to a point must contribute a term to the sum of forces at that point.

A transfer function can be found for this system, with the applied force $f(t)$ as the input and the displacement of the mass $x(t)$ as the output. We can express the Laplace transform of the preceding system equation as

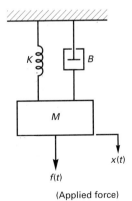

$f(t)$
(Applied force)

Figure 2.18   Mechanical translational system.

$$Ms^2X(s) + BsX(s) + KX(s) = (Ms^2 + Bs + K)X(s) = F(s)$$

The initial conditions are ignored, since the transfer function is to be derived. Thus the transfer function is given by

$$G(s) = \frac{X(s)}{F(s)} = \frac{1}{Ms^2 + Bs + K} \tag{2-23}$$

Example 2.11

As a second example, consider the mechanical system shown in Figure 2.18.. This system is a simplified model of an automobile suspension system of one wheel, with M_1 the mass of the automobile, B the shock absorber, K_1 the springs, M_2 the mass of the wheel, and K_2 the elastance of the tire. Note that two equations must be written, since two independent displacements exist; that is, knowing displacement $x_1(t)$ does not give us knowledge of displacement $x_2(t)$.

First, forces are summed about mass M_1. This mass is enclosed by a dashed closed curve in Figure 2.19, and every element that penetrates this curve exerts a force on M_1. Hence the equation contains three terms (there is no applied force to the mass).

$$M_1\frac{d^2x_1}{dt^2} = -B\left(\frac{dx_1}{dt} - \frac{dx_2}{dt}\right) - K_1(x_1 - x_2)$$

Next the forces on M_2 are summed. Three elements are connected to M_2 in addition to the applied force. Hence five terms appear in the force equation.

$$M_2\frac{d^2x_2}{dt^2} = f(t) - B\left(\frac{dx_2}{dt} - \frac{dx_1}{dt}\right) - K_1(x_2 - x_1) - K_2x_2$$

Collecting terms and taking the Laplace transform of these equations yields

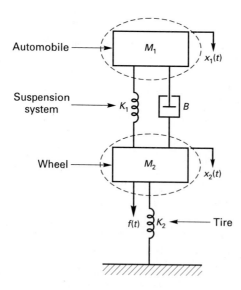

Figure 2.19 Simplified automobile suspension system.

$$M_1 s^2 X_1(s) + B\left[sX_1(s) - sX_2(s)\right] + K_1\left[X_1(s) - X_2(s)\right] = 0$$

$$M_2 s^2 X_2(s) + B\left[sX_2(s) - sX_1(s)\right] + K_1\left[X_2(s) - X_1(s)\right] + K_2 X_2(s) = F(s)$$

Suppose that a transfer function is desired between $F(s)$ and $X_1(s)$, that is, between a force applied to the wheel and the resulting displacement of the car. This transfer function can be found by eliminating $X_2(s)$ in the preceding two equations. The approach used here is to construct a flow graph of the equation and to find the transfer function using Mason's gain formula. This approach is not the shortest one; it is given here for practice in constructing flow graphs (or block diagrams) and in applying Mason's gain formula. There are three variables in these equations; hence the flow graph has three nodes, with the one for $F(s)$ as an input node. The first equation must be solved for one of the unknowns and the second equation for the other unknown. We choose to solve the first equation for $X_1(s)$.

$$X_1(s) = \frac{Bs + K_1}{M_1 s^2 + Bs + K_1} X_2(s) = G_1(s) X_2(s)$$

The rational function (ratio of two polynomials) shown is denoted as $G_1(s)$ for convenience.
 Now the second equation can be solved for $X_2(s)$.

$$X_2(s) = \frac{1}{M_2 s^2 + Bs + K_1 + K_2} F(s) + \frac{Bs + K_1}{M_2 s^2 + Bs + K_1 + K_2} X_1(s)$$

$$= G_2(s) F(s) + G_3(s) X_1(s)$$

The flow graph is constructed as shown in Figure 2.20(a). Thus the transfer function is, from Mason's gain formula,

$$T(s) = \frac{X_1(s)}{F(s)} = \frac{G_1(s) G_2(s)}{1 - G_1(s) G_3(s)}$$

This expression may be evaluated to yield

$$T(s) = \frac{Bs + K_1}{M_1 M_2 s^4 + B(M_1 + M_2) s^3 + (K_1 M_2 + K_1 M_1 + K_2 M_1) s^2 + K_2 Bs + K_1 K_2}$$

The system is seen to be fourth order. The dynamics of the mechanical system are completely described by this transfer function. Given the mass of the car, the mass of the wheel, and the elastance of the tire, the smoothness of the ride is determined by the parameters of the shock absorber and the springs. These parameters are used to tune the system to give a good response. As a physical shock absorber deteriorates, the value of the parameter B changes, which changes the transfer function and hence the quality of the ride. Note also that installing a shock absorber with an incorrect value for the parameter B has the same effect.

The flow graph constructed in the last example is not unique. If the first equation is solved for $X_2(s)$ and the second one for $X_1(s)$, a different flow graph will result. Of course, the transfer function $T(s) = X_1(s)/F(s)$ will not change (see Problem 2.16).

In the last example a signal flow graph was used in the application of Mason's gain formula. A block diagram could have been used instead. The block diagram for the system, constructed from the derived equations, is given in Figure 2.20(b). Note that the geometry of the block diagram is identical to that of the flow graph, since both are constructed in the same manner from the same equations. Mason's gain formula is a function only of the geometry of the flow graph; hence the block diagram can be used as effectively with Mason's gain formula as the flow graph. The block diagram of Figure 2.20(b) yields the same transfer function as does the flow graph of Figure 2.20(a), as the reader can easily show.

A second point should be made concerning the last example. We constructed a block diagram for this example from the system equations as an exercise. Normally, the block diagram is constructed such that each block represents a physical part of the system, as much as is possible. This construction then gives us a better insight into the physical interactions in the system.

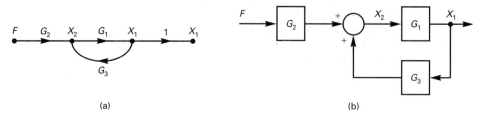

(a)   (b)

Figure 2.20 Models for Example 2.11.

2.6 MECHANICAL ROTATIONAL SYSTEMS

In this section the elements that compose a mechanical linear rotational system are considered. The linear rotational system is analogous (equations are of the same form) to the linear translational system, and the same procedures used for writing the equations for the translational systems may also be used for the rotational systems.

The three elements of the linear rotational system are depicted in Figure 2.21. The first element, *moment of inertia,* is defined by the relationship

$$\tau(t) = J\frac{d^2\theta(t)}{dt^2} = J\frac{d\omega(t)}{dt} \tag{2-24}$$

where $\tau(t)$ is the applied torque, J is the moment of inertia, $\theta(t)$ is the angle of rotation, and $\omega(t)$ is the angular velocity. This equation is analogous to that for mass in a translational system, (2-20). In fact, the moment of inertia of a body is a function of its mass and geometry.

The defining equation for the second element in Figure 2.21, *friction,* is

$$\tau(t) = B\left(\frac{d\theta_1(t)}{dt} - \frac{d\theta_2(t)}{dt}\right) = B[\omega_1(t) - \omega_2(t)] \tag{2-25}$$

where $\tau(t)$ is the applied torque and B is the damping coefficient. It is also assumed that the friction occurs between the two elements shown, and that these elements have zero moments of inertia.

For the third element in Figure 2.21, a *rotational spring,* the defining equation is

$$\tau(t) = K[\theta_1(t) - \theta_2(t)] \tag{2-26}$$

where K is the spring coefficient. Again, the rotating element is assumed to have zero moment of inertia. These equations, along with the principle that the sum of torques around the axis of inertia must be equal to the moment of inertia times its angular acceleration, are utilized to write the equations describing a rotational system. Two examples are given next.

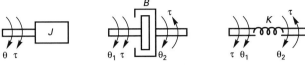

Moment of inertia   Viscous friction   Torsion   **Figure 2.21** Mechanical rotational elements.

Example 2.12

Consider the torsional pendulum modeled in Figure 2.22. One application of this type of pendulum is in clocks that are usually enclosed in glass domes. The moment of inertia of the pendulum bob is represented by J, the friction between the bob and air by B, and the elastance of the brass suspension strip by K. Here it is assumed that the torque is applied at the bob, whereas in a clock the torque is applied by a complex mechanism from the mainspring. Summing torques at the pendulum bob yields

$$J\frac{d^2\theta(t)}{dt^2} = \tau(t) - B\frac{d\theta(t)}{dt} - K\theta(t) \tag{2-27}$$

The transfer function is then easily derived as

$$\frac{\Theta(s)}{T(s)} = \frac{1}{Js^2 + Bs + K} \tag{2-28}$$

Note the similarity of this transfer function to that of the mechanical translational system of Example 2.10. In fact, many different types of physical systems can be accurately modeled by a second-order transfer function of the type given above. This effect is investigated more fully in Section 2.10 on analogous systems.

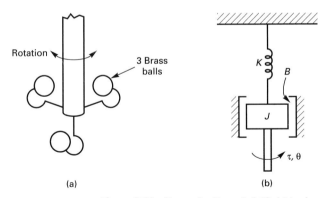

(a)   (b)

Figure 2.22   System for Example 2.12: (a) torsion pendulum; (b) model.

Example 2.13

As a second example of rotational systems, the model of a rigid satellite is derived. The satellite is illustrated in Figure 2.23. The satellite is assumed to be rigid and in a frictionless environment, and to rotate about an axis perpendicular to the page. Torque is applied to the satellite by firing the thrustors shown in Figure 2.23. For example, the firing of the two thrustors, as shown in the figure, tends to increase the angle θ. If the other two thrustors are fired, θ tends to decrease. It is assumed that the thrustor torque $\tau(t)$ is the system input and that the attitude angle $\theta(t)$ is the system output. Then

$$\tau(t) = J\frac{d^2\theta}{dt^2}$$

since the satellite is rigid and no air friction is present. The satellite's moment of inertia is J. The transfer function of the satellite is then

$$G(s) = \frac{\Theta(s)}{T(s)} = \frac{1}{Js^2} \tag{2-29}$$

This is a reasonably accurate model of a rigid satellite and is useful in examples because of its simplicity. Note that if solar panels are attached as appendages to the satellite, the assumption of rigidity is no longer valid (parts of the satellite can move relative to other parts), and the model is much more complex.

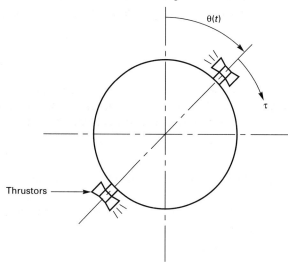

Figure 2.23   Satellite.

2.7 ELECTROMECHANICAL SYSTEMS

In this section the models of two different types of electromechanical systems are developed: the dc generator and the dc servomotor. The dc generator is considered first.

2.7.1 DC Generator

It is assumed that the dc generator is driven by an energy source called a prime mover, which is of sufficient capacity such that the electrical load on the generator does not affect the generator speed. It is further assumed that the generator rotates at a constant speed. A circuit diagram of the generator is given in Figure 2.24. The equation for the field circuit is

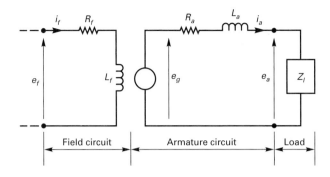

Figure 2.24   DC generator.

$$e_f = R_f i_f + L_f \frac{di_f}{dt} \qquad (2\text{-}30)$$

where the functional dependence of the variables on time is omitted for convenience. In this equation, e_f is the applied field voltage and is considered to be the system input. The field current is i_f, the resistance of the field coil is R_f, and the inductance of the field coil is L_f. The equation for the armature circuit is

$$e_g = R_a i_a + L_a \frac{di_a}{dt} + e_a \qquad (2\text{-}31)$$

where e_g is the generated voltage in the armature circuit, i_a is the armature current, e_a is the armature terminal voltage, and R_a and L_a are the armature resistance and inductance, respectively. The equation that relates the generated voltage e_g to the field flux ϕ is [6]

$$e_g = K\phi \frac{d\theta}{dt} \qquad (2\text{-}32)$$

In this equation, K is a parameter that is determined by the physical structure of the generator and $d\theta/dt$ is the angular velocity of the armature. Since $d\theta/dt$ has been assumed to be constant and since the flux ϕ is directly proportional to the field current i_f, the generated voltage, from (2-32), can be expressed as

$$e_g = K_g i_f \qquad (2\text{-}33)$$

In the following development, all initial conditions are ignored, since we will find the system transfer function. The Laplace transforms of (2-30) and (2-33) yield, respectively,

$$E_f(s) = (sL_f + R_f) I_f(s) \qquad (2\text{-}34)$$

$$E_g(s) = K_g I_f(s) \qquad (2\text{-}35)$$

The equations for the armature circuit are written using the impedance approach, given in Section 2.2. Letting $Z_a(s)$ be the impedance of the armature circuit and $Z_l(s)$ be the generator load impedance, we can write

$$I_a(s) = \frac{E_g(s)}{Z_a(s) + Z_l(s)} = \frac{E_g(s)}{L_a s + R_a + Z_l(s)} \qquad (2\text{-}36)$$

and

$$E_a(s) = I_a(s)Z_l(s) \tag{2-37}$$

In the preceding equations it is assumed that the field voltage, $E_f(s)$, is the system input and that the armature voltage, $E_a(s)$, is the system output. There are four unknowns: $I_f(s)$, $E_g(s)$, $I_a(s)$, and $E_a(s)$. Thus the four equations (2-34), (2-35), (2-36), and (2-37) can be solved for the output as a function of the input, which gives the transfer function. However, we first construct a block diagram from these equations and obtain the transfer function from this block diagram. From (2-34) we solve for $I_f(s)$:

$$I_f(s) = \frac{E_f(s)}{L_f s + R_f} \tag{2-38}$$

and this equation gives us the first block in Figure 2.25,. We obtain the second block in the figure directly from (2-35). The third block is obtained from (2-36), and the fourth comes from (2-37).

Since there are no loops in the block diagram of Figure 2.25, the transfer function is simply the product of the gains transfer functions of the four blocks.

$$G(s) = \frac{E_a(s)}{E_f(s)} = \frac{K_g Z_l(s)}{(L_f s + R_f)\,[L_a s + R_u + Z_l(s)]} \tag{2-39}$$

Note that although we call this device a dc generator, the transfer function gives the manner in which the output voltage varies in time if the field voltage is varied with time. Given $e_f(t)$, we can determine $e_a(t)$ from (2-39). Thus (2-39) gives the *dynamics* of the generator.

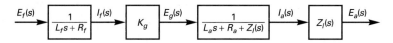

Figure 2.25   Block diagram of generator.

It should be noted that a change in the load conditions on this system affects the system transfer function. This change occurs often in systems and is for some systems a factor that is easy to overlook; hence an engineer should always be alert to loading effects when working with transfer functions.

2.7.2 Servomotor

Next we derive the model of a servomotor. A *servomotor* is a dc motor [7] designed specifically to be used in a closed-loop control system. The circuit diagram of a servomotor is given in Figure 2.26. In this figure $e_a(t)$ is the armature voltage, which is considered to be the system input. The resistance and inductance of the armature circuit are R_m and L_m, respectively. The voltage $e_m(t)$ is the voltage generated in the armature coil because of the motion of the coil in the motor's magnetic field and is usually called the back-EMF. Hence we can write

$$e_m(t) = K\phi \frac{d\theta}{dt} \tag{2-40}$$

where K is a motor parameter, ϕ is the field flux, and θ is the angle of the motor shaft; that is, $d\theta/dt$ is the angular velocity of the shaft. We assume that the flux ϕ remains constant; hence

$$e_m(t) = K_m \frac{d\theta}{dt} \tag{2-41}$$

Note that this assumption is very important; if the flux is a system variable, (2-40) becomes a nonlinear equation, because of the product of two variables. Then the analysis becomes much more difficult and, in particular, the Laplace transform cannot be used. Recall that the Laplace transform of the product of two functions is *not* equal to the product of the transforms.

The Laplace transform of (2-41) yields

$$E_m(s) = K_m s \Theta(s) \tag{2-42}$$

For the armature circuit we can write

$$E_a(s) = (L_m s + R_m) I_a(s) + E_m(s) \tag{2-43}$$

which is solved for $I_a(s)$:

$$I_a(s) = \frac{E_a(s) - E_m(s)}{L_m s + R_m} \tag{2-44}$$

The equation for the developed torque is

$$\tau(t) = K_1 \phi i_a(t) = K_\tau i_a(t) \tag{2-45}$$

since flux is assumed constant. Note that equation would also be nonlinear if flux varied with time. The Laplace transform of this equation yields

$$T(s) = K_\tau I_a(s) \tag{2-46}$$

The final equation is derived from summing torques on the motor armature. In Figure 2.26, the moment of inertia J includes all inertia connected to the motor shaft, and B includes the air friction and the bearings friction. Therefore, the torque equation is

$$J \frac{d^2\theta}{dt^2} = \tau(t) - B \frac{d\theta}{dt} \tag{2-47}$$

and thus we can write

$$T(s) = (Js^2 + Bs) \Theta(s) \tag{2-48}$$

Solving this equation for the motor shaft angle yields

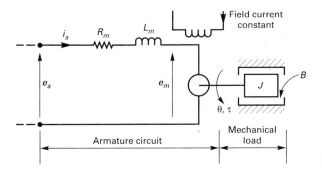

Figure 2.26 Servomotor.

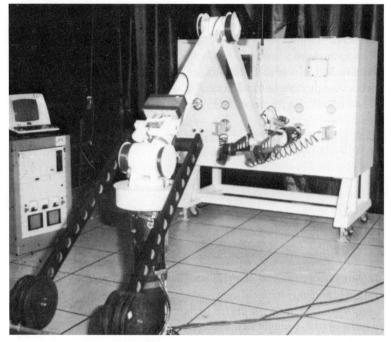

Robot under development by NASA for servicing satellites in orbit. The robot can be "taught" to perform repetitive tasks autonomously. (Courtesy of NASA.)

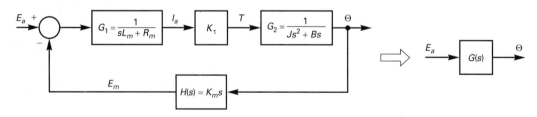

Figure 2.27   Block diagram of servomotor.

$$\Theta(s) = \frac{T(s)}{Js^2 + Bs} \tag{2-49}$$

A block diagram can now be constructed from the four equations (2-42), (2-44), (2-46), and (2-49) and is given in Figure 2.27. From Mason's gain formula we can write the motor transfer function

$$G(s) = \frac{\Theta(s)}{E_a(s)} = \frac{G_1(s) K_\tau G_2(s)}{1 + K_\tau G_1(s) G_2(s) H(s)} \tag{2-50}$$

Evaluating this expression yields

$$G(s) = \frac{K_\tau}{JL_m s^3 + (BL_m + JR_m)s^2 + (BR_m + K_\tau K_m)s} \tag{2-51}$$

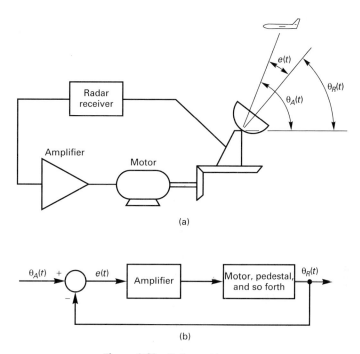

Figure 2.28   Radar tracking system.

An approximation that can often be made for servomotors is to ignore the armature inductance L_m. For the case that L_m is small enough to be ignored, the transfer function is second order and is given by

$$G(s) = \frac{K_\tau}{J R_m s^2 + (B R_m + K_\tau K_m)s} \tag{2-52}$$

Note that this transfer function depends upon the inertia and friction of the load being driven by the motor, as well as the motor parameters. Recall the definition of J and B in (2-47).

The servomotor is used in many examples in this book, with an assumed model as given in either (2-51) or (2-52). A particular servomotor system to be utilized is a radar tracking system, as shown in Figure 2.28. The purpose of this system is to track an aircraft or some equivalent target automatically. In Figure 2.28 the electronics of the radar receiver calculates the error, $e(t)$, between the angle to the target θ_A and the pointing angle of the antenna θ_R. The block in Figure 2.28 labeled motor, pedestal, and so on, would have the model just developed, and the inertia and the friction of the antenna pedestal would be a part of this model.

2.8 SENSORS

In this section we present a brief introduction to sensors used in control systems. For those readers interested in a comprehensive treatment of control system sensors, many good books are available that are devoted entirely to sensors [8–10]. A very good source of material on sensors is the sensor manufacturers (for example, see Ref. 11). In this discussion we consider only sensors for position, velocity, and acceleration. All of these sensors have application, for example, in robotics control.

2.8.1 Position

Perhaps the simplest position sensor is a variable resistor (potentiometer). Resistors can measure both translational and rotational position, as illustrated in Figure 2.29. In each case, R_1 is the resistance from the slider to the common point, and R is the total resistance. The resistance R_1 is then a function of position. The same equation applies for both the lateral position measurement and the angular position measurement, and by voltage division,

$$e(t) = \frac{R_1(t)}{R}E \tag{2-53}$$

where $e(t)$ is the sensor output, E is the applied dc voltage, and $R_1(t)$ is proportional to the displacement. Normally the resistance R_1 is a linear function of displacement, and hence $e(t)$ is directly proportional to displacement. Accordingly,

$$e(t) = K_x x(t) \qquad e(t) = K_\theta \theta(t)$$

where K_x and K_θ are constants, dependent on the construction of the resistors and the dc voltage E.

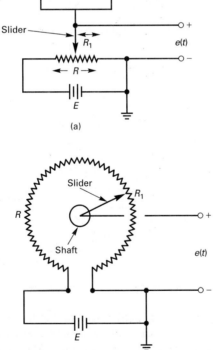

(a)

(b)

Figure 2.29 Resistance position sensors.

The potentiometer as a sensor has the following advantages:

1. It is inexpensive.
2. It can provide a high output voltage.

Some disadvantages are:

  1. The potentiometer becomes noisy with age (poor contact).
  2. It wears out.
  3. Friction loads the mechanical system.
  4. The slider will bounce with fast motions.

A second position sensor is an encoder. An optical *incremental encoder* is illustrated in Figure 2.30. Both translational and rotational encoders are available. In Figure 2.30, a pulse of voltage is generated each time a transparent window passes the light source. With this sensor, electronic circuitry must be available to count the pulses, in order to determine the angle of rotation. This encoder is called an incremental encoder because the generation of a pulse indicates an incremental change in position, and not the actual position.

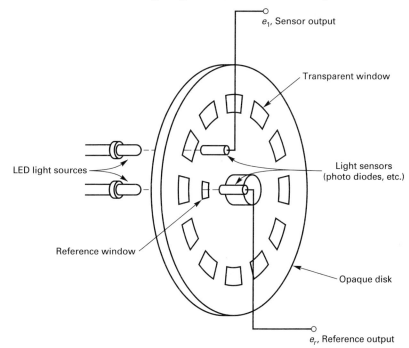

Figure 2.30   Optical incremental encoder.

Usually, a reference window is required to determine absolute position for initializing the system, as shown in Figure 2.30. The reference window is also required in case of temporary loss of power or if noise causes an incorrect count. The count is reinitialized each time the reference window generates a pulse.

The encoder as described above will not indicate a reversal in the direction of motion. One procedure used to determine the direction of rotation is to place a second pick-off at a different window of the encoder track, with the second sensor offset from the first sensor relative to the center of the windows. The second sensor points at the edge of a window when the first sensor points at the center of a window. The result is shown in Figure 2.31, for movement at a constant velocity. Sensor 1 output leading sensor 2 output by 90° indicates one direction of rotation; sensor 1 output lagging sensor 2 output by 90° indicates the other direction of rotation. In Figure 2.31, ω is the frequency of the voltages in rad/s.

More complex encoders are available that will give a binary code as output; the binary code indicates absolute position [8]. This binary code may be transmitted directly to a digital computer for control purposes. This type of encoder, called an *absolute encoder,* will not be discussed here.

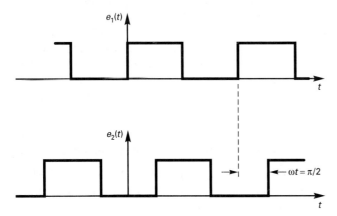

Figure 2.31   Determining the direction of rotation.

2.8.2 Velocity

The encoder of Figure 2.30 can also be used to measure velocity by counting the number of pulses per unit time. A second procedure for determining velocity is to measure the time between adjacent pulses, or the time for a given number of pulses to occur. In either case, the measurements can be quite inaccurate for small velocities, in which the pulses occur at a very low rate.

A special dc generator, called a *tachometer,* can also be used to measure velocity. One type of tachometer has the magnetic field set up by a permanent magnet [8]. Then the dc-generator equation (2-32) can be expressed as

$$e_g(t) = K\phi\frac{d\theta(t)}{dt} = K_\tau\frac{d\theta(t)}{dt} \tag{2-54}$$

where $e_g(t)$ is the generated voltage, ϕ is the magnetic flux, θ is the angle of the shaft, and K is determined by the physical construction of the tachometer. Thus,

$$E_g(s) = K_\tau\Omega(s) \tag{2-55}$$

where $\Omega(s) = \mathcal{L}[d\theta/dt]$ and the units of $\theta(t)$ are radians.

In the armature circuit of the generator of Figure 2.24, the generator output voltage $E_a(s)$ is given by

$$E_a(s) = I_a(s)Z_l(s) = \frac{Z_l(s)}{L_a s + R_a + Z_l(s)} K_\tau \Omega(s)$$

For that frequency range for which the magnitude of the load impedance is much larger than the magnitude of the armature impedance, $E_a(s) \approx K_\tau \Omega(s)$, and the tachometer output voltage is directly proportional to its shaft velocity. The transfer function of the tachometer, with shaft position $\Theta(s)$ as the input, is given by

$$\frac{E_a(s)}{\Theta(s)} = K_\tau s \tag{2-56}$$

Hence we see the equivalent differentiation that appears in the velocity measurement. However, actual differentiation does not occur; the velocity measurement is made directly.

A second procedure for obtaining velocity is to measure position and then differentiate this signal. If the position measurement contains high-frequency noise (the amplitude of the noise changes rapidly), differentiation will increase the amplitude of this noise, since the differentiator output is equal to the rate of change of signal amplitude. If possible, the velocity is always measured, rather than differentiating the position signal to obtain the velocity.

2.8.3 Acceleration

We will now consider the measurement of acceleration. For an isolated rigid mass,

$$f(t) = M\ddot{x}(t) = Ma(t) \tag{2-57}$$

where $f(t)$ is force, $x(t)$ is displacement, $\ddot{x}(t)$ denotes $d^2 x(t)/dt^2$, and $a(t)$ is acceleration. Hence, if the force is measured, the acceleration is given by

$$a(t) = \frac{1}{M} f(t) = K f(t) \tag{2-58}$$

and the acceleration is directly proportional to the force. Accelerometers are based on the measurement of the force on a mass; that is, the force is measured and the acceleration is calculated from (2-58).

Consider the accelerometer illustrated in Figure 2.32. The accelerometer is rigidly attached to the body and is to determine the acceleration of that body. For this accelerometer, the inertial force $f(t)$ is to be determined by generating a force $f_a(t)$ necessary to prevent the mass M from moving relative to the accelerometer case. Then $f_a(t) = f(t)$ and the acceleration of the body is given by

$$a(t) = \frac{1}{M} f_a(t) \tag{2-59}$$

We will not consider the electronic equipment and the methods used for generating the force $f_a(t)$; these topics are beyond the scope of this book. However, a closed loop system is required to maintain a constant position of the mass M relative to the accelerometer case. Thus, for this instrument, a closed-loop control system is required internal to the instrument.

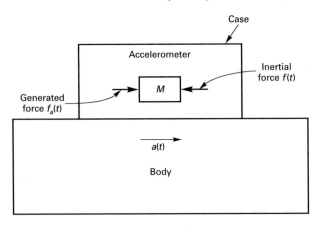

Figure 2.32 Measuring acceleration.

A second type of accelerometer is based on the piezoelectric characteristic of certain crystals. A force applied in a given direction in the structure of the crystal will generate a voltage between two surfaces of the crystal. This voltage is proportional to the applied force.

A diagram of piezoelectric accelerometer is given in Figure 2.33. The acceleration of the inertia mass creates a force on the crystal, which generates a voltage $e(t)$ across the crystal which is proportional to the force. Hence the voltage $e(t)$ is proportional to the acceleration of the mass. The mechanical spring and the mass in Figure 2.33 give the accelerometer a resonance (a peaking in the frequency response). The frequencies in the movement of the mass must be much less than the resonant frequency, or the resulting measurements will be grossly in error. Of course, any sensor will have a limited bandwidth over which the instrument is accurate; that is, any sensor has dynamics.

Let $x(t)$ be the displacement of the mass of the accelerometer. The voltage out of the accelerometer in Figure 2.33 is given by

$$e(t) = K_a f_a(t) = K_a M \ddot{x}(t) \Rightarrow \quad \frac{E(s)}{X(s)} = (K_a M) s^2 \tag{2-60}$$

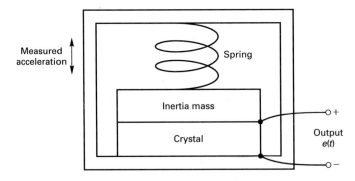

Figure 2.33 Piezoelectric accelerometer.

Note that, in effect, the accelerometer is a second-order differentiator of position. As in the case of a velocity measurement, in general we do not obtain acceleration by differentiating a position signal, because of the amplification of high-frequency noise.

2.9 TEMPERATURE-CONTROL SYSTEM

In this section a specific temperature-control system is modeled. A diagram of the system is shown in Figure 2.34. Liquid at temperature T is flowing out of a tank at some rate and is replaced with a liquid at temperature T_i, where $T_i < T$. The liquid in the tank is heated by an electric heater. A mixer agitates the liquid such that the liquid throughout the tank can be considered to be at the temperature $T(t)$, which in general varies with time.

The following definitions are needed to derive the system model.

$q_e(t)$ = heat flow supplied by the electric heater

$q_l(t)$ = heat flow into the liquid

$q_o(t)$ = heat flow via the liquid leaving the tank

$q_i(t)$ = heat flow via the liquid entering the tank

$q_s(t)$ = heat flow through the tank's surface area

By the principle of the conservation of energy, the heat added to the tank of liquid must equal that transferred from the tank plus that remaining in the tank. Thus

$$q_e + q_i = q_l + q_o + q_s \tag{2-61}$$

where, for convenience, the dependence on t is not shown.

Now

$$q_l = C\frac{dT}{dt} \tag{2-62}$$

where C is the thermal capacity of the liquid; that is, C is a parameter of the system [12]. Letting V equal the flow into the tank and out of the tank (assumed equal) and H equal the specific heat of the liquid, we can write

$$q_i = VHT_i \tag{2-63}$$

and

$$q_o = VHT \tag{2-64}$$

If R is the thermal resistance to that flow through the tank surface and T_a is the ambient air temperature outside the tank, then

$$q_s = \frac{T - T_a}{R} \tag{2-65}$$

Substituting (2-62) through (2-65) into (2-61) yields

$$q_e + VHT_i = C\frac{dT}{dt} + VHT + \frac{T - T_a}{R} \tag{2-66}$$

This is a first-order linear differential equation with T the dependent variable and q_e, T_i, and T_a as forcing functions. In terms of a control system, the control input is the electric heater

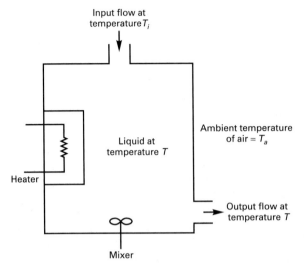

Input flow at
temperature T_i

Liquid at
temperature T

Ambient temperature
of air $= T_a$

Heater

Output flow at
temperature T

Figure 2.34 Thermal System.

Mixer

output q_e. The terms T_i and T_a are called disturbance inputs, or simply disturbances, and are inputs over which we have no control. These inputs generally cause responses that are undesirable, since we cannot control them. Usually a major requirement of a control system design is to minimize the effects of the disturbances on the system.

To continue the discussion of (2-66), the temperature T is the variable to be controlled, and we consider it to be the output. Note that if the flow V is a function of time, (2-66) is a first-order linear time-varying differential equation. We cannot find transfer functions to represent this equation, since, in the second term on the right-hand side of (2-66), both V and T are functions of time. Recall that the Laplace transform of a product of functions is not equal to the product of the transforms.

To simplify the analysis, we assume that the flow V is constant. Taking the Laplace transform of (2-66) and solving for the temperature $T(s)$ yields

$$T(s) = \frac{Q_e(s)}{Cs + VH + (1/R)} + \frac{VHT_i(s)}{Cs + VH + (1/R)} + \frac{(1/R)T_a(s)}{Cs + VH + (1/R)} \tag{2-67}$$

Different configurations may be used to represent (2-67) as a block diagram; one is given in Figure 2.35.

If we ignore the disturbance inputs, the system transfer function is given by

$$G(s) = \frac{T(s)}{Q_e(s)} = \frac{K_1}{\tau s + 1} \tag{2-68}$$

A transfer function of this form is called a *first-order lag*. The reason for the term *lag* will be evident when frequency response is covered. In (2-68) the disturbance terms were ignored; however, as we will see later, at some stage of the design process we must consider these terms. As was just stated, for certain control system designs the minimization of disturbance effects is a major design specification.

One final point should be made. In the preceding derivation we considered the heat energy out of the electric heater to be the system input. However, this energy is usually controlled by controlling the voltage applied to the heater. If the heater is adequately modeled as a resistance, then electric energy per unit time (power) is given by

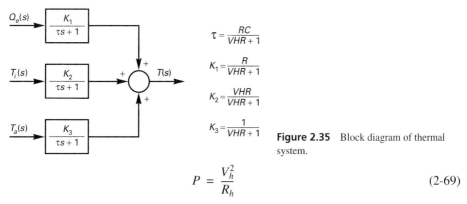

$$\tau = \frac{RC}{VHR + 1}$$

$$K_1 = \frac{R}{VHR + 1}$$

$$K_2 = \frac{VHR}{VHR + 1}$$

$$K_3 = \frac{1}{VHR + 1}$$

Figure 2.35 Block diagram of thermal system.

$$P = \frac{V_h^2}{R_h} \tag{2-69}$$

where P is the power in watts, V_h is the effective (rms) voltage in volts applied to the heater, and R_h is the heater resistance in ohms. Hence the electric energy that is converted to heat is a nonlinear function of the applied voltage, and no transfer function exists from heater voltage input V_h to liquid temperature T.

Note how this system has been simplified by judicious assumptions. First we assumed that the flow into the tank is exactly equal to the flow out, which more than likely will not be the case. Furthermore, the flow was assumed to be constant. Next we ignored disturbance inputs; these must be considered at some point in the design. Then we assumed the input to be the physical variable that gives us a transfer function. Usually, in a practice case, V_h is the input, resulting in a nonlinear system. Nonlinear systems are considered in Chapter 14. As we progress in analysis and design in this text, we will see how difficulties such as these can be overcome, at least to a degree.

2.10 ANALOGOUS SYSTEMS

In the preceding discussion, note that the different types of physical systems are modeled by linear differential equations with constant coefficients. Different types of physical systems that are modeled by the same form of equations are called *analogous* systems. In this section we consider some analogous systems.

We illustrate this concept with the circuit of Figure 2.36. Using the impedance concept, we can write

$$\frac{I(s)}{V(s)} = \frac{1}{Z(s)} = \frac{1}{Ls + R + (1/Cs)} = \frac{Cs}{LCs^2 + RCs + 1}$$

If the voltage across the capacitor is considered to be the output, the transfer function from the voltage input to output is

$$G(s) = \frac{V_c(s)}{V(s)} = \frac{(1/Cs)I(s)}{V(s)} = \frac{1}{LCs^2 + RCs + 1}$$

This transfer function is of a general second-order form

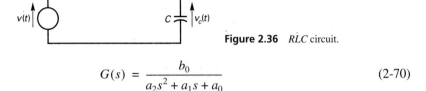

Figure 2.36 *RLC* circuit.

$$G(s) = \frac{b_0}{a_2 s^2 + a_1 s + a_0} \tag{2-70}$$

which is of the same form as the linear translational mechanical system (2-23) of Example 2.10, the torsional pendulum (2-28) of Example 2.12, the satellite (2-29) of Example 2.13, the dc generator (2-39) with resistive load, and the servomotor (2-52). In a particular case, certain coefficients may be zero. These transfer functions are listed in Table 2.1.

System	Equation number	Equation
Mechanical translational	(2-23)	$\dfrac{1}{Ms^2 + Bs + K}$
Torsional pendulum	(2-28)	$\dfrac{1}{Js^2 + Bs + K}$
Satellite	(2-29)	$\dfrac{1}{Js^2}$
Dc generator	(2-39)	$\dfrac{K_g R_l}{(L_f s + R_f)(L_a s + R_a + R_l)}$
Servomotor	(2-52)	$\dfrac{K_\tau}{JR_m s^2 + (BR_m + K_\tau K_m)s}$

Since we are modeling the physical systems with linear differential equations with constant coefficients, naturally the transfer functions will be of the same form. The most general second-order transfer function is of the form

$$G(s) = \frac{b_2 s^2 + b_1 s + b_0}{a_2 s^2 + a_1 s + a_0} \tag{2-71}$$

where one of the nonzero coefficients can be chosen to be unity with no loss in generality (by dividing both numerator and denominator by that coefficient). Usually a_2 is set to unity. This second-order transfer function can model any physical system that is adequately described by a second-order linear differential equation with constant coefficients. Hence all such systems are classified as analogous systems.

The purpose of this section is to introduce the concept of analogous systems. There are many applications of this concept. For example, if we understand vibrations in linear mechanical systems, we can apply this understanding to the response of linear electrical circuits, and vice versa. If we understand the response of linear electrical circuits, we also have an understanding of heat flow in linear heat-transfer systems, even though this is not immediately obvious. An example is given next to illustrate this.

Example 2.14

In the temperature-control system of Section 2.9, the transfer function from electric energy in to temperature out derived in (2-68) to be

$$\frac{T(s)}{Q_e(s)} = \frac{K_1}{\tau s + 1}$$

Consider now the circuit shown in Figure 2.37. The transfer function of this circuit is given by

$$\frac{V_c(s)}{V(s)} = \frac{(1/sC_2)I(s)}{V(s)} = \frac{(1/sC_2)}{R + (1/sC_1) + (1/sC_2)} = \frac{(C_e/C_2)}{RC_e s + 1}$$

where

$$C_e = \frac{C_1 C_2}{C_1 + C_2}$$

If we choose C_e/C_2 equal to K_1 and RC_e equal to τ, the two transfer functions are equal. Suppose a step input is applied to the circuit. From the study of electrical circuits, we know that the voltage on C_2 will rise exponentially to a final value. Hence a step input of electrical power to the heater in the temperature control system will result in an exponential rise in temperature to a final value. A step input of voltage to the heater terminals will give the step input in electrical power. In fact, this procedure of applying a step input in power to a temperature control system is one method that is used to determine the parameters of the model of the system.

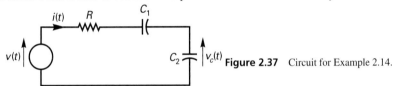

Figure 2.37   Circuit for Example 2.14.

2.11 TRANSFORMERS AND GEARS

This section presents two analogous coupling systems: electrical transformers and mechanical gears. We consider only the ideal cases and see that the describing equations for the two systems are of the same form.

Consider first the electrical transformer. The circuit diagram of the transformer is shown in Figure 2.38. For the ideal transformer, it is assumed that at any time the power into the transformer is equal to the power out of the transformer, giving

$$p(t) = e_1(t)i_1(t) = e_2(t)i_2(t) \tag{2-72}$$

where $p(t)$ is the power transferred by the transformer. It can also be shown that, from consideration of the flux in the transformer core [13],

$$N_1 i_1(t) = N_2 i_2(t) \tag{2-73}$$

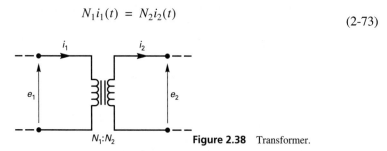

Figure 2.38 Transformer.

where N_1 is the number of turns in the input (primary) coil and N_2 is the number of turns in the output (secondary) coil. Then from (2-72) and (2-73),

$$\frac{e_1(t)}{N_1} = \frac{e_2(t)}{N_2} \tag{2-74}$$

The ratio N_1/N_2 is called the *turns ratio,* and, from (2-72) and (2-74),

$$\frac{N_1}{N_2} = \frac{e_1(t)}{e_2(t)} = \frac{i_2(t)}{i_1(t)} \tag{2-75}$$

Thus the transformer is a coupling device whose purpose is usually to change either voltage levels or current levels. Normally, the ideal transformer model is a reasonable approximation to a physical transformer that has a ferrous core. The ideal transformer model is then given by (2-75). An example is given next.

Example 2.15

Consider the circuit of Figure 2.39. In this circuit the voltage source and the resistance are the Thévenin's equivalent circuit [2] of an amplifier, which is coupled to a load $Z_l(s)$ through the transformer. Now

$$E_i(s) = RI_1(s) + E_1(s)$$

and

$$E_2(s) = Z_l(s)I_2(s)$$

From (2-75),

$$E_1(s) = \frac{N_1}{N_2} E_2(s) \qquad I_1(s) = \frac{N_2}{N_1} I_2(s)$$

Using these two equations and the secondary loop equation given earlier, we can write the primary loop equation as

$$E_i(s) = RI_1(s) + \frac{N_1}{N_2} E_2(s) = \left[R + \left(\frac{N_1}{N_2} \right)^2 Z_l(s) \right] I_1(s)$$

Hence the impedance $Z_s(s)$, as seen by the voltage source, is

$$Z_s(s) = \frac{E_i(s)}{I_1(s)} = R + \left(\frac{N_1}{N_2} \right)^2 Z_l(s)$$

and in the primary circuit, the effect of the secondary load impedance is an equivalent impedance of $(N_1/N_2)^2 Z_l(s)$. Thus the transformer has the effect of converting the secondary impedance from one level to a different level. This characteristic is useful in matching amplifiers to loads, for maximum power transfer [14].

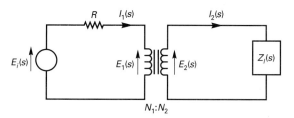

Figure 2.39   Transformer-coupled circuit.

Next, mechanical coupling through the use of gears is considered. Two gear wheels are shown in Figure 2.40. It is assumed that the gears are ideal; that is, the gears are rigid with the gear teeth meshing perfectly, and the gears have no inertia. Then the distance traveled on the circumference of one gear is equal to that traveled on the circumference of the other, or

$$r_1 \theta_1 = r_2 \theta_2 \tag{2-76}$$

where r_1 is the radius of the first gear and θ_1 is the angle rotated by the first gear. The values r_2 and θ_2 are similarly defined. Also, the force exerted by one gear must equal the reaction force of the other at the point of contact. Thus

$$\frac{\tau_1}{r_1} = \frac{\tau_2}{r_2} \tag{2-77}$$

since force is equal to torque divided by radius. In (2-77), τ_i is the torque and r_i is the radius of the ith gear, $i = 1, 2$. Note that these equations are of the same form as those of the ideal transformer, (2-73) and (2-74), with the ratio of radii r_1/r_2 analogous to the turns ratio N_1/N_2. An example is given next.

Example 2.16

Consider the gear system shown in Figure 2.41. Now,

$$\tau_2 = J \frac{d^2 \theta_2}{dt^2}$$

and

$$\tau_1 = \left(\frac{r_1}{r_2} \right) \tau_2 = \left(\frac{r_1}{r_2} \right) J \frac{d^2 \theta_2}{dt^2} = \left[\left(\frac{r_1}{r_2} \right)^2 J \right] \frac{d^2 \theta_1}{dt^2}$$

Thus one effect of the gears is to transform the moment of inertia by the square of the radius ratio, and we see that the characteristics are analogous to those of the transformer.

2.12 ROBOTIC CONTROL SYSTEM

Shown in Figure 2.42 is a line drawing of an industrial robot. In general, the arm of a robot has a number of joints. The current approach to the design of control systems for robot joints

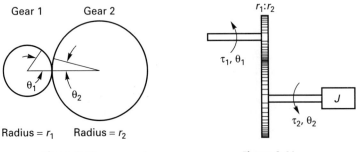

Figure 1:
Gear 1   Gear 2
Radius = r_1 Radius = r_2
θ_1 θ_2

Figure 2.40

Figure 2 (Figure 2.41):
$r_1{:}r_2$
τ_1, θ_1
τ_2, θ_2
J

Figure 2.41

is to treat each joint of the robot arm as a simple joint servomechanism, ignoring the effects of the movements of all other joints. While this approach is simple in terms of analysis and design, the result is often a less than desirable control of the joint [15]. However, in this section we take this approach of considering each joint independently.

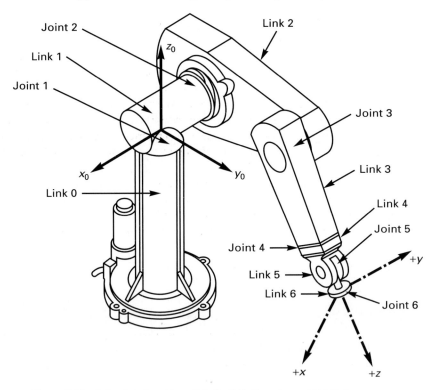

Figure 2.42 Industrial robot. (From K.S. Fu, R.C. Gonzalez, and C.S.G. Lee, *Robotics: Control, Sensing, Visions and Intelligence,* McGraw-Hill Book Company, New York, 1987.)

Figure 2.43 illustrates a single-joint robot arm, with only the joint shown. The actuator is assumed to be an armature-control dc servomotor of the type described in Section 2.7 and Figure 2.27. In addition, it is assumed that the arm is connected to the motor through gears, with a gear ratio of $n = r_1/r_2$ (see Section 2.11).

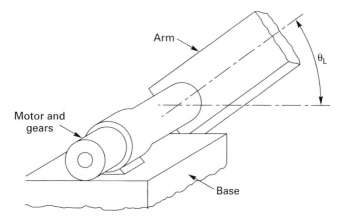

Figure 2.43   Robot arm joint.

In industrial robots, hydraulic or pneumatic actuators may be used rather than dc servomotors. However, since we have already developed the model of a dc servomotor in Section 2.7, we assume the use of a dc servomotor. The block diagram of the robot arm joint is then as given in Figure 2.44. In this figure, the motor parameters are the same as those in Figure 2.27. The motor input voltage is E_a, I_a is the motor armature current, T is the developed torque, θ_m is the motor shaft angle, and θ_L is the angle of the arm. The moment of inertia J represents the total moment of inertia reflected to the motor shaft, including the moment of inertia of the arm. The friction coefficient B represents the total friction reflected to the motor shaft. Example 2.16 illustrates the reflection of moment of inertia through gears.

It is seen from Figure 2.44 that the model of this joint is third order. As in the case of the servomotor of Section 2.7, if the inductance L_m of the motor armature can be ignored, the model of the joint can be reduced to second order. Usually, the motor is a permanent magnet, armature excited, continuous rotation (not stepper) motor, and these motors tend to have a low armature inductance.

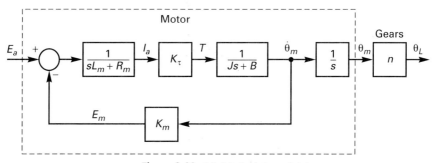

Figure 2.44   Model of robot arm joint.

A feedback control system for a single joint of a robot arm is given in Figure 2.45 [15]. In this diagram, θ_c is the desired, or commanded, angle of the robot arm, and θ_L is the actual angle of the arm. The compensator has the transfer function

$$G_c(s) = K_P + K_D s$$

and is a proportional-plus-derivative (PD) compensator. The design of PD compensators is considered in Chapters 7 and 9. The control system for a single joint of a robot arm, as shown in Figure 2.45, is considered in examples and problems in many of the remaining chapters.

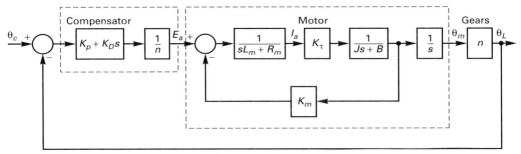

Figure 2.45 Robot joint control system.

2.13 SYSTEM IDENTIFICATION

In the preceding sections the laws of physics were used to obtain models of physical systems. In this section a different approach to obtaining system models is discussed. No methods are developed, since these methods are beyond the scope of this book; however, it is important that the reader be aware that this body of knowledge exists.

The methods to be mentioned here come under the general heading of *system identification* [16,17]. The methods are based on calculating the coefficients of the transfer function from data obtained from measurements on the system input and output. All these procedures require that the order of the transfer function be assumed. The methods may be based on the frequency response (sinusoidal response), the step response, the response to a rectangular pulse, or the response to a more general input. Generally, the methods can be considered to be a type of curve fitting, where the assumed transfer function is fitted to the available data in some optimal manner. This results in the "best" fit of the transfer function to the data but does not necessarily result in a good model. For example, we can fit a second-order transfer function to the input–output data of a highly nonlinear physical system. The result is a best fit of the transfer function to the data, but in this case the best fit is so poor as to be useless. Hence these methods, as with all system-modeling techniques, must be used with great care.

2.14 LINEARIZATION

This chapter presents linear models of several types of physical systems. However, the more accurate models of physical systems are nonlinear. In this section we illustrate a procedure for linearizing some types of nonlinear equations. The Laplace transform cannot be used to solve nonlinear differential equations; it can be used to solve the linearized equations. The procedures for linearization are developed more completely in Chapter 14.

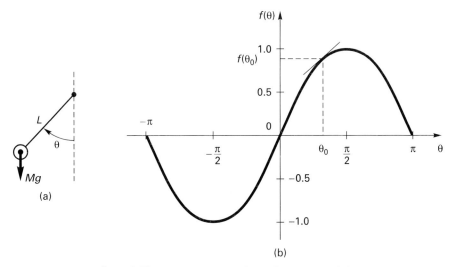

Figure 2.46 Simple pendulum with nonlinear characteristics.

To illustrate linearization, we consider the simple pendulum shown in Figure 2.46(a). The angle of the pendulum is designated as θ, the mass of the pendulum bob is M, and the length of the (weightless) arm from the axis of rotation to the center of the bob is L. The force acting on the bob is then Mg, where g is the gravitational constant. Since we are interested in linearization, we simply state the equation of motion of the pendulum; this equation is derived in any general physics book. The motion of the pendulum is described by the differential equation

$$\frac{L}{g}\frac{d^2\theta(t)}{dt^2} = -\sin\theta(t) \tag{2-78}$$

assuming no friction. This equation is nonlinear because of the $\sin\theta(t)$ term. We cannot find the Laplace transform of this function unless $\theta(t)$ is known.

The nonlinear characteristic for the function $f(\theta) = \sin\theta$ is illustrated in Figure 2.46(b). The usual procedure for linearizing a characteristic such as this is to replace the characteristic with a straight line, which may give a reasonably accurate model in some small region of operation. For example, in Figure 2.46(b), suppose that we wish to linearize the characteristic about the point θ_0. By the Taylor series expansion, we can represent a function $f(\theta)$ about a point θ_0 by the series

$$f(\theta) = f(\theta_0) + \left.\frac{df}{d\theta}\right|_{\theta=\theta_0}(\theta - \theta_0) + \left.\frac{d^2f}{dt^2}\right|_{\theta=\theta_0}\frac{(\theta - \theta_0)^2}{2!} + \cdots \tag{2-79}$$

for those values of θ for which the series converges. For the case that θ is close to θ_0, we may be able to ignore the higher-order derivative terms in this expansion, and

$$f(\theta) \cong f(\theta_0) + \left.\frac{df}{d\theta}\right|_{\theta=\theta_0}(\theta - \theta_0) \tag{2-80}$$

For the simple pendulum modeled in (2-78), $f(\theta) = \sin\theta$, and, in (2-80),

$$\sin \theta \cong \sin \theta_0 + (\cos \theta_0)(\theta - \theta_0) \tag{2-81}$$

Since the pendulum usually operates about the point $\theta = 0°$, we linearize about this point, and with $\theta_0 = 0°$ in (2-81),

$$\sin \theta \cong 0 + (1)(\theta - 0) = \theta \tag{2-82}$$

This, of course, is the usual small-angle approximation for $\sin \theta$. Substituting (2-82) into (2-78) and rearranging terms yields the linear differential equation

$$\frac{d^2\theta(t)}{dt^2} + \frac{g}{L}\theta(t) = 0 \tag{2-83}$$

It is left as an exercise for the reader to show that the solution of this linearized equation yields simple harmonic motion for the pendulum (see Problem 2.23).

To summarize the linearization procedure, suppose that we wish to linearize a nonlinear function $f(x)$ about a point x_0. From the Taylor series expansion (2-79), the linear approximation for $f(x)$ about the point x_0 is

$$f(x) \cong f(x_0) + \frac{df(x)}{dx}\bigg|_{x = x_0} (x - x_0) \tag{2-84}$$

The accuracy of this approximation depends upon the magnitudes of the terms of (2-79) that are ignored.

This section gives a simple example of the linearization of nonlinear differential equations. Since all physical systems are inherently nonlinear, some type of linearization is required to allow us to model these systems with linear differential equations. Linearization procedures are covered more thoroughly in Chapter 14; in that chapter some of the implications of linearization on system stability are discussed.

2.15 SUMMARY

In this chapter we presented methods for obtaining mathematical models of physical systems based on the laws of physics. The chapter covered some electrical, mechanical, electromechanical, and thermal systems. An exhaustive investigation of the modeling of physical systems was not intended. Instead, a sufficient number of different types of systems were described to justify the use of transfer functions in the remainder of this book.

The transfer functions developed are of the form of the ratio of two polynomials in the Laplace transform variable s. Physical systems are usually modeled in this manner for control system analysis and design, at least in the preliminary stages of the work. Consideration of nonlinearities, time-varying parameters, and so forth, may prove necessary in later stages if the linear time-invariant model is found to be inadequate. As a final topic, the linearization of nonlinear differential equations was introduced. Since all physical systems are inherently nonlinear, the linear models that we use in analysis and design are always the result of some type of linearization. The topic of linearization is covered more completely in Chapter 14.

The next chapter presents a method of modeling linear time-invariant systems that is different from the transfer function. This method is the state-variable procedure and is useful in simulation and in modern control analysis and design.

REFERENCES

1. *IEEE Standard Dictionary of Electrical and Electronic Terms.* New York: IEEE, 1984.

2. J. D. Irwin. *Basic Engineering Circuit Analysis,* 5th ed. Upper Saddle River, NJ: Prentice Hall, 1999.

3. V. P. Nelson et al. *Digital Logic Circuit Design and Analysis.* Upper Saddle River, NJ: Prentice Hall, 1995

4. J. L. Agnew and R. C. Knapp. *Linear Algebra with Applications,* 3rd ed. Pacific Grove, CA: Brooks/Cole, 1989.

5. S. J. Mason. "Feedback Theory—Some Properties of Flow Graphs," *Proc. IRE,* 41 (September 1953): 1144–1156.

6. G. McPherson and R. D. Laramore. *An Introduction to Electrical Machines and Transformers.* New York: Wiley, 1991.

7. S. J. Chapman. *Electrical Machinery Fundamentals.* New York: McGraw-Hill, 1984.

8. C. W. deSilva. *Control Sensors and Actuators.* Englewood Cliffs, NJ: Prentice Hall, 1989.

9. M. F. Hordeski. *Design of Microprocessor Sensor & Control Systems.* Reston, VA: Reston Publishing Company, 1985.

10. S. Wolf and R. Smith. *Student Reference Manual for Electronic Instrumentation.* Englewood Cliffs, NJ: Prentice Hall, 1990.

11. E. E. Herceg. *Handbook of Measurement and Control.* Pennsauken, NJ: Schaevitz Engineering, 1976.

12. J. D. Trimmer. *Response of Physical Systems.* New York: Wiley, 1950.

13. W. A. Blackwell and L. L. Grigsby. *Introductory Network Theory.* Boston: Prindle, Weber & Schmidt, 1985.

14. M. E. Van Valkenburg. *Network Analysis.* Englewood Cliffs, NJ: Prentice Hall, 1974.

15. K. S. Fu, R. C. Gonzalez, and C. S. G. Lee. *Robotics: Control, Sensing, Vision, and Intelligence.* New York: McGraw-Hill, 1987.

16. D. Graupe. *Identification of Systems.* Huntington, NY: R. E. Kreiger, 1976.

17. L. Ljung and E. J. Ljung. *System Identification: Theory for the User.* Upper Saddle River, NJ: Prentice Hall, 1998.

PROBLEMS

2.1. For the circuit shown in Figure P2.1, find the voltage transfer function $V_2(s)/V_1(s)$.

2.2. Consider the circuit of Figure P2.2.

    (a) Find the voltage transfer functions $V_2(s)/V_1(s)$.

    (b) Suppose that an indicator L_2 is connected across the output terminals in parallel with R_3. Find the transfer funtion $V_2(s)/V_1(s)$.

    (c) A constant input voltage of 10 V is applied to the circuit. Using the final-value theorem of the Laplace transform (see Appendix B, Section B.2), find the steady-state values of the output voltages for the circuits of (a) and (b).

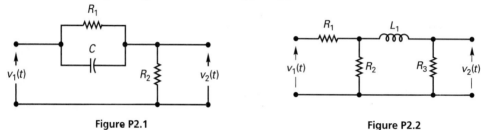

Figure P2.1   Figure P2.2

2.3. (a) Find the voltage transfer functions $V_a(s)/V_i(s)$ for each of the op-amp circuits in Figure P2.3.

    (b) Express each output voltage $v_a(t)$ as a function of the input voltage and the circuit parameters.

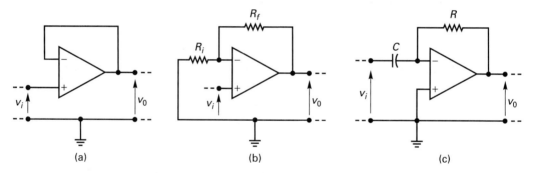

(a)   (b)   (c)

Figure P2.3

2.4. (a) Design an op-amp circuit that realizes the voltage transfer function $-10/s$. Do not use a resistance value of less than 10 kΩ.

    (b) Repeat (a) for a voltage amplifier with a gain of 10 .

    (c) Repeat (b) for the voltage gain of −10.

2.5. (a) Design an op-amp circuit that realizes the voltage transfer function $G_a(s)=1/(0.1s+1)$. Do not use a resistance value of less than 10 kΩ.

    (b) Repeat (a) for the voltage transfer function $G_b(s)=s/(0.1s+1)$.

    (c) Repeat (a) for the voltage transfer function $G_c(s)=-1/(0.1s+1)$.

2.6. Shown in Figure P2.6 is a commonly used op-amp circuit for sensors. In this figure e_s is the output voltage of the transducer, and the −12-V input is used to offset a bias (constant) voltage in e_s. Express v_o as a function of e_s.

2.7. Consider the flow graph given in Figure P2.7.

    (a) Solve for A and B using Mason's gain formula.

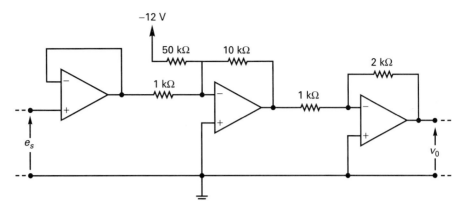

Figure P2.6

(b) Write the equations upon which the flow graph is based.
(c) Solve the equations of (b) by matrix inversion.
(d) Solve the equations of (b) by Cramer's rule.
(e) Solve the equations of (b) using MATLAB.
(f) Verify your solution by direct substitution in the equations in (b).

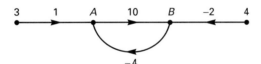

Figure P2.7

2.8. (a) Draw a flow graph for the equations given. Generate the node for the variable A from the first equation, the node for B from the second, and the node for C from the third.

$$A + 2B + C = 4$$
$$A - B - C = 0$$
$$4A - C = 2$$

(b) Use Mason's gain formula to solve these equations.
(c) Verify your solution in (b) by solving the equations by matrix inversion.
(d) Verify your solution in (b) by solving the equations by Cramer's rule.
(e) Verify your solution by direct substitution into the given equations.
(f) Verify your solution using MATLAB.

2.9. (a) Using Mason's gain formula, find the transfer function C/R for the flow graph of Figure P2.9. One of the forward paths is easily overlooked.
(b) The equation for node A, in terms of the labeled nodes, is

$$A = R - G_1A - G_2G_3G_4B - G_3G_4R$$

In a like manner, write the equations for nodes B and C, and solve these equations using Cramer's rule to verify the results in (a).
(c) Solve the equations in (b) using MATLAB.

2.10. (a) For the flow graph Figure P2.10, use Mason's gain formula to find the transfer function $C(s)/R(s)$.
(b) Write three equations in the variables A, B, and C. Then verify the results in (a) using Cramer's rule.

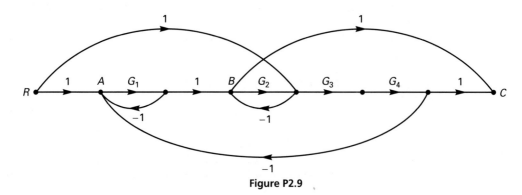

Figure P2.9

(c) Solve the equations in (b) using MATLAB.

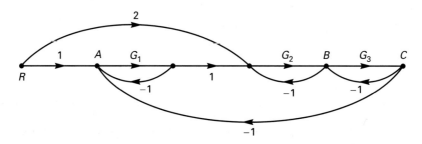

Figure P2.10

2.11. (a) The flow graph of Figure P2.11 is called a simulation diagram and is covered in Chapter 3. These diagrams are very useful in the analysis and design of systems. For this simulation diagram, find the transfer function $C(s)/R(s)$ using Mason's gain formula.

(b) Write three equations in the variables $A(s)$, $B(s)$, and $C(s)$. Then verify the results in (a) using Cramer's rule

(c) Solve the equations in (b) using MATLAB.

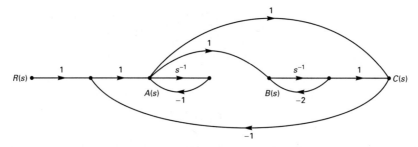

Figure P2.11

2.12. Given the block diagrams of Figure P2.12.

(a) Find the transfer functions G_a and G_b such that the block diagram of (b) is equivalent to that of (a).

(b) Find the transfer functions G_c and G_d such that the block diagram of (c) is equivalent to that of (a).

2.13. Given the block diagrams of Figure P2.13.

(a) Find the transfer functions G_a and G_b such that the block diagram of (b) is equivalent to that of (a).

(b) Find the transfer functions G_c and G_d such that the block diagram of (c) is equivalent to that of (a).

2.14. (a) Write the differential equations for the mechanical system shown in Figure P2.14(a). There are no applied forces; the system is excited only by initial conditions.

(b) A force $f(t)$ is applied downward to the mass M. Find the transfer function from the applied force to the displacement, $x_1(t)$, of the mass; that is, find $X_1(s)/F(s)$.

(c) Repeat (a) for the system of Figure P2.14(b).

(d) A force $f(t)$ is applied downward to the mass M in Figure P2.14(b). Find the transfer function $Y(s)/F(s)$.

2.15. Consider the mechanical system of Figure P2.15.

(a) Write the differential equations that describe this system.

(b) Find the transfer function from the applied force $f(t)$ to the displacement, $y(t)$, of the mass; that is, find $Y(s)/F(s)$.

2.16. Consider the mechanical system of Example 2.11, Section 2.5.

(a) Generate a flow graph for this system by solving the first equation of the example for $X_2(s)$ and the second equation for $X_1(s)$.

(b) Show that the transfer function $X_1(s)/F(s)$ for the flow graph of (a) is the same as that found in the example.

2.17. For the servomotor of Section 2.7, suppose that the shaft cannot be considered rigid but must be modeled as a torsional spring, as shown in Figure P2.17.

(a) Write the system differential equations.

(b) Draw a system block diagram, with $E_a(s)$ as the input and $\Theta_l(s)$ as the output.

(c) Find the transfer function $\Theta_m(s)/E_a(s)$.

(d) Find the transfer function $\Theta_l(s)/E_a(s)$.

(e) Find the transfer function $\Theta_l(s)/\Theta_m(s)$.

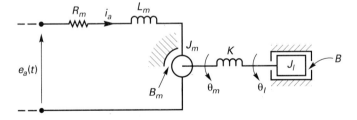

Figure P2.17

2.18. (a) Shown in Figure P2.18 is a fluid-coupled mechanical rotational system. Torque is applied to the outer cylinder having inertia J_1. Energy is coupled in to inertia J_2 through the fluid friction B_1. Write the differential equations for this system.

(b) Use the equations of (a) to find the transfer function $\theta_2(s)/T(s)$, where $T(s) = \mathcal{L}[\tau(t)]$.

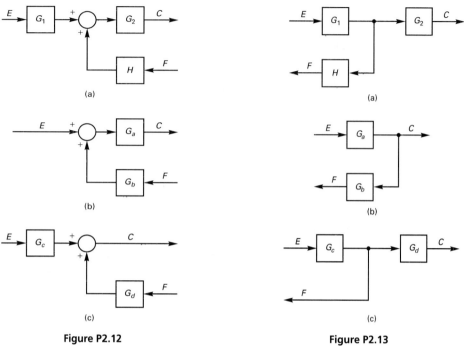

Figure P2.12

Figure P2.13

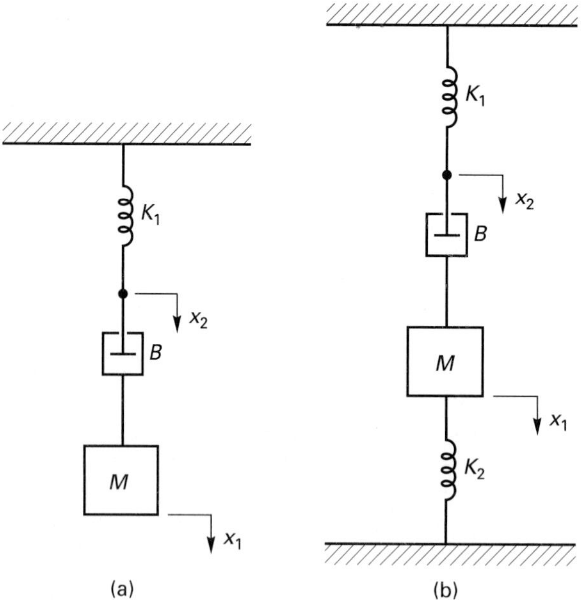

Figure P2.14

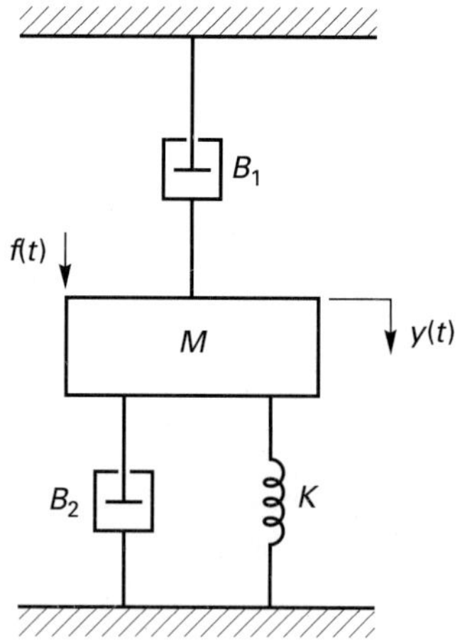

Figure P2.15

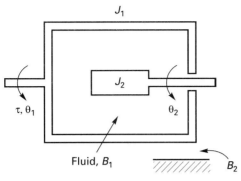

Figure P2.18

2.19. Draw an analogous electrical circuit for the mechanical system of Figure P2.14(b). Let electrical currents be analogous to mechanical velocities.

2.20. Draw an analogous electrical circuit for the automobile suspension-system model of Example 2.11. Let electrical currents be analogous to mechanical velocities.

2.21. Shown in Figure P2.21 is a gear-coupled mechanical system and an equivalent model of this system. Find the element values in the equivalent model such that the relationship of θ_I to τ_1 is the same for both systems.

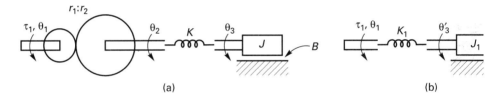

(a)   (b)

Figure P2.21

2.22. Shown in Figure P2.22 is the block diagram of the servo-control system for one of the joints of a robot. This system is described in Section 2.12.
   (a) Find the plant transfer function $\theta_L(s)/E_a(s)$.
   (b) Find the closed-loop system transfer function $\theta_L(s)/\theta_c(s)$.
   (c) Find the transfer function from the system input $\theta_c(s)$ to the motor armature voltage $E_a(s)$ for the closed-loop system.
   (d) Suppose that, in the closed-loop system, the signal $E_a(s)$ is known. Express the output $\theta_L(s)$ as the function of the signal $E_a(s)$.

2.23. Consider the linear model of the simple pendulum developed in Section 2.14.
   (a) Show that if the bob is held stationary at an initial angle θ_1 and then released, the resulting motion is sinusoidal in time.
   (b) Find the frequency in hertz of the sinusoidal motion as a function of the parameters.
   (c) The following MATLAB program solves the differential equation in (a). Run this program, and show that the results verify those in (a).
```
syms x xl L g
x=dsolve('D2x+(g/L)*x=0,Dx(0)=0,x(0)=xl')
```

2.24. Given the differential equation $\dot{x} + f(x) = 0$. Linearize the differential equation about $x = 0$, if the function $f(x)$ is given by

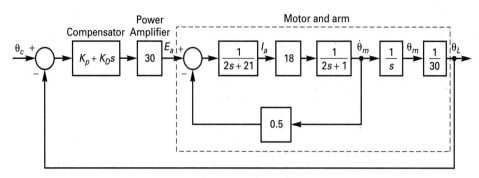

Figure P2.22

(a) $f(x) = x$ (the function is linear)
(b) $f(x) = x^3$
(c) $f(x) = e^{-x}$

3

State-Variable Models

In Chapter 2 two models of linear time-invariant (LTI) analog systems were presented: linear differential equations with constant coefficients and transfer functions. By use of the Laplace transform the transfer function can be derived from the differential equations, and a differential equation model can be derived from the transfer function using the inverse Laplace transform. In this chapter we consider a third type of model: the *state variable model* [1,2]. This model is a differential equation model, but the equations are always written in a specific format. The state-variable model, or state-space model, is expressed as n first-order coupled differential equations. These equations preserve the system's input–output relationship (that of the transfer function); in addition, an internal model of the system is given. Some additional reasons for developing the state model are as follows:

1. Computer-aided analysis and design of state models are performed more easily on digital computers for higher-order systems, while the transfer function approach tends to fail for these systems because of numerical problems.

2. In state-variable design procedures, we feed back more information (internal variables) about the plant; hence we can achieve a more complete control of the system than is possible with the transfer function approach.

3. Design procedures that result in the "best" control system are almost all based on state variable models. By "best" we mean that the system has been designed in such a way as to minimize (or maximize) a mathematical function that expresses the design criteria.

4. Even if we do not implement the state-variable design (some are not practical), we can produce a "best" system response. Then we attempt to approximate this "best" system response using classical design procedures (covered in Chapters 7 and 9).

5. State-variable models are generally required for digital simulation (digital computer solution of differential equations).

3.1 STATE-VARIABLE MODELING

We begin this section by giving an example to illustrate state-variable modeling. The system model used to illustrate state variables, a linear mechanical translational system, is given in Figure 3.1. This same model was utilized in Section 2.5, except here the displacement is denoted as $y(t)$. The differential equation describing this system is given by

$$M\frac{d^2y\,(t)}{dt^2} = f\,(t) - B\frac{dy\,(t)}{dt} - Ky\,(t) \tag{3-1}$$

and the transfer function is given by

$$G(s) = \frac{Y(s)}{F(s)} = \frac{1}{Ms^2 + Bs + K} \tag{3-2}$$

This equation gives a description of the position $y(t)$ as a function of the force $f(t)$. Suppose that we also want information about the velocity. Using the state variable approach, we define the two state variables $x_1(t)$ and $x_2(t)$ as

$$x_1(t) = y(t) \tag{3-3}$$

and

$$x_2\,(t) = \frac{dy(t)}{dt} = \frac{dx_1(t)}{dt} = \dot{x}_1(t) \tag{3-4}$$

Thus $x_1(t)$ is the position of the mass and $x_2(t)$ is its velocity. Then, from (3-1), (3-3), and (3-4), we may write

$$\frac{d^2y(t)}{dt^2} = \frac{dx_2(t)}{dt} = \dot{x}_2(t) = -\left(\frac{B}{M}\right)x_2(t) - \left(\frac{K}{M}\right)x_1(t) + \left(\frac{1}{M}\right)f(t) \tag{3-5}$$

where the overdot denotes the derivative with respect to time. The state-variable model is contained in (3-3), (3-4), and (3-5). However, the model is usually written in a specific format, which is given by rearranging the preceding equations as

$$\dot{x}_1\,(t) = x_2(t)$$

$$\dot{x}_2\,(t) = -\left(\frac{K}{M}\right)x_1(t) - \left(\frac{B}{M}\right)x_2(t) + \left(\frac{1}{M}\right)f(t)$$

$$y\,(t) = x_1(t)$$

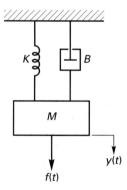

Figure 3.1 Mechanical translational system.

Usually, state equations are written in a vector-matrix format, since this allows the equations to be manipulated much more easily (see Appendix A for a review of matrices). In this format the preceding equations become

$$
\begin{bmatrix} \dot{x}_1(t) \\ \dot{x}_2(t) \end{bmatrix} = \begin{bmatrix} 0 & 1 \\ -\dfrac{K}{M} & -\dfrac{B}{M} \end{bmatrix} \begin{bmatrix} x_1(t) \\ x_2(t) \end{bmatrix} + \begin{bmatrix} 0 \\ \dfrac{1}{M} \end{bmatrix} f(t)
$$

$$
y(t) = \begin{bmatrix} 1 & 0 \end{bmatrix} \begin{bmatrix} x_1(t) \\ x_2(t) \end{bmatrix}
$$

We will now define the state of a system.

Definition   The state of a system at any time t_0 is the amount of information at t_0 that, together with all inputs for $t \geq t_0$, uniquely determines the behavior of the system for all $t \geq t_0$.

It will be shown that the state vector of the standard form of the state variable equations, to be developed next, satisfies this definition.

The standard form of the state equations of a linear time-invariant (LTI) analog system is given by

$$
\begin{aligned}
\dot{\mathbf{x}}(t) &= \mathbf{A}\mathbf{x}(t) + \mathbf{B}\mathbf{u}(t) \\
\mathbf{y}(t) &= \mathbf{C}\mathbf{x}(t) + \mathbf{D}\mathbf{u}(t)
\end{aligned}
\tag{3-6}
$$

where the vector, $\dot{\mathbf{x}}(t)$, is the time derivative of the vector $\mathbf{x}(t)$. In these equations,

$\mathbf{x}(t) =$ state vector $= (n \times 1)$ vector of the states of an nth-order system

$\mathbf{A} = (n \times n)$ system matrix

$\mathbf{B} = (n \times r)$ input matrix

$\mathbf{u}(t) =$ input vector $= (r \times 1)$ vector composed of the system input functions

$\mathbf{y}(t) =$ output vector $= (p \times 1)$ vector composed of the defined outputs

$\mathbf{C} = (p \times n)$ output matrix

$\mathbf{D} = (p \times r)$ matrix to represent direct coupling between input and output

The signals in expanded form are given by

$$
\dot{\mathbf{x}}(t) = \begin{bmatrix} \dot{x}_1(t) \\ \dot{x}_2(t) \\ \vdots \\ \dot{x}_n(t) \end{bmatrix} \qquad \mathbf{x}(t) = \begin{bmatrix} x_1(t) \\ x_2(t) \\ \vdots \\ x_n(t) \end{bmatrix} \qquad \mathbf{u}(t) = \begin{bmatrix} u_1(t) \\ u_2(t) \\ \vdots \\ u_r(t) \end{bmatrix} \qquad \mathbf{y}(t) = \begin{bmatrix} y_1(t) \\ y_2(t) \\ \vdots \\ y_p(t) \end{bmatrix}
$$

We refer to the two matrix equations of (3-6) as the *state-variable equations* of the system. The first equation is called the *state equation*, and the second one is called the *output equation*. The first equation, the state equation, is a first-order matrix differential equation, and the state vector, $\mathbf{x}(t)$, is its solution. Given knowledge of $\mathbf{x}(t)$ and the input vector $\mathbf{u}(t)$, the output equation yields the output $\mathbf{y}(t)$. Usually the matrix $\mathbf{D}$ is zero, since in physical systems, dynamics appear in all paths between the inputs and the outputs. A nonzero value of

D indicates at least one direct path between the inputs and the outputs, in which the path transfer function can be modeled as a pure gain.

In the equation for the state variables $\mathbf{x}(t)$ in (3-6), only the first derivatives of the state variables may appear on the left side of the equation, and no derivatives may appear on the right side. No derivatives may appear in the output equation. Valid equations that model a system may be written without following these rules; however, those equations will not be in the standard format.

The general form of the state equations just given allows for more than one input and more than one output; these systems are called *multivariable systems.* For the case of one input, the matrix **B** is a column vector, and the vector $\mathbf{u}(t)$ is a scalar. For the case of one output, the vector $\mathbf{y}(t)$ is a scalar, and the matrix **C** is a row vector. In this book we do not differentiate notationally between a column vector and a row vector. The use of the vector implies the type. An example is now given to illustrate a multivariable system.

Example 3.1

Consider the system described by the coupled differential equations

$$\ddot{y}_1 + k_1 \dot{y}_1 + k_2 y_1 = u_1 + k_3 u_2$$
$$\dot{y}_2 + k_4 y_2 + k_5 \dot{y}_1 = k_6 u_1$$

where u_1 and u_2 are inputs, y_1 and y_2 are outputs, and k_i, $i = 1, \ldots, 6$ are system parameters. The notational dependence of the variables on time has been omitted for convenience. We may define the states as the outputs and, where necessary, the derivatives of the outputs.

$$x_1 = y_1 \qquad x_2 = \dot{y}_1 = \dot{x}_1 \qquad x_3 = y_2$$

From the system differential equations we write

$$\dot{x}_2 = -k_2 x_1 - k_1 x_2 + u_1 + k_3 u_2$$
$$\dot{x}_3 = -k_5 x_2 - k_4 x_3 + k_6 u_1$$

We rewrite the differential equations in the following order:

$$\dot{x}_1 = x_2$$
$$\dot{x}_2 = -k_2 x_1 - k_1 x_2 + u_1 + k_3 u_2$$
$$\dot{x}_3 = -k_5 x_2 - k_4 x_3 + k_6 u_1$$

with the output equations

$$y_1 = x_1$$
$$y_2 = x_3$$

These equations may be written in matrix form as

$$\dot{\mathbf{x}} = \begin{bmatrix} 0 & 1 & 0 \\ -k_2 & -k_1 & 0 \\ 0 & -k_5 & -k_4 \end{bmatrix} \mathbf{x} + \begin{bmatrix} 0 & 0 \\ 1 & k_3 \\ k_6 & 0 \end{bmatrix} \mathbf{u}$$

$$\mathbf{y} = \begin{bmatrix} 1 & 0 & 0 \\ 0 & 0 & 1 \end{bmatrix} \mathbf{x}$$

In the preceding example suppose that the equations model a mechanical translational system. Suppose further that both y_1 and y_2 are displacements in the system. Then x_2, which is $\dot{y}_1$, is a velocity in the system. Hence all the state variables represent physical variables within the system; x_1 and x_3 are displacements and x_2 is a velocity. Generally, we prefer that the state variables be physically identifiable variables, but this is not necessary; this topic is developed further in Section 3.5. Next we consider a method for obtaining the state model directly from the system transfer function. In general, this method does not result in the state variables being physical variables.

3.2 SIMULATION DIAGRAMS

In Section 3.1 we presented an example of finding the state model of a system directly from the system differential equations. The procedure in that example is very useful and is employed in many practical situations. However, if a system-identification technique (introduced in Section 2.13) is used to obtain the system model, only a transfer function may be available to describe the system. For this and other reasons, it is advantageous to have a method available to obtain a state model directly from a transfer function.

The method considered here is based on the use of simulation diagrams. A *simulation diagram* is a certain type of either a block diagram or a flow graph that is constructed to have a specified transfer function or to model a specified set of differential equations. Given the transfer function, the differential equations, or the state equations of a system, we can construct a simulation diagram of the system. The simulation diagram is aptly named, since it is useful in constructing either digital computer or analog computer simulations of a system.

The basic element of the simulation diagram is the integrator. Figure 3.2 shows the block diagram of an integrating device. In this figure

$$y(t) = \int x(t)\, dt$$

and the Laplace transform of this equation yields

$$Y(s) = \frac{1}{s}X(s) \tag{3-7}$$

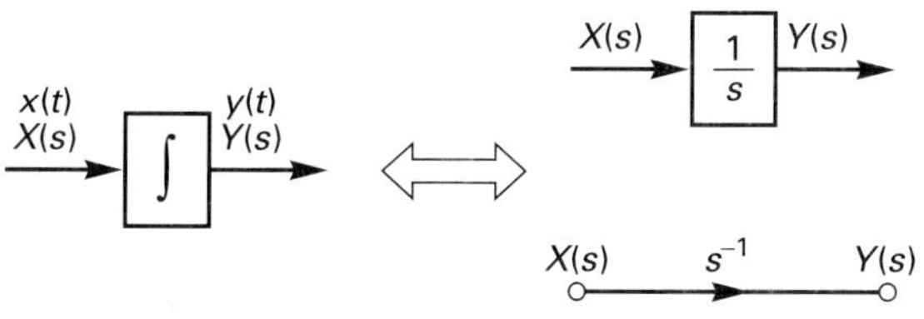

Figure 3.2 Integrating device.

Hence the transfer function of a device that integrates a signal is $1/s$, as shown in Figure 3.2. We are interested only in the transfer function; thus we have ignored the initial condition on $y(t)$.

If the output of an integrator is labeled as $y(t)$, the input to the integrator must be dy/dt. We use this characteristic of the integrator to aid us in constructing simulation diagrams. For example, two integrators are cascaded in Figure 3.3(a). If the output of the second integrator is $y(t)$, the input to this integrator must be $\dot{y}(t)$. In a like manner, the input to the first integrator must be $\ddot{y}(t)$, where $\ddot{y}(t) = d^2y(t)/dt^2$.

We can use these two integrators to construct a simulation diagram of the mechanical system of Figure 3.1. The input to the cascaded integrators in Figure 3.3(a) is $\ddot{y}(t)$, and the equation that $\ddot{y}(t)$ must satisfy for the mechanical system is obtained from (3-1).

$$\ddot{y}(t) = -\frac{B}{M}\dot{y}(t) - \frac{K}{M}y(t) + \frac{1}{M}f(t) \tag{3-8}$$

Hence a summing junction and appropriate gains can be added to the block diagram of Figure 3.3(a) to satisfy this equation, as shown in Figure 3.3(b). Figure 3.3(b) is called a *simulation diagram* of the mechanical system.

In the approach just described, a simulation diagram of integrators, gains, and summing junctions was constructed that satisfied a given differential equation. This is the general approach for the construction of simulation diagrams. Note that, by Mason's gain formula, the transfer function of the simulation diagram in Figure 3.3(b) is given by

$$G(s) = \frac{(1/M)\,s^{-2}}{1 + (B/M)\,s^{-1} + (K/M)\,s^{-2}} = \frac{1/M}{s^2 + (B/M)\,s + (K/M)} \tag{3-9}$$

This transfer function checks that derived in (3-2).

Note that we can reverse the last procedure. We can begin with the given transfer function and, through the reverse of the given steps, construct the simulation diagram. If the

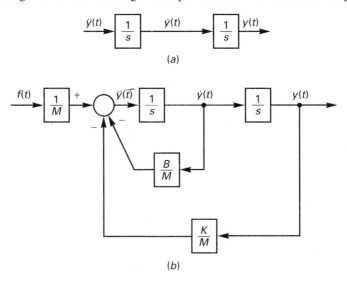

(a)

(b)

Figure 3.3   Simulation diagrams.

simulation diagram is constructed from the system differential equations, the simulation diagram will usually be unique. However, if the transfer function is used to construct the simulation diagram, the simulation diagram can be of many different forms, that is, the simulation diagram is not unique. Two common, useful forms of the simulation diagram are now presented.

The two different simulation diagrams to be given realize the general transfer function

$$G(s) = \frac{b_{n-1}s^{-1} + b_{n-2}s^{-2} + \cdots + b_0 s^{-n}}{1 + a_{n-1}s^{-1} + a_{n-2}s^{-2} + \cdots + a_0 s^{-n}}$$

$$= \frac{b_{n-1}s^{n-1} + b_{n-2}s^{n-2} + \cdots + b_0}{s^n + a_{n-1}s^{n-1} + a_{n-2}s^{n-2} + \cdots + a_0} \qquad (3\text{-}10)$$

The first simulation diagram, called the *control canonical form*, is given in Figure 3.4 for the case $n = 3$, that is, for

$$G(s) = \frac{b_2 s^2 + b_1 s + b_0}{s^3 + a_2 s^2 + a_1 s + a_0} \qquad (3\text{-}11)$$

The second simulation diagram, called the *observer canonical form*, is given in Figure 3.5 for (3-11). The reasons for these particular names are from modern control theory and will become evident later. It is seen from Mason's gain formula that the two simulation diagrams have the transfer function of (3-11). It is noted that the state vector $\mathbf{x}(t)$ in Figure 3.4 is *not* equal to state vector $\mathbf{x}(t)$ in Figure 3.5.

Note that in the transfer function of (3-10), the order of the numerator must be at least one less than the order of the denominator. Of course, any coefficient a_i or b_i may be zero. The general transfer function of a physical system is always assumed to be of this form; the

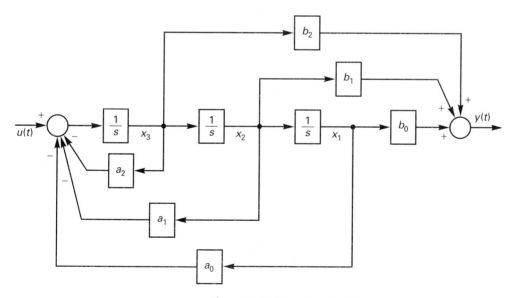

Figure 3.4   Control-canonical form.

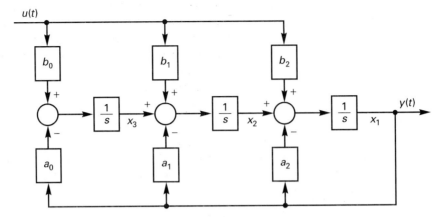

Figure 3.5 Observer-canonical form.

reason for this assumption will become evident when the frequency response of systems is covered. However, simulation diagrams can be constructed for transfer functions for which the order of the numerator is equal to that of the denominator [see Problem 3.6(d)].

Once a simulation diagram of a transfer function is constructed, a state model of the system is easily obtained. The procedure to do this has two steps:

1. Assign a state variable to the output of each integrator.
2. Write an equation for the input of each integrator and an equation for each system output. These equations are written as functions of the integrator outputs and the system inputs.

This procedure yields the following state equations for the control canonical form of Figure 3.4, where the states are identified in the figure. The state equations are, from Figure 3.4,

$$\dot{\mathbf{x}} = \begin{bmatrix} 0 & 1 & 0 \\ 0 & 0 & 1 \\ -a_0 & -a_1 & -a_2 \end{bmatrix} \mathbf{x} + \begin{bmatrix} 0 \\ 0 \\ 1 \end{bmatrix} u$$

$$y = \begin{bmatrix} b_0 & b_1 & b_2 \end{bmatrix} \mathbf{x}$$

(3-12)

These equations are easily extended to the nth-order system (see Problem 3.25). Suppose, in (3.12), $y(t)$ is position for a position-control system. Then, in general, $x_1(t)$ is not position, $x_2(t)$ is not velocity, and so on. For the observer canonical form of Figure 3.5, the state equations are written as

$$\dot{\mathbf{x}} = \begin{bmatrix} -a_2 & 1 & 0 \\ -a_1 & 0 & 1 \\ -a_0 & 0 & 0 \end{bmatrix} \mathbf{x} + \begin{bmatrix} b_2 \\ b_1 \\ b_0 \end{bmatrix} u$$

$$y = \begin{bmatrix} 1 & 0 & 0 \end{bmatrix} \mathbf{x}$$

(3-13)

Suppose again that $y(t)$ is position for a position-control system. Then $x_1(t)$ is also position, but $x_2(t)$ in general is not velocity, and so on. Note that the coefficients that appear in the transfer function also appear directly in the matrices of the state equations. In fact, it is evident that the state equations for these two standard forms can be written directly from the transfer function, without drawing simulation diagrams.

To illustrate further the procedure for writing state equations from the simulation diagram, consider the following example.

Example 3.2

Consider again the mechanical system of Figures 3.1 and 3.3(b). In the simulation diagram of Figure 3.3(b), which is repeated in Figure 3.6, a state has been assigned to each integrator output. With this assignment, the input to the rightmost integrator is $\dot{x}_1$; thus the equation for this integrator input is

$$\dot{x}_1 = x_2$$

For the other integrator the input is $\dot{x}_2$. Thus the equation for this integrator input is

$$\dot{x}_2 = -\frac{K}{M}x_1 - \frac{B}{M}x_2 + \frac{1}{M}f$$

and the system output equation is

$$y = x_1$$

These equations may be written in the standard state-variable matrix format as

$$\dot{\mathbf{x}}(t) = \begin{bmatrix} 0 & 1 \\ -\dfrac{K}{M} & -\dfrac{B}{M} \end{bmatrix} \mathbf{x}(t) + \begin{bmatrix} 0 \\ \dfrac{1}{M} \end{bmatrix} f(t)$$

$$y(t) = \begin{bmatrix} 1 & 0 \end{bmatrix} \mathbf{x}(t)$$

These equations are the same as those developed in the first section of this chapter, and the state model is seen to be in the control canonical form.

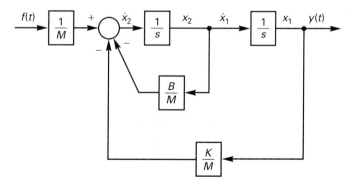

Figure 3.6 Simulation diagram.

Example 3.3

In this example we develop a state-variable model of a dc motor used in rolling steel. Figure 3.7(a) illustrates steel rolling. The purpose of the rolling is to reduce the thickness of the steel, and $w_3 < w_2 < w_1$. Two sets of rollers are shown in Figure 3.7(a); in general, several more sets will appear in a steel-rolling mill. As the steel is rolled, the thickness decreases, with a corresponding increase in the length of the steel sheet. In Figure 3.7(a) the second set of roller must rotate at a greater speed than the first set. The speed of the second set of rollers is determined by the speed of the first set and the reduction in the thickness of the steel in the first set. Hence the speed of the second set must be controlled accurately. A block diagram of a motor speed control system is given in Figure 3.7(b), and a dc motor is depicted in Figure 3.7(c). From Section 2.7, the equations of the motor are

[eq. 2-41]
$$e_m(t) = K_m \frac{d\theta(t)}{dt}$$

[eq. 2-43]
$$e_a(t) = L_m \frac{di_a(t)}{dt} + R_m i_a(t) + e_m(t)$$

[eq. 2-45]
$$\tau(t) = K_\tau i_a(t)$$

[eq. 2-47]
$$J \frac{d^2\theta(t)}{dt^2} + B \frac{d\theta(t)}{dt} = \tau(t)$$

In these equations the variables are

$$e_a(t) = \text{armature voltage (the input)}$$
$$e_m(t) = \text{back-EMF}$$
$$i_a(t) = \text{armature current}$$
$$\tau(t) = \text{developed torque}$$
$$\theta(t) = \text{motor shaft angle}$$
$$d\theta(t)/dt = \omega(t) = \text{shaft speed (the output)}$$

In a speed-control system, the motor speed $\omega(t)$ is controlled by varying the armature voltage $e_a(t)$. Hence $e_a(t)$ is the input variable and $\omega(t)$ is the output variable.

We choose as the state variables $x_1(t) = \omega(t) = d\theta(t)/dt$ and $x_2(t) = i_a(t)$. Note that both state variables can be measured easily, and this is one reason for the choice. The measurements of the states of a system are shown to be very important in the state variable design procedures of Chapter 10.

The state equations will now be derived. From (2-45) and (2-47),

$$\frac{d^2\theta(t)}{dt^2} = -\frac{B}{J} \frac{d\theta(t)}{dt} + \frac{K_\tau}{J} i_a(t)$$

or

$$\dot{x}_1(t) = -\frac{B}{J} x_1(t) + \frac{K_\tau}{J} x_2(t).$$

From (2-41) and (2-43),

$$\frac{di_a(t)}{dt} = \dot{x}_2(t) = -\frac{K_m}{L_m} x_1(t) - \frac{R_m}{L_m} x_2(t) + \frac{1}{L_m} e_a(t)$$

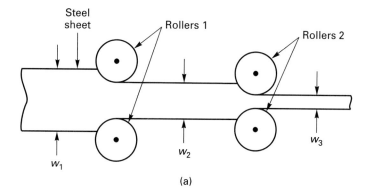

(a)

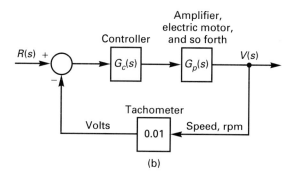

(b)

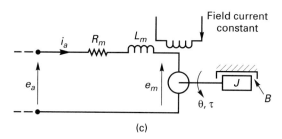

(c)

Figure 3.7   (a) Steel rolling; (b) dc motor speed control for one set of rollers; (c) model for the motor.

The output equation is $y(t) = d\theta(t)/dt = x_1(t)$. Hence the state equations are

$$\dot{\mathbf{x}}(t) = \begin{bmatrix} -\dfrac{B}{J} & \dfrac{K_\tau}{J} \\[2mm] -\dfrac{K_m}{L_m} & -\dfrac{R_m}{L_m} \end{bmatrix} \mathbf{x}(t) + \begin{bmatrix} 0 \\[2mm] \dfrac{1}{L_m} \end{bmatrix} u(t)$$

$$y(t) = \begin{bmatrix} 1 & 0 \end{bmatrix} \mathbf{x}(t)$$

where the input $u(t) = e_a(t)$. The simulation diagram for these equations is shown in Figure 3.8, where $\Omega(s) = L[d\theta(t)/dt]$. This example develops a state model that is used in practical speed-control design; the model is neither of the canonical forms presented earlier.

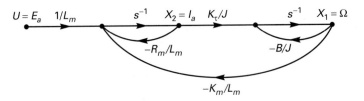

Figure 3.8   Simulation diagram for Example 3.3.

In this and the preceding section, we examined some specific ways to represent a system with state variables. These methods are by no means exhaustive, since the selection of the state variables is not a unique process. As will be shown in Section 3.5, there are actually an unbounded number of different correct state variable representations for any system.

The number of state variables required to model a system will always be equal to the order of the system. While this number is apparent for single-input, single-output systems described by a single nth-order differential equation or by a transfer function, such is sometimes not the case for multivariable systems such as the one in Example 3.1. For that example, the number of state variables is equal to the number of dependent variables plus the number of their derivatives up to $(r - 1)$, where r is the order of the highest derivative of a given dependent variable.

3.3 SOLUTION OF STATE EQUATIONS

We have developed procedures for writing the state equations of a system, given either the system differential equations or the system transfer function. In this section we present two methods for finding the solution of the state equations.

3.3.1 Laplace Transform Solution

The standard form of the state equation is given by

$$\dot{\mathbf{x}}(t) = \mathbf{A}\mathbf{x}(t) + \mathbf{B}\mathbf{u}(t) \qquad (3\text{-}14)$$

This equation will now be solved using the Laplace transform. Consider the first equation in the set (3-14):

$$\dot{x}_1 = a_{11}x_1 + a_{12}x_2 + \cdots + a_{1n}x_n + b_{11}u_1 + \cdots + b_{1r}u_r \qquad (3\text{-}15)$$

where a_{ij} and b_{ij} are the appropriate elements of the matrices $\mathbf{A}$ and $\mathbf{B}$, and the dependence on time is omitted for convenience. The Laplace transform of this equation yields

$$
\begin{aligned}
sX_1(s) - x_1(0) = {} & a_{11}X_1(s) + a_{12}X_2(s) + \cdots + a_{1n}X_n(s) \\
& + b_{11}U_1(s) + \cdots + b_{1r}U_r(s)
\end{aligned}
\qquad (3\text{-}16)
$$

where the initial condition is included, since we will find the complete solution. The Laplace transform of the second equation in (3-14) yields

$$sX_2(s) - x_2(0) = a_{21}X_1(s) + a_{22}X_2(s) + \cdots + a_{2n}X_n(s)$$
$$+ b_{21}U_1(s) + \cdots + b_{2r}U_r(s) \tag{3-17}$$

The Laplace transform of the remaining equations in (3-14) yields equations of the same form. These equations may be written in matrix form as

$$s\mathbf{X}(s) - \mathbf{x}(0) = \mathbf{A}\mathbf{X}(s) + \mathbf{B}\mathbf{U}(s)$$

where

$$\mathbf{x}(0) = \begin{bmatrix} x_1(0) & x_2(0) & \cdots & x_n(0) \end{bmatrix}^T$$

We would like to solve this equation for $\mathbf{X}(s)$; to do this we collect all terms containing $\mathbf{X}(s)$ on the left side of the equation:

$$s\mathbf{X}(s) - \mathbf{A}\mathbf{X}(s) = \mathbf{x}(0) + \mathbf{B}\mathbf{U}(s) \tag{3-18}$$

It is necessary to factor $\mathbf{X}(s)$ in the left side to solve this equation. First the term $s\mathbf{X}(s)$ must be written as $s\mathbf{I}\mathbf{X}(s)$, where $\mathbf{I}$ is the identity matrix. Then

$$s\mathbf{I}\mathbf{X}(s) - \mathbf{A}\mathbf{X}(s) = (s\mathbf{I} - \mathbf{A})\mathbf{X}(s) = \mathbf{x}(0) + \mathbf{B}\mathbf{U}(s) \tag{3-19}$$

This additional step is necessary since the subtraction of the matrix $\mathbf{A}$ from the scalar s is not defined; we cannot factor $\mathbf{X}(s)$ directly in (3-18). Equation (3-19) may now be solved for $\mathbf{X}(s)$:

$$\mathbf{X}(s) = (s\mathbf{I} - \mathbf{A})^{-1}\mathbf{x}(0) + (s\mathbf{I} - \mathbf{A})^{-1}\mathbf{B}\mathbf{U}(s) \tag{3-20}$$

and the state vector $\mathbf{x}(t)$ is the inverse Laplace transform of this equation.

To obtain a general relationship for the solution, we define the *state transition matrix* $\mathbf{\Phi}(t)$ as

$$\mathbf{\Phi}(t) = \mathcal{L}^{-1}[(s\mathbf{I} - \mathbf{A})^{-1}] \tag{3-21}$$

This matrix is also called the *fundamental matrix*. The matrix $(s\mathbf{I} - \mathbf{A})^{-1}$ is called the *resolvant* of $\mathbf{A}$ [2]. Note that for an nth-order system, the state transition matrix is an $(n \times n)$ matrix. The inverse Laplace transform of a matrix is defined as the inverse Laplace transform of the elements of the matrix. Finding the inverse Laplace transform indicated in (3-21) in general is difficult, time consuming, and prone to errors. A more practical procedure for calculating the state vector $\mathbf{x}(t)$ is by way of a computer simulation. Both digital computer simulations and analog computer simulations are discussed later in this chapter.

As a final point, note from (3-20) that the state transition matrix is the solution to the differential equation

$$\dot{\mathbf{x}}(t) = \mathbf{A}\mathbf{x}(t) \Rightarrow \mathbf{x}(t) = \mathbf{\Phi}(t)\mathbf{x}(0) \tag{3-22}$$

Now an example is presented to illustrate the calculation of a state transition matrix using (3-21).

Example 3.4

As an example, consider the system described by the transfer function

$$G(s) = \frac{Y(s)}{U(s)} = \frac{1}{s^2 + 3s + 2} = \frac{s^{-2}}{1 + 3s^{-1} + 2s^{-2}}$$

The observer-canonical form is used to develop a state model for this example, and the flow graph form of the simulation diagram is given in Figure 3.9. The state equations are then

$$\dot{\mathbf{x}}(t) = \begin{bmatrix} -3 & 1 \\ -2 & 0 \end{bmatrix} \mathbf{x}(t) + \begin{bmatrix} 0 \\ 1 \end{bmatrix} u(t)$$

$$y(t) = \begin{bmatrix} 1 & 0 \end{bmatrix} \mathbf{x}(t)$$

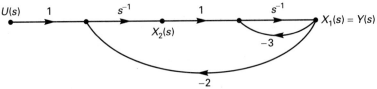

Figure 3.9   System for Example 3.4.

It is seen that Mason's gain formula yields the correct transfer function. In addition, a MATLAB program that calculates the transfer function is

```
e3p4=ss([-3 1;-2 0],[0; 1],[1 0],0);
tf(e3p4)
```

To find the state transition matrix, we first calculate the matrix $(s\mathbf{I} - \mathbf{A})$.

$$s\mathbf{I} - \mathbf{A} = s\begin{bmatrix} 1 & 0 \\ 0 & 1 \end{bmatrix} - \begin{bmatrix} -3 & 1 \\ -2 & 0 \end{bmatrix} = \begin{bmatrix} s+3 & -1 \\ 2 & s \end{bmatrix}$$

To find the inverse of this matrix, we calculate its adjoint matrix.

$$\text{Adj}\,(s\mathbf{I} - \mathbf{A}) = \begin{bmatrix} s & 1 \\ -2 & s+3 \end{bmatrix}$$

The determinant of the matrix is

$$\det(s\mathbf{I} - \mathbf{A}) = s^2 + 3s + 2 = (s+1)(s+2)$$

and the inverse matrix is then the adjoint matrix divided by the determinant.

$$(s\mathbf{I} - \mathbf{A})^{-1} = \begin{bmatrix} \dfrac{s}{(s+1)(s+2)} & \dfrac{1}{(s+1)\,(s+2)} \\ \dfrac{-2}{(s+1)(s+2)} & \dfrac{s+3}{(s+1)\,(s+2)} \end{bmatrix}$$

$$= \begin{bmatrix} \dfrac{-1}{s+1} + \dfrac{2}{s+2} & \dfrac{1}{s+1} + \dfrac{-1}{s+2} \\ \dfrac{-2}{s+1} + \dfrac{2}{s+2} & \dfrac{2}{s+1} + \dfrac{-1}{s+2} \end{bmatrix}$$

The state transition matrix is the inverse Laplace transform of this matrix.

$$\Phi(t) = \begin{bmatrix} -e^{-t} + 2e^{-2t} & e^{-t} - e^{-2t} \\ -2e^{-t} + 2e^{-2t} & 2e^{-t} - e^{-2t} \end{bmatrix}$$

Hence we see that the state transition matrix for a second-order system is a (2×2) matrix. In a like manner, the state transition matrix for an nth-order system is $(n \times n)$. A MATLAB program that solves directly for the elements of $\Phi(t)$ using (3-22) is given by

```
S=dsolve('Dx1=-3*x1+x2, Dx2=-2*x1, x1(0)=x10,x2(0)=x20')
s.x1
s.x1
```

MATLAB can also be used to find the inverse Laplace transforms of each of the element of the state transition matrix, from (3-21).

With the definition of the state transition matrix in (3-21), the equation for the complete solution of the state equations, (3-20), which is repeated next, can be found.

[eq. (3-20)] $\mathbf{X}(s) = (s\mathbf{I} - \mathbf{A})^{-1}\mathbf{x}(0) + (s\mathbf{I} - \mathbf{A})^{-1}\mathbf{B}U(s)$

We illustrate the complete solution of this equation using an example before giving the general form of the solution.

Example 3.5

Consider the same system as described in Example 3.4. State equations were derived to be

$$\dot{\mathbf{x}}(t) = \begin{bmatrix} -3 & 1 \\ -2 & 0 \end{bmatrix} \mathbf{x}(t) + \begin{bmatrix} 0 \\ 1 \end{bmatrix} u(t)$$

with the Laplace transform of the transition matrix given by

$$\Phi(s) = \mathcal{L}\left[\Phi(t)\right] = (s\mathbf{I} - \mathbf{A})^{-1} = \begin{bmatrix} \dfrac{s}{(s+1)(s+2)} & \dfrac{1}{(s+1)(s+2)} \\ \dfrac{-2}{(s+1)(s+2)} & \dfrac{s+3}{(s+1)(s+2)} \end{bmatrix}$$

Suppose that a unit step is applied as an input. Then $U(s) = 1/s$, and the second term in (3-20) becomes

$$(s\mathbf{I} - \mathbf{A})^{-1}\mathbf{B}U(s) = \begin{bmatrix} \dfrac{s}{(s+1)(s+2)} & \dfrac{1}{(s+1)(s+2)} \\ \dfrac{-2}{(s+1)(s+2)} & \dfrac{s+3}{(s+1)(s+2)} \end{bmatrix} \begin{bmatrix} 0 \\ 1 \end{bmatrix} \dfrac{1}{s}$$

$$= \begin{bmatrix} \dfrac{1}{s(s+1)(s+2)} \\ \dfrac{s+3}{s(s+1)(s+2)} \end{bmatrix} = \begin{bmatrix} \dfrac{\frac{1}{2}}{s} + \dfrac{-1}{s+1} + \dfrac{\frac{1}{2}}{s+2} \\ \dfrac{\frac{3}{2}}{s} + \dfrac{-2}{s+1} + \dfrac{\frac{1}{2}}{s+2} \end{bmatrix}$$

The inverse Laplace transform of this term is

$$\mathcal{L}^{-1}\left((s\mathbf{I} - \mathbf{A})^{-1}\mathbf{B}U(s)\right) = \begin{bmatrix} \left(\dfrac{1}{2}\right) - e^{-t} + \left(\dfrac{1}{2}\right)e^{-2t} \\ \left(\dfrac{3}{2}\right) - 2e^{-t} + \left(\dfrac{1}{2}\right)e^{-2t} \end{bmatrix}$$

A MATLAB program that solves for this term is given by

```
S=dsolve('Dx1=-3*x1+x2, Dx2=-2*x1+1, x1(0)=0,x2(0)=0')
s.x1
s.x2
```

The state transition matrix was derived in Example 3.4. Hence the complete solution of the state equations is given by

$$\mathbf{x}(t) = \mathcal{L}^{-1}((s\mathbf{I} - \mathbf{A})^{-1}\mathbf{x}(0) + (s\mathbf{I} - \mathbf{A})^{-1}\mathbf{B}U(s))$$

$$= \begin{bmatrix} -e^{-t} + 2e^{-2t} & e^{-t} - e^{-2t} \\ -2e^{-t} + 2e^{-2t} & 2e^{-t} - e^{-2t} \end{bmatrix} \begin{bmatrix} x_1(0) \\ x_2(0) \end{bmatrix} + \begin{bmatrix} \left(\frac{1}{2}\right) - e^{-t} + \left(\frac{1}{2}\right)e^{-2t} \\ \left(\frac{3}{2}\right) - 2e^{-t} + \left(\frac{1}{2}\right)e^{-2t} \end{bmatrix}$$

and the state variables are given by

$$x_1(t) = (-e^{-t} + 2e^{-2t})x_1(0) + (e^{-t} - e^{-2t})x_2(0) + \left(\frac{1}{2}\right) - e^{-t} + \left(\frac{1}{2}\right)e^{-2t}$$

and

$$x_2(t) = (-2e^{-t} + 2e^{-2t})x_1(0) + (2e^{-t} - e^{-2t})x_2(0) + \left(\frac{3}{2}\right) - 2e^{-t} + \left(\frac{1}{2}\right)e^{-2t}$$

Finding the complete solution of the state equations is long and involved, even for a second-order system. The necessity for reliable machine solutions, such as a digital computer simulation, is evident.

A general form of the solution of the state equations of (3-20) is developed next. The second term in the right member of this equation is a product of two terms in the Laplace variable s. Thus the inverse Laplace transform of this term can be expressed as a convolution integral (see Appendix **B**). The inverse Laplace transform of (3-20) is then

$$\mathbf{x}(t) = \Phi(t)\mathbf{x}(0) + \int_0^t \Phi(t - \tau)\mathbf{B}\mathbf{u}(\tau)\, d\tau \tag{3-23}$$

By the convolution theorem, this solution can also be expressed as

$$\mathbf{x}(t) = \Phi(t)\mathbf{x}(0) + \int_0^t \Phi(\tau)\mathbf{B}\mathbf{u}(t - \tau)\, d\tau \tag{3-24}$$

Note that the solution is composed of two terms. The first term is often referred to as either the *zero-input part* or the *initial-condition part* of the solution, and the second term is called either the *zero-state part* or the *forced part*. Equations (3-23) and (3-24) are sometimes called the *convolution solution*.

We see that the state transition matrix is central to the solution of the state equations. The solution in this form is quite difficult to calculate except for the simplest of systems. The preceding example is used to illustrate this form of the solution.

Example 3.6

In the preceding example, the input was a unit step. Thus, in (3-23),

$$\int_0^t \Phi(t-\tau)\mathbf{B}\mathbf{u}(\tau)\,d\tau = \int_0^t \begin{bmatrix} -e^{-(t-\tau)} + 2e^{-2(t-\tau)} & e^{-(t-\tau)} - e^{-2(t-\tau)} \\ -2e^{-(t-\tau)} + 2e^{-2(t-\tau)} & 2e^{-(t-\tau)} - e^{-2(t-\tau)} \end{bmatrix} \begin{bmatrix} 0 \\ 1 \end{bmatrix} d\tau$$

$$= \begin{bmatrix} \displaystyle\int_0^t (e^{-(t-\tau)} - e^{-2(t-\tau)})\,d\tau \\ \displaystyle\int_0^t (2e^{-(t-\tau)} - e^{-2(t-\tau)})\,d\tau \end{bmatrix}$$

$$= \begin{bmatrix} \left(e^{-t}e^{\tau} - \left(\frac{1}{2}\right)e^{-2t}e^{2\tau}\right)_0^t \\ \left(2e^{-t}e^{\tau} - \left(\frac{1}{2}\right)e^{-2t}e^{2\tau}\right)_0^t \end{bmatrix} = \begin{bmatrix} (1 - e^{-t}) - \left(\frac{1}{2}\right)(1 - e^{-2t}) \\ 2(1 - e^{-t}) - \left(\frac{1}{2}\right)(1 - e^{-2t}) \end{bmatrix}$$

$$= \begin{bmatrix} \left(\frac{1}{2}\right) - e^{-t} + \left(\frac{1}{2}\right)e^{-2t} \\ \left(\frac{3}{2}\right) - 2e^{-t} + \left(\frac{1}{2}\right)e^{-2t} \end{bmatrix}$$

This result checks that of Example 3.5. The result derived here is only the forced part of the solution. The initial-condition part of the solution is obtained from the term $\Phi(t)\mathbf{x}(0)$ in (3-23). This term was evaluated in Example 3.5 and is not repeated here.

The complete solution to the state equations was derived in this section. This solution may be evaluated either by the Laplace transform or by a combination of the Laplace transform and the convolution integral. Either procedure is long, time consuming, and prone to errors. The practical method for evaluating the time response of a system is through simulation.

3.3.2 Infinite Series Solution

As just shown, the state transition matrix can be evaluated using the Laplace transform. An alternative procedure for this evaluation is now developed.

One method of solution of differential equations is to assume as a solution an infinite series with unknown coefficients. The infinite series is then substituted into the differential equation to evaluate the unknown coefficients. This method is now used to find the state transition matrix. For the case that all inputs to a system are zero, the state equations may be written as

$$\dot{\mathbf{x}}(t) = \mathbf{A}\mathbf{x}(t) \tag{3-25}$$

with the solution, from (3-23),

$$\mathbf{x}(t) = \Phi(t)\mathbf{x}(0) \tag{3-26}$$

Since we are solving for a vector $\mathbf{x}(t)$, the solution must be assumed to be of the form

$$\dot{\mathbf{x}}(t) = (\mathbf{K}_0 + \mathbf{K}_1 t + \mathbf{K}_2 t^2 + \mathbf{K}_3 t^3 + \cdots)\mathbf{x}(0) = \sum_{i=0}^{\infty} \mathbf{K}_i t^i \mathbf{x}(0) \tag{3-27}$$

$$= \Phi(t)\mathbf{x}(0)$$

where the $(n \times n)$ matrices $\mathbf{K}_i$ are unknown and t is the scalar time. Differentiating this expression yields

$$\dot{\mathbf{x}}(t) = (\mathbf{K}_1 + 2\mathbf{K}_2 t + 3\mathbf{K}_3 t^2 + \cdots)\mathbf{x}(0) \tag{3-28}$$

Substituting (3-27) and (3-28) into (3-25) yields

$$(\mathbf{K}_1 + 2\mathbf{K}_2 t + 3\mathbf{K}_3 t^2 + \cdots)\mathbf{x}(0) = \mathbf{A}(\mathbf{K}_0 + \mathbf{K}_1 t + \mathbf{K}_2 t^2 + \cdots)\mathbf{x}(0) \tag{3-29}$$

We next perform the following operations. First evaluate (3-29) at $t = 0$. Then differentiate (3-29) and evaluate the result at $t = 0$. Differentiate again and evaluate at $t = 0$. Repeat this operation, and each evaluation results in an equation in the unknown matrices $\mathbf{K}_i$. The effect of this procedure is to equate the coefficients of t^i in (3-29). The resulting equations are

$$\mathbf{K}_1 = \mathbf{A}\mathbf{K}_0$$
$$2\mathbf{K}_2 = \mathbf{A}\mathbf{K}_1$$
$$3\mathbf{K}_3 = \mathbf{A}\mathbf{K}_2 \tag{3-30}$$
$$\vdots$$

Evaluating (3-27) at $t = 0$ shows that $\mathbf{K}_0 = \mathbf{I}$. Then the other matrices are evaluated from (3-30) as

$$\mathbf{K}_1 = \mathbf{A}$$
$$\mathbf{K}_2 = \frac{\mathbf{A}^2}{2!}$$
$$\mathbf{K}_3 = \frac{\mathbf{A}^3}{3!} \tag{3-31}$$
$$\vdots$$

Hence, from (3-27) the state transition matrix may be expressed as

$$\Phi(t) = \mathbf{I} + \mathbf{A}t + \mathbf{A}^2\frac{t^2}{2!} + \mathbf{A}^3\frac{t^3}{3!} + \cdots \tag{3-32}$$

Because of the similarity of (3-32) and the Taylor's series expansion of the scalar exponential,

$$e^{kt} = 1 + kt + k^2\frac{t^2}{2!} + k^3\frac{t^3}{3!} + \cdots \tag{3-33}$$

the state transition matrix is often written, for notational purposes only, as the matrix exponential

$$\Phi(t) = e^{\mathbf{A}t} \tag{3-34}$$

The *matrix exponential* is defined by (3-32) and (3-34). An example illustrating this technique is presented next.

Example 3.7

To give an example for which the series in (3-32) has a finite number of nonzero terms, the satellite model of Section 2.6 is used. From Example 2.13, the transfer function of the satellite is

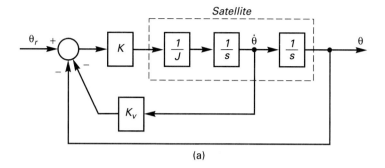

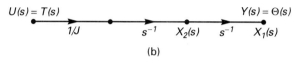

(b)

Figure 3.10 Attitude control system for satellite.

$$G(s) = \frac{\Theta(s)}{T(s)} = \frac{1}{Js^2}$$

where $T(s) = \mathcal{L}\,[\tau(t)]$ is the torque applied to the satellite and $\theta(s)$ is the attitude angle of the satellite. A block diagram for a control system for the satellite is given in Figure 3.10(a). In this control system, both position $\theta(t)$ and rate $\dot{\theta}(t)$ are measured and fed back to control the attitude of the satellite. The simulation diagram for the satellite only is given in Figure 3.10(b). We write the state equations for the satellite directly from this simulation diagram:

$$\dot{\mathbf{x}}(t) = \mathbf{A}\mathbf{x}(t) + \mathbf{B}u(t) = \begin{bmatrix} 0 & 1 \\ 0 & 0 \end{bmatrix}\mathbf{x}(t) + \begin{bmatrix} 0 \\ \frac{1}{J} \end{bmatrix}u(t)$$

Thus in (3-32),

$$\mathbf{A} = \begin{bmatrix} 0 & 1 \\ 0 & 0 \end{bmatrix}$$

$$\mathbf{A}^2 = \begin{bmatrix} 0 & 1 \\ 0 & 0 \end{bmatrix}\begin{bmatrix} 0 & 1 \\ 0 & 0 \end{bmatrix} = \begin{bmatrix} 0 & 0 \\ 0 & 0 \end{bmatrix}$$

$$\mathbf{A}^3 = \mathbf{A}\mathbf{A}^2 = \mathbf{0}$$

In a like manner,

$$\mathbf{A}^n = \mathbf{A}^2\mathbf{A}^{n-2} = \mathbf{0}; \qquad n \geq 3$$

The state transition matrix is then, from (3-32),

$$\Phi(t) = \mathbf{I} + \mathbf{A}t = \begin{bmatrix} 1 & 0 \\ 0 & 1 \end{bmatrix} + \begin{bmatrix} 0 & 1 \\ 0 & 0 \end{bmatrix}t = \begin{bmatrix} 1 & t \\ 0 & 1 \end{bmatrix}$$

For this example, the state transition matrix is easily calculated using the infinite series expansion. In general this will not be the case.

The series expansion of $\Phi(t)$ is well suited to evaluation on a digital computer if $\Phi(t)$ is to be evaluated at only a few instants of time. The series expansion is also very useful in the analysis of digital control systems [3,4]. However, as a practical matter, the time response of a system should be evaluated by simulation.

3.4 TRANSFER FUNCTIONS

A procedure was given in Section 3.2 for writing the state equations of a system if the transfer function is known. In this section we are interested in the inverse of this procedure: Given the state equations, find the transfer function. One approach is the use of the state equations to construct a simulation diagram and the use of Mason's gain formula to find the transfer function of the simulation diagram. This approach is direct but tends to be long and tedious except for the simplest of systems. In addition, finding all the loops and all the forward paths in a simulation diagram is an almost impossible task for high-order systems. In this section we present a matrix procedure for finding the transfer function from the state equations; this procedure can be implemented as a digital computer program [2]. This procedure will now be developed.

The standard form of the state equations is given by

$$\dot{\mathbf{x}}(t) = \mathbf{A}\mathbf{x}(t) + \mathbf{B}u(t)$$
$$y(t) = \mathbf{C}\mathbf{x}(t) \tag{3-35}$$

for a single-input, single-output system. The matrix $\mathbf{D}$ of (3-6) has been omitted, since it is normally the null matrix. The Laplace transform of the differential equation in (3-35) yields [see (3-18)]

$$s\mathbf{X}(s) = \mathbf{A}\mathbf{X}(s) + \mathbf{B}U(s) \tag{3-36}$$

ignoring initial conditions. This equation may be written as

$$(s\mathbf{I} - \mathbf{A})\mathbf{X}(s) = \mathbf{B}U(s) \tag{3-37}$$

and is easily solved for $\mathbf{X}(s)$:

$$\mathbf{X}(s) = (s\mathbf{I} - \mathbf{A})^{-1}\mathbf{B}U(s) \tag{3-38}$$

The Laplace transform of the output equation in (3-35) yields

$$Y(s) = \mathbf{C}\mathbf{X}(s) \tag{3-39}$$

Substitution of (3-38) into (3-39) gives the desired transfer function,

$$Y(s) = \mathbf{C}(s\mathbf{I} - \mathbf{A})^{-1}\mathbf{B}U(s) = G(s)U(s) \tag{3-40}$$

by the definition of the transfer function. Hence the transfer function of the system is

$$G(s) = \mathbf{C}(s\mathbf{I} - \mathbf{A})^{-1}\mathbf{B} = \mathbf{C}\Phi(s)\mathbf{B} \tag{3-41}$$

For the case that D is not zero, the transfer function is found to be

$$G(s) = \mathbf{C}(s\mathbf{I} - \mathbf{A})^{-1}\mathbf{B} + D \tag{3-42}$$

The interested reader may derive this relationship (see Problem 3.27). An example is now given to illustrate this procedure.

Example 3.8

For this example the system of Example 3.4 is used. The transfer function for that example was given as

$$G(s) = \frac{Y(s)}{U(s)} = \frac{1}{s^2 + 3s + 2}$$

and the state equations were found to be

$$\dot{\mathbf{x}}(t) = \begin{bmatrix} -3 & 1 \\ -2 & 0 \end{bmatrix} \mathbf{x}(t) + \begin{bmatrix} 0 \\ 1 \end{bmatrix} u(t)$$

$$y(t) = \begin{bmatrix} 1 & 0 \end{bmatrix} \mathbf{x}(t)$$

The matrix $(s\mathbf{I} - \mathbf{A})^{-1}$ was calculated in Example 3.4. Then, from (3-41) and Example 3.4, the transfer function is given by

$$G(s) = \mathbf{C}(s\mathbf{I} - \mathbf{A})^{-1}\mathbf{B} = \begin{bmatrix} 1 & 0 \end{bmatrix} \begin{bmatrix} s+3 & -1 \\ 2 & s \end{bmatrix}^{-1} \begin{bmatrix} 0 \\ 1 \end{bmatrix}$$

$$= \begin{bmatrix} 1 & 0 \end{bmatrix} \begin{bmatrix} \dfrac{s}{(s+1)(s+2)} & \dfrac{1}{(s+1)(s+2)} \\ \dfrac{-2}{(s+1)(s+2)} & \dfrac{s+3}{(s+1)(s+2)} \end{bmatrix} \begin{bmatrix} 0 \\ 1 \end{bmatrix}$$

$$= \begin{bmatrix} 1 & 0 \end{bmatrix} \begin{bmatrix} \dfrac{1}{(s+1)(s+2)} \\ \dfrac{s+3}{(s+1)(s+2)} \end{bmatrix} = \dfrac{1}{(s+1)(s+2)}$$

This transfer function checks the one given. A MATLAB program that verifies the transfer function calculation is given by

```
e3p8=([-3 1;-2 0],[0; 1], [1 0], 0);
tf (e3p8)
```

where n gives the numerator coefficients for $G(s)$ and d the denominator coefficients.

For the multivariable case (more than one input and/or more than one output), $\mathbf{U}(s)$ is an $(r \times 1)$ vector and $\mathbf{Y}(s)$ is a $(p \times 1)$ vector. The preceding derivation remains valid, with (3-40) becoming

$$\mathbf{Y}(s) = \mathbf{C}(s\mathbf{I} - \mathbf{A})^{-1}\mathbf{B}\mathbf{U}(s) = \mathbf{G}(s)\mathbf{U}(s) \qquad (3\text{-}43)$$

where $\mathbf{G}(s)$ is now a $(p \times r)$ matrix. The elements of $\mathbf{G}(s)$, $G_{ij}(s)$, are each transfer functions, where, by superposition,

$$G_{ij}(s) = \frac{Y_i(s)}{U_j(s)} \qquad (3\text{-}44)$$

with all other inputs equal to zero, that is, with each $U_k(s)$, $k \neq j$, equal to zero. An example of a multivariable system is the temperature control system in Section 2.9.

3.5 SIMILARITY TRANSFORMATIONS

In this chapter procedures for finding a state-variable model from either the system differential equations or the system transfer function have been presented. It has been shown that a unique state model does not exist. Two general state models, the control canonical form and the observer canonical form, can always be found. A single-input single-output system has only one input–output model (transfer function), but the number of internal models (state models) is unbounded, as we now show.

The state model of a single-input, single-output model is

$$\dot{\mathbf{x}}(t) = \mathbf{A}\mathbf{x}(t) + \mathbf{B}u(t) \tag{3-45}$$

$$y(t) = \mathbf{C}\mathbf{x}(t) + Du(t) \tag{3-46}$$

and the transfer function is given by

[eq. (3-42)] $$\frac{Y(s)}{U(s)} = G(s) = \mathbf{C}(s\mathbf{I} - \mathbf{A})^{-1}\mathbf{B} + D$$

There are many combinations of the matrices $\mathbf{A}$, $\mathbf{B}$, $\mathbf{C}$, and D that will satisfy (3-42) for a given $G(s)$.

Suppose that we are given a state model of a system as in (3-45) and (3-46). Now, define a different state vector $\mathbf{v}(t)$ that is of the same order as $\mathbf{x}(t)$, such that the elements of $\mathbf{v}(t)$ are linear combinations of the elements of $\mathbf{x}(t)$, that is,

$$v_1(t) = q_{11}x_1(t) + q_{12}x_2(t) + \cdots + q_{1n}x_n(t)$$
$$v_2(t) = q_{21}x_1(t) + q_{22}x_2(t) + \cdots + q_{2n}x_n(t)$$
$$\cdot$$
$$\cdot \tag{3-47}$$
$$\cdot$$
$$v_n(t) = q_{n1}x_1(t) + q_{n2}x_2(t) + \cdots + q_{nn}x_n(t)$$

This equation can be written in matrix form:

$$\mathbf{v}(t) = \mathbf{Q}\mathbf{x}(t) = \mathbf{P}^{-1}\mathbf{x}(t) \tag{3-48}$$

where the matrix $\mathbf{Q}$ has been defined as the inverse of the matrix $\mathbf{P}$, to satisfy common notation. Hence, the state vector $\mathbf{x}(t)$ can be expressed as

$$\mathbf{x}(t) = \mathbf{P}\mathbf{v}(t) \tag{3-49}$$

where the matrix $\mathbf{P}$ is called a transformation matrix, or simply a transformation. An example is now given.

Example 3.9

Consider the system of Example 3.4, which has the transfer function

$$G(s) = \frac{Y(s)}{U(s)} = \frac{1}{s^2 + 3s + 2}$$

and the state equations

$$\dot{\mathbf{x}}(t) = \begin{bmatrix} -3 & 1 \\ -2 & 0 \end{bmatrix} \mathbf{x}(t) + \begin{bmatrix} 0 \\ 1 \end{bmatrix} u(t)$$

$$y(t) = \begin{bmatrix} 1 & 0 \end{bmatrix} \mathbf{x}(t)$$

For example, suppose that the elements of $\mathbf{v}(t)$ are arbitrarily defined as

$$v_1(t) = x_1(t) + x_2(t)$$
$$v_2(t) = x_1(t) + 2x_2(t)$$

as shown in Figure 3.11. Thus

$$\mathbf{v}(t) = \mathbf{Q}\mathbf{x}(t) = \begin{bmatrix} 1 & 1 \\ 1 & 2 \end{bmatrix} \mathbf{x}(t)$$

and

$$\mathbf{P}^{-1} = \mathbf{Q} = \begin{bmatrix} 1 & 1 \\ 1 & 2 \end{bmatrix}, \qquad \therefore \mathbf{P} = \begin{bmatrix} 2 & -1 \\ -1 & 1 \end{bmatrix}$$

Thus the components of $\mathbf{x}(t)$ can be expressed as functions of the components of $\mathbf{v}(t)$.

$$x_1(t) = 2v_1(t) - v_2(t)$$
$$x_2(t) = -v_1(t) + v_2(t)$$

It is seen from this exercise that, given the vector $\mathbf{x}(t)$, we can solve for the vector $\mathbf{v}(t)$. Or, given the vector $\mathbf{v}(t)$, we can solve for the vector $\mathbf{x}(t)$. The matrix inversion can be verified by the MATLAB program

```
Q = [1 1; 1 2];
P = inv(Q)
```

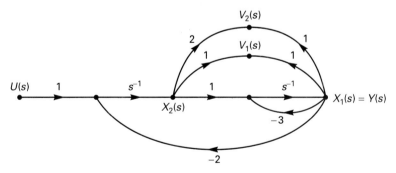

Figure 3.11 System for Example 3.9.

The preceding example illustrates transforming from one state vector to a different state vector. This transformation alters the internal model of the system (the state model) but does not alter the input–output model of the system (the transfer function). This type of transformation is called a *similarity transformation*. The details of similarity transformations are developed next.

Assume that we are given a state model of a *multivariable* system of the form of (3-45) and a similarity transformation of the form of (3-49).

[eq. (3-45)] $$\dot{\mathbf{x}}(t) = \mathbf{A}\mathbf{x}(t) + \mathbf{B}\mathbf{u}(t)$$

[eq. (3-49)] $$\mathbf{x}(t) = \mathbf{P}\mathbf{v}(t)$$

Substituting (3-49) into (3-45) yields

$$\mathbf{P}\dot{\mathbf{v}}(t) = \mathbf{A}\mathbf{P}\mathbf{v}(t) + \mathbf{B}u(t) \tag{3-50}$$

Solving this equation for $\dot{\mathbf{v}}(t)$ results in the state model for the variables $\mathbf{v}(t)$:

$$\dot{\mathbf{v}}(t) = \mathbf{P}^{-1}\mathbf{A}\mathbf{P}\mathbf{v}(t) + \mathbf{P}^{-1}\mathbf{B}u(t) \tag{3-51}$$

The output equation for a multivariable system

[eq. (3-46)] $$y(t) = \mathbf{C}\mathbf{x}(t) + Du(t)$$

becomes, using (3-49),

$$y(t) = \mathbf{C}\mathbf{P}\mathbf{v}(t) + Du(t) \tag{3-52}$$

Therefore, we have the state equations expressed as a function of the state vector $\mathbf{x}(t)$ in (3-45) and (3-46) and as a function of the transformed state vector $\mathbf{v}(t)$ in (3-51) and (3-52).

The state equations as a function of $\mathbf{v}(t)$ can be expressed in the standard format for a multivariable system as

$$\dot{\mathbf{v}}(t) = \mathbf{A}_v\mathbf{v}(t) + \mathbf{B}_v\mathbf{u}(t) \tag{3-53}$$

$$\mathbf{y}(t) = \mathbf{C}_v\mathbf{v}(t) + D_v u(t) \tag{3-54}$$

where the subscripts indicates the transformed matrices. The matrices for the vector $\mathbf{x}(t)$ will not be subscripted. Comparing (3-53) with (3-51) and (3-54) with (3-52), we see that the equations for the transformed matrices are

$$\begin{aligned} \mathbf{A}_v &= \mathbf{P}^{-1}\mathbf{A}\mathbf{P} & \mathbf{B}_v &= \mathbf{P}^{-1}\mathbf{B} \\ \mathbf{C}_v &= \mathbf{C}\mathbf{P} & \mathbf{D}_v &= \mathbf{D} \end{aligned} \tag{3-55}$$

An example will be presented now to illustrate similarity transformations.

Example 3.10

As an example, consider the system of Example 3.9,

$$\dot{\mathbf{x}}(t) = \mathbf{A}\mathbf{x}(t) + \mathbf{B}u(t) = \begin{bmatrix} -3 & 1 \\ -2 & 0 \end{bmatrix}\mathbf{x}(t) + \begin{bmatrix} 0 \\ 1 \end{bmatrix}u(t)$$

$$y(t) = \mathbf{C}\mathbf{x}(t) = \begin{bmatrix} 1 & 0 \end{bmatrix}\mathbf{x}(t)$$

with the transformation

$$\mathbf{P}^{-1} = \begin{bmatrix} 1 & 1 \\ 1 & 2 \end{bmatrix} \qquad \mathbf{P} = \begin{bmatrix} 2 & -1 \\ -1 & 1 \end{bmatrix}$$

From (3-55), the system matrices for the transformed system become

$$\mathbf{A}_v = \mathbf{P}^{-1}\mathbf{A}\mathbf{P} = \begin{bmatrix} 1 & 1 \\ 1 & 2 \end{bmatrix}\begin{bmatrix} -3 & 1 \\ -2 & 0 \end{bmatrix}\begin{bmatrix} 2 & -1 \\ -1 & 1 \end{bmatrix}$$

$$= \begin{bmatrix} -5 & 1 \\ -7 & 1 \end{bmatrix}\begin{bmatrix} 2 & -1 \\ -1 & 1 \end{bmatrix} = \begin{bmatrix} -11 & 6 \\ -15 & 8 \end{bmatrix}$$

$$\mathbf{B}_v = \mathbf{P}^{-1}\mathbf{B} = \begin{bmatrix} 1 & 1 \\ 1 & 2 \end{bmatrix}\begin{bmatrix} 0 \\ 1 \end{bmatrix} = \begin{bmatrix} 1 \\ 2 \end{bmatrix}$$

$$\mathbf{C}_v = \mathbf{C}\mathbf{P} = \begin{bmatrix} 1 & 0 \end{bmatrix}\begin{bmatrix} 2 & -1 \\ -1 & 1 \end{bmatrix} = \begin{bmatrix} 2 & -1 \end{bmatrix}$$

The transformed state equations are then

$$\dot{\mathbf{v}}(t) = \mathbf{A}_v\mathbf{v}(t) + \mathbf{B}_v u(t) = \begin{bmatrix} -11 & 6 \\ -15 & 8 \end{bmatrix}\mathbf{v}(t) + \begin{bmatrix} 1 \\ 2 \end{bmatrix}u(t)$$

$$y(t) = \mathbf{C}_v\mathbf{v}(t) = \begin{bmatrix} 2 & -1 \end{bmatrix}\mathbf{v}(t)$$

The calculations in this example can be verified by the MATLAB program

```
A  = [-3 1; -2 0]; B = [0; 1]; C = [1 0]; D = 0; T = [1 1; 1 2];
sys=ss(A,B,C,D)
sys=ss2ss(sys, T)
```

The preceding example gives two state models of the same system. If a different transformation matrix **P** had been chosen, a third model would result. In fact, for each different transformation **P** that has an inverse, a different state model results. Thus there are an unlimited number of state models for a given system transfer function. The choice of the state model for a given system is usually based on using the natural state variables (position, velocity, and so on), on ease of design, and so forth.

To check the state model developed in the preceding example, we derive the transfer function of this model by two different procedures.

Example 3.11

To obtain the transfer function for the transformed system equations of Example 3.10 using Mason's gain formula, the simulation diagram is drawn from the state equations,

$$\dot{\mathbf{v}}(t) = \begin{bmatrix} -11 & 6 \\ -15 & 8 \end{bmatrix}\mathbf{v}(t) + \begin{bmatrix} 1 \\ 2 \end{bmatrix}u(t)$$

$$y(t) = \begin{bmatrix} 2 & -1 \end{bmatrix}\mathbf{v}(t)$$

It is shown in Figure 3.12. Note that the transfer function of this simulation diagram must be

$$G(s) = \frac{1}{s^2 + 3s + 2}$$

To verify this transfer function, Mason's gain formula is used. Note in Figure 3.12 that the simulation diagram has three loops and that two of these loops do not touch. Thus for Mason's gain formula,

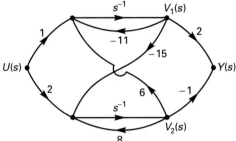

Figure 3.12   System for Example 3.11.

$$\Delta = 1 - [-11s^{-1} + 8s^{-1} + (-15) s^{-1} (6) s^{-1}] + [-11s^{-1}] [8s^{-1}]$$
$$= 1 + 3s^{-1} + 2s^{-2}$$

There are four forward paths; as an exercise the reader might try to find these before reading further. The forward path gains and the corresponding Δ_i are

$$M_1 = 2s^{-1} \qquad\qquad \Delta_1 = 1 - 8s^{-1}$$
$$M_2 = 15s^{-2} \qquad\qquad \Delta_2 = 1$$
$$M_3 = -2s^{-1} \qquad\qquad \Delta_3 = 1 + 11s^{-1}$$
$$M_4 = (2)(6)(2) s^{-2} \qquad \Delta_4 = 1$$

The transfer function is then

$$G(s) = \frac{M_1\Delta_1 + M_2\Delta_2 + M_3\Delta_3 + M_4\Delta_4}{\Delta}$$

$$= \frac{(2s^{-1})(1 - 8s^{-1}) + (15s^{-2})(1) + (-2s^{-1})(1 + 11s^{-1}) + (24s^{-2})(1)}{1 + 3s^{-1} + 2s^{-2}}$$

$$= \frac{s^{-2}}{1 + 3s^{-1} + 2s^{-2}} = \frac{1}{s^2 + 3s + 2}$$

This verifies the transfer function. Note the difficulty in identifying all loops and forward paths. This transfer function can also be checked using the matrix approach

[eq. (3-42)] $$G(s) = C_v (sI - A_v)^{-1}B_v$$

Now

$$sI - A_v = \begin{bmatrix} s + 11 & -6 \\ 15 & s - 8 \end{bmatrix}$$

and thus

$$\det (sI - A_v) = s^2 + 3s - 88 + 90 = s^2 + 3s + 2$$

The adjoint of $sI - A_v$ is given by

$$\text{Adj} (sI - A_v) = \begin{bmatrix} s-8 & 6 \\ -15 & s + 11 \end{bmatrix}$$

Thus the inverse of this matrix is

$$(s\mathbf{I} - \mathbf{A}_v)^{-1} = \frac{\text{Adj}\,(s\mathbf{I} - \mathbf{A}_v)}{\det\,(s\mathbf{I} - \mathbf{A}_v)} = \begin{bmatrix} \dfrac{s-8}{s^2+3s+2} & \dfrac{6}{s^2+3s+2} \\ \dfrac{-15}{s^2+3s+2} & \dfrac{s+11}{s^2+3s+2} \end{bmatrix}$$

From (3-42), the system transfer function is given by

$$G(s) = \mathbf{C}_v\,(s\mathbf{I} - \mathbf{A}_v)^{-1}\mathbf{B}_v = \begin{bmatrix} 2 & -1 \end{bmatrix} \begin{bmatrix} \dfrac{s-8}{s^2+3s+2} & \dfrac{6}{s^2+3s+2} \\ \dfrac{-15}{s^2+3s+2} & \dfrac{s+11}{s^2+3s+2} \end{bmatrix} \begin{bmatrix} 1 \\ 2 \end{bmatrix}$$

$$= \begin{bmatrix} \dfrac{2s-1}{s^2+3s+2} & \dfrac{-s+1}{s^2+3s+2} \end{bmatrix} \begin{bmatrix} 1 \\ 2 \end{bmatrix} = \frac{1}{s^2+3s+2}$$

The transfer function is checked, with much less probability of error when compared to using Mason's gain formula. Recall also that the matrix procedure can be implemented in a digital computer program, as illustrated in the MATLAB program of Example 3.8.

Similarity transformations have been demonstrated through examples. Certain important properties of these transformations are derived next. Consider first the determinant of $s\mathbf{I} - \mathbf{A}_v$. From (3-55),

$$\det(s\mathbf{I} - \mathbf{A}_v) = \det(s\mathbf{I} - \mathbf{P}^{-1}\mathbf{AP}) = \det(s\mathbf{P}^{-1}\mathbf{IP} - \mathbf{P}^{-1}\mathbf{AP}) \qquad (3\text{-}56)$$

where $\det(\cdot)$ denotes the determinant. For two square matrices

$$\det \mathbf{R}_1\mathbf{R}_2 = \det \mathbf{R}_1 \det \mathbf{R}_2$$

Then (3-56) becomes

$$\det(s\mathbf{I} - \mathbf{A}_v) = \det \mathbf{P}^{-1} \det(s\mathbf{I} - \mathbf{A}) \det \mathbf{P} \qquad (3\text{-}57)$$

Since, for a matrix square $\mathbf{R}$,

$$\mathbf{R}^{-1}\mathbf{R} = \mathbf{I}$$

then

$$\det \mathbf{R}^{-1} \det \mathbf{R} = \det \mathbf{R}^{-1}\mathbf{R} = \det \mathbf{I} = 1 \qquad (3\text{-}58)$$

Thus (3-57) becomes

$$\det(s\mathbf{I} - \mathbf{A}_v) = \det(s\mathbf{I} - \mathbf{A}) \qquad (3\text{-}59)$$

The zeros of $\det(s\mathbf{I} - \mathbf{A})$ are the characteristic values, or eigenvalues, of $\mathbf{A}$ (see Appendix A). Thus the characteristic values of $\mathbf{A}_v$ are equal to the characteristic values of $\mathbf{A}$. This is the first property of similarity transformations.

A second property can be derived as follows. From (3-55),

$$\det \mathbf{A}_v = \det \mathbf{P}^{-1}\mathbf{AP} = \det \mathbf{P}^{-1} \det \mathbf{A} \det \mathbf{P} = \det \mathbf{A} \qquad (3\text{-}60)$$

Then the determinant of $\mathbf{A}_v$ is equal to the determinant of $\mathbf{A}$. This property can also be seen from the fact that the determinant of a matrix is equal to the product of its characteristic values (see Appendix A). As shown earlier, the characteristic values of A_v are equal to those of A; thus the determinants must be equal.

The third property of a similarity transformation can also be seen from the fact that the eigenvalues of A_v are equal to those of A. Since the trace of a matrix is equal to the sum of the eigenvalues,

$$\text{tr}\,A_v = \text{tr}\,A$$

To summarize the properties of similarity transformations, let $\lambda_1, \lambda_2, \ldots, \lambda_n$ be the eigenvalues of the matrix A. Then for the similarity transformation

$$A_v = P^{-1}AP$$

1. The eigenvalues of A are equal to those of A_v; or

$$\det(sI - A) = \det(sI - A_v) = (s - \lambda_1)(s - \lambda_2)\cdots(s - \lambda_n) \tag{3-61}$$

2. The determinant of A is equal to the determinant of A_v:

$$\det A = \det A_v = \lambda_1\lambda_2\cdots\lambda_n \tag{3-62}$$

3. The trace of A is equal to the trace of A_v:

$$\text{tr}\,A = \text{tr}\,A_v = \lambda_1 + \lambda_2 + \cdots + \lambda_n \tag{3-63}$$

4. The following transfer functions are equal:

$$C(sI - A)^{-1}B = C_v(sI - A_v)^{-1}B_v$$

The proof of the fourth property is left as an exercise (see Problem 3.27); this property is illustrated in Example 3.11. The first three properties are illustrated by the following example.

Example 3.12

The similarity transformation of Example 3.10 is used to illustrate the three properties just developed. From Example 3.10, the matrices A and the A_v are given by

$$A = \begin{bmatrix} -3 & 1 \\ -2 & 0 \end{bmatrix} \qquad A_v = \begin{bmatrix} -11 & 6 \\ -15 & 8 \end{bmatrix}$$

Then

$$\det(sI - A) = \begin{vmatrix} s+3 & -1 \\ 2 & s \end{vmatrix} = s^2 + 3s + 2$$

and

$$\det(sI - A_v) = \begin{vmatrix} s+11 & -6 \\ 15 & s-8 \end{vmatrix} = s^2 + 3s - 88 + 90 = s^2 + 3s + 2$$

Thus the two determinants are equal.

Next, the eigenvalues are found from

$$\det(sI - A) = s^2 + 3s + 2 = (s+1)(s+2)$$

and $\lambda_1 = -1$, $\lambda_2 = -2$. The determinants of the two matrices are

$$\det \mathbf{A} = \begin{vmatrix} -3 & 1 \\ -2 & 0 \end{vmatrix} = 2 \qquad \det \mathbf{A}_v = \begin{vmatrix} -11 & 6 \\ -15 & 8 \end{vmatrix} = -88 + 90 = 2$$

and both determinants are equal to the products of the eigenvalues. The traces of the two matrices are both equal to −3, which is the sum of the eigenvalues. The results in this example can be verified with the MATLAB program

```
A = [-3 1; -2 0]; Av = [-11 6; -15 8];
detsImA = poly(A),  detsImAv = poly(Av)
evA = eig(A),       evAv = eig(Av)
detA = det(A),      detAv = det(Av)
trA = trace(A),     trAv = trace(Av)
```

3.6 DIGITAL SIMULATION

It has been stated previously that the practical procedure for finding the time response of a system is through simulation, rather than by directly solving the differential equations or by using the Laplace transform. The Laplace transform is used throughout this book because of the ease in analyzing low-order systems. However, for higher-order systems, simulations are necessary.

In this section we consider the numerical solution of differential equations by *integration algorithms*. Many algorithms are available for numerical integration [5].

Integration algorithms differ in execution speed, accuracy, programming complexity, and so forth. In this section we consider a very simple algorithm—*Euler's method*. Euler's method is seldom used in practical situations because more efficient and more accurate algorithms are available. It is presented here because of the simplicity and its ease of programming.

The process of numerically integrating a time function using Euler's method is illustrated in Figure 3.13. We wish to integrate $z(t)$ numerically; that is, we wish to find $x(t)$, where

$$x(t) = \int_0^t z(\tau)\, d\tau + x(0) \tag{3-64}$$

Suppose that we know $x(t)$ at $t = (k-1)H$; that is, we know $x[(k-1)H]$ and we want to calculate $x(kH)$, where H is called the numerical integration increment and is the step size of the algorithm. Euler's algorithm is obtained by assuming $z(t)$ constant at the value $z[(k-1)H]$ for $(k-1)H \le t < kH$. Then

$$x(kH) = x[(k-1)H] + Hz[(k-1)H] \tag{3-65}$$

In (3-65), $x(kH)$ is only an approximation for $x(t)$ in (3-64) evaluated at $t = kH$. Note in Figure 3.13 that we are approximating the area under the $z(t)$ curve for $(k-1)H \le t < kH$ with the area of the rectangle shown. For this reason, this method is also called the *rectangular rule*.

In simulation, we are interested in the integration of differential equations [that is, in (3-64), $z(t)$ is the derivative of $x(t)$]. We illustrate this case with a simple example. Suppose that we wish to solve the differential equation

$$\dot{x}(t) + x(t) = 0 \qquad x(0) = 1 \tag{3-66}$$

The solution is obviously

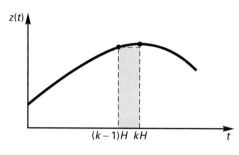

Figure 3.13   Euler rule for numerical integration.

$$x(t) = e^{-t}, \quad t \geq 0$$

However, by numerical integration in (3-65),

$$x(kH) = x[(k-1)H] + H\{-x[(k-1)H]\} \tag{3-67}$$

since, from (3-66), for this differential equation,

$$z(t) = \dot{x}(t) = -x(t)$$

Suppose that we choose $H = 0.1$ s. Then, solving (3-67) iteratively starting with $k = 1$ [we know $x(0)$],

$$x(0.1) = x(0) - Hx(0) = 1.0 - (0.1)(1.0) = 0.9$$
$$x(0.2) = x(0.1) - Hx(0.1) = 0.9 - 0.09 = 0.81$$

$$\vdots$$

$$x(1.0) = x(0.9) - Hx(0.9) = 0.3487$$

Since, for this example, we know the solution, we calculate $x(t)$ at $t = 1.0$ s as

$$x(1.0) = e^{-1.0} = 0.3679$$

and we can see the error due to the numerical integration. If we choose $H = 0.01$ s, the value of $x(1.0)$ is calculated to be 0.3660, and the error is much less.

In the preceding example, if we choose H larger, the error is larger. If we decrease H, the error decreases, as shown. In fact, as H decreases, the error will decrease to a minimum value. Then a further decrease in H results in an increase in the error, due to round-off in the computations. With H equal to 0.1 s, 10 iterations are required to calculate $x(1)$. With H equal to 0.001 s, 1000 iterations are required to calculate $x(1)$. The round-off errors in the computations are larger for the latter case, since more calculations are made. If H is made sufficiently smaller, this round-off error becomes appreciable. Thus, if H is chosen to be too large, the errors are large due to the algorithm being a poor approximation for the integral. If H is chosen to be too small, the errors are also large because of the round-off in the computations.

An acceptable value of H is usually found by experimenting with the algorithm, that is, by varying H and noting the effect on the solution. This can be done by choosing a value of H and calculating the solution. Then H is reduced and the solution is calculated again. If the solution has changed very little, we can conclude that the first value of H is small enough. If not, H is reduced again and the same test applied. The value of H can also be increased, with the same test used.

Next we consider the simulation of a system described by the state equations

$$\dot{\mathbf{x}}(t) = \mathbf{A}\mathbf{x}(t) + \mathbf{B}\mathbf{u}(t) \tag{3-68}$$

To develop Euler's rule for this matrix equation, consider the first element of the state vector, $x_1(t)$. This element can be expressed as

$$x_1(t) = \int_0^t \dot{x}_1(\tau)\, d\tau + x_1(0) \tag{3-69}$$

Comparing this equation with (3-64), from (3-65) we write

$$x_1(kH) = x_1[(k-1)H] + H\dot{x}_1[(k-1)H] \tag{3-70}$$

The same development applies for any element of the state vector, $x_i(t)$. Hence we can write

$$x_i(kH) = x_i[(k-1)H] + H\dot{x}_i[(k-1)H] \qquad 1 \le i \le n \tag{3-71}$$

This equation may be written in terms of vectors:

$$\mathbf{x}(kH) = \mathbf{x}[(k-1)H] + H\dot{\mathbf{x}}[(k-1)H] \tag{3-72}$$

where, from (3-68), $\dot{\mathbf{x}}[(k-1)H]$ is given by

$$\dot{\mathbf{x}}[(k-1)H] = \mathbf{A}\mathbf{x}[(k-1)H] + \mathbf{B}\mathbf{u}[(k-1)H] \tag{3-73}$$

The numerical integration algorithm is then

1. Let $k = 1$.
2. Evaluate $\dot{\mathbf{x}}[(k-1)H]$ in (3-73).
3. Evaluate $\mathbf{x}(kH)$ in (3-72).
4. Let $k = k + 1$.
5. Go to step 2.

This algorithm is particularly easy to program but in general requires a much smaller step size than does some of the more complex algorithms. The interested reader may consult Ref. 5 for additional discussion of numerical integration. An example of the integration of state equations is given next.

Example 3.13

In this example we illustrate the numerical integration of state equations by Euler's rule. For the example we use the state equations of Example 3.5, since the complete analytical solution of the equations was derived in Example 3.5. From that example, the state equations are given by

$$\dot{\mathbf{x}}(t) = \begin{bmatrix} -3 & 1 \\ -2 & 0 \end{bmatrix} \mathbf{x}(t) + \begin{bmatrix} 0 \\ 1 \end{bmatrix} u(t)$$

$$y(t) = \begin{bmatrix} 1 & 0 \end{bmatrix} \mathbf{x}(t)$$

and the solution is, for a unit step input,

$$x_1(t) = (-e^{-t} + 2e^{-2t})x_1(0) + (e^{-t} - e^{-2t})x_2(0) + (\tfrac{1}{2}) - e^{-t} + (\tfrac{1}{2})e^{-2t}$$

$$x_2(t) = (-2e^{-t} + 2e^{-2t})x_1(0) + (2e^{-t} - e^{-2t})x_2(0) + (\tfrac{2}{3}) - 2e^{-t} + (\tfrac{1}{2})e^{-2t}$$

Suppose we choose $H = 0.01$ s. In addition, let $\mathbf{x}(0) = \mathbf{0}$ to simplify the calculations somewhat. From step 2 of the algorithm and (3-73),

$$\dot{\mathbf{x}}(0) = \mathbf{A}\mathbf{x}(0) + \mathbf{B}u(0) = \mathbf{B}u(0) = \begin{bmatrix} 0 \\ 1 \end{bmatrix}(1) = \begin{bmatrix} 0 \\ 1 \end{bmatrix}$$

Then, from step 3 and (3-72),

$$\mathbf{x}(0.01) = \mathbf{x}(0) + H\dot{\mathbf{x}}(0) = H\dot{\mathbf{x}}(0) = (0.01)\begin{bmatrix} 0 \\ 1 \end{bmatrix} = \begin{bmatrix} 0 \\ 0.01 \end{bmatrix}$$

For the second iteration of the algorithm,

$$\dot{\mathbf{x}}(0.01) = \mathbf{A}\mathbf{x}(0.01) + \mathbf{B}u(0.01) = \begin{bmatrix} -3 & 1 \\ -2 & 0 \end{bmatrix}\begin{bmatrix} 0 \\ 0.01 \end{bmatrix} + \begin{bmatrix} 0 \\ 1 \end{bmatrix} = \begin{bmatrix} 0.01 \\ 1 \end{bmatrix}$$

and from (3-72),

$$\mathbf{x}(0.02) = \mathbf{x}(0.01) + H\dot{\mathbf{x}}(0.01) = \begin{bmatrix} 0 \\ 0.01 \end{bmatrix} + (0.01)\begin{bmatrix} 0.01 \\ 1 \end{bmatrix} = \begin{bmatrix} 0.0001 \\ 0.02 \end{bmatrix}$$

An additional iteration yields

$$\dot{\mathbf{x}}(0.02) = \begin{bmatrix} -3 & 1 \\ -2 & 0 \end{bmatrix}\begin{bmatrix} 0.0001 \\ 0.02 \end{bmatrix} + \begin{bmatrix} 0 \\ 1 \end{bmatrix} = \begin{bmatrix} 0.0197 \\ 0.9998 \end{bmatrix}$$

$$\mathbf{x}(0.03) = \begin{bmatrix} 0.0001 \\ 0.02 \end{bmatrix} + (0.01)\begin{bmatrix} 0.0197 \\ 0.9998 \end{bmatrix} = \begin{bmatrix} 0.00030 \\ 0.03000 \end{bmatrix}$$

The exact values of the state vector can be calculated from the analytical solution just given and are computed with the simulation values in Table 3.1. Although giving values of the state vector over this short period of time does not give a good indication of the accuracy of the algorithm for the given value of H, this example does indicate the nature of numerical integration.

TABLE 3.1   NUMERICAL INTEGRATION RESULTS

	Simulation		Exact	
t	$x_1(t)$	$x_2(t)$	$x_1(t)$	$x_2(t)$
0	0	0	0	0
0.01	0	0.01	0.00005	0.00999
0.02	0.0001	0.02	0.0002	0.01999
0.03	0.0003	0.03	0.0004	0.02999

The iterative nature of numerical integration makes it well suited for digital computation. A MATLAB program that implements the Euler rule for this example is given by.

```
A = [-3 1; -2 0]; B = [0; 1]; C = [1 0];
H = 0.01;
xk = [0;0];
for k=1:4
    t = (k-1)*H;
    t
    xk
    xdotkp1 = A*xk + B;
    xkp1 = xk + H*xdotkp1;
    xk = xkp1;
end
```

The MATLAB control system toolbox offers a simple and convenient procedure for numerically solving linear constant-coefficient differential equations. One procedure is illustrated in the following example.

Example 3.14

We consider the system of Example 3.13. The state equations and the system output $y(t) = x_1(t)$ are given in that example. One MATLAB program that simulates this system is given by

```
A = [-3 1; -2 0]; B = [0; 1]; C = [1 0]; D = 0;
e3p14=ss(A, B, C, D)
step(e3p14)
```

The first statement enters the systems matrices, the second one identifies the system e3p14 as a system in state space, and the third one calculates the system step response $y(t)$. This response is plotted, with the result shown in Figure 3.14. Note that the final value of $y(t)$ is 0.5, which is checked by the response given in Example 3.13.

Almost all numerical-integration routines in MATLAB choose the step size automatically. Several routines use a variable-step-size routine, in which the step size varies as the calculated signals vary.

As a second example, the total response of the system of Example 3.13 to a step input and nonzero initial conditions is now calculated by simulation.

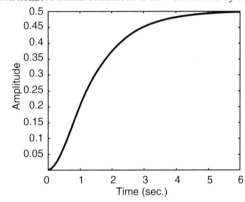

Figure 3.14   Simulation step response for Example 3.14.

Example 3.15

We now modify the program given in the last example to calculate the system step response with the initial conditions $\mathbf{x}(0) = [1\ 1]^T$. The solution to this set of equations is given in Example 3.5, with $y(t)$ given by

$$y(t) = x_1(t) = 0.5 - e^{-t} + 1.5e^{-2t} \qquad t \geq 0$$

The program required is

```
A = [-3 1;-2 0]; B = [0; 1]; C = [1 0]; D = 0;
e3p14=ss(A, B, C, D)
t = 0:0.1:6;
u = stepfun(t,0);
x0 = [1; 1];
lsim(e3p14, u, t, x0)
```

The third statement sets the initial time to zero, the numerical integration increment to 0.1 s, and the final time to 6 s. The fourth statement identifies the input as a unit step function, and the fifth statement enters the initial conditions. The sixth statement performs the simulation. The response is polotted in Figure 3.15. If the last statement is replaced with

$$[y,t] = \text{lsim}(\text{e3p14, u, t, x0})y',t'$$

the values of $y(t)$ and t are listed. For example,

$$y(t)|_{t=1} = (0.5 - e^{-t} + 1.5e^{-2t})|_{t=1} = 0.3351$$

Running the simulation gives the same value, which verifies the solution found in Example 3.5.

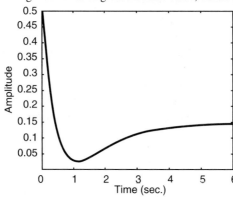

Figure 3.15 Simulation response for Example 3.15.

Example 3.16

In this example, we illustrate MATLAB simulation using a transfer function. It is shown in Example 3.4 that the transfer function for the system of Example 3.14 is

$$G(s) = \frac{1}{s^2 + 3s + 2}$$

A MATLAB program that finds the system unit step response is as follows:

```
num = [0 0 1]; den = [1 3 2];
e3p16 = tf(num, den);
step(e3p16)
```

This simulation results in a plot that is identical to that of Figure 3.14.

Simulation using MATLAB offers many options not covered here. Some of these options will be covered as required. In addition, for more complex systems, SIMULINK®, a simulation program based on MATLAB, is available, and will be introduced at appropriate points in this book.

3.7 CONTROLS SOFTWARE

Many commercial software packages are available for the analysis, design, and simulation of feedback control systems. In general, we use the student versions of MATLAB and SIMULINK in this book. These packages contain all the required control-system analysis and simulations programs. Some of the design techniques introduced in this book are not included in MATLAB. These techniques are implemented as MATLAB m-files in the sections on design, and can be run with the MATLAB package. See the Preface for instructions for downloading from the Internet all MATLAB programs given in this book.

3.8 ANALOG SIMULATION

The most common method of simulation, digital simulation, was discussed in the last section. A different method of simulation, analog simulation, is considered in this section.

Before the general availability of fast digital computers, analog computer simulation was the principal method of simulation. This method has declined in importance over the past few years; however, it is still necessary in certain situations. If the system to be simulated contains some high-frequency terms in the transient response, the numerical integration increment required in a digital simulation is very small (see Section 3.6). Hence the execution time of the simulation is large, and the simulation is very expensive. In this case an analog computer simulation may be more practical.

Also, a simulation can be made more accurate by including some of the physical hardware in the simulation. This method is used extensively in space and missile-system simulations. These simulations are called *hardware-in-the-loop simulations* and must operate in *real time*; that is, 1 s simulation time must equal 1 s actual time, or else the hardware will not respond properly. If the simulated part of the system contains some high-frequency terms in the transient response, a digital simulation may not operate in real time, thus requiring that an analog computer be used. Generally, a hardware-in-the-loop simulation will contain both analog simulation and digital simulation, in addition to the physical-system hardware. A combined analog-digital simulation is called a *hybrid simulation*. A hybrid computer is a unit in which an analog computer and a digital computer are interfaced to give the unit the best characteristics of each computer.

As mentioned in Section 2.2, the basic elements of the analog computer are integrators and amplifiers. Examples of the block diagram of these two devices are given in Figure 3.16. Each device shown in the figure has two inputs; an actual analog computer device would probably have four or more. The integrator in Figure 3.16(a) satisfies the relationship

$$y = -\int (x_1 + 10x_2)\, dt \qquad (3\text{-}74)$$

An equivalent simulation diagram is also given in the figure. The amplifier in Figure 3.16(b) satisfies the relationship

$$y = -(x_1 + 10x_2) \qquad (3\text{-}75)$$

A simulation diagram for the amplifier is also given.

The third element of the analog computer discussed here is the potentiometer, shown in Figure 3.16(c). The potentiometer is used in conjunction with integrators and amplifiers to realize the system parameters. The potentiometer is a voltage divider that satisfies the equation

$$y = \frac{R_1}{R} x \qquad (3\text{-}76)$$

Since $R_1 \leq R$, the gain of a potentiometer is less than or equal to unity.

We now construct, as an example, an analog-computer diagram for a second-order system. The construction of these diagrams is very similar to the construction of simulation diagrams as given in Section 3.2. The difference can be seen in the simulation diagrams of the analog computer components given in Figure 3.16. The analog computer diagram must be constructed only from those components.

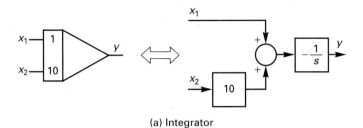

(a) Integrator

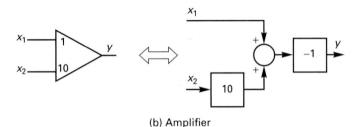

(b) Amplifier

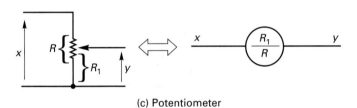

(c) Potentiometer

Figure 3.16   Analog computer components.

As an example, consider the transfer function

$$T(s) = \frac{C(s)}{R(s)} = \frac{2}{s^2 + 3s + 2} = \frac{2s^{-2}}{1 - (-3s^{-1} - 2s^{-2})} \tag{3-77}$$

The analog computer diagram for this transfer function is given in Figure 3.17(a), and an equivalent simulation diagram is shown in Figure 3.17(b). The simulation diagram is constructed, using Mason's gain formula, to have the transfer function (3-77). However, recall that the simulation diagram components for this example must be limited to those given in Figure 3.16.

The programming of older analog computers requires the actual interconnection of the components using wires specifically constructed for the computer. The computer contains a plugboard of connectors, with the connectors to each computer component labeled on the board. The programmer connects the integrators, amplifiers, and potentiometers as required. For example, in Figure 3.17(a), each connecting line shown would be a physical wire. For a complex simulation, debugging the simulation can be very difficult. The plugboard will appear as a "rat's nest" of wires. In general, debugging a digital simulation is much easier than debugging an analog simulation. However, some of the modern analog computers may be programmed using a simulation language, with the wiring of the components performed electronically.

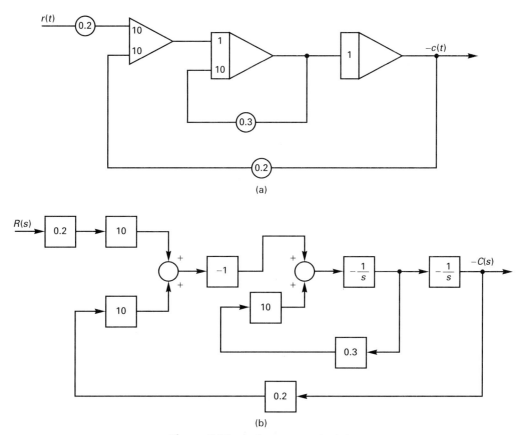

Figure 3.17 Analog computer simulation.

3.9 SUMMARY

In this chapter we presented the state-variable method of modeling physical systems. This method of modeling is required for the simulation of a system. Simple digital and analog simulation was also covered in this chapter. State variable models are also required in the area called modern control analysis and design, which is introduced in Chapter 10. The advantage of the state model of a system is that knowledge of the internal structure as well as the input–output characteristics of the system are available, whereas the transfer function model gives us only input–output information. Generally we make use of both model forms in analysis and design.

REFERENCES

1. P. M. De Russo, R. J. Roy, and C. M. Close. *State Variables for Engineers.* New York: Wiley, 1965.
2. B. Friedland. *Control System Design.* New York: McGraw-Hill, 1986.
3. G. F. Franklin, J. D. Powell, and M. Workman. *Digital Control of Dynamic Systems*, 3rd ed. Reading, MA: Addison-Wesley, 1998.
4. C. L. Phillips and H. T. Nagle. *Digital Control System Analysis and Design*, 3rd ed. Englewood Cliffs, NJ: Prentice Hall, 1995.
5. S. D. Conte and C. deBoor. *Elementary Numerical Analysis: An Algorithmic Approach.* New York: McGraw-Hill, 1982.

PROBLEMS

3.1. (a) Letting $x_1(t) = y(t)$ and $x_2(t) = \dot{y}(t)$, write the state equations for the system described in the differential equation. Express these equations in matrix form.

$$\frac{d^2y(t)}{dt^2} + 4\frac{dy(t)}{dt} + 3y(t) = 2u(t)$$

(b) Draw a simulation diagram for the system in (a).

(c) Write the state equations for the system described by the differential equation

$$\dot{y}(t) + 3y(t) = 2u(t)$$

(d) Draw both the control-canonical and the observer-canonical simulation diagrams.

3.2. (a) Draw a simulation diagram of the system described by the transfer function

$$\frac{Y(s)}{U(s)} = G(s) = \frac{7}{s^2 + 9s + 8}$$

(b) From the simulation diagram, write a set of state equations for the system of (a).

(c) Repeat (a) and (b) for the transfer function

$$\frac{Y(s)}{U(s)} = G(s) = \frac{6}{s + 2}$$

3.3. (a) Draw a simulation diagram for the system described by the transfer function

$$\frac{Y(s)}{U(s)} = G(s) = \frac{s^2}{s^3 + 1}$$

(b) From the simulation diagram write a set of state equations for the system of (a).

3.4. (a) Consider the closed-loop control system in Figure P3.4(a), and its simulation diagram in part (b) of the figure. Write the state equations for the control system, from the simulation diagram.

(b) Find $G_c(s)$, the compensator transfer function, and $G_p(s)$, the plant transfer function, directly from the simulation diagram.

(c) Find the closed-loop transfer function $Y(s)/U(s)$.

(d) Show that the denominator of the closed-loop transfer function is equal to $\det(s\mathbf{I} - \mathbf{A})$, using $\mathbf{A}$ from (a).

(e) Verify the results of (d) by finding the Δ of Mason's gain formula. Note that, in this problem the denominator of the closed-loop transfer function has been calculated by three different procedures.

(f) Modify the MATLAB program of Example 3.4 to verify the closed-loop transfer function.

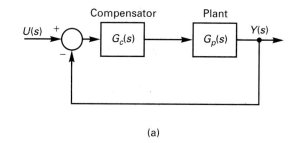

(a)

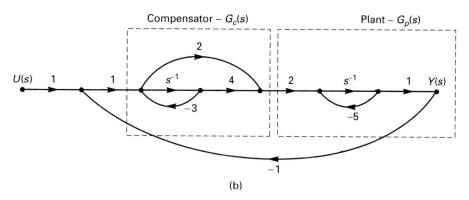

(b)

Figure P3.4   Control system for Problem 3.4.

3.5. Repeat Problem 3.4 for the case that the compensator is a gain K, that is, $G_c(s) = K$.

3.6. **(a)** Draw a simulation diagram of the system described by the transfer function

$$\frac{Y(s)}{U(s)} = G(s) = \frac{10s + 5}{s^2 + 5s + 6}$$

(b) From the simulation diagram, write a set of state equations.

(c) Repeat (a) and (b) for the transfer function

$$\frac{Y(s)}{U(s)} = G(s) = \frac{10s}{s + 2}$$

(d) Repeat (a) and (b) for the transfer function

$$G(s) = \frac{s^2 + 10s + 5}{s^2 + 5s + 6}$$

(e) In (a), (c), and (d), use Mason's gain formula to verify the transfer functions.

(f) Modify the MATLAB program of Example 3.4 to verify the transfer functions in (e).

3.7. Consider the motor in the steel-rolling system of Example 3.3. In this problem we use the same motor for a position control system; that is, the motor is not used for speed control.

(a) Choosing the state variables to be $x_1 = i_a$, $x_2 = \theta$, $x_3 = d\theta/dt$, and the output $y = \theta$, write the state equations for the motor.

(b) Draw a simulation diagram for (a). Note that when using the laws of physics, the state model is neither the control canonical nor the observer canonical form.

(c) Using Mason's gain formula and the simulation diagram of (b), verify the transfer function $\Theta(s)/E_a(s)$ developed in Section 2.7.

3.8. The equations for the circuit of Figure P3.8 are developed in Example 2.2, Section 2.2.

(a) Express these equations in the state variable format, using the variables $x(t) = q(t) = \int i(t)\,dt$, $u(t) = v_1(t)$, and $y(t) = v_2(t)$.

(b) Draw a simulation diagram for these state equations.

(c) Verify the transfer function calculated in Example 2.3, using Mason's gain formula; that is, show that

$$\frac{Y(s)}{U(s)} = \frac{R_2 Cs + 1}{(R_1 + R_2)\,Cs + 1}$$

(d) Verify the transfer funtion in (c) using (3-42).

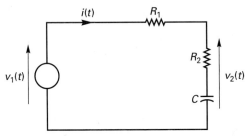

Figure P3.8

3.9. Consider the rigid satellite of Figure P3.9, which was discussed in Example 2.13 of Section 2.6. The thrustors develop a torque τ, and the resultant rotation is modeled by $\tau(t) = J\,d^2\theta(t)/dt^2$.

(a) Develop a state model for this system, with the torque τ as the input and the angle θ as the output.

(b) Draw a simulation diagram for the system.

(c) Using Mason's gain formula, write the system transfer function from the simulation diagram.

(d) Letting $J = 1$, use MATLAB to verify the transfer function in (c).

3.10. A thermal test chamber is illustrated in Figure P3.10(a). This chamber, which is a large room, is used to test large devices under various thermal stresses. The chamber is heated with steam, which is controlled by an electrically activated valve. The temperature of the chamber is measured by a sensor based on a thermistor, which is a semiconductor resistor whose resistance varies with temperature. Opening the door of the chamber affects the temperature of the chamber and hence must be considered as a disturbance.

A model of the thermal chamber is shown in Figure P3.10(b). The control input is the voltage $m(t)$, which controls the valve in the steam line, as shown in the figure. A step function $d(t) = 6u(t)$ is used to model the opening of the chamber door. With the door closed, $d(t) = 0$.

(a) Find a second-order state model of the system with two inputs, $m(t)$ and $d(t)$, and one output, $c(t)$.

(b) Find a first-order state model of the system with the same inputs and outputs as in (a). This model is possible because the characteristics equations are identical for both blocks in Figure P3.10.

(c) Find the transfer function matrix $\mathbf{G}(s)$, where

$$C(s) = \mathbf{G}(s)\mathbf{U}(s)$$

and where $\mathbf{G}(s)$ is of order 1×2 and $\mathbf{U}(s) = [M(s)\ \ D(s)]^{\mathrm{T}}$.

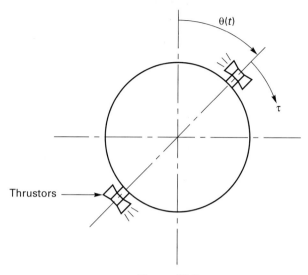

Figure P3.9

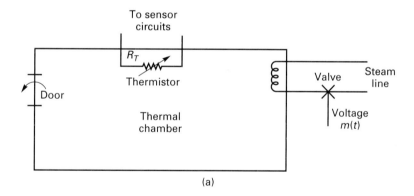

(a)

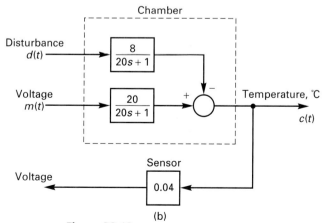

(b)

Figure P3.10 System for Problem 3.10.

3.11. Shown in Figure P3.11 is the block diagram of a simplified cruise-control system for an automobile. The actuator controls the throttle position, the carburetor is modeled as a first-order lag
with a 1-s time constant, and the engine and load is also modeled as a first-order lag with a 3-s
time constant. For this problem, assume that the disturbance is zero.

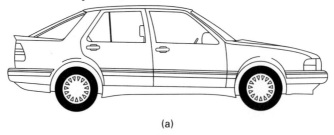

(a)

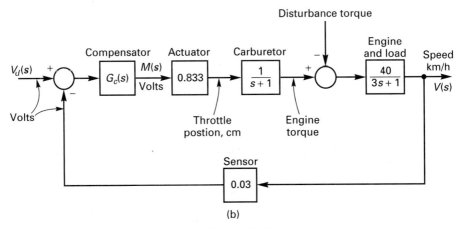

(b)

Figure P3.11

(a) Calculate the transfer function

$$G_p(s) = \frac{V(s)}{M(s)}$$

Mason's gain formula cannot be used directly to solve for this transfer function. Why?

(b) Find a state model for the plant in (a), with one state associated with the carburetor and the
other state associated with the total load on the engine.

(c) Let $G_c(s) = 1$. Find the state model of the closed-loop system such that the states are the
same as those in (b). *Hint:* Draw a simulation diagram.

(d) Find the closed-loop transfer function using

$$T(s) = \frac{G_c(s)G_p(s)}{1 + G_c(s)G_p(s)H}$$

(e) Verify the results in (d) using MATLAB and the state model.

3.12. Consider the mechanical translational system shown in Figure P3.12. All parameters are in
consistent units.

(a) Write a set of state equations for this system.

(b) From these equations, construct a simulation diagram for the system.

(c) Using Mason's gain formula and the simulation diagram of (b), find the transfer function $Y(s)/F(s)$.

(d) Verify the results of (c) using MATLAB and the state equations of (a).

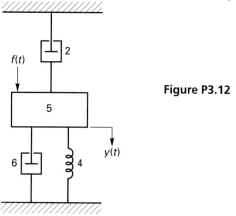

Figure P3.12

3.13. A first-order system is modeled by the state equations

$$\dot{x}(t) = -3x(t) + 4u(t)$$

$$y(t) = x(t)$$

(a) Find the Laplace transform of the state transition matrix.

(b) Find the state transition matrix.

(c) If the input $u(t)$ is a unit step function, with $x(0) = 0$, find $y(t)$, $t > 0$, using (3-24).

(d) If the input $u(t)$ is a unit step function, with $x(0) = -1$, find $y(t)$, $t > 0$, using (3-24). The results of (b) and (c) are useful.

(e) Verify the results of (c), using the transfer function approach.

(f) Verify the results in (d), using the Laplace transform of the state equation.

(g) Use MATLAB to verify the results in (d).

3.14. Repeat all parts of Problem 3.13 for the first-order system

$$\dot{x}(t) = -x(t) + u(t)$$

$$y(t) = x(t) + u(t)$$

3.15. Given the state equations

$$\dot{\mathbf{x}}(t) = \begin{bmatrix} 0 & 2 \\ -2 & -5 \end{bmatrix} \mathbf{x}(t) + \begin{bmatrix} 0 \\ 1 \end{bmatrix} u(t)$$

$$y(t) = \begin{bmatrix} 1 & 0 \end{bmatrix} \mathbf{x}(t)$$

(a) Find the Laplace transform of the state transition matrix.

(b) Find the state transition matrix.

(c) If the input $u(t)$ is a unit step function, with $x_1(0) = x_2(0) = 0$, find $\mathbf{x}(t)$, $t \geq 0$, using (3-24).

(d) If the input $u(t)$ is a unit step function, with $x_1(0) = 1$, $x_2(0) = 2$, find $\mathbf{x}(t)$, $t \geq 0$. The results of (c) are useful.

(e) Verify the calculations of $x_1(t)$ in (c), using a transfer function approach.

(f) Modify the MATLAB program of Example 3.4 to verify the results in (d).

3.16. Given the state equations

$$\dot{x}(t) = \begin{bmatrix} 0 & 2 \\ -1 & -3 \end{bmatrix} x(t) + \begin{bmatrix} 0 \\ 1 \end{bmatrix} u(t)$$

$$y(t) = \begin{bmatrix} 1 & 0 \end{bmatrix} x(t)$$

(a) Find the Laplace transform of the state transition matrix.
(b) Find the state transition matrix.
(c) If the input $u(t)$ is a unit step function, with $x_1(0) = x_2(0) = 0$, find $x(t)$, $t \geq 0$, using (3-24).
(d) If the input $u(t)$ is a unit step function, with $x_1(0) = 1$, $x_2(0) = -1$, find $x(t)$, $t \geq 0$. The results of (c) are useful.
(e) Verify the calculation of $x_1(t)$ in (c), using a transfer function approach.
(f) Modify the MATLAB program of Example 3.4 to verify the results in (d).

3.17. Use the state equations and (3-42), verify the transfer functions found in
(a) Problem 3.6(e)
(b) Problem 3.9(c)
(c) Problem 3.12(d)
(d) Problem 3.13(e)
(e) Problem 3.14(e)

3.18. For the satellite of Problem 3.9 with $J = 0.1$, a state model is given by

$$\dot{x}(t) = \begin{bmatrix} 0 & 1 \\ 0 & 0 \end{bmatrix} x(t) + \begin{bmatrix} 0 \\ 10 \end{bmatrix} u(t)$$

$$y(t) = \begin{bmatrix} 1 & 0 \end{bmatrix} x(t)$$

with $u(t) = \tau(t)$ and $y(t) = \theta(t)$.
(a) Find the Laplace transform of the state transition matrix.
(b) Find the state transition matrix.
(c) If the input $u(t)$ is a unit step function, with $x(0) = 0$, find $x(t)$, $t > 0$, using (3-24).
(d) If the input $u(t)$ is a unit step function, with $x(0) = [0 \ 1]^T$, find $x(t)$, $t > 0$. The results of (b) and (c) are useful.
(e) Verify the results of (c), using the transfer function approach.
(f) Modify the MATLAB program of Example 3.4 to verify the results in (d).
(g) Describe the response of the physical system with $u(t) = 0$ and the initial conditions in (d); that is, what would you observe if you were watching the satellite?

3.19. Consider the rigid satellite of Figure P3.9 which was discussed in Example 2.13 of Section 2.6. The thrustors develop a torque τ, and the resultant rotation is modeled by $\tau(t) = J \, d^2\theta(t)/dt^2$.
(a) Develop a state model for this system, with the torque τ as the input, the angle θ as the output, and θ and $\dot{\theta}$ and as the states.
(b) Draw a simulation diagram for the system.
(c) A similarity transformation is defined as [see (3-49)]

$$\dot{x}(t) = Pv(t) = \begin{bmatrix} 2 & 0 \\ 1 & 1 \end{bmatrix} v(t)$$

Express the state model in terms of the states $v(t)$.
(d) Draw a simulation diagram for the state equations in (c), and use Mason's gain formula to verify the transfer function.

(e) Modify the simulation diagram of (b) to show the states $\mathbf{v}(t)$, as in figure 3.11. Compare the transfer functions from the input τ to the states $v_1(t)$ and $v_2(t)$, for the two simulation diagrams.

(f) Simulate the two state models with unit step inputs, with $J = 1$. The two responses will be equal, indicating that the two models have the same input–output characteristics.

(g) With $J = 1$, verify the resuilts in (c) using MATLAB.

3.20. For the two state models of Problem 3.19, show that:

(a) $|s\mathbf{I} - \mathbf{A}| = |s\mathbf{I} - \mathbf{A}_v|$.

(b) $|\mathbf{A}| = |\mathbf{A}_v|$.

(c) $\text{tr } \mathbf{A} = \text{tr } \mathbf{A}_v$.

(d) $\mathbf{C}(s\mathbf{I} - \mathbf{A})^{-1}\mathbf{B} = \mathbf{C}_v(s\mathbf{I} - \mathbf{A}_v)^{-1}\mathbf{B}_v$.

3.21. Consider the system

$$\dot{\mathbf{x}}(t) = \begin{bmatrix} 0 & 2 \\ -2 & -5 \end{bmatrix} \mathbf{x}(t) + \begin{bmatrix} 0 \\ 1 \end{bmatrix} u(t)$$

$$y(t) = \begin{bmatrix} 1 & 0 \end{bmatrix} \mathbf{x}(t)$$

A similarity transformation is defined as [see(3-49)]

$$\dot{\mathbf{x}}(t) = \mathbf{P}\mathbf{v}(t) = \begin{bmatrix} 1 & 1 \\ 1 & 2 \end{bmatrix} \mathbf{v}(t)$$

(a) Express the state model in terms of the states $\mathbf{v}(t)$.

(b) Draw a simulation diagram for the state model in $\mathbf{x}(t)$ and one for the state model in $\mathbf{v}(t)$.

(c) Show by Mason's gain formula that the transfer functions of the two simulation diagrams in (b) are equal.

(d) Simulate the two state models with unit step inputs. The two responses will be equal, indicating that the two models have the same input–output characteristics.

(e) Verify the results in (a) using MATLAB.

3.22. For the two state models of Problem 3.21, show that:

(a) $|s\mathbf{I} - \mathbf{A}| = |s\mathbf{I} - \mathbf{A}_v|$

(b) $|\mathbf{A}| = |\mathbf{A}_v|$

(c) $\text{tr } \mathbf{A} = \text{tr } \mathbf{A}_v$

(d) $\mathbf{C}(s\mathbf{I} - \mathbf{A})^{-1}\mathbf{B} = \mathbf{C}_v(s\mathbf{I}_v - \mathbf{A}_v)^{-1}\mathbf{B}_v$

3.23. Use a MATLAB simulation to verify the responses in (d) of

(a) Problem 3.13

(b) Problem 3.14

(c) Problem 3.15

(d) Problem 3.16

(e) Problem 3.18

3.24. Construct an analog computer diagram for each of the following transfer functions.

(a) $G(s) = \dfrac{8.7}{s + 5}$

(b) $G(s) = \dfrac{15s + 10.5}{s^2 + 9.7s + 83.4}$

3.25. For the general nth-order system with the transfer function

$$G(s) = \frac{b_{n-1}s^{n-1} + \cdots + b_1 s + b_0}{s^n + a_{n-1}s^{n-1} + \cdots + a_1 s + a_0}$$

(a) Write the state equations for this system in control-canonical form.

(b) Write the state equations for this system in observer-canonical form.

3.26. Given the state equations $\dot{\mathbf{x}}(t) + \mathbf{A}\mathbf{x}(t) + \mathbf{B}u(t)$ and $y(t) = \mathbf{C}\mathbf{x}(t) + Du(t)$, derive the transfer function $G(s) = Y(s)/U(s)$ in (3-42); that is, derive $G(s) = \mathbf{C}(s\mathbf{I} - \mathbf{A})^{-1}\mathbf{B} + D$.

3.27. Show that $\mathbf{C}(s\mathbf{I} - \mathbf{A})^{-1}\mathbf{B} = \mathbf{C}_v(s\mathbf{I} - \mathbf{A}_v)^{-1}\mathbf{B}_v$, where the various matrices are related by

$$\mathbf{A}_v = \mathbf{P}^{-1}\mathbf{A}\mathbf{P} \qquad \mathbf{B}_v = \mathbf{P}^{-1}\mathbf{B} \qquad \mathbf{C}_v = \mathbf{C}\mathbf{P}$$

4

System Responses

This chapter is concerned with the response of linear time-invariant systems. The Laplace transform is used to investigate the response of first- and second-order systems. Then it is shown that the response of higher-order systems can be considered to be the sum of the responses of first- and second-order systems; therefore, much emphasis is placed on understanding the responses of these low-order systems.

The actual time signals that appear in control systems normally are not analytical signals; that is, we cannot express these signals as mathematical expressions. Hence we cannot generally find the Laplace transform of these signals. Even if the Laplace transform of the signals can be found, it is often not worth the effort. However, there are exceptions to these statements. One very common input to control systems is the step function. For example, if we activate an electric heater by applying a constant voltage through the operation of a switch or a relay, we have applied a step function of voltage to the heater. Generally, any time that we activate a system, we are applying a step function. For another example, if we command a constant temperature, it is frequently done by adding a step function to the input of the system. In addition, the step-function input is important because the step response of a system displays important characteristics of that system, as will be shown in this chapter.

A less common input to control systems is the ramp function. However, such inputs do appear. For example, in the automatic landing of aircraft, the aircraft is commanded to follow a ramp for the glide slope; this glide slope is commonly approximately 3°. Also, in the heat treatment of steel, it often necessary for the temperature of the steel to be increased at a constant rate for a period of time. Thus the input to the temperature control system must be a ramp function. Because of cases such as these and for other reasons that are discussed later, we are interested in the ramp response of control systems.

A third input of major importance is the sinusoidal function. The importance of this function is not obvious, since we usually do not require a control system to follow (track) a sinusoid. However, if we know the response of a linear time-invariant system to sinusoids of all frequencies, we have a *complete description* of the system; we can design compensators for the system based on this description alone. This is of immense practical value, since we can often measure the sinusoidal response of a physical system.

For the reasons just given, we investigate the response of systems to step inputs, ramp inputs, and sinusoidal inputs in this chapter. The value of the results obtained will become more evident later.

4.1 TIME RESPONSE OF FIRST-ORDER SYSTEMS

In this section the time response of first-order systems is investigated. The transfer function of a general first-order system can be written as

$$G(s) = \frac{C(s)}{R(s)} = \frac{b_0}{s + a_0} \qquad (4\text{-}1)$$

where $R(s)$ is the input function and $C(s)$ is the output function; this notation is common.

A more common notation for the first-order transfer function is

$$G(s) = \frac{C(s)}{R(s)} = \frac{K}{\tau s + 1} \qquad (4\text{-}2)$$

since physical meaning can be given to both K and τ. We call (4-2) the standard form of the first-order system. Of course, in (4-1) and (4-2),

$$a_0 = \frac{1}{\tau} \qquad b_0 = \frac{K}{\tau} \qquad (4\text{-}3)$$

Recall that initial conditions are ignored in the calculation of transfer functions. We now show how the initial condition can be included in (4-2). First we rewrite (4-2) as

$$\left(s + \frac{1}{\tau}\right)C(s) = \frac{K}{\tau}R(s) \qquad (4\text{-}4)$$

The differential equation for the transfer function of (4-2) is the inverse Laplace transform of (4-4).

$$\dot{c}(t) + \frac{1}{\tau}c(t) = \frac{K}{\tau}r(t) \qquad (4\text{-}5)$$

Now we take the Laplace transform of (4-5) and include the initial condition term.

$$sC(s) - c(0) + \frac{1}{\tau}C(s) = \frac{K}{\tau}R(s) \qquad (4\text{-}6)$$

Solving for $C(s)$ yields

$$C(s) = \frac{c(0)}{s + (1/\tau)} + \frac{(K/\tau)R(s)}{s + (1/\tau)} \qquad (4\text{-}7)$$

and we have the effect of $c(0)$ on $C(s)$. This equation can be represented in block diagram form, as shown in Figure 4.1(a); however, initial conditions usually are not shown as inputs on the system block diagram. Note that the initial condition as an input has a Laplace transform of $c(0)$, which is a constant. The inverse Laplace transform of a constant is an impulse function. Hence, *for this representation,* the initial condition as an input appears as the

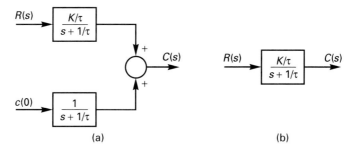

Figure 4.1 First-order system.

impulse function $c(0)\delta(t)$. Here we can see that the impulse function has a practical meaning, even though the impulse function is not a physically realizable signal. Since we usually ignore initial conditions in block diagrams, we normally use the system block diagram as shown in Figure 4.1(b).

4.1.1 System Step Response

We now find the unit step response for the standard first-order system of (4-2). For this case, $R(s) = 1/s$ and, from (4-2),

$$C(s) = \frac{K/\tau}{s\,[s + (1/\tau)]} = \frac{K}{s} + \frac{-K}{s + (1/\tau)} \tag{4-8}$$

The inverse Laplace transform of this expression yields

$$c(t) = K(1 - e^{-t/\tau}), \, t > 0 \tag{4-9}$$

The response can also be expressed as

$$c(t) = K(1 - e^{-t/\tau})u(t)$$

Using (4-5), we verify this response with the MATLAB program

```
dsolve('Dc=-c/tau + K/tau, c(0) = 0')
```

The first term in the response (4-9) originates in the pole of the input $R(s)$ and is called the *forced response;* since this term does not go to zero with increasing time, it is also called the *steady-state response.* The second term in (4-9) originates in the pole of the transfer function $G(s)$ and is called the *natural response;* since this term goes to zero with increasing time, it is also called the *transient response.*

The step response of a first-order system as given in (4-9) is plotted in Figure 4.2. The two components of the response are plotted separately, along with the complete response. Note that the exponentially decaying term has an initial slope of K/τ; that is,

$$\frac{d}{dt}(-Ke^{-t/\tau})_{t=0} = \frac{K}{\tau}e^{-t/\tau}\bigg|_{t=0} = \frac{K}{\tau} \tag{4-10}$$

Mathematically, the exponential term does not decay to zero in a finite length of time. However, if the term continued to decay at its initial rate, it would reach a value of zero in τ seconds. The parameter τ is called the *time constant* and has the units of seconds.

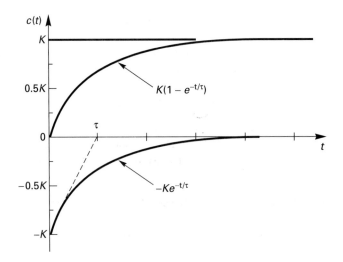

Figure 4.2 Step response of first-order system.

TABLE 4.1

t	$e^{-t/\tau}$
0	1
τ	0.3679
2τ	0.1353
3τ	0.0498
4τ	0.0183
5τ	0.0067

The decay of an exponential function is illustrated in Table 4.1 as a function of the time constant τ. It is seen from this table that the exponential function has decayed to less than 2 percent of its initial value in four time constants and to less than 1 percent of its initial value in five time constants. In a practical sense, we often consider an exponential term to have decayed to zero in four to five time constants. In this book we somewhat arbitrarily consider an exponential term to be zero after four time constants. Other authors sometimes use other definitions. The output of a physical system that is modeled by the first-order transfer function will not continue to vary for all finite time. Thus it is reasonable to make the approximation that the response becomes constant after a number of time constants. In addition, the identification of time constants with exponential terms allows us to compare the decay of one term relative to another. For example, if we were designing a position-control system for the pen of a plotter for a digital computer, we know immediately that a time constant of 1 s for the control system is much too slow, whereas values in the region of 0.1 s and faster (smaller) are more reasonable.

Consider again the response of (4-9) as plotted in Figure 4.2. Note that

$$\lim_{t \to \infty} c(t) = K \qquad (4\text{-}11)$$

We call this limit the *final value,* or the *steady-state value,* of the response. Hence in the general first-order transfer function

$$G(s) = \frac{K}{\tau s + 1}$$

τ is the system time constant and K is the steady-state response to a unit step input. Thus both parameters in the transfer function have meaning, and this accounts for the popularity of writing the transfer function in this manner. An example is now given.

Example 4.1

We wish to find the unit step response of a system with the transfer function

$$G(s) = \frac{2.5}{0.5s + 1} = \frac{5}{s + 2}$$

Then

$$C(s) = G(s)R(s) = \frac{5}{s + 2}\frac{1}{s} = \frac{5/2}{s} + \frac{-5/2}{s + 2}$$

and

$$c(t) = (5/2)\,(1 - e^{-2t})$$

Hence the pole of the transfer function at $s = -2$ gives the time constant $\tau = 1/2$ s. The steady-state value of the response is 5/2, and since the system constant is 0.5 s, the output reaches steady state in about 2 s. A MATLAB program that calculates both the partial-fraction expansion and the unit step response is given by

```
Cnum = [0 0 5]; Cden = [1 2 0];
[r,p,k] = residue(Cnum,Cden)
pause
G=tf ([0 5],[1 2]);
step (G)
```

4.1.2 System dc Gain

A general procedure for finding the steady-state response to a unit step input for a system of *any order* is now developed. From Appendix B, the final-value theorem of the Laplace transform is

$$\lim_{t \to \infty} c(t) = \lim_{s \to 0} sC(s) = \lim_{s \to 0} sG(s)R(s) \qquad (4\text{-}12)$$

provided that $c(t)$ has a final value. For the case that the input is a unit step, $R(s)$ is equal to $1/s$ and

$$\lim_{t \to \infty} c(t) = \lim_{s \to 0} sG(s)\frac{1}{s} = \lim_{s \to 0} G(s) \qquad (4\text{-}13)$$

If $c(t)$ has a final value, the right-hand limit in this equation exists. If $c(t)$ does not have a final

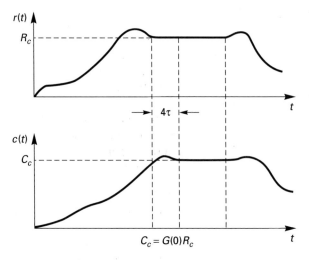

Figure 4.3 Input and output illustrating dc gain.

value, the right-hand limit may still exist but has no meaning. Assume that $c(t)$ has a final value. Since the steady-state input is unity, then the output is also the gain in the steady state; that is, $G(0)$ is the system steady-state gain for the input constant. This is true independent of the order of the system. This gain is of such importance that it deserves a name. Since a constant electrical signal is called a dc signal, we will call $G(0)$ the system dc gain.

System dc gain. The *system dc gain* is the steady-state gain to a constant input for the case that the output has a final value, and it is equal to the system transfer function evaluated at $s = 0$.

Note that for the dc gain of a system to have meaning, it is not necessary that a step function be applied to the system. It is only necessary that the input function $r(t)$ be constant for a period longer than four time constants. This point is illustrated in Figure 4.3. The input becomes constant at a value R_c, and the output then becomes constant 4τ seconds later at the value C_c. Then

$$\frac{C_c}{R_c} = \lim_{s \to 0} G(s)$$

4.1.3 System Ramp Response

We now consider ramp responses. For the input equal to a unit ramp function, $r(t) = t$ and $R(s) = 1/s^2$. Then, from (4-2),

$$C(s) = G(s)R(s) = \frac{K/\tau}{s^2[s + (1/\tau)]} = \frac{K}{s^2} + \frac{-K\tau}{s} + \frac{K\tau}{s + (1/\tau)} \qquad (4\text{-}14)$$

Therefore,

$$c(t) = Kt - K\tau + K\tau e^{-t/\tau} \qquad (4\text{-}15)$$

This response is verified with the following MATLAB program:

```
syms s F K Tau
F = (K/Tau) / (s^2*(s+1/Tau));
ilaplace (F)
```

Thus the ramp response is composed of three terms: a ramp, a constant, and an exponential. We wish to make two points relative to this response.

First, the exponential has the same time constant as in the step response. Of course, this is true, regardless of the input. However, the amplitude of the exponential is different in the ramp response as compared to the step response. The amplitude is different by the factor τ; if τ is large, the exponential can have a major effect on the system response. This can be a problem in designing systems to follow ramp inputs.

Next, the steady-state response $c_{ss}(t)$ is given by

$$c_{ss}(t) = Kt - K\tau \qquad (4\text{-}16)$$

Here we define the steady-state response to be composed of those terms that do not approach zero as time increases. The *system error* for a control system is defined later; however, it is a measure of the difference between the desired system output and the actual system output. In general, we wish this error to be zero in the steady state for a control system. For this example, if we desire that the output follow a ramp, the error will include the constant term $K\tau$. A design specification might be to alter the system such that the steady-state system error for a ramp input is zero or at least is very small. Problems such as this are covered in the chapters on design.

4.2 TIME RESPONSE OF SECOND-ORDER SYSTEMS

In this section we investigate the response of second-order systems to certain inputs. We assume that the system transfer function is of the form

$$G(s) = \frac{C(s)}{R(s)} = \frac{b_0}{s^2 + a_1 s + a_0} \qquad (4\text{-}17)$$

However, as in the first-order case, the coefficients are generally written in a manner such that they have physical meaning. The *standard form* of the second-order transfer function is given by

$$G(s) = \frac{\omega_n^2}{s^2 + 2\zeta\omega_n s + \omega_n^2} \qquad (4\text{-}18)$$

where ζ is defined to be the dimensionless *damping ratio* and ω_n is defined to be the *natural frequency,* or *undamped frequency.* Note also that the dc gain, $G(0)$, is unity. Later we consider both the case that the dc gain is other than unity and the case that the numerator is other than a constant. Note also that *all* system characteristics of the standard second-order system are functions of only ζ and ω_n, since ζ and ω_n are the only parameters that appear in the transfer function (4-18).

Consider first the unit step response of this second-order system:

$$C(s) = G(s)R(s) = \frac{\omega_n^2}{s(s^2 + 2\zeta\omega_n s + \omega_n^2)} \qquad (4\text{-}19)$$

The inverse Laplace transform is not derived here (see Problem 4.8); however, assuming for the moment that the poles of $G(s)$ are complex, the result is

$$c(t) = 1 - \frac{1}{\beta} e^{-\zeta \omega_n t} \sin(\beta \omega_n t + \theta) \tag{4-20}$$

where $\beta = \sqrt{1 - \zeta^2}$ and $\theta = \tan^{-1}(\beta/\zeta)$. In this response, $\tau = 1/\zeta\omega_n$ is the time constant of the exponentially damped sinusoid in seconds (we can usually ignore this term after approximately four time constants). Also, $\beta\omega_n$ is the frequency of the damped sinusoid.

We wish now to show typical step responses for a second-order system. The step response given by (4-20) is a function of both ζ and ω_n. If we specify ζ, we still cannot plot $c(t)$ without specifying ω_n. To simplify the plots, we give $c(t)$ for a specified ζ as a function of $\omega_n t$. A family of such curves for various values of ζ is very useful and is given in Figure 4.4 for $0 \le \zeta \le 2$. Note that for $0 < \zeta < 1$, the response is a damped sinusoid. For $\zeta = 0$, the sinusoid is undamped, or of sustained amplitude. For $\zeta \ge 1$, the oscillations have ceased. It is apparent from (4-20) that for $\zeta < 0$, the response grows without limit. We consider only the case that $\zeta \ge 0$ in this chapter. A MATLAB program that calculates some of the step responses of Figure 4.4 is given by

```
zeta = [0.2 0.5 1 2];
for k = 1:4
  G =tf ([1],[1 2*zeta(k) 1]);
    step(G)
    hold on
end
hold off
```
The two poles of the transfer function $G(s)$ in (4-18) occur at

$$s = -\zeta\omega_n \pm j\omega_n \sqrt{1 - \zeta^2}$$

For $\zeta > 1$, these poles are real and unequal, and the damped sinusoid portion of $c(t)$ is replaced by the weighted sum of two exponential functions; that is,

$$c(t) = 1 + k_1 e^{-t/\tau_1} + k_2 e^{-t/\tau_2} \tag{4-21}$$

where $\tau_1 = 1/(\zeta\omega_n + \omega_n\sqrt{\zeta^2 - 1})$, $\tau_2 = 1/(\zeta\omega_n - \omega_n\sqrt{\zeta^2 - 1})$ are the two system time constants. For $\zeta = 1$, the poles of $G(s)$ are real and equal, so that

$$c(t) = 1 + k_1 e^{-t/\tau} + k_2 t e^{-t/\tau}, \tau = 1/\omega_n$$

For $0 < \zeta < 1$, the system is said to be *underdamped*, and for $\zeta = 0$ it is said to be *undamped*. For $\zeta = 1$, the system is said to be *critically damped*, and for $\zeta > 1$, the system is *over-damped*.

For a linear time-invariant system,

$$C(s) = G(s)R(s) \tag{4-22}$$

For the case that $r(t)$ is a unit impulse function, $R(s) = 1$ and

$$c(t) = \mathcal{L}^{-1}[G(s)] = g(t) \tag{4-23}$$

where $g(t)$ is the *unit impulse response,* or *weighting function,* of a system with the transfer function $G(s)$. Then, by the convolution integral (see Appendix B), for a general

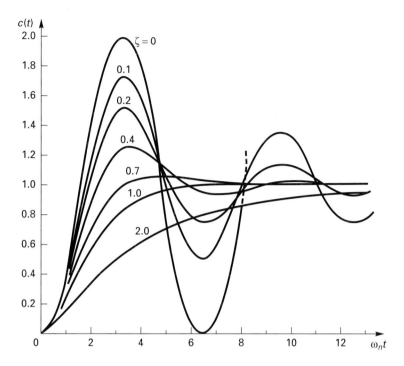

Figure 4.4   Step response for second-order system (4-20).

input $r(t)$,

$$c(t) = \int_0^t g(\tau) r(t - \tau) \, d\tau \tag{4-24}$$

from (4-22). [In (4-24), τ is the variable of integration and is not related to the time constant.] Hence, all response information for a general input is contained in the impulse response $g(t)$.

Recall also from Section 4.1 that an initial condition on a first-order system can be modeled as an impulse function input. While the initial condition excitation of higher-order systems cannot be modeled as simply as that of the first-order system, the impulse response of any system does give an indication of the nature of the initial-condition response, and thus the transient response, of the system. The unit-impulse response of the second-order system (4-18) is given in Figure 4.5. This figure is a plot of the function

$$c(t) = \mathcal{L}^{-1}\left(\frac{\omega_n^2}{s^2 + 2\zeta\omega_n s + \omega_n^2} \right) = \frac{\omega_n}{\beta} e^{-\zeta\omega_n t} \sin\beta\omega_n t = g(t) \tag{4-25}$$

Compare Figure 4.4 with Figure 4.5 and note the similarity of the information. In fact, the unit impulse response of a system is the derivative of the unit step response (see Problem 4.9). The impulse response of the second-order system can also be considered to be the response to certain initial conditions, with $r(t) = 0$ (see Problem 4.9).

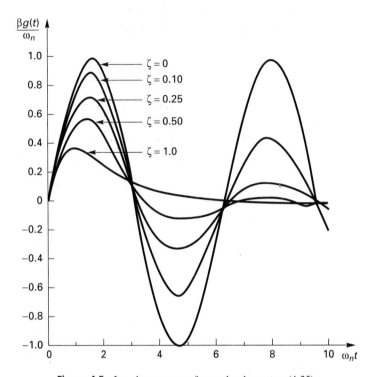

Figure 4.5   Impulse response of second-order system (4-25).

4.3 TIME RESPONSE SPECIFICATIONS IN DESIGN

Before a control system is designed, specifications must be developed that describe the characteristics that the system should possess. For example, some of these specifications may be written in terms of the system step response. Control system specifications in the time domain for the standard second-order system are developed in this section.

A typical unit-step response of a standard second-order system is shown in Figure 4.6. Some characteristics that describe this response are now developed. The rise time of the response can be defined in different ways; we define *rise time* T_r, as the time required for the response to rise from 10 percent of the final value to 90 percent of the final value, as shown in the figure. The peak value of the step response is denoted by M_{pt}, the time to reach this peak value is T_p, and the *percent overshoot* is defined by the equation

$$\text{percent overshoot} = \frac{M_{pt} - c_{ss}}{c_{ss}} \times 100 \tag{4-26}$$

where c_{ss} is the steady-state, or final, value of $c(t)$. In Figure 4.6 $c_{ss} = 1$ [see (4-20)].

The *settling time,* T_s, is the time required for the output to settle to within a certain percent of its final value. Two common values used are 5 percent and 2 percent. For the second-order response of (4-20), approximately four time constants are required for $c(t)$ to settle to within 2 percent of the final value (see Table 4.1). Regardless of the percentage used, the settling time will be directly proportional to the time constant τ for a standard second-order underdamped system; that is,

$$T_s = k\tau = \frac{k}{\zeta\omega_n} \tag{4-27}$$

where k is determined by the defined percentage. As stated previously, we use k equal to four in this book.

Whereas all the preceding parameters are defined on the basis of the underdamped response, T_r, T_s, and c_{ss} are all equally meaningful for the overdamped and the critically damped responses. Of course, M_{pt}, T_p, and percent overshoot have no meaning for these cases.

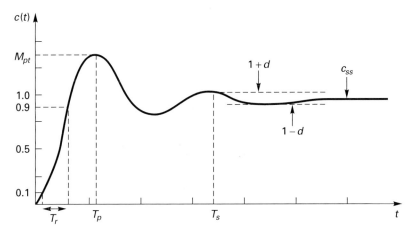

Figure 4.6   Typical step response.

We would like to develop analytical relationships for percent overshoot and T_p in Figure 4.6. The impulse response of the second-order system is given in (4-25),

$$g(t) = \frac{\omega_n}{\beta}e^{-\zeta\omega_n t}\sin\beta\omega_n t \tag{4-28}$$

As just stated, this response is the derivative of the step response of Figure 4.6. Thus the first value of time greater than zero for which this expression is zero is that time for which the slope of the step response is zero, or that time T_p. Therefore,

$$\beta\omega_n T_p = \pi \qquad \beta = \sqrt{1-\zeta^2}$$

or

$$T_p = \frac{\pi}{\omega_n\sqrt{1-\zeta^2}} \tag{4-29}$$

The peak value of the step response occurs at this time, and evaluating the sinusoid in (4-20) at $t = T_p$ yields

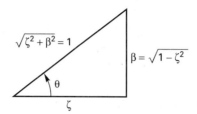

Figure 4.7 Relationship of θ, ζ, and β.

$$\sin(\beta\omega_n t + \theta)\big|_{\beta\omega_n t = \pi} = -\sin\theta = -\sin\left(\tan^{-1}\frac{\beta}{\zeta}\right) = -\frac{\beta}{1} \tag{4-30}$$

(see Figure 4.7). Thus from (4-20) and (4-30),

$$c(t)\big|_{t = T_p} = M_{pt} = 1 + e^{-\zeta\pi/\sqrt{1-\zeta^2}} \tag{4-31}$$

Therefore, since $c_{ss} = 1$, the percent overshoot is, from (4-26),

$$\text{percent overshoot} = e^{-\zeta\pi/\sqrt{1-\zeta^2}} \times 100 \tag{4-32}$$

The percent overshoot is thus a function only of ζ and is plotted versus ζ in Figure 4.8.
We can express (4-29) as

$$\omega_n T_p = \frac{\pi}{\sqrt{1-\zeta^2}} \tag{4-33}$$

Thus the product $\omega_n T_p$ is also a function of only ζ, and this is also plotted in Figure 4.8.
Since T_p is an approximate indication of the rise time, Figure 4.8 also roughly indicates rise time. An example is now given to indicate the usefulness of these curves.

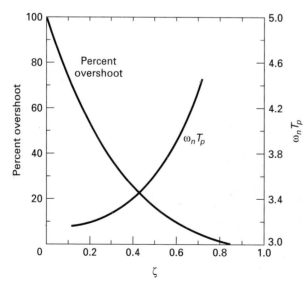

Figure 4.8 Relationship of percent overshoot, ζ, ω_n, and T_p for a second-order system.

Example 4.2

We will consider the design of the servo system of Figure 4.9. Normally, the signal out of the difference circuit that implements the summing junction does not have sufficient power to drive the motor, and a power amplifier must be used. The transfer function of the system is

$$T(s) = \frac{0.5K_a/[s(s+2)]}{1+0.5K_a/[s(s+2)]} = \frac{0.5K_a}{s^2+2s+0.5K_a} = \frac{\omega_n^2}{s^2+2\zeta\omega_n s+\omega_n^2}$$

where K_a is the gain of the power amplifier. We assume that K_a is the only parameter in the system that can be varied. Hence, in designing this system, we can set only one characteristic of the system; that is, we can set only ζ, or only ω_n, or only a single function of the two. We cannot set ζ and ω_n independently by choosing the power amplifier gain K_a.

Suppose that the servo controls the pen position of a plotter in which we can allow no overshoot for a step input. From Figure 4.4, $\zeta = 1$ gives the fastest response with no overshoot for a given ω_n, and we thus choose this value. From the derived transfer function,

$$2\zeta\omega_n = 2(1)\omega_n = 2$$

Therefore, $\omega_n = 1$. Also, from the derived transfer function,

$$\omega_n^2 = (1)^2 = 0.5K_a$$

and therefore $K_a = 2.0$. Hence setting the power amplifier gain to 2.0 results in the fastest response with no overshoot for the plotter, but in this design we have no control over the actual speed of response.

Note that the settling time will be

$$T_s = \frac{4}{\zeta\omega_n} = 4 \text{ s}$$

This plotter would not be fast enough for most applications. To improve the plotter, first we should choose a different motor, one that responds faster (see Problem 4.10). Then it would probably be necessary to add a compensator in the closed-loop system, a topic that is covered in the chapters on design.

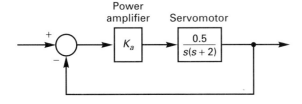

Figure 4.9   System for Example 4.2.

4.3.1 Time Response and Pole Locations

The parameters of the step response of the standard second-order system can also be related to the pole locations of the transfer function. Since the transfer function is given by

$$G(s) = \frac{\omega_n^2}{s^2+2\zeta\omega_n s+\omega_n^2}$$

the poles are calculated to occur at $s = -\zeta\omega \pm j\omega_n\sqrt{1-\zeta^2}$. The poles are shown in Figure 4.10(a). The settling time T_s is related to these poles by (4-27), which is repeated here:

[eq. (4-27)] $$T_s = k\tau = \frac{k}{\zeta\omega_n}$$

where k is often chosen to be equal to 4. The settling time is then inversely related to the real part of the poles. If in design the settling is specified to be less than or equal to some value T_{sm}, $\zeta\omega_n \geq k/T_{sm}$, and the pole locations are then restricted to the region of the s-plane indicated in Figure 4.10(b). Hence the speed of response is increased by moving the poles to the left in the s-plane.

The angle α in Figure 4.10(a) satisfies the relationship

$$\alpha = \tan^{-1}\frac{\sqrt{1-\zeta^2}}{\zeta} = \cos^{-1}\zeta \qquad (4\text{-}34)$$

The percent overshoot is given by

[eq. (4-32)] $\qquad$ percent overshoot $= e^{-\zeta\pi/\sqrt{1-\zeta^2}} \times 100$

Hence this equation can be expressed as

$$\text{percent overshoot} = e^{-\pi/\tan\alpha} \times 100$$

Decreasing the angle α reduces the percent overshoot. Hence, specifying the percent overshoot to be less than a particular value restricts the pole locations to the region of the s-plane, as shown in Figure 4.10(c). These relationships will now be illustrated with an example.

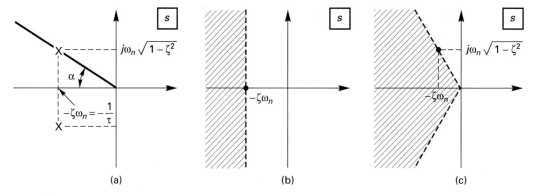

Figure 4.10   Pole locations.

Example 4.3

As an example, suppose that, in the design of a second-order system, the percent overshoot in the step response of a second-order system is limited to 4.32 percent, which is the value that results from $\zeta = 0.707$ (a commonly used value). Thus, in (4-34), $\alpha = \tan^{-1} 1 = \cos^{-1} 0.707 = 45°$. Suppose that the design specifications require a maximum settling time of 2 s. Then $\tau \leq 0.5$, and

$$\zeta\omega_n = \frac{1}{\tau} \leq 2$$

Hence, the transfer function pole locations of the second-order system are limited to the regions of the s-plane shown in Figure 4.11. The pole locations that exactly satisfy the limits of the specifications are $s = -2 \pm j2$.

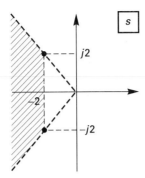

Figure 4.11 Poles for the example.

In summary, for the standard second-order system, the characteristics of the step response are related to the transfer function parameters by the following equations:

[eq. (4-27)] $$T_s = k\tau = \frac{k}{\zeta\omega_n}$$

where we let $k = 4$, and

[eq. (4-32)] $$\text{percent overshoot} = e^{-\zeta\pi/\sqrt{1-\zeta^2}} \times 100$$

[eq. (4-33)] $$\omega_n T_p = \frac{\pi}{\sqrt{1-\zeta^2}}$$

The angle α of the pole locations in the s-plane, defined in Figure 4.10, is given by

[eq. (4-34)] $$\alpha = \tan^{-1}\frac{\sqrt{1-\zeta^2}}{\zeta} = \cos^{-1}\zeta$$

4.4 FREQUENCY RESPONSE OF SYSTEMS

In the preceding sections, the time responses of first- and second-order systems were considered. In this section we give meaning to the steady-state response of systems to sinusoidal inputs, which is called the *frequency response*. We show later that the frequency response has meaning far beyond the calculation of the time response to sinusoids.

Suppose that the input to a system with the transfer function $G(s)$ is the sinusoid

$$r(t) = A \cos\omega_1 t \tag{4-35}$$

Then

$$R(s) = \frac{As}{s^2 + \omega_1^2} \tag{4-36}$$

and

$$C(s) = G(s)R(s) = G(s)\frac{As}{(s - j\omega_1)(s + j\omega_1)} \qquad j = \sqrt{-1} \qquad (4\text{-}37)$$

We can expand this expression into partial fractions of the form

$$C(s) = \frac{k_1}{s - j\omega_1} + \frac{k_2}{s + j\omega_1} + C_g(s) \qquad (4\text{-}38)$$

where $C_g(s)$ is the collection of all the terms in the partial-fraction expansion that originate in the denominator of $G(s)$. It is assumed that the system is such that the terms in $C_g(s)$ will decay to zero with increasing time. Therefore, only the first two terms in (4-38) contribute to the steady-state response. The result is that the steady-state response to a sinusoid is also a sinusoid of the same frequency.

From (4-37) and (4-38),

$$k_1 = \frac{j\omega_1}{j2\omega_1}AG(j\omega_1) = \frac{1}{2}AG(j\omega_1) \qquad (4\text{-}39)$$

$$k_2 = \frac{-j\omega_1}{-j2\omega_1}AG(-j\omega_1) = \frac{1}{2}AG(-j\omega_1) \qquad (4\text{-}40)$$

and k_2 is seen to be the complex conjugate of k_1.

Since, for a given value of ω_1, $G(j\omega_1)$ is a complex number, it is convenient to express $G(j\omega_1)$ as

$$G(j\omega_1) = |G(j\omega_1)|e^{j\phi\,(j\omega_1)} \qquad (4\text{-}41)$$

where $|G(j\omega_1)|$ is the magnitude and ϕ is the angle of the complex number. Then from (4-38) through (4-41), the sinusoidal steady-state value of $c(t)$ is

$$\begin{aligned}
c_{ss}(t) &= k_1 e^{j\omega_1 t} + k_2 e^{-j\omega_1 t} \\
&= \frac{A}{2}|G(j\omega_1)|e^{j\phi}e^{j\omega_1 t} + \frac{A}{2}|G(j\omega_1)|e^{-j\phi}e^{-j\omega_1 t} \\
&= A|G(j\omega_1)|\frac{e^{j\,(\omega_1 t + \phi)} + e^{-j\,(\omega_1 t + \phi)}}{2} = A|G(j\omega_1)|\cos(\omega_1 t + \phi)
\end{aligned} \qquad (4\text{-}42)$$

since

$$|G(-j\omega_1)| = |G(j\omega_1)| \qquad (4\text{-}43)$$

In (4-42), $\phi = \phi(j\omega_1)$. We see then that the *steady-state gain* of a system for a sinusoidal input is the *magnitude* of the transfer function evaluation at $s = j\omega_1$, and the *phase shift* of the output sinusoid relative to the input sinusoid is the angle of $G(j\omega_1)$. This proof is seen to be the same as that for ac circuit analysis, in which we work with the impedance $Z(j\omega_1)$ [1]. Sinusoidal steady-state response is now illustrated with an example.

Example 4.4

Consider a system with the transfer function

$$G(s) = \frac{5}{s + 2}$$

and an input of $7\cos 3t$. Then

$$G(s)|_{s=j3} = \frac{5}{2+j3} = 1.387e^{-j56.3°} = 1.387\angle{-56.3°}$$

and the steady-state output is given by

$$c_{ss}(t) = (1.387)(7)\cos(3t - 56.3°) = 9.709\cos(3t - 56.3°)$$

Since the system time constant is 0.5 s, the output would reach steady state approximately 2 s after the application of the input signal. The complex gain $G(j3)$ is evaluated with the MAT-LAB program

```
G = tf([5],[1, 2]);
Gj3 = evalfr(G, j*3);
magGj3 = abs(Gj3)
phaseGj3 = angle(Gj3)*180/pi
```

Recall that the sinusoidal steady-state response is determined from the system transfer function, $G(s)$, evaluated at $s = j\omega_1$, where ω_1 is the frequency of the sinusoid. We define the *frequency response function* to be the function $G(j\omega)$, $0 \le \omega < \infty$. For a given value of ω, $G(j\omega)$ is a complex number. Thus the function $G(j\omega)$ is a complex function.

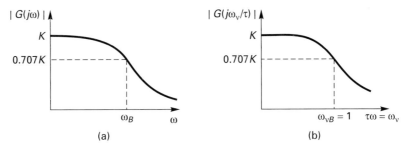

Figure 4.12   Illustration of bandwidth.

4.4.1 First-Order Systems

Consider first the frequency response of a first-order system. Now, the standard form of the first-order system is given by

[eq. (4-27)]
$$G(s) = \frac{K}{\tau s + 1}$$

The frequency response function of this system is then

$$G(j\omega) = \frac{K}{1 + j\tau\omega} = |G(j\omega)|e^{j\phi(\omega)} \tag{4-44}$$

where the magnitude and the phase of the frequency response are given by

$$|G(j\omega)| = \frac{K}{[1+\tau^2\omega^2]^{1/2}} \qquad \phi(\omega) = -\tan^{-1}\tau\omega \tag{4-45}$$

In this section we are interested primarily in the magnitude of the frequency response. A plot of $|G(j\omega)|$ for (4-45) is given in Figure 4.12(a).

In Figure 4.12(a) the frequency ω_B denotes the frequency at which the gain is equal to $1/\sqrt{2}$ times the gain at very low frequencies; this frequency is called the *system bandwidth*. The factor $1/\sqrt{2}$ originated in the study of amplifiers and is the frequency at which the power output of an amplifier has decreased to one-half the maximum low-frequency value. In this book we use this definition for bandwidth. For the first-order system of (4-43), the bandwidth is found from

$$\frac{K}{(1 + \tau^2 \omega_B^2)^{1/2}} = \frac{K}{\sqrt{2}} \tag{4-46}$$

or $\omega_B = 1/\tau$. Hence we see that the time constant τ also has meaning in the frequency domain.

It is convenient at this point to normalize frequency by the factor τ; that is, we define normalized frequency ω_v to be equal to $\tau\omega$. Then, the normalized frequency response $G_n(j\omega_v)$ is given by

$$G_n(j\omega_v) = G(j\omega)|_{\omega = \omega_v/\tau} = G(j\omega_v/\tau) = \frac{K}{(1 + \omega_v^2)^{1/2}} e^{j\phi} \tag{4-47}$$

A plot of the magnitude of the normalized frequency response is given in Figure 4.12(b). The bandwidth in normalized frequency is $\omega_{vB} = 1$, which checks the previous calculation for the bandwidth.

In a like manner, if we normalize time by $t_v = t/\tau$, the step response in normalized time, $c_n(\tau_v)$, is given by

$$c_n(t_v) = c(t)|_{t = \tau t_v} = K(1 - e^{-t_v}) \tag{4-48}$$

from (4-9), and this normalized step response is independent of τ. A plot of this normalized step response is given in Figure 4.13.

Suppose that for a given first-order system, it is desired to decrease the rise time by a factor of 2. The new time constant required is then $\tau/2$, Figure 4.13. Then, from (4-46), the bandwidth is increased by a factor of 2. In fact, for a first-order system, to decrease the rise time by any factor, the bandwidth must be increased by the same factor.

4.4.2 Second-Order Systems

We now consider second-order systems with the standard second-order transfer function

$$G(s) = \frac{\omega_n^2}{s^2 + 2\zeta\omega_n s + \omega_n^2} = \frac{1}{(s/\omega_n)^2 + 2\zeta(s/\omega_n) + 1} \tag{4-49}$$

The frequency response is given by

$$G(j\omega) = \frac{1}{[1 - (\omega/\omega_n)^2] + j2\zeta(\omega/\omega_n)} \tag{4-50}$$

For this transfer function we define normalized frequency as $\omega_v = \omega/\omega_n$ and the magnitude of the frequency response is given by

$$|G_n(j\omega_v)| = |G(j\omega)||_{\omega = \omega_n\omega_v} = \frac{1}{((1 - \omega_v^2)^2 + (2\zeta\omega_v)^2)^{1/2}} \tag{4-51}$$

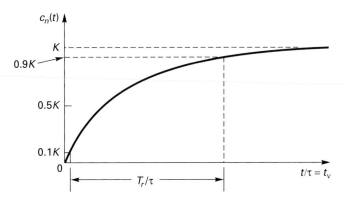

Figure 4.13 Normalized step response.

Plots of this frequency response for various values of ζ are given in Figure 4.14. A MAT-LAB program that calculates some of the frequency responses in Figure 4.14 is given by

```
w = 0:0.05:3;
z = [0.25 0.5 0.707 1];
for k = 1:4
    Gnum = [1]; Gden = [1 2*z(k) 1];
    Gjomega = freqs(Gnum,Gden,w);
    Gmag = abs(Gjomega);
    plot(w,Gmag)
        title('Frequency Response of G(s)')
        xlabel('Omega')
        ylabel('G(j*Omega)')
        grid
    hold on
end
hold off
```

Note from Figure 4-14 that, for a given ζ, the ratio ω_B/ω_n is equal to a constant. Hence, increasing ω_n causes a bandwidth to increase by the same factor with ζ constant. As a second point, consider (4-33):

[eq. (4-33)]
$$\omega_n T_p = \frac{\pi}{\sqrt{1 - \zeta^2}}$$

Then, for ζ constant, increasing ω_n decreases T_p, and hence the rise time T_r, by the same factor. Therefore, for ζ constant, we have the same result as for the first-order system. Increasing the bandwidth by a given factor decreases rise time by exactly the same factor for the first-and second-order systems discussed here.

The result derived in the preceding paragraph applies exactly to the two systems considered but only approximately to systems in general. To increase the speed of response of a system, it is necessary to increase the system bandwidth. For a particular system, an approximate relationship can be derived (see [2]), and is given by

$$\omega_B T_r \cong \text{constant} \tag{4-52}$$

where this constant has a value in the neighborhood of 2. This relationship can be plotted exactly for the standard second-order transfer function of (4-49) and (4-51). The rise time T_r can be obtained directly from Figure 4.4 for various values of ζ. Note that the actual rise time in seconds is $T_r = T_{rv}/\omega_n$, where T_{rv} is the normalized rise time in the units of $\omega_n t$, from Figure 4.4. The bandwidth can be obtained from Figure 4.14, and the actual bandwidth is given by $\omega_B = \omega_{Bv}\omega_n$, where ω_{Bv} is in the normalized bandwidth in the units of Figure 4.14. Hence the bandwidth–rise time product is

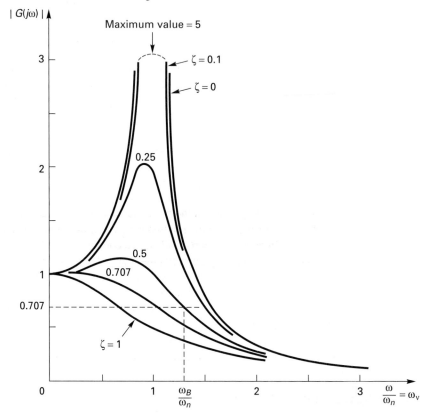

Figure 4.14   Frequency response of a second-order system (4-50).

$$\omega_B T_r = (\omega_{Bv}\omega_n)\frac{T_{rn}}{\omega_n} = \omega_{Bv} T_{rn}$$

This product is independent of ω_n and is plotted in Figure 4.15 as a function of ζ for $0 = \zeta \le 2$. Note that as ζ increases, the bandwidth–rise time product approaches the constant value of 2.20, which is the value for a first-order system (see Problem 4.18).

It is also of value to correlate the peak in the magnitude of the frequency response, Figure 4.14, with the peak in the step response, Figure 4.4. It has been shown that the peak value in the step response is a function only of the damping ratio ζ. The magnitude of the frequency response is given by (4-51), in the normalized frequency variable $\omega_v = \omega/\omega_n$. To

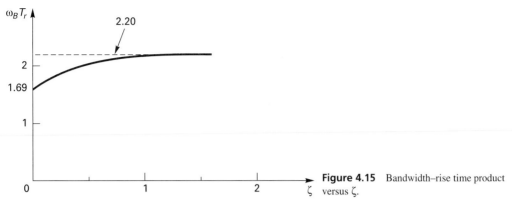

Figure 4.15 Bandwidth–rise time product versus ζ.

find the peak value of this magnitude, we differentiate (4-51) and set this result to zero. Replacing ω_v with ω/ω_n and solving the resulting equation yields the frequency at which the peak occurs,

$$\omega_r = \omega_n\sqrt{1-2\zeta^2} \qquad \zeta < 0.707 \qquad (4\text{-}53)$$

The maximum magnitude of the frequency response, denoted by $M_{p\omega}$, is then, from (4-51),

$$M_{p\omega} = |G(j\omega_r)| = \frac{1}{2\zeta\sqrt{1-\zeta^2}} \qquad (4\text{-}54)$$

Thus the peak value of the magnitude of the frequency response is a function only of ζ, as is the peak value of the step response. These relationships are shown graphically in Figure 4.16.

The peaking in a frequency response indicates a condition called *resonance,* and ω_r in Figure 4.16 is called the *resonant frequency.* For the standard second-order system, the peak in the frequency response is directly related to the amount of overshoot in the step response (and hence in the transient response). For higher-order systems, generally the higher the peak in the magnitude of the frequency response (the more resonance present), the more the overshoot that will occur in the transient response. For this reason, we prefer that control systems not have significant resonances. In design, this restriction is usually specified by giving a maximum allowable value for $M_{p\omega}$. The phenomenon of resonance is discussed in greater detail in Chapter 8.

4.5 TIME AND FREQUENCY SCALING

We now present a general development relative to time scaling and frequency scaling. The purpose of this development is to show that while most transfer functions that appear in this book have time constants in the neighborhood of unity, these transfer functions can be considered to be time scaled from transfer functions with *any values* for time constants.

We consider first the effects of *time scaling.* Let t denote time before scaling, t_v denote scaled time, and σ denote the scaling parameter. Note that for this development, both t and t_v denote time. We define the time scaling by the relationship

$$t = \sigma t_v \qquad \Rightarrow \qquad t_v = \frac{t}{\sigma} \qquad (4\text{-}55)$$

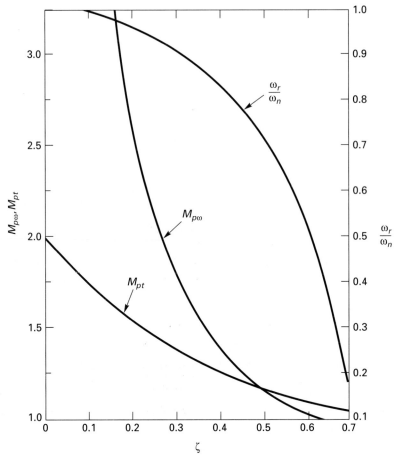

Figure 4.16 Second-order system characteristics.

Let $c(t)$ be the function to be scaled and $c_s(t_v)$ be the scaled function (the subscript s denotes a scaled function), such that

$$c_s(t_v) = c(t)\big|_{t = \sigma t_v} = c(\sigma t_v) \tag{4-56}$$

An example of time scaling for the case that $\sigma = 10$ is now presented. Figure 4.17 (4-56) gives a plot of a function $c(t)$ versus t, and we wish to generate a plot of $c_s(t_v) = c(10t_v)$ versus t_v. The t_v-axis, determined from $t_v = 0.1t$ in (4-55), is shown directly below the t-axis; hence the function shown is also $c_s(t_v)$ when considered to be plotted against the t_v-axis. We see from this example that $\sigma > 1$ yields a faster signal; that is, the rate of change of the signal at any time is increased by a factor of σ. For $\sigma < 1$, the signal is slower. In neither case is the amplitude of the signal changed.

Consider next the effects of time scaling on frequency. For the sinusoidal signal $c(t) = \cos(t)$, time scaling yields

$$c_s(t_v) = \cos \omega t\big|_{t = \sigma t_v} = \cos(\omega \sigma t_v) = \cos[(\sigma \omega)t_v] = \cos \omega_v t_v \tag{4-57}$$

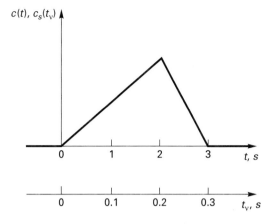

Figure 4.17   Time-scaled function.

where ω_v denotes scaled frequency. Hence, effectively, the frequency has been changed by the factor σ; once again, for $\sigma > 1$, we see that the effect is to speed up the signal. The time-scaling function $t = \sigma t_v$ affects the frequency domain as $\omega_v = \sigma\omega$, or $\omega = \omega_v/\sigma$. *Hence time scaling also results in frequency scaling.* These scaling effects were shown in Section 4.4 for first- and second-order system responses.

We now consider the effects of time scaling on linear differential equations with constant coefficients, which is a model used for linear time-invariant analog systems. Consider the ith derivative of the function $c(t)$. Then, from (4-55),

$$\left.\frac{d^i c(t)}{dt^i}\right|_{t=\sigma t_v} = \frac{d^i c(\sigma t_v)}{d(\sigma t_v)^i} = \frac{1}{\sigma^i}\frac{d^i c_s(t_v)}{dt_v^i} \tag{4-58}$$

Hence, for the general nth-order differential-equation model with input $r(t)$ and output $c(t)$ and with $a_n = 1$,

$$\frac{d^n c(t)}{dt^n} + a_{n-1}\frac{d^{n-1}c(t)}{dt^{n-1}} + \cdots + a_1\frac{dc(t)}{dt} + a_0 c(t)$$

$$= b_m\frac{d^m r(t)}{dt^m} + b_{m-1}\frac{d^{m-1}r(t)}{dt^{m-1}} + \cdots + b_1\frac{dr(t)}{dt} + b_0 r(t) \tag{4-59}$$

the effect of the time scaling $t = \sigma t_v$, from (4-58), is given by

$$\frac{1}{\sigma^n}\frac{d^n c_s(t_v)}{dt_v^n} + \frac{a_{n-1}}{\sigma^{n-1}}\frac{d^{n-1}c_s(t_v)}{dt_v^{n-1}} + \cdots + \frac{a_1}{\sigma}\frac{dc_s(t_v)}{dt_v} + a_0 c_s(t_v)$$

$$= \frac{b_m}{\sigma^m}\frac{d^m r_s(t_v)}{dt_v^m} + \frac{b_{m-1}}{\sigma^{m-1}}\frac{d^{m-1}r_s(t_v)}{dt_v^{m-1}} + \cdots + \frac{b_1}{\sigma}\frac{dr_s(t_v)}{dt_v} + b_0 r_s(t_v) \tag{4-60}$$

The transfer function of the system before time scaling is given by, from (4-59),

$$\frac{C(s)}{R(s)} = G(s) = \frac{b_m s^m + b_{m-1}s^{m-1} + \cdots + b_1 s + b_0}{s^n + a_{n-1}s^{n-1} + \cdots + a_1 s + a_0} \tag{4-61}$$

The transfer function after time scaling is given by, from (4-60),

$$\frac{C_s(s_v)}{R_s(s_v)} = G_s(s_v) = \frac{\dfrac{b_m}{\sigma^m}s_v^m + \dfrac{b_{m-1}}{\sigma^{m-1}}s_v^{m-1} + \cdots + \dfrac{b_1}{\sigma}s_v + b_0}{\dfrac{1}{\sigma^n}s_v^n + \dfrac{a_{n-1}}{\sigma^{n-1}}s_v^{n-1} + \cdots + \dfrac{a_1}{\sigma}s_v + a_0} \tag{4-62}$$

$$= \frac{\sigma^{n-m}b_m s_v^m + \sigma^{n-m+1}b_{m-1}s_v^{m-1} + \cdots + \sigma^{n-1}b_1 s_v + \sigma^n b_0}{s_v^n + \sigma a_{n-1}s_v^{n-1} + \cdots + \sigma^{n-1}a_1 s_v + \sigma^n a_0}$$

where s_v is the Laplace-transform variable for time t_v. The effect of the time scaling $t = \sigma t_v$ on a system transfer function is to replace the general numerator coefficient b_i with $\sigma^{n-i}b_i$ and the general denominator coefficient a_i with $\sigma^{n-i}a_i$. An example is now given.

Example 4.5

Consider a first-order system with the transfer function

$$G(s) = \frac{8}{4s+1} = \frac{2}{s+0.25}$$

Hence the dc gain is 8 and the time constant is 4 s. We want to time scale this transfer function such that the scaled system has the same form (amplitude) of response, but with the time rate of change of the response increased by a factor of 1000 (the time constant decreased to 0.004 s, or 4 ms). Thus $t = 1000t_v$ and $\sigma = 1000$. From (4-62), with $n = 1$,

$$G_s(s_v) = \frac{(1000)2}{s_v + (1000)0.25} = \frac{2000}{s_v + 250} = \frac{8}{0.004s_v + 1}$$

Hence the time constant is now 4 ms, while the amplitude of the response to any input is unchanged. Consider next a second-order system with the transfer

$$G(s) = \frac{9}{s^2 + 0.6s + 9} = \frac{(3)^2}{s^2 + 2(0.1)(3)s + (3)^2}$$

We see that $\zeta = 0.1$, $\omega_n = 3$, and the time constant $\tau = 1/(\zeta\omega_n) = 3.33$ s. Once again, we time scale this transfer function such that the scaled system has the same form of response but is sped up by a factor of 1000. Thus $t = 1000t_v$ and $\sigma = 1000$. From (4-62), with $n = 2$,

$$G_s(s_v) = \frac{(3)^2(1000)^2}{s_v^2 + 2(0.1)(3)(1000)s_v + (3)^2(1000)^2}$$

$$= \frac{9,000,000}{s_v^2 + 600s_v + 9,000,000}$$

We see then that for the time scaled system, ζ is unchanged, $\omega_{nv} = 3000$, and the time constant $\tau_v = 1/\zeta\omega_{nv} = 0.00333$. The form of the response is unchanged, but the time rate of change is increased by a factor of 1000. Note that for this second-order system, the range of the coefficients is from 1 to 9 million when the time constant is on the order of milliseconds.

We see from the developments in this section that the time scaling $t_v = t/\sigma$ affects the system response in the following manner:

1. The amplitude (form) of the response is unaffected.

2. The scaled time constants τ_v are given by $\tau_{vi} = \tau_i/\sigma$, where τ_i are the time constants prior to scaling.

3. The damping ratios ζ_i are unaffected.

4. Frequency is scaled by $\omega_v + \sigma\omega$

4.6 RESPONSE OF HIGHER-ORDER SYSTEMS

Thus far we have considered only standard first- and second-order systems. In this section we extend the results developed for these systems to higher-order systems.

However, first we must consider a second-order system with nonunity dc gain. A transfer function for this case can be expressed as

$$G_K(s) = KG(s) = \frac{K\omega_n^2}{s^2 + 2\zeta\omega_n s + \omega_n^2} \tag{4-63}$$

where K is the dc gain and $G(s)$ is the standard second-order transfer function. Obviously the frequency response of $G(s)$, plotted in Figure 4.14, applies also to $G_K(s)$ if the magnitude axis is multiplied by the factor K. Hence the shape of the frequency response remains the same. The step response of this system is given by

$$C_K(s) = G_K(s)\frac{1}{s} = KG(s)\frac{1}{s} = KC(s) \tag{4-64}$$

where $C(s)$ is the step response of the unity dc gain system, as plotted in Figure 4.6. Thus the step response of the nonunity dc gain system is the step response of the unity dc gain system multiplied by the value of the dc gain, and the shape of the step response remains the same. These conclusions apply only to the specific transfer function given, (4-63), and not to systems in general.

Now we investigate the response of higher-order systems. Consider first a third-order transfer function, expanded into partial fractions.

$$G(s) = \frac{K\omega_n^2/\tau}{(s + 1/\tau)(s^2 + 2\zeta\omega_n s + \omega_n^2)} = \frac{k_1}{s + 1/\tau} + \frac{k_3 s + k_2}{s^2 + 2\zeta\omega_n s + \omega_n^2} \tag{4-65}$$

Note that the numerator is chosen such that the dc gain is K. The system output can be expressed as

$$
\begin{aligned}
C(s) &= G(s)R(s) = \frac{k_1}{s + 1/\tau}R(s) + \frac{k_3 s + k_2}{s^2 + 2\zeta\omega_n s + \omega_n^2}R(s) \\
&= \frac{k_1}{s + 1/\tau}R(s) + \frac{k_2}{s^2 + 2\zeta\omega_n s + \omega_n^2}R(s) + \frac{k_3 s}{s^2 + 2\zeta\omega_n s + \omega_n^2}R(s) \\
&= C_1(s) + C_2(s) + C_3(s)
\end{aligned}
\tag{4-66}
$$

Thus the response of this third-order system can be modeled as the sum of the responses of a first-order system and two second-order systems, as shown in Figure 4.18.

Suppose that the input $R(s)$ in (4-66) is a unit step. The first term, $C_1(s)$, is the step response of a first-order system of dc gain $k_1\tau$ [see (4-9)]. The third term in (4-66), $C_3(s)$, can be expressed as

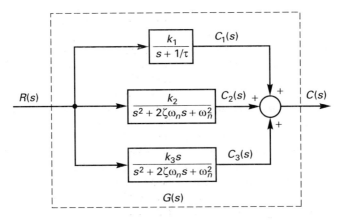

Figure 4.18 Response of a third-order system.

$$C_3(s) = \frac{k_3 s}{s^2 + 2\zeta\omega_n s + \omega_n^2}R(s) = \frac{k_3}{k_2}[sC_2(s)] \tag{4-67}$$

and thus

$$c_3(t) = \frac{k_3}{k_2}\frac{dc_2(t)}{dt} \tag{4-68}$$

where $c_2(t)$ is the step response of the standard second-order system, (4-20), but with the dc gain equal to k_2/ω_n^2. Hence the step response of this third-order system is composed of three terms:

1. The step response of a first-order system of known dc gain and time constant τ.
2. The step response of a standard second-order system with known dc gain, damping ratio ζ, and natural frequency ω_n.
3. The derivative of the response in 2, with the same ζ and ω_n, but with a different dc gain.

It is seen from this derivation that the step response of a third-order system is a complex function of all the parameters of the transfer function. However, some general comments can be made. For the preceding system, the time constant of the first-order term is known, as is that of the second-order term. Since the time response is the sum of terms dependent on these time constants, the settling time of the system will approximately equal four times the slower time constant; that is, the slow term will normally determine the system settling time.

However, the estimation of the percent overshoot and the rise time is more complex. The damping ratio, ζ, of the second-order term gives an approximation of the percent overshoot, using the standard second-order curves of Figure 4.8. The rise time is even more difficult to estimate. Perhaps the best estimate is obtained from the bandwidth, such as is given in (4-52). Of course, (4-52) applies to any order system. An example is given to illustrate this discussion.

Example 4.6

Consider the third-order system with the transfer function

$$G(s) = \frac{8}{(s+2.5)(s^2+2s+4)} = \frac{1.524}{s+2.5} + \frac{-1.524s+0.762}{s^2+2s+4}$$

For the first-order transfer function, the time constant is 0.4 s. For the second-order transfer function the damping ratio is $\zeta = 0.5$ and the natural frequency $\omega_n = 2$. Thus the time constant is 1.0 s, and we estimate the system settling time to be approximately 4 s, or four times the slower time constant. From Figure 4.8, ζ equal to 0.5 indicates a system overshoot of approximately 17 percent. The step response of this third-order system was obtained by simulation and is plotted in Figure 4.19(a). The actual settling time was found to be 3.2 s and the actual overshoot was found to be 11 percent.

The magnitude of the frequency response was calculated using a digital computer program and is plotted in Figure 4.19(b). The bandwidth was found to be $\omega_B = 2.13$ rad/s. From the simulation the actual rise time was found to be 1.05 s, and hence the bandwidth–rise time product is 2.24. For example, suppose that system specifications require a rise time of 0.5 s. We see then from (4-52) that it would be necessary to increase the system bandwidth to approximately 4.0 rad/s. Methods to increase the bandwidth of a control system are covered in the chapters on design.

The frequency response in Figure 4.19(b) can be calculated with the MATLAB program

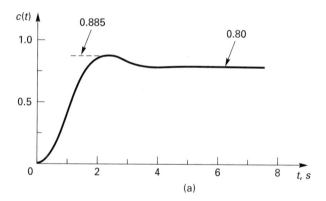

(a)

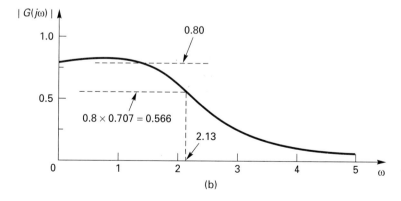

(b)

Figure 4.19   Example 4.6.

```
w = logspace(-1,1);
Gnum = [0 0 0 8]; Gden1 = [1 2.5]; Gden2 = [1 2 4];
Gden = conv(Gden1,Gden2);
   Gjomega = freqs(Gnum,Gden,w);
   Gmag = abs(Gjomega);
      plot(w,Gmag)
      title('Frequency Response of G(s)')
      xlabel('Omega')
      ylabel('G(j*omega)')
      grid
```

For systems of order higher than three, the preceding approach can be used. The system can be considered to be a number of first-and second-order systems connected in parallel. An examination of time constants gives an indication of the settling time. The various damping ratios indicate the type of system transient response, and the combination of the damping ratios and the natural frequencies indicate the rise time. However, the results are only approximate and in some cases are very much in error. The contribution of some of the subsystems to the output may be so small that these subsystems should be ignored. This, of course, reduces the order of the model, which is the topic of the next section.

4.7 REDUCED-ORDER MODELS

In this chapter we initially considered first- and second-order models. Then we saw how high-order models can be considered to be first-and second-order models connected in parallel, and estimates of the system characteristics can be based on the characteristics of the first- and second-order models. These estimates may or may not be accurate; no general statements can be made.

In this section we consider the topic of order reduction. In some cases we may be able to reduce a high-order model to a lower-order model, perhaps to a first- or second-order model, and still obtain reasonable accuracy from the lower-order model. If a model is specified for a physical system, we can consider that model to be a reduced-order model, since the model of a physical system can always be made more accurate by increasing the order of the model. *Hence any model of a physical system will be ignoring certain system characteristics and can be considered to be a reduced-order model.*

The question naturally arises as to why we would purposely reduce the accuracy of our model. Some reasons are as follows:

1. Both analysis and design are much simpler for low-order models (systems).
2. Numerical accuracy for computer computations is better for low-order models.
3. If the low-order model is first- or second-order, a wealth of information is available relating to analysis and design.
4. Since high-order models are in themselves inaccurate, in some cases the low-order model may give results that are almost as accurate as those of the high-order model. Adding orders to a model may not significantly increase the accuracy of the model.

5. Insight is developed more easily for a low-order model. The importance of insight in practical design, gained through working with both the model and the physical system, cannot be overemphasized.

In reducing the order of a system, the *s*-plane may be partitioned as shown in Figure 4.20[3]. This figure is approximate, and it depends on the particular transfer function being considered. The dominant poles generally give transient-response terms that are slow. The insignificant poles generally give transient-response terms that are much faster; the poles may be ignored in many cases. As a rule of thumb, the ratio *b/a* is 5 to 10 or greater. Once again, this rule of thumb does not apply to every transfer function. Reducing the order of a system will be mathematically justified in Chapter 7 (using root-locus justification) and Chapter 8 (using frequency-response justification).

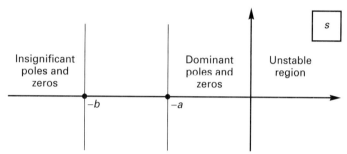

Figure 4.20   Regions of the *s*-plane.

4.8 SUMMARY

In this chapter we considered the response of the linear time-invariant analog systems, that is, systems described by linear differential equations with constant coefficients. We can consider the response to be characterized in one of two ways. First, if we know the input function, we can solve for the output as a function of time. The response is composed of two parts: the transient part (natural part), whose nature is determined entirely by the system, and the forced (steady-state) part, whose nature is determined by the nature of the input function.

The second method for characterizing the response of a system is the frequency response, that is, the response of the system to input sinusoids. This response is general in that if the complete frequency response is known, we have a complete description of the system. In fact, we see in Chapter 9 that knowledge of the plant frequency response allows us to design the closed-loop control system; we do not require a knowledge of the plant transfer function. The advantage of this approach is that the frequency response can often be obtained by measurements on the physical system.

In the examples in this book, time constants are generally on the order of 1 s, for ease in calculations. Time scaling is introduced to show the effects of scaling time constants on the coefficients of transfer functions. It is shown how the results of the examples can be applied to systems with time constants of any values.

The emphasis of this chapter is on the response of first- and second-order systems. However, it is shown that higher-order systems can be represented as first- and second-order systems in parallel. Finally, the topic of reducing the order of a model is mentioned.

REFERENCES

1. J. D. Irwin. *Basic Engineering Circuit Analysis,* 5th ed. Upper Saddle River, NJ: Prentice Hall, 1996.

2. M. E. Van Valkenburg. *Network Analysis.* Englewood Cliffs, NJ: Prentice Hall, 1974.

3. B. C. Kuo. *Automatic Control Systems,* 7th ed. Upper Saddle River, NJ: Prentice Hall, 1996.

PROBLEMS

4.1. **(a)** The first-order plant of Figure P4.1(a) has the unit step response given in Figure P4.1(b). Find the parameters of the transfer function.

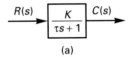

(a)

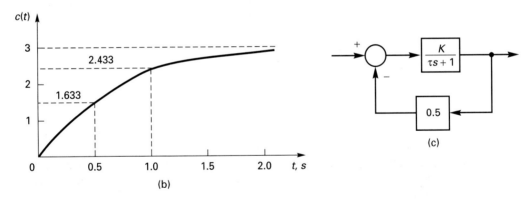

(c)

Figure P4.1

   (b) The plant is connected into the closed-loop system as shown in Figure P4.1(b). Sketch the unit stepresponse of the closed-loop system.

4.2. The closed-loop first-order system of Figure P4.2(a) has the unit step response given in Figure P4.2(b). The steady state-state value of $c(t)$ is 0.96. Find the parameters K and τ.

4.3. **(a)** Sketch the unit step response of the system of Figure P4.3(a), giving approximate values on two axis.
   (b) Run a simulation of the system, and compare the response with the results of (a).
   (c) Repeat (a) for the system of Figure P4.3(b).
   (d) Repeat (b) for the system of Figure P4.3(b)

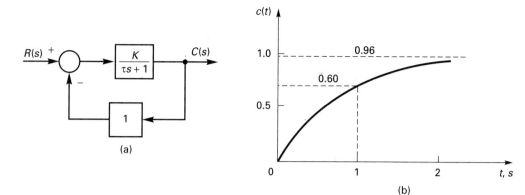

Figure P4.2

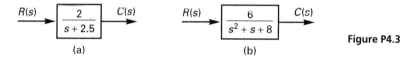

Figure P4.3

4.4. (a) For the system shown in Figure P4.4, sketch the unit step response of the system without mathematically solving for the time response $c(t)$. Indicate approximate numerical values on both the amplitude axis and the time axis.

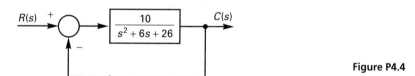

Figure P4.4

(b) Run SIMULINK simulation of the system, constructed in the exact form of Figure P4.4, and compare the results to those in (a).

4.5. Shown in Figure P4.5 is a satellite attitude control system. The model of the satellite is developed in Example 2.13, Section 2.6. The moment inertia J of the satellite has been normalized to unity.

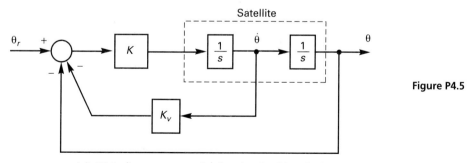

Figure P4.5

(a) Write the system transfer function for this attitude control system.

(b) The system is commanded to assume an attitude of $10°$ $[\theta_r(t)=10u(t)]$. After the transients die out (the system reaches steady state), what will be the attitude angle of the satellite $[\theta_{ss}(t)]$?

(c) The closed-loop system is to respond to a step input in minimum time with no overshoot, which requires that $\zeta = 1$. Find K_v as a function of K such that this specification is satisfied.

(d) The system of (c) is to reach steady state approximately 6 s after a command to change the attitude angle (after the application of an input). Find the value of K that satisfies this specification.

(e) Verify the results of (b), (c), and (d) with a MATLAB simulation of the system.

(f) The rate signal is measured using a rate gyro. Suppose that the rate gyro fails, such that no signal appears in the rate path (effectively $K_v = 0$). What is the nature of the system response in the failure mode? (This failure occurred on a space lab mission of NASA, with the predictable result.)

(g) Simulate the system with $K_v = 0$ to verify the results of (f).

4.6. **(a)** Develop a SIMULINK simulation for the closed-loop satellite control system of Figure P4.5, in the exact form given in this figure. Let $K = 0.444$ and $K_v = 3$, from the Problem 4.5(d).

(b) The values of K and K_v result in $\zeta = 1$. Run the simulation to verify this design.

4.7. Given the system of Figure P4.7.

(a) If $r(t) = 5u(t)$, find the steady-state value of $c(t)$.

(b) Approximately how many seconds are required for the system to reach steady state?

(c) Do you expect the transient response to be oscillatory? Justify your answer.

(d) Verify your results by obtaining a step response by simulation.

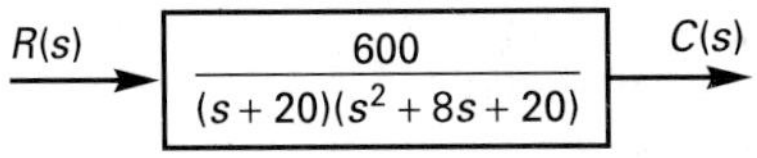

$$R(s) \longrightarrow \boxed{\dfrac{600}{(s+20)(s^2+8s+20)}} \longrightarrow C(s)$$

Figure P4.7

4.8. Show that the inverse Laplace transform of (4-19) is (4-20); that is,

$$\mathcal{L}^{-1}\left[\frac{\omega_n^2}{s(s^2+2\zeta\omega_n s+\omega_n^2)}\right] = 1-\frac{1}{\beta}e^{-\zeta\omega_n t}\sin(\beta\omega_n t+\theta)$$

where $\beta = \sqrt{1-\zeta^2}$ and $\theta = \tan^{-1}(\beta/\zeta)$.

4.9. **(a)** Show that for the system with the transfer function $G(s)$, the unit impulse response is the derivative of the unit step response.

(b) Show that the unit step response of the system of (4-18), given in (4-25), is the derivative of the unit step response, given in (4-20).

(c) Show that for a system with the transfer function $G(s)$, the unit step response is the derivative of the unit ramp response.

4.10. The digital plotter of Example 4.2, Section 4.3, is to have a settling time of 0.25 s. The motor specified in the example cannot achieve this specification.

(a) A motor is to be chosen with the transfer function $G(s) = K/s(\tau_m s+1)$ and connected into the system of Figure P4.10. Design the system to meet the settling-time specification and the result in $\zeta = 1$ to eliminate overshoot for step inputs, by finding values for K, K_a, and τ_m.

(b) Simulate the system to verify the design.

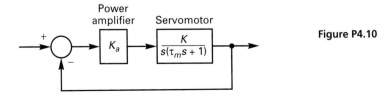

Figure P4.10

4.11. Given the system of Figure P4.11.
   (a) Find the range of K for which the system is
      (i) underdamped
      (ii) critically damped
      (iii) over damped
   (b) Find the value of K that will result in the system having minimum settling time.

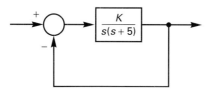

Figure P4.11

4.12. The middle-C string (f = 256 Hz) of a guitar is plucked, and the sound becomes inaudible after approximately 2 s.
   (a) Find the approximate value of the damping ratio ζ for this string.
   (b) Find the approximate value of the natural frequency ω_n for this string.
   (c) Give the transfer function of the string. What is the physical input and what is the physical output for this transfer function?
   (d) List all the assumptions made in modeling the string.

4.13. Consider the system with the transfer function

$$\frac{C(s)}{R(s)} = \frac{K_1}{s + \alpha}$$

   (a) Find the region of allowable s-plane pole locations such that the system settling time is less than 10 s.
   (b) Solve for the allowable ranges of K_1 and α for (a).

4.14. Consider the system with the transfer function

$$\frac{C(s)}{R(s)} = \frac{\omega_n^2}{s^2 + 2\zeta\omega_n s + \omega_n^2}$$

   (a) Find the region of allowable s-plane pole locations such that the system settling time is less than 2 s and an overshoot for a step response is less than 10 percent.
   (b) Solve for the allowable ranges of ζ and ω_n for (a).

4.15. Consider the control system of a satellite given in Figure P4.5.
   (a) Find the closed-loop transfer function.
   (b) Find the closed-loop dc gain.
   (c) If $K_v = 0$, find the closed-loop system gain (magnitude of the closed-loop frequency response) at resonance.
   (d) Design specifications for the system are that the peak closed-loop gain cannot be greater than 1.25 and that the system time constant $\tau = 1$ s. Design the system by finding K and K_v such that the specifications are satisfied. Note that the damping of the system has been increased by the velocity (also called rate and derivative) feedback.
   (e) Verify the results of (d) by plotting the frequency response using MATLAB.

4.16. Consider the system of Example 4.6, Section 4.6, which has the transfer function

$$G(s) = \frac{8}{(s + 2.5)(s^2 + 2s + 4)}$$

Both the system unit step response and the system frequency response are plotted in Figure 4.19. The final value of the unit step response is seen to be 0.8, and the initial value of the frequency response is also 0.8. Will these two values always have the same magnitude for any stable system? Give a proof for your answer.

4.17. Show that for a first-order system with the transfer function

$$G(s) = \frac{K}{\tau s + 1}$$

the product of bandwidth and rise time is equal to 2.197.

4.18. For the system of Figure P4.3(a), the input $r(t) = 3 \cos 2t$ is applied at $t = 0$.
   (a) Find the steady-state system response.
   (b) Find the range of time t for which the system is in steady-state.
   (c) Find the steady-state response for the input $r(t) = 3 \cos 8t$.
   (d) Why is the amplitude of the response in (a) much greater than that in (c), with the amplitides of the input signals equal?
   (e) Construct and run a SIMULINK simulation that verifies the results of (a) and (b).

4.19. Given the system of Figure P4.19.
   (a) Find the closed-loop transfer function.
   (b) Find the closed-loop dc gain.
   (c) If $K_v = 0$, find the closed-loop system gain (magnitude of the closed-loop transfer function) at resonance.
   (d) A design specification is that the peak closed-loop gain can be no greater than 1.26 at any frequency. Find K_v such that this specification is satisfied. Note that the damping of the system has been increased by velocity (also called rate and derivative) feedback.
   (e) Verify the results of (d) using MATLAB.

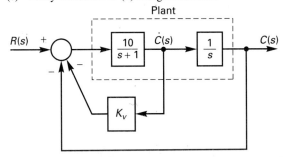

Figure P4.19

4.20. Given a system described by the standard second-order transfer function

$$G(s) = \frac{25}{s^2 + 4s + 25}$$

   (a) Sketch the system unit step response. Any figures in this chapter may be used.
   (b) Find the transfer function if this system is time scaled with $t_v = 0.01t$.
   (c) From the transfer function in (b), find ζ_v and ω_{nv} for the scaled system.
   (d) Repeat (a) for the time scaled system.
   (e) Verify the step responses in (a) and (d) using MATLAB simulation.
   (f) Given the frequency scaling $s_v = as$, where the parameter a is real. Find the effect of this scaling on the time response of a standard second-order system.

4.21. (a) Given a system described by the transfer function

$$G(s) = \frac{500}{s + 200}$$

Time scale this transfer function such that the time constant of the scaled system is $\tau = 1$ s.

(b) Repeat (a) for the system transfer function

$$G(s) = \frac{0.00016}{s^2 + 0.016s + 0.000256}$$

(c) For verification, calculate the time constants directly from the scaled transfer functions, for (a) and (b).

(d) Verify the results in (a) and (b) using the step responses from a SIMULINK simulation.

4.22. Given a system described by the third-order transfer function

$$G(s) = \frac{b_2 s^2 + b_1 s + b_0}{(s + 2)(s^2 + 20s + 200)}$$

Note that the coefficients in the numerator are not specified.

(a) Estimate the system rise time, settling time, and percent overshoot in the step response.

(b) Using a MATLAB simulation, find the step responses and compare the actual parameters of the step responses with the estimated values for the transfer-function numerator poly nomials

(i) $b_2 s^2 + b_1 s + b_0 = 2s^2 + 240s + 800$

(ii) $b_2 s^2 + b_1 s + b_0 = 2s^2 + 50s + 420$

(iii) $b_2 s^2 + b_1 s + b_0 = 0.1s^2 + 402s + 820$

5

Control System Characteristics

In this chapter we cover certain desired characteristics that a closed-loop control system should have. Although certain control systems should have characteristics in addition to those listed in this chapter, it is generally desirable that all control systems should have characteristics introduced in this section; these characteristics are covered more completely in the following sections.

1. In all cases we require that a control system respond in some controlled manner to applied inputs and initial conditions. We call this characteristic *stability*. Generally, a system is not usable if it is unstable; hence we require a control system to be stable when operated in a prescribed manner.

2. Since an exact model of a physical system is never available, the characteristics of a physical closed-loop control system should be reasonably insensitive to the parameters of the mathematical model used in the design of the system. In addition, the characteristics of a physical plant will possibly change with time and with environmental conditions such as temperature, humidity, and altitude. We would also like the closed-loop physical system's characteristics to be insensitive to these changes. Hence we are interested in the *sensitivity* of the closed-loop system to parameter changes in the system.

3. All physical systems have unwanted inputs, called *disturbances*, in addition to the inputs used to control the systems. In general we desire that a system not respond in a significant manner to these disturbances. Thus the *disturbance rejection* capabilities of a system are important.

4. The steady-state errors present in a control system when certain specified inputs are applied are of critical importance in some control systems. This topic is called *steady-state accuracy*.

5. In Section 5.2 we define the *natural response* of a system as that part of the response that is always present, independent of the input signal. If this response goes to zero with increasing time, the natural response is also called the *transient response*. We considered the transient response in Chapter 4; we discuss it further here.

6. As a final topic in this chapter, the relation of the characteristics just given to the *closed-loop frequency response* is considered.

5.1 CLOSED-LOOP CONTROL SYSTEM

In this section the structure of a single-loop closed-loop control system is given. The required components of the system are defined, and the functions of these components are discussed.

Consider the control system shown in Figure 5.1(a). The *plant* is the physical system, or process, to be controlled, and in this case it is considered to include all power amplifiers, actuators, gears, and so on. The *sensor* is the physical instrumentation that measures the output signal and converts this measurement to a usable signal at the summing junction. The *compensator* is a dynamic physical system purposely added to the loop to enhance the closed-loop system characteristics.

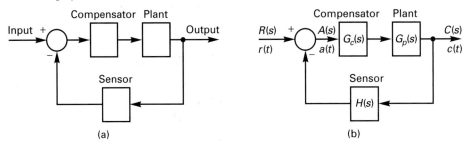

Figure 5.1 Closed-loop control system.

The transfer functions (mathematical models) of the components of the control system are shown in Figure 5.1(b). By Mason's gain formula, the system transfer function is

$$T(s) = \frac{C(s)}{R(s)} = \frac{G_c(s)G_p(s)}{1 + G_c(s)G_p(s)H(s)} \tag{5-1}$$

In general we denote the closed-loop transfer function by $T(s)$, a transfer function in a forward path by $G(s)$, and a transfer function in a feedback path by $H(s)$, with subscripts as needed.

Usually the instrumentation system (sensor) responds much faster than the plant. Another way of saying this is that the sensor bandwidth is much wider than that of the plant. For this case the sensor can be modeled as a pure gain, which we denote as H_k. As an example, suppose that the input to a certain temperature sensor is temperature in degrees Celsius and that the sensor output changes 50 mV for a change in temperature of 1°C. Then the transfer function of the sensor is the gain

$$H_k = 50 \text{ mV}/1°C = 0.05 \text{ V}/°C$$

Modern military aircraft utilize many feedback control systems. (Courtesy of McDonnel-Douglas Corporation.)

within the sensor bandwidth. The transfer function (5-1), for the case that the sensor can be considered to be a pure gain, is then

$$T(s) = \frac{C(s)}{R(s)} = \frac{G_c(s)G_p(s)}{1 + H_k G_c(s) G_p(s)} \tag{5-2}$$

Before proceeding further, it is necessary that we clearly define the components of the closed-loop system that we are considering. For any single-input, single-output feedback control system having the form shown in Figure 5.1, the following are true:

1. The system will have a variable, $c(t)$, that we wish to control. This is known as the *controlled variable* or the *system output.*

2. The system will have an *input variable* or *system input, $r(t)$*, the value of which is a measure or indication of (but not necessarily equal to in magnitude or units) the *desired value of the system output, $c_d(t)$*. Note that $c_d(t)$ is not a variable that appears in the system of Figure 5.1(b).

3. The difference between the desired value of the output and the actual output is the *system error, $e(t)$*:

$$\text{system error} = e(t) = c_d(t) - c(t) \tag{5-3}$$

4. The output of the summing junction in the system of Figure 5.1(b), $a(t)$, is the *actuating signal*, since this signal brings about response in the plant.

Note from Figure 5.1 that:

1. The *system error* is not a signal within the system.

2. In general, the system input $r(t)$ is not the same as $c_d(t)$, the desired system output.

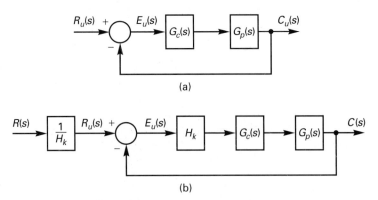

Figure 5.2   Systems having unity feedback.

We now define a unity feedback system, as shown in Figure 5.2(a). A *unity feedback system* is one in which the gain in the feedback path is unity. For unity feedback systems, we consider the following always to be true:

1. The units of the input, $r_u(t)$, are the same as those of the output, $c_u(t)$.
2. The input signal, $r_u(t)$, is also the desired value of the output, $c_d(t)$.
3. Hence the system error is also the actuating signal, that is,

$$e_u(t) \;=\; r_u(t) - c_u(t) \tag{5-4}$$

If the sensor model can be considered to be a pure gain, as in (5-2), the block diagram of Figure 5.1 can be manipulated into an equivalent system having unity feedback. The resultant block diagram is given in Figure 5.2(b). Note that the transfer function of Figure 5.2(b) is (5-2). The equivalent forward path gain within the loop is $H_k G_c(s) G_p(s)$, and the input to the unity-feedback portion of the system is now $R_u(s) = R(s)/H_k$. We see then that if we employ the modified input $R_u(s)$, the system can be represented by the system of Figure 5.2(a). We call this model the *unity feedback model* of the system (this model is used often in this book), and we see that the following assumptions apply:

1. For the application considered, the sensor can be modeled as a pure gain H_k.
2. The equivalent forward path transfer function within the loop is $H_k G_c(s) G_p(s)$.
3. The input of the unity feedback model, $R_u(s)$, is the physical system input $R(s)$ multiplied by $1/H_k$.

Note that the effect of this manipulation is to amplitude scale the equivalent system input at the summing junction to have the same units as the system output; in the physical system the system input has the units of the sensor output. *In any system in which we use a unity feedback model, the above assumptions apply.* If these assumptions are not valid, the unity feedback model cannot be employed.

We see from this discussion that, for the unity feedback model, the units of the input $r_u(t)$ and of the output $c_u(t)$ must be the same; that is, if the output units are in degrees Celsius, the input must also be in degrees Celsius. However, if the input to the physical system is $R(s)$, the input to the unity feedback is $R_u(s) = R(s)/H_k$. Hence the physical-system input is given by

$$\text{Input to the physical system} = H_k r_u(t)$$

For the preceding example, with $H_k = 0.05$ V/°C, to command a temperature of 60°C for the system (that is, the desired value of the output is 60°C), we would apply a voltage of

$$r(t) = H_k r_u(t) = 0.05 \times 60 = 3.0 \ V$$

to the input of the physical system. As shown by this example, the input of the physical system must be scaled to match the units of the sensor output.

Example 5.1

Consider the system shown in Figure 5.3, which is a temperature-control system. The compensator is a proportional-plus-integral (PI) compensator, with two gains K_P and K_I, to be determined by the design process. This compensator is one of the most commonly used types and is covered in detail in the design chapters. Suppose that the physical plant is a large chamber used to test devices under various thermal stresses. The plant input is an electrical signal (in volts) that operates a valve to control the flow of steam, which heats the chamber. The output is the chamber temperature (measured at specified location in the chamber) in degrees Celsius.

The sensor is a measurement system based on a thermistor, which is a semiconductor device whose resistance varies with temperature. The sensor, *in this application*, can be modeled as a pure gain of $H_k = 0.05$ V/°C. Note that the bandwidth of the chamber is $\omega_B = 0.1$ rad/s (or, $f_B = \omega_B/2\pi = 0.016$ Hz). Hence the temperature of the chamber will change very slowly (the time constant of the chamber is 10 s). By assuming that the sensor is a pure gain, we are assuming that the thermistor temperature can change very quickly compared to that of the chamber. Thus the bandwidth of the thermistor is very large compared to that of the chamber.

By the preceding development, an equivalent block diagram of the system is as shown in Figure 5.3(c). Both the inputs and outputs of Figure 5.3(b) and (c) are the same. In addition, the signal $E(s)$ in Figure 5.3(c) is the Laplace transform of the system error in degrees Celsius.

We can simplify the equivalent system further, as shown in Figure 5.3(d), by defining an equivalent input $R_u(s)$ to be

$$R_u(s) = \frac{1}{H_k} R(s) = 20 R(s)$$

We usually prefer to work with a block diagram in this form [Fig. 5.3(d)], since the signal $E(s)$ is the system error and the input function is in degrees Celsius, the units of both the output and the error.

5.2 STABILITY

One of the most difficult questions to answer with respect to a *physical* system is that of a general stability. Usually, by *stability*, we mean that a stable system remains under control. Hence a stable system will respond in some appropriate manner to an applied input. For an unstable system there is little apparent relation between the system input and the system output. An example of an unstable physical system is an automobile that is hydroplaning on a

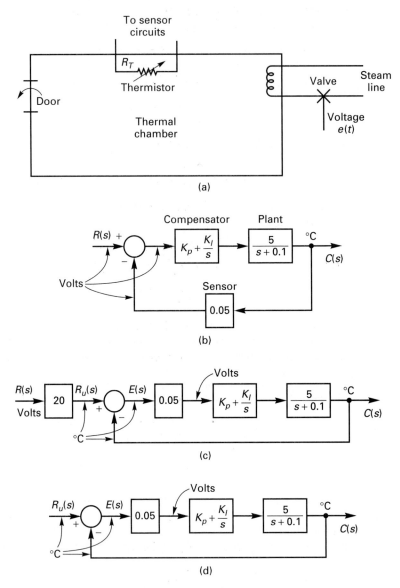

Figure 5.3 System for Example 5.1.

wet road. There is little relation between the driver inputs of the positions of the steering wheel and the brake pedal, and the position and the orientation of the car.

While there is considerable difficulty in determining the stability of a physical system under all conditions, the stability question is not difficult for *linear time-invariant systems* (*models*). For these systems, we use the *bounded-input, bounded-output* (BIBO) definition of stability:

Bounded-input, bounded-output stability   A system is *bounded-input, bounded-output stable,* if, for every bounded input, the output remains bounded for all time.

We now develop the criteria for the BIBO stability of linear time-invariant systems. Suppose that we have a control system with the configuration of Figure 5.1(b). By letting $G(s) = G_c(s)G_p(s)$, the transfer function is readily written from (5-1) as

$$T(s) = \frac{C(s)}{R(s)} = \frac{G(s)}{1 + G(s)H(s)} \tag{5-5}$$

Recalling that $G(s)$ and $H(s)$ are rational functions of s, let

$$G(s) = \frac{N_G(s)}{D_G(s)} \qquad H(s) = \frac{N_H(s)}{D_H(s)}$$

where $N_G(s)$, $D_G(s)$, $N_H(s)$, and $D_H(s)$ are all polynomials in s. Substitution of these transfer functions into (5-5) yields

$$
\begin{aligned}
T(s) &= \frac{N_G(s)/D_G(s)}{1 + N_G(s)N_H(s)/D_G(s)D_H(s)} \\
&= \frac{N_G(s)D_H(s)}{D_G(s)D_H(s) + N_G(s)N_H(s)} = \frac{P(s)}{Q(s)}
\end{aligned}
\tag{5-6}
$$

where $P(s)$ and $Q(s)$ are polynomials in s. Hence, $T(s)$ is a rational function of s.

We define the system *characteristic equation* as the denominator polynomial of a system transfer function set equal to zero. We justify this definition below. For the system of (5-6), the characteristic equation is

$$Q(s) = D_G(s)D_H(s) + N_G(s)N_H(s) = 0 \tag{5-7}$$

Although this relationship is the system characteristic equation, we usually prefer to express the characteristic equation in a different manner. If the denominator of (5-5) is set to zero,

$$1 + G(s)H(s) = 1 + \frac{N_G(s)N_H(s)}{D_G(s)D_H(s)} = 0 \tag{5-8}$$

(5-7) results. Hence we see that the system characteristic equation for a system of this configuration is given by either (5-7) or (5-8). We usually express the system characteristic equation as in (5-8).

Let the characteristic equation of (5-7) be represented in factored form as

$$Q(s) = a_n \prod_{i=1}^{n} (s - p_i) = a_n(s - p_1)(s - p_2)\cdots(s - p_n) = 0 \tag{5-9}$$

where a_n is a constant. Since the roots of the characteristic equation are the same as the poles of the closed-loop transfer function, we use the expressions *system roots* and *system poles* interchangeably in referring to them.

From the preceding development we can express the output as

$$C(s) = T(s)R(s) = \frac{P(s)}{a_n \prod\limits_{i=1}^{n}(s - p_i)} R(s)$$

$$= \frac{k_1}{s - p_1} + \frac{k_2}{s - p_2} + \cdots + \frac{k_n}{s - p_n} + C_r(s)$$

where $C_r(s)$ is the sum of the terms, in the partial-fraction expansion, that originate in the poles of $R(s)$. Thus $C_r(s)$ is the forced response. We have assumed that the characteristic equation has no repeated roots; that case is considered below. The inverse Laplace transform of $C(s)$ yields

$$c(t) = k_1 e^{p_1 t} + k_2 e^{p_2 t} + \cdots + k_n e^{p_n t} + c_r(t) = c_n(t) + c_r(t) \qquad (5\text{-}10)$$

We define the terms of $c_n(t)$ to be the *natural response* terms, since these terms originate in the poles of the transfer function and the functional forms are independent of the input. If $r(t)$ is bounded, all terms in $c_r(t)$ will remain bounded, since $c_r(t)$ is of the functional form of $r(t)$. Thus if the output becomes unbounded, it is because at least one of the natural-response terms, $k_i e^{p_i t}$, has become unbounded. This unboundness cannot occur if the real part of each of the roots p_i of the characteristic equation is negative.

We see from the preceding discussion that the requirement for a linear time-invariant system to be stable is that all roots of the characteristic equation (poles of the closed-loop transfer function) *must lie in the left half of the s-plane*. We will call a system *not stable*, or *nonstable*, if not all roots are in the left half-plane. If the characteristic equation has roots on the imaginary axis ($j\omega$-axis) with all other roots in the left half-plane, the steady-state output will be sustained oscillations for a bounded input, unless the input is a sinusoid (which is bounded) whose frequency is equal to the magnitude of the $j\omega$-axis roots (as shown in Example 5.2 below). For this case, the output becomes unbounded. Such a system is called *marginally stable*, since only certain bounded inputs (sinusoids of the frequency of the poles) will cause the output to become unbounded. For an *unstable* system, the characteristic equation has at least one root in the right half of the *s*-plane; for this case the output will become unbounded for any input.

The expression for $c(t)$ in (5-10) applies only for the case that the roots of the characteristic equation are distinct, that is, no repeated roots. For the case of an *m*th-order repeated root, the partial-fraction expansion of $C(s)$ yields terms of the form

$$\mathcal{L}^{-1}\left[\frac{k}{(s - p_i)^m}\right] = \frac{k}{(m - 1)!}t^{m-1}e^{p_i t} \qquad (5\text{-}11)$$

This term is bounded provided that the real part of p_i is negative, which is the same condition developed above. Hence,

> A linear time-invariant system is bounded-input, bounded-output stable provided all roots of the system characteristic equation (poles of the closed-loop transfer function) lie in the left half of the *s*-plane.

An example is given next.

Example 5.2

A system with the closed-loop transfer function

$$T(s) = \frac{2}{(s+1)(s+2)}$$

is stable, since the characteristic equation is given by

$$Q(s) = (s+1)(s+2) = 0$$

and the roots of this equation, -1, and -2, are in the left half of the s-plane. The natural-response terms for this system are $k_1 e^{-t}$ and $k_2 e^{-2t}$.

A different system, with the transfer function

$$T(s) = \frac{10s+24}{s^3+2s^2-11s-12} = \frac{10(s+2.4)}{(s+1)(s-3)(s+4)}$$

is unstable, because the pole at $s=3$ is a right half-plane pole. The natural-response terms are $k_1 e^{-t}$, $k_2 e^{3t}$, and $k_3 e^{-4t}$. The system is unstable because the term $k_2 e^{3t}$ becomes unbounded with increasing time. The poles of this transfer function can be found with the MATLAB program

```
p = [1 2 -11 -12];
r = roots(p)
```

A third system has the transfer function

$$T(s) = \frac{s}{s^2+1}$$

and is marginally stable, since the poles are $s = \pm j$. The natural-response term can be expressed as $k \sin(t + \theta)$, which is bounded. However, if the input $\sin t$ is applied, the output is

$$C(s) = T(s)R(s) = \frac{s}{s^2+1}\frac{1}{s^2+1} = \frac{s}{(s^2+1)^2}$$

and

$$c(t) = t \sin t$$

which is not bounded. As stated earlier, the type of system that has a bounded natural response but an unbounded output for certain bounded inputs is called a marginally stable system.

As just shown, the system characteristic equation can be obtained from the denominator of the closed-loop transfer function. If this denominator is expressed as a polynomial, this denominator is the *characteristic polynomial*. Since the denominator of the closed-loop transfer function is denoted as Δ in Mason's gain formula (see Section 2.4), the characteristic polynomial can also be obtained from this Δ. Thus the system characteristic equation can also be expressed as

$$\Delta(s) = 0 \tag{5-12}$$

where $\Delta(s)$ is the Δ of Mason's gain formula.

The characteristic equation can also be obtained from the state variable model

$$\dot{\mathbf{x}}(t) = \mathbf{A}\mathbf{x}(t) + \mathbf{B}u(t)$$
$$y(t) = \mathbf{C}\mathbf{x}(t) \tag{5-13}$$

Since the transfer function is given by

[eq. (3-41)] $$T(s) = \mathbf{C}(s\mathbf{I} - \mathbf{A})^{-1}\mathbf{B}$$

the denominator of this transfer function is given by $\det(s\mathbf{I} - \mathbf{A})$, and the system characteristic equation is then

$$\det(s\mathbf{I} - \mathbf{A}) = 0 \tag{5-14}$$

In summary, the characteristic equation for a linear time-invariant system can be expressed in three ways.

$$1 + G_c(s)G_p(s)H(s) = 0 \tag{5-15}$$

[eq. (5-12)] $$\Delta(s) = 0$$

[eq. (5-14)] $$\det(s\mathbf{I} - \mathbf{A}) = 0$$

Equation (5-15) applies to the single-loop system of Figure 5.1(b), and (5-12) and (5-14) apply to any linear time-invariant system.

5.3 SENSITIVITY

In this section we introduce a concept that is a primary reason for employing feedback control. Much of the theory of classical control was developed as a result of the effort to implement electronic amplifiers with nonvarying characteristics [1]. It was necessary for the amplifier characteristics not to vary appreciably over a long period of time, even though the characteristics of some of the major components would change significantly. The general topic of system characteristics changing with system parameter variations is called *sensitivity*.

In order to measure sensitivity adequately, we must have a mathematical definition for it. We can begin by considering the ratio of the percent change in the system transfer function to the percent change in a parameter b of the transfer function and letting this function be a measure of the sensitivity of the transfer function to the parameter b. We denote this ratio as S.

$$S = \frac{\Delta T(s)/T(s)}{\Delta b/b} = \frac{\Delta T(s)}{\Delta b}\frac{b}{T(s)} \tag{5-16}$$

In this equation $\Delta T(s)$ is the change in the transfer function $T(s)$ due to the change of the amount Δb in the parameter b. By definition, the *sensitivity function* is (5-16) evaluated in the limit as Δb approaches zero. Hence, the sensitivity function is given by

$$S_b^T = \lim_{\Delta b \to 0} \frac{\Delta T(s)}{\Delta b}\frac{b}{T(s)} = \frac{\partial T(s)}{\partial b}\frac{b}{T(s)} \tag{5-17}$$

To be mathematically correct, $T(s)$ should be written as $T(s, b)$, since the transfer function depends on both s and b.

The general sensitivity function of a characteristic W with respect to the parameter b is

$$S_b^W = \frac{\partial W}{\partial b}\frac{b}{W} \tag{5-18}$$

In general, the sensitivity function is a function of the Laplace transform variable s, which makes the sensitivity function very difficult to interpret. However, if we replace s in (5-17) with $j\omega$, we have the sensitivity as a frequency response. Then we can assign meaning to the sensitivity for frequencies in the bandwidth of the system. Since the system will not transmit frequencies outside its bandwidth (an approximation), the sensitivity for frequencies much greater than the system bandwidth is usually of little interest.

We now derive some useful sensitivity functions. We consider the control system of Figure 5.1(b), which has the transfer function

$$T(s) = \frac{G_c(s)G_p(s)}{1 + G_c(s)G_p(s)H(s)}$$

Consider first the sensitivity of the system transfer function $T(s)$ to the plant transfer function $G_p(s)$. From (5-18),

$$S_{G_p}^T = \frac{\partial T}{\partial G_p} \frac{G_p}{T} = \frac{(1 + G_c G_p H)G_c - G_c G_p(G_c H)}{(1 + G_c G_p H)^2} \frac{G_p}{G_c G_p/(1 + G_c G_p H)}$$

$$= \frac{1}{1 + G_c G_p H} \tag{5-19}$$

where the functional dependency on the Laplace transform variable s is understood. Evaluating (5-19) as a function of frequency yields

$$S_{G_p}^T(j\omega) = \frac{1}{1 + G_c(j\omega)G_p(j\omega)H(j\omega)}$$

The term $G_c G_p H$ evaluated at a specified frequency is called the *loop gain* at that frequency. This term is the transfer function of the loop in the control system with the minus sign at the summing junction ignored. Hence, at frequencies within the system bandwidth, we would like for the loop gain to be as large as possible, to reduce the sensitivity of the system characteristics to the parameters within the plant. Generally one of the purposes of the compensator $G_c(s)$ is to allow the loop gain to be increased without destabilizing the system.

Next we derive the sensitivity of the system transfer function to parameter changes in the sensor $H(s)$. Hence

$$S_H^T = \frac{\partial T}{\partial H} \frac{H}{T} = \frac{-G_c G_p(G_c G_p)}{(1 + G_c G_p H)^2} \frac{H}{G_c G_p/(1 + G_c G_p H)} = \frac{-G_c G_p H}{1 + G_c G_p H} \tag{5-20}$$

The minus sign indicates that an increase in H results in a decrease in T. Thus, for the system sensitivity with respect to the sensor to be small, the loop gain must be small. However, a small loop gain makes the system sensitive to variations in the plant. We see then that the system cannot be insensitive to both the plant and the sensor. To solve this problem, we generally can choose high-quality stable components for the sensor. We may not be able to do this for the plant. In addition, many plants, by their very nature, have parameters that vary extensively during operation. An example is an aircraft, whose parameters vary over very wide ranges with speed, altitude, and so on. Hence aircraft autopilot (compensator) designs must satisfy rigorous specifications for sensitivity to plant variations.

Thus far we have considered sensitivity to G_p or H. However, the transfer functions G_p or H will vary because of the variation of some parameter within the transfer functions. Let this parameter be b. The sensitivity of the system transfer function $T(s)$ to a parameter b in $G_p(s)$ can be expressed as

$$S_b^T = \frac{\partial T}{\partial b} \frac{b}{T} = \frac{\partial T}{\partial G_p} \frac{\partial G_p}{\partial b} \frac{b}{T} \tag{5-21}$$

Two examples are now given to illustrate the use of sensitivity functions.

Example 5.3

In this example we consider certain aspects of the design of the temperature-control system of Example 5.1, with the change that the compensator is a proportional (P) type, with $G_c(s) = K_P$, a pure gain. Also, let the plant transfer function be

$$G_p(s) = \frac{K}{s + 0.1}$$

where K has a nominal value of 5.0. Furthermore, the sensor is modeled as a pure gain, H_k, which has a nominal value of 0.05. We first find the sensitivity of $T(s)$ with respect to K, the numerator of $G_p(s)$. From (5-21), since

$$T(s) = \frac{K_p G_p(s)}{1 + K_p G_p(s) H_k}$$

then

$$S_K^T = \frac{\partial T}{\partial G_p} \frac{\partial G_p}{\partial K} \frac{K}{T} = \frac{K_P}{(1 + K_P G_p H_k)^2} \frac{G_p}{K} \frac{K}{K_P G_p/(1 + K_P G_p H_k)} = \frac{1}{1 + K_P G_p H_k}$$

Thus the sensitivity about the nominal value of K, as a function of frequency, is

$$S_K^T(j\omega) = \frac{1}{1 + K_P [5/(0.1 + j\omega)] (0.05)} = \frac{0.1 + j\omega}{0.1 + 0.25 K_P + j\omega}$$

In a like manner, from (5-20),

$$S_H^T(j\omega) = \frac{-K_P G_p(j\omega) H_k}{1 + K_P G_p(j\omega) H_k} = \frac{-K_P [5/(0.1 + j\omega)] (0.05)}{1 + K_P [5/(0.1 + j\omega)] (0.05)} = \frac{-0.25 K_P}{0.1 + 0.25 K_P + j\omega}$$

Figure 5.4 shows these sensitivity functions plotted as a function of frequency for values of the compensator gain, K_P, of 1 and 10. These values were chosen somewhat arbitrarily.

Note that the sensitivity of the system transfer function to the gain K is smaller at low frequencies. The bandwidth of the system can be obtained from the system transfer function:

$$T(j\omega) = \frac{K_P G_p(j\omega)}{1 + K_P G_p(j\omega) H_k} = \frac{5 K_P}{0.1 + 0.25 K_P + j\omega}$$

Note that the system gain is reduced to $0.707T(j0)$ at the frequency $\omega = (0.1 + 0.25 K_P)$, and thus the bandwidth is $\omega_B = (0.1 + 0.25 K_P)$. For $K_P = 1$, the bandwidth is $\omega_B = 0.35$, and for $K_P = 10$, $\omega_B = 2.60$ rad/s. These bandwidths are indicated in Figure 5.4.

We can draw the following conclusions from Figure 5.4:

1. The sensitivity of the system to K decreases with increasing open-loop gain, whereas the sensitivity to H increases with increasing open-loop gain.

2. The system is very sensitive to K outside the system bandwidth, which in general is not significant, and is very sensitive to H inside the system bandwidth, which is significant. Hence the sensor must be constructed of high-quality components.

Two of the frequency responses in Figure 5.4 can be calculated with the MATLAB program

```
Kp = [1 10];  w = logspace (-1,1);
for k = 1:2
  Gnum = [1 0.1]; Gden = [1 0.1+0.25*Kp(k)];
    Gjomega = freqs(Gnum,Gden,w);Gmag = abs(Gjomega);
      plot (w,Gmag)
      title ('Sensitivity Functions')
      xlabel ('Omega'),ylabel ('Sensitivity')
      grid
      hold on
end
hold off
```

Example 5.4

Design aspects of the system of Example 5.3 are considered further in this example. The compensator is now a PI type, with the transfer function

$$G_c(s) = K_P + \frac{K_I}{s}$$

Since none of the differentiations in the sensitivity functions were with respect to G_c, these sensitivity functions are unchanged from the preceding example, except that K_P is replaced with $K_P + K_I/j\omega$. Hence

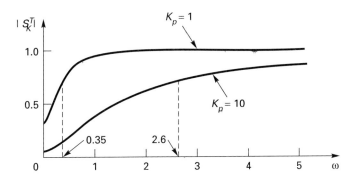

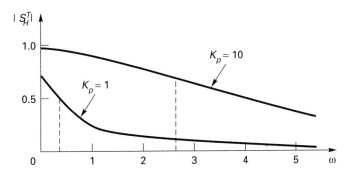

Figure 5.4   Results of Example 5.3.

$$S_K^T(j\omega) = \frac{1}{1 + [(K_I + j\omega K_P)/j\omega] \, [5/(0.1 + j\omega)] \, (0.05)}$$

$$= \frac{j\omega(0.1 + j\omega)}{0.25 K_I - \omega^2 + j(0.1 + 0.25 K_P)\omega}$$

and

$$S_H^T(j\omega) = \frac{[-(K_I + j\omega K_P)/j\omega] \, [5/(0.1 + j\omega)] \, (0.05)}{1 + [(K_I + j\omega K_P)/j\omega] \, [5/(0.1 + j\omega)] \, (0.05)}$$

$$= \frac{-(0.25 K_I + j0.25 K_P\omega)}{0.25 K_I - \omega^2 + j(0.1 + 0.25 K_P)\omega}$$

These sensitivity functions are more complex and will not be plotted, because effects other than sensitivity enter the analysis. However, note that the sensitivity of T with respect to K is zero at dc ($\omega = 0$). This indicates that if the input is constant and the system is in steady state, the system gain is independent of K (provided that K is not zero). Thus we see an important reason for the popularity of the PI controller.

5.4 DISTURBANCE REJECTION

In this section we consider the effects of disturbance inputs on the response of a control system. In a control system we have the input to the plant that is used to control the plant, and this input, called the *manipulated variable*, is denoted by $m(t)$. However, any physical control system will have other inputs that influence the plant output and that, in general, we do not control. We call these inputs *disturbances* and usually attempt to design the control system such that these disturbances have a minimal effect on the system.

An example of disturbance inputs is now given. Suppose that we wish to design a control system for a certain radar tracking antenna. The parabolic antenna is rigidly mounted on a pedestal, and the pedestal is rotated by the application of a torque by a dc servomotor, as described in Section 2.7 and Figure 2.28. The block diagram of a servomotor is shown in Figure 5.5 (see Figure 2.27), where $G_2(s)$ contains the inertia of both the antenna and pedestal. The wind, of course, also exerts a torque on the antenna. Hence the total torque on the pedestal is the sum of the motor torque and the wind torque. The wind-torque input, denoted as $D_\tau(s)$, has been added to Figure 5.5. By superposition, the system output is

$$C(s) = \frac{K_\tau G_1(s)G_2(s)}{1 + K_\tau G_1(s)G_2(s)H(s)}M(s) + \frac{G_2(s)}{1 + K_\tau G_1(s)G_2(s)H(s)}D_\tau(s)$$

$$= G_p(s)M(s) + G_d(s)D_\tau(s) \qquad (5\text{-}22)$$

where $G_p(s)$ is the transfer function from the control input $M(s)$ (the armature voltage) to the output and $G_d(s)$ is the transfer function from the disturbance input $D_\tau(s)$ (the wind torque) to the output. The control system design should minimize the effects of the wind on the position (pointing direction) of the antenna while allowing the antenna to respond in some prescribed manner to the system input.

We can also develop disturbance models using the state variable approach. Consider the state variable model of a plant with a single output and two inputs.

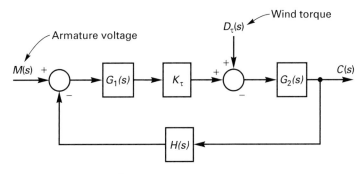

Figure 5.5 Servomotor model.

$$\dot{\mathbf{x}}(t) = \mathbf{A}\mathbf{x}(t) + \mathbf{B}\mathbf{u}(t)$$

$$y(t) = \mathbf{C}\mathbf{x}(t)$$

where $\mathbf{u}(t)$ is of order 2×1 and $y(t)$ is a scalar. The input vector $\mathbf{u}(t)$ is given by

$$\mathbf{u}(t) = \begin{bmatrix} m(t) \\ d(t) \end{bmatrix}$$

where $m(t)$ is the control input and $d(t)$ is the disturbance input. From (3-43), the transfer function matrix of the system is

$$\mathbf{G}(s) = \mathbf{C}(s\mathbf{I} - \mathbf{A})^{-1}\mathbf{B} = [G_p(s)\ G_d(s)] \tag{5-23}$$

In this equation $G_p(s)$ is the transfer function from the control input $M(s)$ to the output, and $G_d(s)$ is the transfer function from the disturbance input $D(s)$ to the output. Note that if there are more than two inputs, (5-23) still applies but with the matrix $\mathbf{G}(s)$ having more than two components.

We now investigate the problems involved in disturbance rejection. The block diagram of a system with a disturbance input can be drawn as shown in Figure 5.6. Of course, for a particular system, the equations of the system can often be manipulated to give a block diagram of the form of Figure 5.5. However, the form of Figure 5.6 is general and is used in many practical situations; hence we use this one.

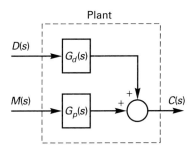

Figure 5.6 Plant with a disturbance input.

A closed-loop control system with a disturbance input is shown in Figure 5.7. By superposition and Mason's gain formula we can write the output expression

$$C(s) = \frac{G_c(s) G_p(s)}{1 + G_c(s) G_p(s) H(s)} R(s) + \frac{G_d(s)}{1 + G_c(s) G_p(s) H(s)} D(s)$$

$$= T(s) R(s) + T_d(s) D(s)$$

In this expression $T(s)$ is the transfer function from the reference input $R(s)$ to the output, and $T_d(s)$ is the transfer function from the disturbance input $D(s)$ to the output.

If we wish to reject the disturbance input, then, in some sense, $T_d(s)D(s)$ must be small. We use a frequency response approach to investigate disturbance rejection. In general, $T_d(j\omega)$ cannot be made small for all frequencies ω. However, through the design process for $G_c(j\omega)$, we may be able to make the gain $T_d(j\omega)$ small over a significant portion of the system bandwidth. Recall from the preceding section that the loop gain, $G_c(j\omega)G_p(j\omega)H(j\omega)$, must be made large to reduce sensitivity to plant variations. We have the same requirement here. For the case that the loop gain is large,

$$T(j\omega) = \frac{G_c(j\omega) G_p(j\omega)}{1 + G_c(j\omega) G_p(j\omega) H(j\omega)} \cong \frac{G_c(j\omega) G_p(j\omega)}{G_c(j\omega) G_p(j\omega) H(j\omega)} = 1 \qquad (5\text{-}24)$$

for $H(j\omega) = 1$, and the output tracks the reference input very well over the frequency band for which the loop gain is large.

For the disturbance transfer function,

$$T_d(j\omega) = \frac{G_d(j\omega)}{1 + G_c(j\omega) G_p(j\omega) H(j\omega)} \cong \frac{G_d(j\omega)}{G_c(j\omega) G_p(j\omega) H(j\omega)} \qquad (5\text{-}25)$$

Hence this ratio must be small for good disturbance rejection. An obvious way to accomplish this is to make $G_d(j\omega)$ small. However, the control engineer may not be able to influence the design of the plant. For example, the control engineer may be required to use a commercially available radar antenna to implement a radar tracking system. He or she will not be able to specify the structure of the antenna to reduce wind forces on the antenna. However, when allowed, the control engineer's influence on the design of the plant can significantly reduce the effects of disturbances.

A second method for reducing disturbance effect is to increase the loop gain by increasing the gain of $G_c(j\omega)$, since the control engineer does specify the compensation. Note that increasing the plant gain will also increase the loop gain, but increasing the plant

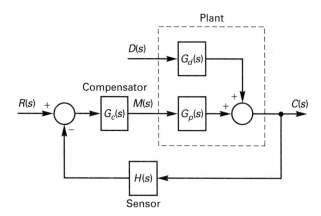

Figure 5.7   Closed-loop system with a disturbance input.

gain may also increase the gain of $G_d(j\omega)$; hence in this case disturbance rejection is not improved. Thus to reject disturbances, the loop gain must be increased in such a manner that the gain from the disturbance input to the system output is not increased. This increase is usually accomplished by increasing the compensator gain.

A third method of disturbance rejection is to reduce the magnitude of the disturbance. Suppose, for example, that the disturbance is an electrical noise coupled into the sensitive sensor circuits. Careful layout, shielding, and proper grounding of these circuits usually reduces, or even eliminates, this noise. This method of reducing disturbance effects is the most effective solution to the disturbance problems. However, it cannot be applied in all cases.

A fourth method of disturbance rejection is called *feedforward* and can be applied if the disturbance can be measured by a sensor. Feedforward disturbance rejection is illustrated in Figure 5.8. In this system the disturbance $D(s)$ is measured and transmitted to the system summing junction through the transfer function $G_{cd}(s)$, which is a compensator for the disturbance input. The addition of this compensator does not affect the transfer function from the reference input $R(s)$ to the output, which is given by (5-24). However, the transfer function from the disturbance input to the output is now

$$T_d(s) = \frac{G_d(s) - G_{cd}(s)\, G_c(s)\, G_p(s)}{1 + G_c(s)\, G_p(s)\, H(s)} \tag{5-26}$$

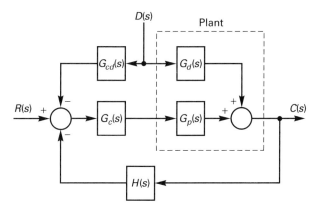

Figure 5.8   Feedforward compensation.

If the product $G_{cd}G_c$ can be chosen such that the numerator is small, then good disturbance rejection will occur. If the numerator can be made equal to zero, that is, if

$$G_d(s) - G_{cd}(s)\, G_c(s)\, G_p(s) = 0$$

or if

$$G_{cd}(s)\, G_c(s) = \frac{G_d(s)}{G_p(s)} \tag{5-27}$$

the disturbance will be rejected completely. The effect of satisfying (5-27) is to produce a component in the output of $G_p(s)$, due to $D(s)$, which is equal and opposite to the output of $G_d(s)$. If (5-27) cannot be satisfied exactly for all frequencies, good disturbance rejection will occur over the frequency band for which it is approximately satisfied. However, remember that a transfer function is only an approximate model of a physical system. Hence, if (5-27) is satisfied exactly, the quality of the disturbance rejection in the physical system will depend on the accuracy of the models $G_p(s)$ and $G_d(s)$.

In summary, disturbances may be rejected in the following ways:

1. Reduce the gain $G_d(j\omega)$ between the disturbance input and the output.
2. Increase the loop again $G_c G_p H(j\omega)$ without increasing the gain of $G_d(j\omega)$. This is usually accomplished by the choice of the compensator $G_c(j\omega)$.
3. Reduce the magnitude of the disturbance $d(t)$. This should always be attempted, if reasonable.
4. Use feedforward compensation, if the disturbance can be measured.

Good disturbance rejection usually results from a combination of these procedures plus (possibly) others. An example is given next.

Example 5.5

The design of the temperature-control system of Example 5.1 is considered further in this example. Recall that the model given represents a thermal test chamber. We assume that the disturbance to be considered is the opening of a door into the chamber, which will certainly affect the chamber temperature. Consider that the model given in Figure 5.9(a) is an adequate model. It is assumed that the disturbance of opening the chamber door can be represented in this model by a unit step input for $d(t)$. Thus the effect on the plant in Figure 5.9(a) of the door remaining open for a long period of time (at least 40 seconds) is a 6°C drop in temperature, since $G_d(0) = -6$.

We now consider the closed-loop system. The PI control system is shown in Figure 5.9(b). The transfer function from the disturbance input to the system output is given by

$$T_d(s) = \frac{G_d(s)}{1 + G_c(s)G_p(s)H(s)}$$

$$= \frac{-0.6/(s + 0.1)}{1 + (K_P + K_I/s)\,[5/(s + 0.1)]\,(0.05)}$$

or

$$T_d(j\omega) = \frac{-j0.6\omega}{0.25K_I - \omega^2 + j(0.1 + 0.25K_P)\omega}$$

This function may be plotted versus frequency to determine the quality of the disturbance rejection for a given PI controller. In this example we have assumed that the disturbance can be modeled as a unit step function. Thus *in the steady state* the disturbance has *no effect*, since $T_d(0) = 0$. This is a direct result of the integrator in the compensator having an infinite gain at dc. Note however that opening the chamber door will have a transient effect, and the values of K_P and K_I will determine the characteristics of this transient effect.

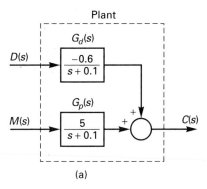

(a)

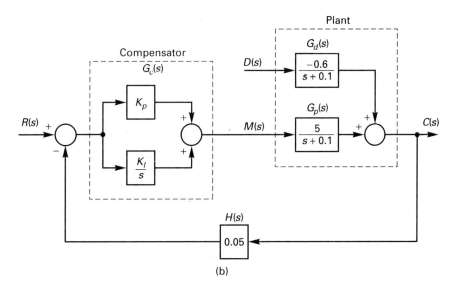

(b)

Figure 5.9   System for Example 5.5.

5.5 STEADY-STATE ACCURACY

Many control system are designed to operate with the reference input $r(t)$ equal to a constant value. An example is the temperature-control system used in homes and in work areas. The reference input (the setting on the thermostat) is a constant value, and disturbance inputs, such as thermal leakage through outside surface areas, tend to keep the system output from maintaining the commanded value of temperature for any length of time. Of course, the reference input may be changed from one constant value to another constant value from time to time. We refer to this type of change as a step change; it is illustrated in Figure 5.10. Here we have arbitrarily chosen $t = 0$ to be the time of the change. This signal can be written as

$$r(t) = R_0 + (R_1 - R_0)u(t) \tag{5-28}$$

where $u(t)$ is the unit step. By superposition, the system response can be separated into two independent parts. The first part is the response to the constant input R_0, and the second part is the response to the step of amplitude $(R_1 - R_0)$. Hence the system step response is also the response to a step change in the input, and we see then another reason that the step response of a system is very important.

We first consider the unity feedback system of Figure 5.11 and consider nonunity feedback systems later. *In all the following derivations, it is assumed that the system is stable.* If the system is not stable, none of the results derived in this section have meaning. For the system of Figure 5.11, the output is given by

$$C(s) = \frac{G_c(s)G_p(s)}{1 + G_c(s)G_p(s)} R(s) \tag{5-29}$$

For convenience we express $G_c(s)G_p(s)$ as

$$G_c(s)G_p(s) = \frac{F(s)}{s^N Q_1(s)} \tag{5-30}$$

where neither of the polynomials $F(s)$ nor $Q_1(s)$ has a zero at $s = 0$. Since the transfer function of an integrator is $1/s$, then N is the number of free integrators in the transfer function $G_c(s)G_p(s)$. We define free integrators as integrators that are not part of any loop in a simulation diagram of the transfer function. The number of free integrators, N, is called the *system type*, and we now demonstrate its importance.

As was defined in Section 5.1, the system error for the system of Figure 5.11 is the difference between the system input and the system output, that is,

$$\text{system error} = e(t) = r(t) - c(t) \tag{5-31}$$

The *steady-state system error* is, by definition, the steady-state value of $e(t)$. Denoting the steady-state error by e_{ss}, then, by the final theorem of the Laplace transform (see Appendix B),

$$e_{ss} = \lim_{s \to 0} sE(s) \tag{5-32}$$

provided that $e(t)$ has a final value. For the system of Figure 5.11, from Mason's gain formula,

$$E(s) = \frac{R(s)}{1 + G_c(s)G_p(s)} \tag{5-33}$$

and

$$e_{ss} = \lim_{s \to 0} \frac{sR(s)}{1 + G_c(s)G_p(s)} \tag{5-34}$$

The steady-state error is next calculated for three different input functions.

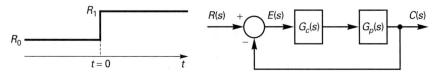

Figure 5.10   Step change in the input.   **Figure 5.11**   Unity feedback system.

5.5.1 Step Response

We first calculate the steady-state error for the unit step input. For the unit step, $R(s) = 1/s$ and from (5-34).

$$e_{ss} = \lim_{s \to 0} \frac{1}{1 + G_c(s)G_p(s)} = \frac{1}{1 + \lim_{s \to 0} G_c(s)G_p(s)} = \frac{1}{1 + K_p} \qquad (5\text{-}35)$$

where K_p is called the *position error constant* and is given by

$$K_p = \lim_{s \to 0} G_c(s)G_p(s) \qquad (5\text{-}36)$$

For the case that $N \geq 1$ in (5-30), K_p is unbounded and the steady-state error is zero. Hence, for a system that is type 1 or higher, the steady-state error for a unit step input is zero. For a type 0 system, the steady-state error is not zero and is given by (5-35).

For the case that a step of amplitude A is applied, that is, if $R(s) = A/s$, then from (5-34),

$$e_{ss} = \frac{A}{1 + K_p} \qquad (5\text{-}37)$$

where K_p is defined in (5-36). Since in the steady state a step input is a constant input, (5-37) applies for any case that the system is in steady state with a constant input of A units.

5.5.2 Ramp Response

We next consider the steady-state error for a unit ramp input for which $r(t) = tu(t)$ and thus $R(s) = 1/s^2$. The steady-state error is then, from (5-34),

$$e_{ss} = \lim_{s \to 0} \frac{1}{s + sG_c(s)G_p(s)} = \frac{1}{\lim_{s \to 0} sG_c(s)G_p(s)} = \frac{1}{K_v} \qquad (5\text{-}38)$$

In this equation, K_v is called the *velocity error constant*, since commanding a position control system with a ramp function will give a constant velocity output in the steady state. The velocity error constant is, from (5-30) and (5-38),

$$K_v = \lim_{s \to 0} sG_c(s)G_p(s) = \lim_{s \to 0} \frac{F(s)}{s^{N-1}Q_1(s)} \qquad (5\text{-}39)$$

Hence, for a type 2 and higher system, the steady-state error for a ramp input is zero, since K_v is unbounded. For a type 1 system, the steady-state error is finite and nonzero and is given by (5-38). For a type 0 system, K_v is zero and the steady-state error is unbounded.

These errors are illustrated in Figure 5.12. In Figure 5.12(a), for a type 0 system the output is a ramp in the steady state, but the output ramp has a different slope from the input ramp. Hence the error continues to grow without limit. In Figure 5.12(b), for a type 1 system the slopes of the input and output ramps are equal, but the output ramp is offset from the input ramp by the amount of the steady-state error. In Figure 5.12(c), for a type 2 and higher system, there is no error in the steady state.

If the input is the ramp function $Atu(t)$, the steady-state error is seen to be

$$e_{ss} = \frac{A}{K_v} \qquad (5\text{-}40)$$

where K_v is given by (5-39).

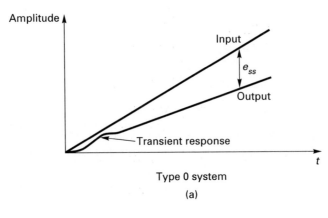

Type 0 system

(a)

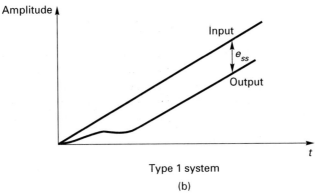

Type 1 system

(b)

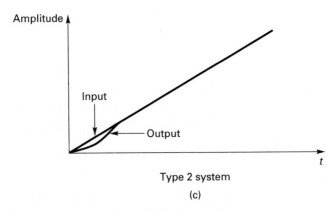

Type 2 system

(c)

Figure 5.12   Ramp responses.

5.5.3 Parabolic Input

The final input that we consider is the parabolic input $r(t) = (t^2/2)u(t)$, which has the Laplace transform $R(s) = 1/s^3$. This input causes a position control system to respond with a constant acceleration in the steady state. The steady-state error, from (5-34), is

$$e_{ss} = \lim_{s \to 0} \frac{1}{s^2 + s^2 G_c(s) G_p(s)} = \frac{1}{\lim_{s \to 0} s^2 G_c(s) G_p(s)} = \frac{1}{K_a} \tag{5-41}$$

where K_a is called the *acceleration error constant*. Thus

$$K_a = \lim_{s \to 0} s^2 G_c(s) G_p(s) = \lim_{s \to 0} \frac{F(s)}{s^{N-2} Q_1(s)} \tag{5-42}$$

For a type 3 and higher system, the steady-state error is zero. For a type 2 system, the steady-state error is finite and nonzero. For type 0 and type 1 systems, the steady-state error is unbounded.

The results of these derivations are given in Table 5.1. From Table 5.1 we see that the higher the system type, the more accurate is the steady-state response. We can show this as follows. For a well-behaved input, the input function can be expanded into a Taylor's series [2]:

$$r(t) = r(0) + \frac{dr}{dt}\bigg|_{t=0} t + \frac{1}{2!} \frac{d^2 r}{dt^2}\bigg|_{t=0} t^2 + \cdots \tag{5-43}$$

Thus the input function can be expressed as a step plus a ramp plus a parabolic function, and so forth. Higher system types will result in the errors from more of these terms being zero. But, as we will see later, it is quite difficult to stabilize a system of type 2 and higher. In addition, the transient responses of these type systems tend to be very poor. Note also from Table 5.1 that increasing the loop gain also reduces the steady-state errors, provided the system remains stable.

TABLE 5.1   STEADY-STATE ERROR CONSTANTS

N	$1/s$	$1/s^2$	$1/s^3$	Error constants
		$R(s)$		
0	$\dfrac{1}{1 + K_p}$	∞	∞	$K_p = \lim\limits_{s \to 0} G_c G_p$
1	0	$\dfrac{1}{K_v}$	∞	$K_v = \lim\limits_{s \to 0} s G_c G_p$
2	0	0	$\dfrac{1}{K_a}$	$K_a = \lim\limits_{s \to 0} s^2 G_c G_p$

An example is given next to illustrate the use of these results.

Example 5.6

Consider the system of Figure 5.13, which is the model of a position-control system. This motor has a time constant of 0.1 s; thus it can reach full speed from standstill in approximately 0.4 s when operated open loop. There is no dynamic compensation in the system; however, the gain K of the power amplifier is assumed variable. The steady-state error constants, from Table 5.1, are

$$K_p = \lim_{s \to 0} KG_p(s) \to \infty$$

$$K_v = \lim_{s \to 0} sKG_p(s) = 10K$$

$$K_a = \lim_{s \to 0} s^2 KG_p(s) = 0$$

Since the system is type 1, we could have stated the values of K_p and K_a without calculations. The steady-state error for a unit ramp input is $e_{ss} = 1/10K$. For example, suppose that the output units are centimeters and K has been set to a value of 10. Then, in the steady-state, the velocity error is 0.01 cm/s. However, we do not know the time required to reach steady state. The time constant just given is that of the open-loop system. It would be necessary to solve for the poles of the closed-loop transfer function to determine the time constant of the closed-loop system. The transient response of closed-loop systems is discussed in Section 5.6.

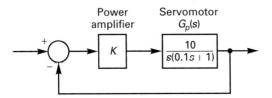

Figure 5.13   System for Example 5.6.

5.5.4 Nonunity-Gain Feedback

For a system in which the feedback gain is not unity, the units of the input are usually different from the units of the output. Typically the sensor output is a voltage, and thus the physical system input at the summing junction must be a voltage. The physical system's output may be, for example, position in millimeters, temperature in degrees Celsius, or angle in degrees. Therefore, there is no meaning in subtracting volts from millimeters; we must first transform the system into an equivalent unity feedback system such that the input and output are in the same units. The system steady-state error will then be in these units.

Given in Section 5.1 is a method for transforming a nonunity feedback system into a unity feedback system for the case that the sensor transfer function is a pure gain. In calculating steady-state errors, we consider transfer functions only under the conditions that the Laplace variable s approaches zero (see Table 5.1). In general, sensors that are measuring a constant (dc) signal have an output that is constant, in the steady state. Hence the sensor (dc) gain is given by

$$\lim_{s \to 0} H(s) = H_k \tag{5-44}$$

where H_k is finite and nonzero. Thus, for steady-state analysis, generally the method of Section 5.1 can be used to transform a nonunity feedback system into a unity feedback system. It will be noted that the input to the unity feedback model, $r_u(t)$, is also the desired value of the output, $c_d(t)$, so that the output of the model summing junction is indeed the system error.

The nonunity feedback system is shown in Figure 5.14(a), and the unity feedback model is shown in Figure 5.14(b), where H_k is given by (5-44). If this system is compared with that of Figure 5.11 (the system for which Table 5.1 applies), the two systems are the same if $G_c(s)G_p(s)$ in Figure 5.11 is replaced with $H_kG_c(s)G_p(s)$ in Figure 5.14. Thus the equations for the error coefficients in Table 5.1 apply to the nonunity feedback system if the given substitution is made. Then, from Table 5.1, the error coefficients are

$$K_p = \lim_{s \to 0} G_cG_pH$$

$$K_v = \lim_{s \to 0} sG_cG_pH \tag{5-45}$$

$$K_a = \lim_{s \to 0} s^2 G_cG_pH$$

To calculate the steady-state error coefficients for the nonunity feedback system, the open-loop function $G_c(s)G_p(s)H(s)$ must be used, not simply the forward path transfer function. The system type is determined by the open-loop function (however, the sensor will have no poles at $s = 0$). Equations (5-45) are general, since, for the unity feedback case, H is equal to unity. With these error coefficients used in Table 5.1, the steady-state errors are in the units of the physical system output. An example is now given to illustrate these results.

Example 5.7

Further aspects in the design of the temperature-control system of Example 5.1 are considered in this example. Both the control system and the unity feedback model are shown in Figure 5.15 [see Figure 5.14(b)]. For the case that the compensator is a proportional type with $G_c(s) = K$,

$$G_c(s)G_p(s)H_k = \frac{0.25K}{s + 0.1}$$

and we see that the system is type 0. From (5-36), the position error constant is given by

$$K_p = \lim_{s \to 0} G_c(s)G_p(s)H_k = 2.5K$$

and the steady-state error for a constant input of A °C is

$$e_{ss} = \frac{A}{1 + K_p} = \frac{A}{1 + 2.5K} \, ^\circ\text{'C}$$

For example, if the reference input is a constant value of 50°C and the compensator gain is chosen as $K = 10$, the steady-state error is equal to $50/26 = 1.92°C$. The steady-state chamber temperature is 48.08°C.

Suppose that a PI compensator is designed. Then the open-loop function is

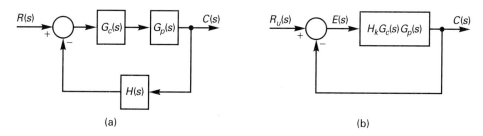

(a)   (b)

Figure 5.14   Steady-state analysis for nonunity feedback system.

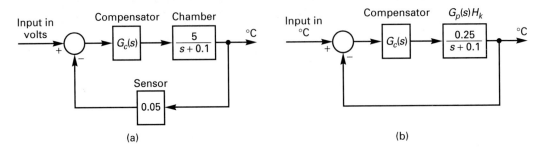

Figure 5.15   System for Example 5.7.

$$G_c(s)G_p(s)H_k = \left(K_P + \frac{K_I}{s}\right)\frac{0.25}{s+0.1} = \frac{0.25(K_I + K_P s)}{s(s+0.1)}$$

Since the system is now type 1, the steady-state error is zero for a constant input. In the preceding equation if the parameters of the plant (represented by the numbers 0.25 and 0.1) change, the steady-state error remains zero. Here we see two of the principal reasons for employing the PI compensator in a control system of this type: (1) the steady-state error for a constant input is zero, and (2) this error is insensitive to the parameters of the plant model. Note also that if a unit ramp is applied at the input, the steady-state error is obtained from the velocity error constant.

$$K_v = \lim_{s \to 0} s G_c(s) G_p(s) H_k = 2.5 K_I$$

and the error is

$$e_{ss} = \frac{1}{K_v} = \frac{0.4}{K_I}$$

Of course, all these results are valid only if the closed-loop system is stable.

5.5.5 Disturbance Input Errors

In this section we consider the *steady-state errors* resulting from disturbance inputs. In general, *any* contribution of disturbance inputs to the system output constitutes error; that is, disturbance inputs are unwanted effects, and in the ideal case these inputs do not contribute at all to the system output.

As discussed in Section 5.4, a control system with one disturbance input can be modeled as shown in Figure 5.16. Let $T(s)$ be the transfer function from the input $R(s)$ to the output $C(s)$, and $T_d(s)$ the transfer function from the disturbance input $D(s)$ to $C(s)$. Then, by superposition,

$$C(s) = T(s)R(s) + T_d(s)D(s) = C_r(s) + C_d(s) \tag{5-46}$$

where $c_r(t)$ is the component of the response from the input $r(t)$, and $c_d(t)$ is the component from $d(t)$. The system error is then

$$e(t) = r(t) - c(t) = [r(t) - c_r(t)] - c_d(t) = e_r(t) + e_d(t) \tag{5-47}$$

where $e_r(t)$ is the error considered earlier in this section. We prefer that disturbance error $e_d(t) = -c_d(t)$ be zero. Assuming that the steady-state value of $e_d(t)$, denoted as e_{dss}, exists, we calculate this steady-value to be

$$e_{dss} = \lim_{s \to 0} sC_d(s) = \lim_{s \to 0} sT_d(s)D(s) \qquad (5\text{-}48)$$

where

$$T_d(s) = \frac{G_d(s)}{1 + G_c(s)G_p(s)} \qquad (5\text{-}49)$$

for the system of Figure 5.16.

We will first consider the case that the disturbance input is modeled as a step function, with $D(s) = B/s$, where B is the amplitude of the step. Examples of a constant disturbance input are constant wind against a radar antenna, or constant heat-energy loss in a temperature-control system. From (5-48) and (5-49), the steady-state error is given by

$$e_{dss} = \lim_{s \to 0} T_d(s)B = \lim_{s \to 0} \frac{G_d(s)B}{1 + G_c(s)G_p(s)} \qquad (5\text{-}50)$$

To draw general conclusions, we must consider the type number of $G_p(s)$, $G_c(s)$, and $G_d(s)$. Often $G_p(s)$ and $G_d(s)$ are of the same type number, since these two transfer functions are from different inputs to the output of the same system.

Consider that all three transfer functions in (5-50) are type 0. The steady-state error for a step disturbance input is then

$$e_{dss} = \frac{G_d(0)B}{1 + G_c(0)G_p(0)} \qquad (5\text{-}51)$$

One combination of type numbers for zero steady-state error is for $G_c(s)$ to be type one (a PI compensator) with $G_d(s)$ and $G_p(s)$ type 0. For this case, $\lim_{s \to 0} G_c(s)$ is unbounded and e_{dss} is zero. Other combinations of type numbers will also result in e_{dss} being zero (see Problem 5.25).

For the case that the disturbance input is modeled as the ramp function $D(s) = B/s^2$, the steady-state error in (5-48) is given by

$$e_{dss} = \lim_{s \to 0} \frac{T_d(s)B}{s} = \lim_{s \to 0} \frac{G_d(s)B}{s[1 + G_c(s)G_p(s)]}$$

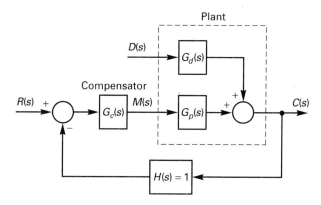

Figure 5.16 Closed-loop system with disturbance input.

For example, if $G_d(s)$, $G_p(s)$, and $G_c(s)$ are all type 0, e_{dss} is unbounded. If $G_c(s)$ is type 2 and $G_p(s)$ and $G_d(s)$ are type 0, $e_{dss} = 0$. Of course, there are many other combinations that will result in the steady-state error being zero (see Problem 5.25).

Note that the derivations are independent of the assumption that $H(s) = 1$. Hence all results apply for the $H(s) \neq 1$, with the denominator of $T_d(s)$ replaced with $G_c(s)G_p(s)H(s)$. An example will now be given.

Example 5.8

Additional aspects of the design of the temperature-control system of Examples 5.1, 5.3, 5.4, 5.5, 5.6, and 5.7 will be considered. For this case, $G_p(s)$ and $G_d(s)$ are both type 0, and $G_c(s)$ is type 1 (a PI compensator). As shown in Example 5.5,

$$T_d(s) = \frac{-0.6s}{s^2 + (0.1 + 0.25K_P)s + 0.25K_I}$$

From (5-50), for $D(s) = 1/s$, a unit step function,

$$e_{dss} = \lim_{s \to 0} T_d(s) = \lim_{s \to 0} \frac{-0.6s}{s^2 + (0.1 + 0.25K_P)s + 0.25K_I} = 0$$

and we see one advantage of the PI compensator. If the disturbance were modeled as a unit ramp function $D(s) = 1/s^2$, the steady-state error is, from (5-52),

$$e_{dss} = \lim_{s \to 0} \frac{T_d(s)B}{s} = \lim_{s \to 0} \frac{-0.6s}{s[s^2 + (0.1 + 0.25K_P)s + 0.25K_I]} = \frac{-2.4}{K_I}$$

5.6 TRANSIENT RESPONSE

In Section 5.2 we discussed stability and the natural response of a system. In that section we represented the closed-loop system transfer function in (5-6) and (5-9) as

$$T(s) = \frac{P(s)}{Q(s)} = \frac{P(s)}{a_n \prod_{i=1}^{n} (s - p_i)} \tag{5-52}$$

where the p_i are the poles of the transfer function (zeros of the characteristic equation). The system response for an input $R(s)$ is given by

$$C(s) = T(s)R(s) = \frac{P(s)}{a_n \prod_{i=1}^{n} (s - p_i)} R(s)$$

$$= \frac{k_1}{s - p_1} + \frac{k_2}{s - p_2} + \cdots + \frac{k_n}{s - p_n} + C_r(s) \tag{5-53}$$

where $C_r(s)$ is the sum of the terms, in the partial-fraction expansion, that originate in the poles of $R(s)$. The inverse Laplace transform of this equation yields

$$c(t) = k_1 e^{p_1 t} + k_2 e^{p_2 t} + \cdots + k_n e^{p_n t} + c_r(t) \tag{5-54}$$

In this expression, $c_r(t)$ is the *forced response* and is of the same functional form as the input $r(t)$. The first n terms of (5-54) form the *natural response*, and these terms approach zero as time increases if the system is stable. A usable control system will be stable. However, in many control systems, the manner in which the natural response terms approach zero is very important. For a stable system, the natural response will go to zero as time increases; for this case the natural response is often called the *transient response*.

Assuming that the system of (5-53) is stable, the transient response in (5-54) is given by

$$\text{transient response} = k_1 e^{p_1 t} + k_2 e^{p_2 t} + \cdots + k_n e^{p_n t} \tag{5-55}$$

The factors $e^{p_i t}$, $i = 1, 2, \ldots, n$, are called the *modes* of the system. The nature, or form, of each term is determined by the pole location p_i. The amplitude of each term is k_j, which is determined by

$$k_j = \left. \frac{P(s)}{a_n \displaystyle\prod_{\substack{i=1 \\ i \neq j}}^{n} (s - p_i)} R(s) \right|_{s = p_j} \tag{5-56}$$

Thus the amplitude of each term is determined by all other pole locations, the numerator polynomial (zeros of the transfer function), and the input function.

For a high-order system, it is difficult to make general assertions concerning the transient response; however, we can consider the nature of each term of this response. A time constant is associated with each real pole with the value

$$\tau_i = \frac{-1}{p_i} = \frac{1}{|p_i|} \tag{5-57}$$

The negative sign is required, since, for a stable system, the values of the real poles are negative.

For each set of complex-conjugate poles of values

$$p_i = -\zeta_i \omega_{ni} \pm j \omega_{ni} \sqrt{1 - \zeta_i^2} \tag{5-58}$$

there is associated a damping ratio ζ_i, a natural frequency ω_{ni}, and a time constant $\tau_i = 1/\zeta_i \omega_{ni}$ (see Section 4.3). If one pole of the transfer function dominates (we then call this pole the *dominant pole* of the system), the system responds essentially as a first-order system and the transient-response characteristics may be determined as shown in Section 4.1. If two poles dominate (the two poles are called the *dominant poles*), the system responds essentially as a second-order system and the transient-response characteristics may be determined as shown in Section 4.3. For higher-order systems, the rise time, the overshoot, and the settling time are all complicated functions of the poles and zeros of the closed-loop transfer function, as discussed in Section 4.5.

5.7 CLOSED-LOOP FREQUENCY RESPONSE

In Chapter 4 the important topic of frequency response was introduced. In earlier sections of this chapter the necessity of maintaining a high open-loop gain $G_c(j\omega)G_p(j\omega)H(j\omega)$ was demonstrated for low sensitivity to plant-parameter variations, for good disturbance rejection, and for low system errors. Note that these comments refer to *closed-loop characteristics* that are determined from the *open-loop gain*.

The input-output characteristics of a system are determined by the closed-loop frequency response. For the single-loop system that we have considered in this chapter, the closed-loop frequency response is given by

$$T(j\omega) = \frac{G_c(j\omega)G_p(j\omega)}{1 + G_c(j\omega)G_p(j\omega)H(j\omega)} \tag{5-59}$$

The value of this function at $\omega = 0$ is the system dc gain, that is, the steady-state gain for a constant input. This value determines the steady-state error for a constant input. To obtain the actual steady-state error in units of the output, it is necessary to multiply (5-59) by H_k in (5-44) to convert the system to a unity feedback model, as described in the section above.

If we evaluate the magnitude of (5-59) for small ω, we see how the system will track a slowly varying input. If the magnitude has values that are approximately unity at low frequencies, the system will closely track an input that is composed principally of low frequencies. The closed-loop bandwidth can be obtained from (5-59). Since, for a given system, the product of rise time and bandwidth is approximately constant [see (4-52)], a large bandwidth indicates a fast system response.

The presence of any peaks in the plot of the magnitude of the closed-loop frequency response, which denotes resonances, will give an indication of the overshoot, or decaying oscillations, in the transient response. Hence a plot of (5-59) will give useful indications of the characteristics of the closed-loop system.

In summary, a plot of the magnitude of the *closed-loop frequency response T(jω)* will give the following information:

1. From the value at $\omega = 0$, we can determine the steady-state error for a constant input. The low-frequency response indicates the response for slowly varying inputs.

2. The bandwidth gives us an indication of the rise time.

3. The presence of any peaks indicates overshoot in the transient response—the greater the peaks, the more overshoot that will occur.

These characteristics were discussed in Chapter 4 and are investigated in greater detail in Chapter 8.

5.8 SUMMARY

In this chapter the characteristics of linear time-invariant analog control systems were investigated. The purpose of this chapter is to illustrate the types of responses that may be obtained from systems of this type. If a control system is to be designed, the response specifications must be of the types listed in this chapter.

First of all, a control system must be stable. Then we may specify that a system be relatively insensitive to the variations of certain parameters of the plant that are known to change during the operation of the system. If the system is subjected to significant disturbances, we may specify limits of the response to these disturbances, in some manner. If the types of inputs to the system are known, we may specify that the response must follow these inputs in the steady state with a certain accuracy. The manner in which a system responds to certain inputs, in terms of the speed of response, the overshoot in the response, the settling time, and so forth, may be of importance. This response is called the transient response.

The specifications of a closed-loop control system may be based on the above characteristics and may in certain cases include other characteristics. In the following chapters we will see how systems can be designed to satisfy specified characteristics, at least to an extent. Certain specifications may impose conflicting requirements on the system and all cannot be satisfied. In these cases, trade-offs between specified characteristics are necessary such that no characteristic is satisfied exactly but all are satisfied to a degree.

REFERENCES

1. H. W. Bode. *Network Analysis and Feedback Design.* New York: Van Nostrand, 1945.
2. G. B. Thomas, Jr. and R. L. Finney. *Calculus and Analytic Geometry.* Reading, MA: Addison-Wesley, 1996.

PROBLEMS

5.1. A speed-control system of an electric motor has the block diagram given in Figure P5.1. The input signal is a voltage and represents the desired speed.
   (a) The gain of the tachometer is $H_k = 0.02$. What are the units on this gain?
   (b) If the desired motor speed is 300 rpm, find the (constant) input voltage, $r(t)$, that should be applied.
   (c) It is desired that the motor speed increase linearly (as a ramp) from 0 rpm to 500 rpm in 20 s. The desired speed is then to step back to 200 rpm and remain at that speed. Sketch the required input signal $r(t)$, in volts.
   (d) Give the block diagram of the unity feedback model for this system, such that both the input and output have the units of rpm.

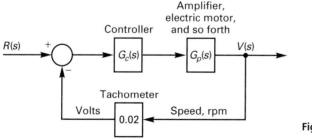

Figure P5.1

5.2. In a player for a compact disc, the laser beam is maintained on the disk track by a position control system. Let the block diagram given in Figure P5.2(a) represent one such control system. The output position can vary ±200 mm, which results in a sensor output-signal variation of ±5 V; that is, the sensor gain is 0.04 V/mm.
   (a) If the desired position is +140 mm, find the (constant) input voltage, $r(t)$, that should be applied.

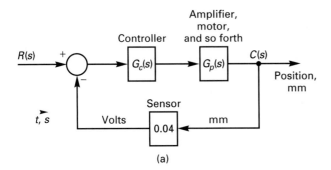

(a)

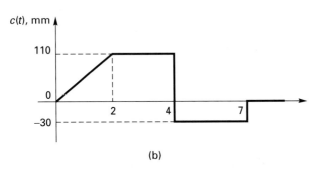

(b)

Figure P5.2

(b) A test signal for the system input is to vary the output position with time as shown in Figure 5.2(b). Sketch the required input signal $r(t)$ in volts.

(c) Give the block diagram of the unity feedback model for this system such that both the input and output are in millimeters.

(d) Complete the diagram for the unity feedback system, and verify the input signal found in (b).

5.3. Consider the control system of Figure P5.3.

(a) Determine if the plant $G_p(s)$ is stable.

(b) In the closed-loop system, velocity (commonly called rate) feedback is employed. The gain of the rate sensor is a. For this part of the problem, let $a = 0$; that is, the rate sensor is removed. Determine if the resulting closed-loop system is stable.

(c) Repeat (b) for rate feedback in the system, with $a = 1$.

(d) Find the range of a for which the system is stable.

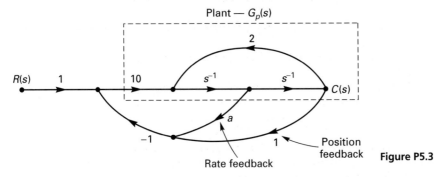

Figure P5.3

5.4. Comment on the stability of a system with the closed-loop transfer function given by each of the following.

(a) $T(s) = \dfrac{5(s+2)}{(s+1)(s^2+s+1)}$

(b) $T(s) = \dfrac{5(-s+2)}{(s+1)(s^2+s+1)}$

(c) $T(s) = \dfrac{5}{(s+1)(s^2+s+1)}$

(d) $T(s) = \dfrac{5}{(s-1)(s^2+3)}$

(e) $T(s) = \dfrac{(s^2+3)}{(s+1)(s^2-s+1)}$

(f) $T(s) = \dfrac{5}{(s+1)(s^2+3)}$

5.5. A closed-loop system has the state equations

$$\dot{\mathbf{x}}(t) = \begin{bmatrix} -1 & 0 & 0 \\ 0 & 2 & 0 \\ -1 & -2 & -3 \end{bmatrix} \mathbf{x}(t) + \begin{bmatrix} 0 \\ 0 \\ 1 \end{bmatrix} u(t)$$

$$y(t) = \begin{bmatrix} 1 & 0 & 0 \end{bmatrix} \mathbf{x}(t)$$

(a) Calculate the system characteristic equation.

(b) Use the results in (a) to determine system stability.

(c) Using MATLAB, verify the results in (b) by two different methods.

5.6. Consider the model of a temperature-control system of Figure P5.6. The nominal value of K is 30, of α is 2, and of β is 1.

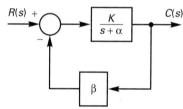

Figure P5.6

(a) Find the sensitivity of the closed-loop transfer function to K, as a function of $s = j\omega$; that is, find $S_K^T(j\omega)$.

(b) Find the sensitivity of the closed-loop transfer function to α, as a function of $s = j\omega$; that is, find $S_\alpha^T(j\omega)$.

(c) Find the sensitivity of the closed-loop transfer function to β, as a function of $s = j\omega$; that is, find $S_\beta^T(j\omega)$.

(d) Calculate the sensitivities of (a), (b), and (c) at dc, that is, at $s = j\omega = j0$.

(e) Compare the three sensitivities by plotting $\left|S_K^T(j\omega)\right|$, $\left|S_\alpha^T(j\omega)\right|$, and $\left|S_\beta^T(j\omega)\right|$.

(f) Give the effect on each sensitivity of increasing K.

(g) Verify the plot of $\left|S_K^T(j\omega)\right|$ using MATLAB.

5.7. Consider the system of Figure P5.7. The parameter α has a nominal value of 10.

(a) With $H = 0$ (that is, the system is open-loop), find the sensitivity of $T(s)$ to α; that is find S_α^T about the nominal value of α. $T(s)$ is given by

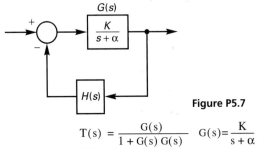

Figure P5.7

$$T(s) = \frac{G(s)}{1 + G(s)\,G(s)} \qquad G(s) = \frac{K}{s + \alpha}$$

(b) Repeat (a) for $H(s) = 1$.

(c) Sketch the magnitude of the sensitivity functions of (a) and (b) as a function of frequency for $K = 1$. Indicate the system's bandwidths.

(d) Repeat (c) for $K = 100$, and note the effects on sensitivity of (i) closed-loop versus open-loop; (ii) high loop-gain versus low loop-gain for the closed-loop system.

(e) Verify the sketches in (c) and (d) using MATLAB.

5.8. A closed-loop system is described by the state equations

$$\dot{\mathbf{x}}(t) = \begin{bmatrix} 0 & 1 \\ -(5+a) & -8 \end{bmatrix} \mathbf{x}(t) + \begin{bmatrix} 0 \\ 1 \end{bmatrix} u(t)$$

$$y(t) = \begin{bmatrix} 6 & 0 \end{bmatrix} \mathbf{x}(t)$$

Find the transfer function $T(s) = Y(s)/U(s)$ to the parameter α.

5.9. Given the system of Figure P5.9. Assume that the system is stable.

(a) For the case that $R(s)$ is zero and $D(s)$ is not zero, find the system characteristic equation.

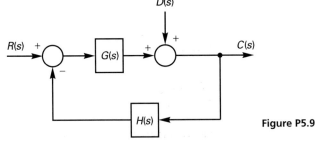

Figure P5.9

(b) For the case that $R(s)$ is not zero and $D(s)$ is zero, find the system characteristic equation.

(c) For the case that both $R(s)$ and $D(s)$ are zero, and the initial conditions on the plant are not zero, find the system characteristic equation.

(d) For the case that $D(s)$ is zero, give the characteristics of $G(s)$ that will cause the steady-state error to be zero for $R(s) = 1/s^2$; that is, for the case that $r(t)$ is a unit ramp function.

(e) For the case that $R(s)$ is zero, give the characteristic of $G(s)$ that will cause the steady-state value of $c(t)$ to be zero, for $D(t) = 1/s$: that is, for the case that the disturbance $d(t)$ is constant.

5.10. Consider the control system of Figure P5.10. It is assumed that the sensor modeled as the gain H_k is perfect; that is, the signal out of H_k is the perfect measurement of $c(t)$. The inaccuracies of the physical sensor are represented by the disturbance $d(t)$, and the sum of the perfect measurement and $d(t)$ is the output of the physical sensor. This is a commonly used model for sensor inaccuracies.

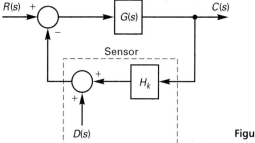

Figure P5.10

(a) Express $C(s)$ as a function of both the system input and the disturbance input.
(b) Assume that the input $r(t)$ is constant. What is the property required of $G(s)$ such that the steady-state gain from $r(t)$ to $c(t)$ is unity? Let $H_k = 1$.
(c) Assume that $G(s)$ has the property found in (b), and $H_k = 1$. Assume that the sensor inaccuracy $d(t)$ is modeled as a constant signal. Find the steady-state gain from $d(t)$ to $c(t)$. We see from this problem why the sensor should be made as accurate as possible.

5.11. In the system of Figure P5.11, let $G_1(s) = 10K/(5s + 1)$, $G_2(s) = 1$, and $H(s) = 1$.

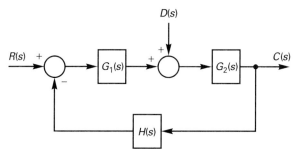

Figure P5.11

(a) Suppose that $d(t) = 5u(t)$. Find the value of K that will limit the steady-state component in $c(t)$ due to $d(t)$ to 1 percent of the value of $d(t)$. Note that this specification limits the dc gain from $D(s)$ to $C(s)$ to the value of 0.01.
(b) If $r(t) = 10u(t)$, find the steady-state error in $c(t)$ due to $r(t)$, for the value of K found in (a). Note that the solutions for (a) and (b) are based on superposition.

5.12. In (a) and (b) of Problem 5.11, the percentage steady-state errors for both the disturbance input signal and the control input signal are equal. Will this always be true? Prove your answer.

5.13. Consider the temperature control system depicted in Figure P5.13. Note the two sensors used to measure the temperature and the rate of change of temperature for feedback. It is shown in Section 9.10 that this connection of sensors realizes a proportional-plus-derivative (PD) compensator. The design of this system involves choosing K to meet certain criteria.
(a) For sensor 2, the minor-loop sensor, express the sensor output $v(t)$ as a function of the output $c(t)$.
(b) Let the disturbance signal $d(t)$ be constant. Design the system (calculate K) such that the steady-state error in $c(t)$ due to $d(t)$ is less than 1 percent of $d(t)$.

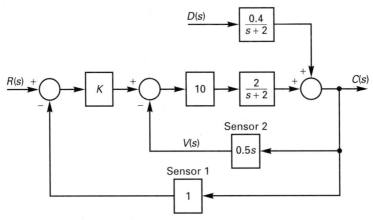

Figure P5.13

(c) Find the system type for this system, with the system error given by $e(t) = r(t) - c(t)$.

(d) What is the effect of sensor 2 on the system steady-state error? The minor loop is added to improve the transient response.

(e) Find the system type if the input signal is a ramp function.

5.14. Consider the system shown in Figure P5.14. For each case given, find the steady-state error for (i) a unit step input; (ii) a unit ramp input. Assume in each case that the closed-loop system is stable.

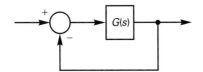

Figure P5.14

(a) $G_a(s) = \dfrac{10}{(s+1)(s+3)}$

(b) $G_b(s) = \dfrac{10}{s(s+1)(s+6)}$

(c) $G_c(s) = \dfrac{7(s+2)}{s^2(s+6)}$

(d) $G_d(s) = \dfrac{6s^2 + 2s + 10}{s(s^2+4)}$

5.15. Consider the dc generator control system of Figure P5.15. Assume that the system is stable for all parts of this problem. The units of both the input and output signals are volts. In each part of this problem, give the units of the error signal.

(a) For $G_c(s) = 1$, find the steady-state error, in the units of the output, for the following inputs. Assume that the inputs given are those to the unity feedback model: (i) input = unit step; (ii) input = unit ramp.

(b) A PI controller is designed with $G_c(s) = 1.0 + 0.1/s$. Repeat (a) for this design.

(c) A PD controller is designed with $G_c(s) = 1.0 + 0.3s$. Repeat (a) for this design.

(d) What is the effect of the integral term in a PI controller on the steady-state error? Why? In (b), the term $0.1/s$ is the integral term.

(e) What is the effect of the derivative term in a PD controller on the steady-state error? Why? In (c), the term $0.3s$ is the derivative term.

(f) In this problem, the unit step input is commanding the system output to what voltage?

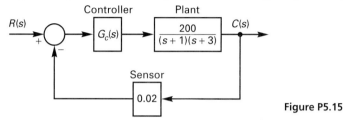

Figure P5.15

5.16. Suppose that in the system of Figure 5.15, the controller is a proportional type; that is, $G_c(s) = K$.

(a) Find the transient-response terms for the case that $K = 0.025$.

(b) Find the transient-response terms for the case that $K = 0.25$.

(c) Find the value of K for which the transient response will have the minimum settling time with no overshoot.

(d) Verify all root calculations with MATLAB.

5.17. Shown in Figure P5.17 is the block diagram of the lateral control system of an automatic aircraft landing system. The output $Y(s)$ is the aircraft lateral position, and the input $Y_c(s)$ is the desired aircraft position. The aircraft position is determined by radar, which is modeled as unity gain with an added noise signal $D_r(s)$. This noise signal represents the inaccuracies of the radar. The signal $D_w(s)$ represents the wind disturbance on the aircraft.

(a) The aircraft is commanded to fly along the extended centerline of the runway during landing. Hence the input $Y_c(s)$ is constant. Assuming that the system is stable, give the characteristic of $G_c(s)G_p(s)$ that will result in the aircraft following the extended centerline exactly in the steady state, provided that all disturbances are zero.

(b) The transfer functions $G_p(s)$ and $G_d(s)$ each have two poles at $s = 0$ in the actual aircraft models. Hence, if the compensator has no poles at $s = 0$, the system is type 2. Suppose that the wind on the aircraft is constant, such that $d_w(t)$ is modeled as a constant signal. A system design criterion requires the steady-state effects of this wind on the aircraft to be zero. What characteristic should $G_c(s)$ have such that this criterion is satisfied?

(c) From (a) and (b), what is the system type for this control system? This system is one of the very few that have a system type this high.

5.18. In the automatic aircraft landing system of Figure P5.17, two disturbance inputs are shown. The disturbance inputs $D_w(s)$ represents wind and $D_r(s)$ represents radar noise (radar inaccuracies).

(a) Write an expression for $Y(s)$ in terms of all inputs.

(b) Find the conditions on the system open-loop function $G_c(s)G_p(s)$ such that the effects of the wind disturbance are small.

(c) Find the conditions on the system open-loop function $G_c(s)G_p(s)$ such that the effects of the radar inaccuracies (noise) are small.

(d) Note that both disturbance effects cannot be made small; that is, the required conditions are conflicting. As an engineer, what would be your approach to solving this problem?

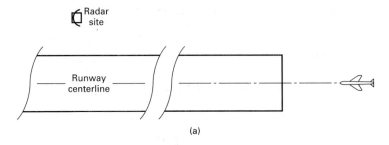

(a)

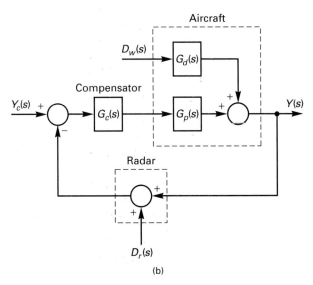

(b)

Figure P5.17

5.19. Consider the closed-loop system transfer functions given. In each case, give the terms that appear in each of the system's natural responses. Underline all factors which indicate that the system is not stable, and encircle all terms which indicate that the system is unstable.

(a) $T_a(s) = \dfrac{10}{(s+1)\,(s-2)}$

(b) $T_b(s) = \dfrac{10}{s\,(s+1)\,(s+6)}$

(c) $T_c(s) = \dfrac{5\,(s+2)}{s^2\,(s+6)}$

(d) $T_d(s) = \dfrac{6s^2 + 2s + 10}{s\,(s^2+3)}$

(e) $T_e(s) = \dfrac{5\,(s+2)}{s\,(s+6)^2}$

(f) $T_f(s) = \dfrac{6s^2 + 2s + 10}{s\,(s^2+3)^2}$

5.20. Consider the closed-loop transfer functions of Problem 5.19.
  (a) Indicate those transfer functions for which the system is marginally stable.
  (b) For those system indicated in (a), give a bounded input signal for which the output signal is not bounded.

5.21. Given the unity feedback system of Figure P5.14 with the transfer function

$$G(s) = \frac{4(s+3)}{s^2 - 2s + 10}$$

  (a) Consider the open-loop system, that is, the system with the feedback path removed. Find the characteristic equation of the open-loop system, and determine its roots. What do you conclude about the stability of this system?
  (b) Consider now the closed-loop system. Find the characteristic equation of the closed-loop system, and determine its roots. What do you conclude about the stability of this system?
  (c) Repeat (a) and (b) with the transfer function

$$G(s) = \frac{4(s-3)}{s^2 + 2s + 10}$$

  Note that one system is open-loop unstable and closed-loop stable, while the other is open-loop stable and closed-loop unstable.
  (d) Find the step responses, using SIMULINK, to verify the results on the closed-loop systems.

5.22. (a) Consider the closed-loop system of Figure P5.14, with the transfer function given by $G(s) = 10/(s^2 + 4s - 6)$. Suppose that, in addition, a sensor in the feedback has the gain H_k. Write the system characteristic equation, and determine the range of the sensor gain H_k for which the system is stable.
  (b) Set H_k to the lower limit found in (a). List the terms that appear in the system response.
  (c) Verify the unstable term in (b) using SIMULINK.

5.23. Consider the steady-state error resulting from a disturbance input, given in (5-48). Suppose that the transfer functions of Figure 5.16 are expressed as

$$G_c(s) = \frac{1}{s^{N_c}} G_{c1}(s) \qquad G_p(s) = \frac{1}{s^N} G_{p1}(s) \qquad G_d(s) = \frac{1}{s^N} G_{d1}(s)$$

  That is, the compensator $G_c(s)$ is of type N_c, and $G_p(s)$ and $G_d(s)$ are each of type N.
  (a) A design criterion requires that the steady-state error for $D(s) = 1/s$, a unit step, be zero. What can you determine about N_c and N such that this criterion is satisfied?
  (b) Repeat (a) if the design criterion requires that the steady-state error for $D(s) = 1/s^2$, a unit ramp, be zero.

6

Stability Analysis

In this chapter we investigate the important topic of determining the stability of a linear time-invariant (LTI) analog system. As was stated in Chapter 5, the bounded-input bounded-output (BIBO) definition of stability is used when investigating the stability of this type of system. The BIBO definition of stability is

> A system is BIBO stable if, for *every* bounded input, the output remains bounded with increasing time.

For a linear time-invariant system, this definition requires that all poles of the closed-loop transfer function (all roots of the system characteristic equation) lie in the left half of the complex plane. This property was proved in Section 5.2. Hence any stability analysis requires determining if the closed-loop transfer function has any poles either on the imaginary axis or in the right half of the s-plane. Recall that a system is *marginally stable* if all poles are in the left half of the complex plane, except for simple poles on the imaginary axis.

The first method of stability analysis to be covered, the Routh–Hurwitz criterion, determines if any roots of a polynomial lie outside the left half of the complex plane. However, this method does not find the exact locations of the roots. The second procedure covered involves the calculation of the exact location of the roots. For first- and second-order systems, these roots can be found analytically. For higher-order systems a digital computer program should be used. The third procedure covered is simulation. This procedure applies to all systems; in addition, it is the only general procedure available for complex nonlinear systems. Stability of LTI systems can also be determined by the root-locus technique, presented in Chapter 7, and by the Nyquist criterion, developed in Chapter 8.

All these methods are based on the (inexact) system model. The *only* method to determine with certainty the stability of a physical system is to operate the system. Then the stability is determined only for those operating conditions tested.

Before the stability analysis techniques are presented, we consider some general properties of polynomials that will prove to be useful. We *always* assume that all polynomial coefficients are real. Consider first the second-order polynomial

$$Q_2(s) = s^2 + a_1 s + a_0 = (s - p_1)(s - p_2) = s^2 - (p_1 + p_2)s + p_1 p_2 \qquad (6\text{-}1)$$

and the third-order polynomial

$$Q_3(s) = s^3 + a_2 s^2 + a_1 s + a_0 = (s - p_1)(s - p_2)(s - p_3)$$

$$= [s^2 - (p_1 + p_2)s + p_1 p_2](s - p_3) \qquad (6\text{-}2)$$

$$= s^3 - (p_1 + p_2 + p_3)s^2 + (p_1 p_2 + p_1 p_3 + p_2 p_3)s - p_1 p_2 p_3$$

Extending this expansion to the nth-order polynomial

$$Q_n(s) = s^n + a_{n-1}s^{n-1} + \cdots + a_1 s + a_0 \qquad (6\text{-}3)$$

we see that the coefficients are given by

a_{n-1} = negative of the sum of all roots

a_{n-2} = sum of the products of all possible combinations of roots taken two at a time

a_{n-3} = negative of the sum of the products of all possible combinations of roots taken three at a time

$$\vdots$$

$a_0 = (-1)^n$ multiplied by the product of all the roots

Suppose first that all roots of a polynomial are real and in the left half-plane. Then all p_i in the examples of (6-1) and (6-2) are real and negative. Therefore, all polynomial coefficients are positive; this characteristic also applies to the general case of (6-3). The only case for which a coefficient can be negative is that there be at least one root in the right half-plane. Note also that if all roots are in the left half-plane, no coefficient can be zero.

If any roots of the polynomials above are complex, the roots must appear in complex-conjugate pairs, since the polynomial coefficients are assumed real. Then, in the rules given for forming the polynomial coefficients, all imaginary parts of the products will cancel. Therefore, if all roots occur in the left half-plane, all coefficients of the general polynomial of (6-3) will be positive. Conversely, if not all coefficients of (6-3) are positive, the polynomial will have at least one root that is not in the left half-plane (that is, on the $j\omega$ axis or in the right half-plane).

In summary, given a polynomial as in (6-3),

1. If any coefficient a_i is equal to zero, then *not all* the roots are in the left half-plane.

2. If any coefficient a_i is negative, then *at least one root* is in the right half-plane.

The converse of rule 2 is not true; if all coefficients of a polynomial are positive, the roots are not necessarily confined to the left half-plane. We now illustrate this point with an example.

Example 6.1

For the polynomial

$$Q(s) = (s + 2)(s^2 - s + 4) = s^3 + s^2 + 2s + 8$$

all coefficients are positive. The roots are at the locations

$$-2, \qquad \frac{1}{2} \pm \frac{j\sqrt{15}}{2}$$

and we see that two of the three roots are in the right half-plane. The roots of the polynomial can be verified by the MATLAB program

```
p = [1 1 2 8];
roots(p)
```

6.1 ROUTH–HURWITZ STABILITY CRITERION

The Routh–Hurwitz criterion is an analytical procedure for determining if all roots of a polynomial have negative real parts and is used in the stability analysis of linear time-invariant systems. The criterion gives the number of roots with positive real parts, and applies to all LTI systems for which the characteristic equation is a polynomial set to zero. This requirement excludes a system that contains an ideal time delay (transport lag). For this special case, which is covered later, the Routh–Hurwitz criterion cannot be employed.

The Routh–Hurwitz criterion applies to a polynomial of the form

$$Q(s) = a_n s^n + a_{n-1} s^{n-1} + \cdots + a_1 s + a_0 \tag{6-4}$$

where we can assume with no loss of generality that $a_0 \neq 0$. Otherwise, the polynomial can be expressed as a power of s multiplied by a polynomial in which $a_0 \neq 0$. The power of s indicates roots at the origin, the number of which is evident; hence, only the latter polynomial need be investigated using the Routh–Hurwitz criterion. We assume in the following developments that a_0 is not zero.

The first step in the application of the Routh–Hurwitz criterion is to form the array below, called the Routh array, where the first two rows are the coefficients of the polynomial in (6-4).

$$
\begin{array}{c|ccccc}
s^n & a_n & a_{n-2} & a_{n-4} & a_{n-6} & \cdots \\
s^{n-1} & a_{n-1} & a_{n-3} & a_{n-5} & a_{n-7} & \cdots \\
s^{n-2} & b_1 & b_2 & b_3 & b_4 & \cdots \\
s^{n-3} & c_1 & c_2 & c_3 & c_4 & \cdots \\
\vdots & \vdots & \vdots & & & \\
s^2 & k_1 & k_2 & & & \\
s^1 & l_1 & & & & \\
s^0 & m_1 & & & &
\end{array}
$$

The column with the powers of s is included as a convenient accounting method. The b row is calculated from the two rows directly above it, the c row, from the two rows directly above it, and so on. The equations for the coefficients of the array are as follows:

$$b_1 = -\frac{1}{a_{n-1}} \begin{vmatrix} a_n & a_{n-2} \\ a_{n-1} & a_{n-3} \end{vmatrix} \qquad b_2 = -\frac{1}{a_{n-1}} \begin{vmatrix} a_n & a_{n-4} \\ a_{n-1} & a_{n-5} \end{vmatrix}, \ldots$$

$$c_1 = -\frac{1}{b_1} \begin{vmatrix} a_{n-1} & a_{n-3} \\ b_1 & b_2 \end{vmatrix} \qquad c_2 = -\frac{1}{b_1} \begin{vmatrix} a_{n-1} & a_{n-5} \\ b_1 & b_3 \end{vmatrix}, \ldots$$

$$(6\text{-}5)$$

and so on. Note that the determinant in the expression for the ith coefficient in a row is formed from the first column and the $(i + 1)$ column of the two preceding rows.

As an example, the Routh array for a fourth-order polynomial (which has five coefficients) is of the form

$$
\begin{array}{c|ccc}
s^4 & x & x & x \\
s^3 & x & x \\
s^2 & x & x \\
s^1 & x \\
s^0 & x
\end{array}
$$

where each table entry is represented by the symbol x. Hence, for a general array, the final two rows of the array will have one element each, the next two rows above two elements each, the next two above three elements each, and so forth.

The Routh–Hurwitz criterion may now be stated as follows:

> With the Routh array calculated as previously defined, the number of polynomial roots in the right half-plane is equal to the number of sign changes in the first column of the array.

An example is now given.

Example 6.2

Consider again the polynomial given in Example 6.1:

$$Q(s) = (s + 2)(s^2 - s + 4) = s^3 + s^2 + 2s + 8$$

The Routh array is

$$
\begin{array}{c|cc}
s^3 & 1 & 2 \\
s^2 & 1 & 8 \\
s^1 & -6 \\
s^0 & 8
\end{array}
$$

where

$$b_1 = -\frac{1}{1} \begin{vmatrix} 1 & 2 \\ 1 & 8 \end{vmatrix} = -6 \qquad c_1 = -\frac{1}{-6} \begin{vmatrix} 1 & 8 \\ -6 & 0 \end{vmatrix} = 8$$

Since there are two sign changes on the first column (from 1 to −6 and from −6 to 8), there are two roots of the polynomial in the right half-plane. This was shown to be the case in Example 6.1.

This example illustrates the application of the Routh–Hurwitz criterion. Note, however, that only the stability of a system is determined. Since the locations of the roots of the characteristic equation are not found, no information about the transient response of a stable system is derived from Routh–Hurwitz criterion. Also, the criterion gives no information about the steady-state response. We obviously need analysis techniques in addition to the Routh–Hurwitz criterion. These techniques are presented in Chapters 7 and 8.

From the equations for the calculation of the elements of the Routh array, (6-5), we see that the array cannot be completed if the first element in a row is zero. For this case, the calculation of the elements of the following row requires a division by zero. Because of this possibility, we divided the application of the criterion into three cases.

6.1.1 Case 1

This case is the one just discussed. For this case none of the elements in the first column of the Routh array is zero, and no problems occur in the calculation of the array. We do not discuss this case further.

6.1.2 Case 2

For this case, the first element in a row is zero, with at least one nonzero element in the same row. This problem can be solved by replacing the first element of the row, which is zero, with a small number ε, which can be assumed to be either positive or negative. The calculation of the array is then continued, and some of the elements that follow that row will be a function of ε. After the array is completed, the signs of the elements in the first column are determined by allowing ε to approach zero. The number of roots of the polynomial in the right half-plane is then equal to the number of sign changes in this first column, as before. An example illustrates this case.

Example 6.3

As an example of case 2, consider the polynomial

$$Q(s) = s^5 + 2s^4 + 2s^3 + 4s^2 + 11s + 10$$

The Routh array is calculated to be

$$
\begin{array}{c|ccc}
s^5 & 1 & 2 & 11 \\
s^4 & 2 & 4 & 10 \\
s^3 & \phi^\varepsilon & 6 & \\
s^2 & -\dfrac{12}{\varepsilon} & 10 & \\
s^1 & 6 & & \\
s^0 & 10 & &
\end{array}
$$

where

$$b_1 = -\frac{1}{2}\begin{vmatrix} 1 & 2 \\ 2 & 4 \end{vmatrix} = 0 \qquad\qquad b_2 = -\frac{1}{2}\begin{vmatrix} 1 & 11 \\ 2 & 10 \end{vmatrix} = 6$$

$$c_1 = -\frac{1}{\varepsilon}\begin{vmatrix} 2 & 4 \\ \varepsilon & 6 \end{vmatrix} = \frac{1}{\varepsilon}(12 - 4\varepsilon) = -\frac{12}{\varepsilon} \qquad c_2 = -\frac{1}{\varepsilon}\begin{vmatrix} 2 & 10 \\ \varepsilon & 0 \end{vmatrix} = 10 =$$

$$d_1 = \frac{\varepsilon}{12}\begin{vmatrix} \varepsilon & 6 \\ \frac{-12}{\varepsilon} & 10 \end{vmatrix} = \frac{\varepsilon}{12}\left[10\varepsilon + 6\left(\frac{12}{\varepsilon}\right)\right] = 6$$

$$e_1 = -\frac{1}{6}\begin{vmatrix} \frac{-12}{\varepsilon} & 10 \\ 6 & 0 \end{vmatrix} = 10$$

In the preceding calculations, the limits were taken as $\varepsilon \to 0$ at convenient points in the calculations rather than waiting until the array was complete. This procedure simplifies the calculations and the final form of the array, and the final result is the same. From the array we see that there are two sign changes in the first column whether ε is assumed positive or negative. The number of sign changes in the first column is always independent of the assumed sign of ε, which leads to the conclusion that the system that falls under case 2 is always unstable. The results in this example are verified by the MATLAB program

```
p = [1 2 2 4 11 10];
roots(p)

results:    r    0.895 ± 1.4561i    -1.2407 ± 1.0375i    -1.3087
```

As an additional point, note in the preceding example that the final element in each row designated by an even power of s is the same, which in that example is 10. It is easily shown that this is always the case (see Problem 6.15).

We make one further point concerning case 2. This point is illustrated by a second-order polynomial. If we replace s with $1/x$ in the polynomial in (6-1), then

$$Q(1/x) = Q(s)|_{s = 1/x} = \left(\frac{1}{x} - p_1\right)\left(\frac{1}{x} - p_2\right) = p_1 p_2 x^{-2}\left(x - \frac{1}{p_1}\right)\left(x - \frac{1}{p_2}\right)$$

and the roots of $Q(1/x)$ are the reciprocals of the roots of $Q(s)$. Hence the roots do not change half-planes; left half-plane roots remain in the left half-plane, and right half-plane roots remain in the right half-plane. In addition, this substitution results in the polynomial coefficients being reversed in order.

$$Q(1/x) = Q(s)|_{s = 1/x} = \frac{1}{x^2} + \frac{a_1}{x} + a_0 = x^{-2}\left(a_0 x^2 + a_1 x + 1\right)$$

Generally, if $Q(s)$ is case 2, $Q(1/x)$ will be case 1. Since the stability of $Q(1/x)$ is the same as that of $Q(s)$, then the Routh–Hurwitz criterion can be applied to $Q(1/x)$. Applying the Routh–Hurwitz criterion to either $Q(s)$ or $Q(1/x)$ will give the stability of a case 2 polynomial. Solving Example 6.3 using this procedure is given as Problem 6.4.

6.1.3 Case 3

A case 3 polynomial is one for which all elements in a row of the Routh array are zero. The method described for case 2 does not give useful information in this case. A simple example illustrates case 3. Let

$$Q(s) = s^2 + 1$$

For this system the roots of the characteristic equation are on the imaginary axis, and thus the system is marginally stable. The Routh array is then

$$
\begin{array}{c|cc}
s^2 & 1 & 1 \\
s^1 & 0 & \\
s^0 & &
\end{array}
$$

and the s^1 row has no nonzero elements. The array cannot be completed because of the zero element in the first column. A second example is

$$Q(s) - (s+1)(s^2 + 2) = s^3 + s^2 + 2s + 2$$

The Routh array is

$$
\begin{array}{c|cc}
s^3 & 1 & 2 \\
s^2 & 1 & 2 \\
s^1 & 0 & \\
s^0 & &
\end{array}
$$

Again the s^1 row is zero, and the array is prematurely terminated.

A case 3 polynomial contains an even polynomial as a factor. An even polynomial is one in which the exponents of s are even integers or zero only. This even-polynomial factor is called the *auxiliary polynomial* and is evident in each of the examples above. The coefficients of the auxiliary polynomial will always be the elements in the row directly above the row of zeros in the array. The exponent of the highest power in the auxiliary polynomial is the exponent that denotes the row containing its coefficients. In the first example, the row directly above the row of zeros, the s^2 row, contains the elements $(1, 1)$. Hence the auxiliary polynomial is

$$Q_a(s) = s^2 + 1$$

For the second example, the s^1 row is all zeros and the s^2 row contains the coefficients $(1, 2)$. Thus the auxiliary equation is

$$Q_a(s) = s^2 + 2$$

The case 3 polynomial may be analyzed in either of two ways. First, once the auxiliary polynomial is found, it may be factored from the characteristic equation, leaving a second polynomial. The two polynomials may then be analyzed separately. However, difficulties may be encountered in applying the Routh–Hurwitz criterion to an auxiliary polynomial of high order; the second example below illustrates this case. If the auxiliary polynomial is of low order, it may be factored algebraically to give the roots.

The second method for analyzing case 3 polynomials is used in this book. Suppose that the row of zeros in the array is the s^i row. Then the auxiliary polynomial is differentiated with respect to s, with the coefficients of the resulting polynomial used to replace the zeros in the s^i row. The calculation of the array then continues as in the case 1 polynomial. First an example is given to illustrate the construction of the array; then the interpretation of the array is discussed.

Example 6.4

Consider the polynomial

$$Q(s) = s^4 + s^3 + 3s^2 + 2s + 2$$

The Routh array is then

$$
\begin{array}{c|ccc}
s^4 & 1 & 3 & 2 \\
s^3 & 1 & 2 & b_1 = -(2-3) = 1 \\
s^2 & 1 & 2 & b_2 = -(0-2) = 2 \\
s^1 & \phi^2 & & c_1 = -(2-2) = 0 \\
s^0 & 2 & & d_1 = -\left(\dfrac{1}{2}\right)(0-4) = 2
\end{array}
$$

Since the s^1 row contains no nonzero elements, the auxiliary polynomial is obtained from the s^2 row and is given by

$$Q_a(s) = s^2 + 2$$

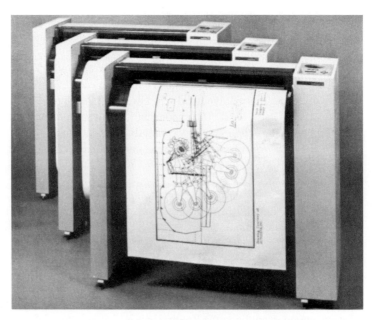

Plotters such as these employ two closed-loop control systems to (1) position the pen in the x-direction and (2) position the paper in the y-direction. (Courtesy of Hewlett-Packard Company.)

Then

$$\frac{dQ_a(s)}{ds} = 2s$$

The coefficient 2 replaces the zero in the s^1 row, and the Routh array is then completed.

The preceding example illustrates completing the array by using the derivative of the auxiliary polynomial. The array itself is interpreted in the usual way; that is, the polynomial in the example has no roots in the right half-plane. However, investigation of the auxiliary polynomial shows that there are roots on the imaginary axis, and thus the system is marginally stable.

The roots of an even polynomial occur in pairs that are equal in magnitude and opposite in sign. Hence, these roots can be purely imaginary, as shown in Figure 6.1(a), purely real, as shown in Figure 6.1(b), or complex, as shown in Figure 6.1(c). Since complex roots must occur in conjugate pairs, any complex roots of an even polynomial must occur in groups of four, which is apparent in Figure 6.1(c). Such roots have quadrantal symmetry; that is, the roots are symmetrical with respect to both the real and imaginary axes. For Figure 6.1(b) and 6.1(c), the Routh array will indicate the presence of roots with positive real parts. If a zero row occurs but the completed Routh array shows no sign changes, roots on the $j\omega$-axis are indicated. Thus, in any case, the presence of an auxiliary polynomial (a row of zeros in the array) indicates a *nonstable* system. An additional example is now given to illustrate further problems with cases 2 and 3.

Example 6.5

Consider the polynomial in Figure 6.1(c):

$$Q(s) = s^4 + 4$$

Recall from Chapter 2 that physical systems generally do not have characteristic polynomials such as this one, but this polynomial does present a good exercise. The Routh array begins with the two rows

$$\begin{array}{c|ccc} s^4 & 1 & 0 & 4 \\ s^3 & 0 & 0 \end{array}$$

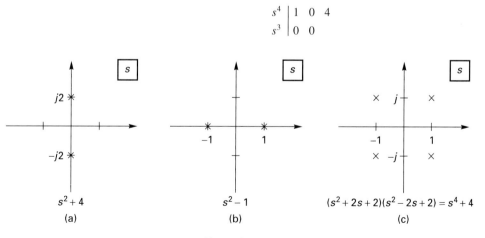

Figure 6.1   Some even polynomials.

and we immediately have a row of zeros. The auxiliary polynomial and its derivative are

$$Q_a(s) = s^4 + 4 \qquad \frac{dQ_a}{ds} = 4s^3$$

Hence the array becomes

$$
\begin{array}{c|l}
s^4 & 1 \quad 0 \quad 4 \\[4pt]
s^3 & \phi^4 \; 0 \qquad b_1 = -\left(\frac{1}{4}\right)(0 - 0) = 0 \\[6pt]
s^2 & \phi^\varepsilon \; 4 \qquad b_2 = -\left(\frac{1}{4}\right)(0 - 16) = 4 \\[6pt]
s^1 & \dfrac{-16}{\varepsilon} \qquad c_1 = -\left(\frac{1}{\varepsilon}\right)(16 - 0) = \dfrac{-16}{\varepsilon} \\[8pt]
s^0 & 4 \qquad d_1 = -\left(\dfrac{-\varepsilon}{16}\right)\left[0 + 4\left(\dfrac{16}{\varepsilon}\right)\right] = 4
\end{array}
$$

The s^2 row has a nonzero element with zero for the first element; the zero is replaced with the small number ε. The array has two sign changes in the first column, indicating two roots with positive real parts. This result agrees with Figure 6.1(c).

  This polynomial is seen to be both case 2 and case 3. The row of zeros in the array indicates the possibility of roots on the $j\omega$-axis. In this example we know that this is not the case. In general it is necessary to factor the auxiliary equation to determine the presence of imaginary roots.

  As a final point for case 3, note that if the given system is marginally stable, the auxiliary polynomial will give the frequency of oscillation of the system. This is seen from Figure 6.1(a), where the roots of the auxiliary polynomial are those poles of the closed-loop transfer function that are located on the $j\omega$-axis.

  Thus far we have used the Routh–Hurwitz criterion only to determine the stability of systems. This criterion can also be used to aid in the *design* of control systems, as shown in the following two examples.

Example 6.6

  The Routh–Hurwitz criterion will be utilized to perform a simple design for the control system shown in Figure 6.2, in which a proportional compensator is employed. The system is type 0 (see Section 5.5), and the steady-state error for a constant input of unity is, from (5-35),

$$e_{ss} = \frac{1}{1 + K_p} = \frac{1}{1 + K}$$

where the position error constant K_p is, for this example,

$$K_p = \lim_{s \to 0} G_c(s) G_p(s) = \lim_{s \to 0} \frac{2K}{s^3 + 4s^2 + 5s + 2} = K$$

Suppose that a design specification is that e_{ss} must be less than 2 percent of a constant input. Thus

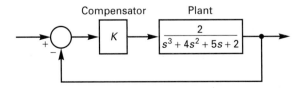

Figure 6.2   System for Example 6.6.

$$e_{ss} = \frac{1}{1 + K_p} = \frac{1}{1 + K} < \frac{1}{50}$$

or, K must be greater than 49. The calculation of steady-state errors is based on the assumption of stability; thus we must ensure that the system is stable for the range of K required. The system characteristic equation is given by

$$1 + G_c(s)G_p(s) = 1 + \frac{2K}{s^3 + 4s^2 + 5s + 2} = 0$$

or

$$Q(s) = s^3 + 4s^2 + 5s + 2 + 2K = 0$$

The Routh array for this polynomial is then

s^3	1	5	$b_1 = -\frac{1}{4}(2 + 2K - 20)$
s^2	4	$2 + 2K$	
s^1	$\dfrac{18 - 2K}{4}$	$\Rightarrow$	$K < 9$
s^0	$2 + 2K$	$\Rightarrow$	$K > -1$

Thus the system is stable only for the compensator gain K greater than -1 but less than 9. The steady-state error criterion cannot be met with the proportional compensator; it will be necessary to use a dynamic compensation such that $G_c(s)$ is a function of s and not simply a pure gain. Of course, a PI compensator with

$$G_c(s) = K_P + \frac{K_I}{s}$$

will make the system type 1, since $G_c(s)G_p(s)$ has a pole at the origin. Hence the steady-state error for a constant input is then zero, *provided* that the compensated system is stable.

Example 6.7

The design in Example 6.6 is continued in this example. We replace the gain K with a PI compensator with the transfer function

$$G_c(s) = K_P + \frac{K_I}{s} = \frac{K_P s + K_I}{s} = 0$$

The system characteristic equation is then

$$1 + G_c(s)G_p(s) = 1 + \frac{2(K_P s + K_I)}{s(s^3 + 4s^2 + 5s + 2)} = 0$$

or

$$Q(s) = s^4 + 4s^3 + 5s^2 + (2 + 2K_P)s + 2K_I = 0$$

The Routh array for this polynomial is then

s^4	1	5	$2K_I$	
s^3	4	$2 + 2K_P$		
s^2	$\dfrac{18 - 2K_P}{4}$	$2K_I$	$\Rightarrow$	$K_P < 9$
s^1	c_1			
s^0	$2K_I$		$\Rightarrow$	$K_I > 0$

where

$$c_1 = \frac{-4}{18 - 2K_P}\left[8K_I - \frac{(2 + 2K_P)(18 - 2K_P)}{4}\right]$$

$$= \frac{4}{18 - 2K_P}[(1 + K_P)(9 - K_P) - 8K_I]$$

From the s^2 row, we see that $K_P < 9$ is required for stability. To simplify the design, we choose $K_P = 3$ and solve for the range of K_I for stability. The element c_1 is then, for $K_P = 3$,

$$c_1 = \frac{24 - 8K_I}{3}$$

Hence, for stability, with $K_P = 3$, the integrator gain K_I must be less than 3.0. Also, from the s^0 row, K_I must be greater than zero. Of course, if the value of K_P is chosen to be different, the range of K_I for stability will also be different.

This design has considered only stability. Usually specifications in addition to stability are given in the design of a control system, and the final values of the gains K_P and K_I are determined by all the specifications.

Example 6.8

This example is a continuation of the preceding example. If we choose the design parameters as $Kp = 3$ and $K_I = 3$, the system is marginally stable. To show this, the Routh array is, from Example 6.7,

$$\begin{array}{c|ccc}
s^4 & 1 & 5 & 6 \\
s^3 & 4 & 8 & \\
s^2 & 3 & 6 & \\
s^1 & \varnothing 6 & & \\
s^0 & 6 & &
\end{array}$$

The auxiliary polynomial is then

$$Q_a(s) = 3s^2 + 6 = 3(s^2 + 2) = 3(s + j\sqrt{2})(s - j\sqrt{2})$$

This system should oscillate with the frequency $\omega = 1.14$ rad/s, and a period of $T = 2\pi/\omega = 4.44$ s, which is confirmed with the SIMULINK simulation depicted in Figure 6.3.

A practical use of the Routh–Hurwitz criterion is illustrated in the last three examples. The Routh–Hurwitz criterion allows us to determine stability as functions of the *parameters* of a system.

In summary, the three cases that occur in the application of the Routh–Hurwitz criterion are as follows:

Case 1. No elements in the first column are zero. There are no problems in completing the array.

Case 2. There is at least one nonzero element in a row, with the first element equal to zero. This always indicates an unstable system. The first element (which is zero) is replaced with the value ε, $|\varepsilon| \ll 1$, and the calculation of the array continues.

Case 3. All elements in a row are zero. This always indicates a system that is not stable, but it may be marginally stable. This case can be analyzed through the use of the auxiliary equation, as described earlier.

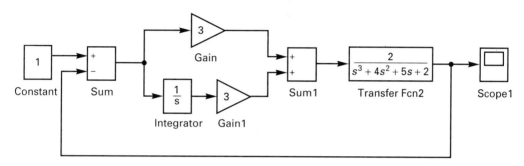

Figure 6.3   SIMULINK simulation for Example 6.8.

6.2 ROOTS OF THE CHARACTERISTIC EQUATION

The Routh–Hurwitz criterion was presented in Section 6.1. This criterion allows us to determine the stability of a linear time-invariant system without finding the roots of the characteristic polynomial. The stability of a system can also be determined by finding the root locations of the characteristic polynomial. For a first-order polynomial, this is trivial. For a second-order polynomial the well-known equations are available. Algebraic methods are available for third-order polynomials but are somewhat difficult to apply. For polynomials of order higher than two, numerical procedures are usually required. These procedures are well suited to the digital computer, and computer programs such as MATLAB are available that can be used to find the roots of a polynomial (see Example 6.1). One *disadvantage* of using numerical techniques is that all parameters in the system must be assigned numerical values. Finding the range of a certain parameter that results in a stable system becomes more difficult when a numerical technique is used to determine the stability. As we saw in Section 6.1, the Routh–Hurwitz criterion is well suited to finding the range of a parameter for stability.

For a first-order system with the characteristic equation

$$Q(s) = a_1 s + a_0 \tag{6-6}$$

both coefficients must be of the same sign for stability. For a second-order system,

$$Q(s) = a_2 s^2 + a_1 s + a_0 \tag{6-7}$$

the Routh array is

$$
\begin{array}{c|cc}
s^2 & a_2 & a_0 \\
s^1 & a_1 \\
s^0 & a_0
\end{array}
$$

Hence, as in the first-order case, the necessary and sufficient condition for stability is that all coefficients in the characteristic equation must be of the same sign. For systems of order higher than two, it is necessary but not sufficient that the coefficients of the characteristic equation be of the same sign. Thus for first- and second-order systems, stability is obvious without solving for the roots of the characteristic polynomial or applying the Routh–Hurwitz criterion.

6.3 STABILITY BY SIMULATION

An obvious method for determining the stability of an LTI system is by simulation. First a simulation of the system, either digital or analog, is constructed. Then the output of the system is observed for typical inputs or simply for an initial-condition excitation. The stability of the system under the conditions simulated will be evident. Often, at some time in the design process, a simulation of the control system is constructed. If the physical system contains either time-varying parameters or significant nonlinearities, a simulation may be the only method available for observing the effects of these phenomena. There are no general methods for analyzing complex nonlinear and/or time-varying systems. Hence accurate simulations of the systems are the only procedures available for determining the characteristics of many of these systems, without actually operating the physical systems.

Numerical integration and digital simulations were introduced in Section 3.6. Digital computer analysis and design software was discussed in Section 3.7. The MATLAB simulation program SIMULINK was demonstrated in some examples.

SIMULINK automatically chooses the numerical increment for numerical integration; however, numerical problems can still occur with an integration increment that is too large. Inaccurate simulations can also occur if an error is made in entering the system model into the computer. One procedure for verifying the system model in a simulation is to employ the Routh–Hurwitz criterion to determine both the gain that results in a marginally stable system, and the frequency of oscillation of this marginally stable system as in Example 6.8. If a simulation of the marginally stable system does not result in a steady-state sinusoidal response at the correct frequency, the simulation is not accurate. The general problem of verifying system simulations is considered further in Chapter 9, after the topic of frequency response is considered.

An introduction to analog-computer simulation was also given in Chapter 3. Before the general availability of fast digital computers, analog-computer simulation was the principal method of simulation. This method has declined in importance over the past few years; however it is still useful in certain situations, as discussed in Chapter 3.

In summary, determining stability by simulation is probably the most common method of determining the stability of systems, other than actually operating the physical systems. This procedure is absolutely necessary for physical systems, such as those used in aerospace applications, that cannot be operated safely in an unstable condition.

6.4 SUMMARY

This chapter was concerned with the very important concept of system stability. For a linear time-invariant system the bounded-input, bounded-output (BIBO) definition of stability was defined. The Routh–Hurwitz criterion was then given as a procedure for testing the BIBO stability of a linear time-invariant system whose characteristic equation is a polynomial in the Laplace variable s. One type of linear time-invariant system does not satisfy the polynomial requirement—a system with ideal time delay. The stability analysis of this type of system is presented in Chapter 8.

Physical systems are generally not linear time-invariant systems. We must consider the nonlinearities of a physical system if we want to develop more-accurate models. It is usually necessary to employ simulations of a system to determine accurately system characteristics, including stability. Hence a brief introduction to stability by simulation concluded this chapter. Simulation of nonlinear systems in covered in Chapter 14.

PROBLEMS

6.1. What conclusions can you reach concerning the root locations of the following polynomials without applying the Routh–Hurwitz criterion or solving for the roots?

(a) $s^3 + 2s^2 + 3s + 1$

(b) $-s^3 - 2s^2 - 3s - 1$

(c) $s^3 + 2s^2 + 3s - 1$

(d) $s^4 + s^3 + s + 2$

(e) $s^4 + s^2 + 1$

(f) $s^4 - 1$

6.2. (a) Apply the Routh–Hurwitz criterion to each of the polynomials of Problem 6.1 to verify your conclusions.

(b) Verify all results using MATLAB to find the root locations.

6.3. Use the Routh–Hurwitz criterion to determine the number of roots in the left half-plane, in the right half-plane, and on the imaginary axis for the given characteristic equations.

(a) $s^5 + 2s^4 + 5s^3 + 4s^2 + 6s = 0$

(b) $s^4 + s^3 + s + 0.5 = 0$

(c) $s^4 + s^3 + 5s^2 + 5s + 2 = 0$

(d) $s^4 + 2s^3 + 3s^2 + 2s + 5 = 0$

(e) $s^4 + s^3 + 5s^2 + 2s + 4 = 0$

(f) $s^4 + 2s^2 + 1 = 0$

(g) Verify all results using MATLAB to find the root locations.

6.4. Consider the polynomial of Example 6.3:

$$Q(s) = s^5 + 2s^4 + 2s^3 + 4s^2 + 11s + 10$$

(a) Apply the Routh–Hurwitz criterion to $Q(1/x)$, where $Q(1/x) = Q(s)\,|_{s=1/x}$, and compare the results with those of Example 6.3.

(b) Verify your results with MATLAB by finding the root locations of both $Q(s)$ and $Q(1/x)$.

6.5. Consider the control system of Figure P6.5. Note the PI compensator and the rate sensor (sensor 2).

(a) What is the low-frequency ($s = j\omega \simeq j0$) gain of the position sensor (sensor 1)?

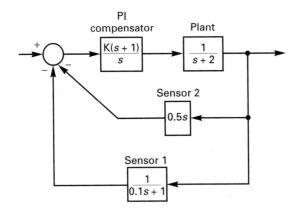

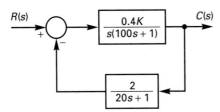

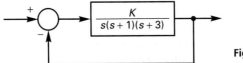

Figure P6.5

(b) What is the low-frequency ($s = j\omega \simeq j0$) gain of the rate sensor (sensor 2)?

(c) From (a) and (b), we see that at low frequencies the system is a unity feedback system. What is the type number of this system?

(d) From (c), we see that the steady-state error for a constant input is zero, *provided that the system is stable*. For what range of K is the system stable?

6.6. Given the system of Figure P6.6.

Figure P6.6

(a) Find the value of K that will result in the system being marginally stable.

(b) With K equal to the value found in (a), find the frequency, ω_0, of oscillation. Also, find the period $T = 2\pi/\omega_0$.

(c) In (b), why is the period of the sinusoid so large?

(d) Use SIMULINK with a step input to verify the period in (b).

6.7. Use the system of Figure P6.7.

Figure P6.7

(a) Find the steady-state error for a constant input of unity, assuming that the system is stable.

(b) Find the range of K for which the result of (a) applies.

(c) Determine the value of K ($K \neq 0$) that will result in the system being marginally stable.

(d) Find the location of all roots of the system characteristic equation for that value of K found in (c).

(e) Verify the value of K in (c) and the frequency of oscillation in (d) by simulation.

6.8. Consider the system of Figure P6.8.

(a) Determine the range of K for stability. Consider both positive and negative values of K.

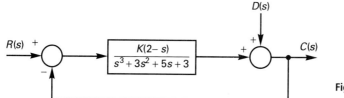

Figure P6.8

(b) Suppose *just for the purposes of this question,* that the system accurately models a ship-steering system, with the closed-loop system input the wheel position controlled by the person at the helm and the output the position of the ship. The disturbance $D(s)$ models the effects of ocean currents on the ship's position. In testing the physical system, the value of gain K is set somewhat lower than the *lower* limit found in (a) for a short period of time. Give a sketch of a typical path that the ship would travel.

(c) Repeat (b) for the case that the value of the gain K is set somewhat higher than the *upper* limit found in (a). Even though the system is unstable in each case, note the difference in the nature of the instability.

(d) Verify the results in (b) and (c) using SIMULINK. Let the input be zero and the disturbance a unit step.

6.9. Given the closed-loop system described by the state equations

$$\dot{\mathbf{x}}(t) = \begin{bmatrix} 1 & 1 \\ -3 & -K \end{bmatrix} \mathbf{x}(t)$$

(a) Determine the range of K for which the system is stable.

(b) The gain K is set to one of the values found in (a) such that the system oscillates. Find the frequency of oscillation and the period of oscillation.

(c) Use MATLAB to find the eigenvalues of A, to verify the results in (a).

(d) Check the results in (b) by simulation. Use a step input.

6.10. We consider further the design of the PI-compensated system of Example 6.7. The characteristic equation is given by

$$1 + G_c(s)G_p(s) = 1 + \frac{2(K_P s + K_I)}{s(s^3 + 4s^2 + 5s + 2)} = 0$$

(a) Find the ranges of the compensator gains K_P and K_I, with both gains positive, such that the closed-loop system is stable.

(b) Is the system stable for $K_I = 0.5$ and $K_P = 1$?

(c) Verify the result in (b) by:
   (i) Calculating the characteristic-equation roots with MATLAB.
   (ii) Finding the step response with SIMULINK.

(d) Repeat (b) and (c) for $K_I = 2$ and $K_P = 1$.

6.11. Shown in Figure P6.11 is the control system for one joint of a robot arm. The controller is a PD compensator. This control system is described in Section 2.12.

(a) Show that the plant transfer function is given by

$$\frac{\Theta_L(s)}{E_a(s)} = \frac{0.15}{s(s+1)(s+5)}$$

(b) Find the ranges of the compensator gains K_P and K_D, with these gains positive, such that the closed-loop system is stable.

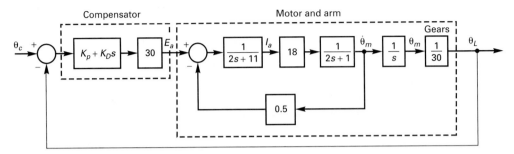

Figure P6.11

(c) Let $K_D = 1$. Find K_P such the system will have a steady-state oscillation, and the period of that oscillation.

(d) Verify the results in (c) using SIMULINK.

6.12. Shown in Figure P6.12 is the temperature-control system of a large test chamber, which is the type considered in Example 5.5. The controller is PI (proportional-plus-integral), which is the type commonly employed for temperature-control systems. The transfer function of the controller is given by $G_c(s) = (K_P + K_I/s)$, where K_P is the proportional gain and K_I is the integral gain. The disturbance $D(s)$ models the opening of the door of the chamber and is a unit step when the door is opened. If the door is closed, $D(s) = 0$.

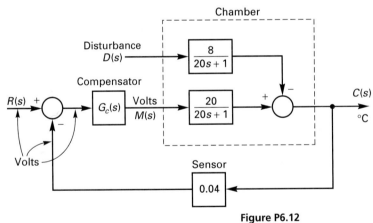

Figure P6.12

(a) Assume that the system is open–loop [$G_c(s) = 0$]. If the door is opened and remains open, find the steady-state change in the chamber temperature in degrees Celsius.

(b) Repeat (a) for the closed-loop system, with $K_I = 0$ and $K_P > 0$. The result is a function of K_P.

(c) Repeat (a) for the closed-loop system, with $K_I > 0$ and $K_P > 0$.

(d) The results of (b) and (c) are based on the assumption that the system is stable. Find the ranges of K_P and K_I, with both gains positive, such that the system is stable.

(e) Let $K_I = 1$. Find K_p such that the system is critically damped.

(f) Verify the results in (c) and (e) by running the system step response with SIMULINK.

(g) Why does the step response for (e) have overshoot, since the system is critically damped?

6.13. Shown in Figure P6.13 is the block diagram of the servo-control system for one of the axes of a digital plotter. The input θ_r is the output of a digital computer, and the output θ_p is the position of the servomotor shaft. It is assumed that the pen-positioning system connected to the motor shaft is rigid (no dynamics) within the system bandwidth.

(a) If $K_d = 1$, find the range of K_v for which the system is stable.
(b) If $K_v = 0.3$, find the range of K_d for which the system is stable.
(c) Consider that K_d is plotted along the *x*-axis (abscissa) and K_v along the *y*-axis (ordinate) of a plane (called the parameter plane). Show the regions of this plane in which the system is stable.
(d) Let $K_d = 1$. Find both the value of K_v that makes the system marginally stable and the period of the resulting oscillation.
(e) Verify the results in (d) with the system step response from SIMULINK. Program the system with the blocks as shown in Figure P6.13.

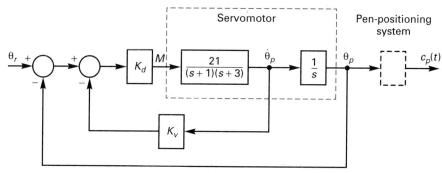

Figure P6.13

6.14. Shown in Figure P6.14 is the simplified block diagram for a cruise-control system for an automobile. The actuator controls the throttle position, the carburetor model is first order with a 1–s time constant, and the engine and load is also modeled as a first-order transfer function with a 3-s time constant. The disturbance models the grade (slope) of the road. The cruise control is to be operable from 70 to 120 km/hr. The compensator is PI (proportional–plus–integral) with the transfer function $G_c(s) = (K_P + K_I s)$.

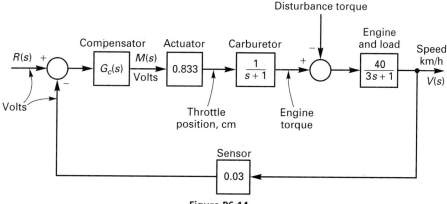

Figure P6.14

(a) Find the ranges of K_P and K_I for which the system is stable.

(b) Suppose that K_I is set to a value in the range found in (a) for stability, but that K_P is set somewhat lower than the limit found in (a). Hence the system is unstable. If you were driving this automobile and activated the cruise control, exactly how would the automobile respond? Be specific, but do not solve for numerical values.

(c) Simulate the system of (b) to verify your results.

6.15. The Routh array has the form

$$
\begin{array}{c|ccc}
\vdots & \vdots & & \\
s^4 & x & x & \textcircled{x} \\
s^3 & x & x & \\
s^2 & x & \textcircled{x} & \\
s^1 & x & & \\
s^0 & \textcircled{x} & &
\end{array}
$$

Show that the values of the elements circled (the final elements in the rows designated by even powers of s) are always equal.

7

Root-Locus Analysis and Design

In this chapter we introduce one of the major analysis and design methods discussed in this book. The method is the root-locus procedure; it indicates to us the characteristics of a control system's natural response. In the following two chapters we use the frequency response to analyze and design control systems. The frequency-response procedures give us information different from that of root locus. We shall see that root-locus procedures and frequency-response procedures complement each other and in many practical design problems both methods are employed; controls engineers must be familiar with both methods.

7.1 ROOT-LOCUS PRINCIPLES

We introduce the root locus through an example. Consider the radar tracking system depicted in Figure 7.1(a) (see Section 2.7). Figure 7.1(b) and (c) show a block diagram of the system and the transfer functions, respectively. The closed-loop system transfer function is given by

$$T(s) = \frac{KG_p(s)}{1 + KG_p(s)} = \frac{\dfrac{K}{s(s+2)}}{1 + \dfrac{K}{s(s+2)}} = \frac{K}{s^2 + 2s + K} \tag{7-1}$$

Hence the characteristic equation, which is the denominator of the closed-loop transfer function set to zero, is

$$s^2 + 2s + K = 0 \tag{7-2}$$

We see, since the polynomial is second order, that the system is stable for all positive values of K.

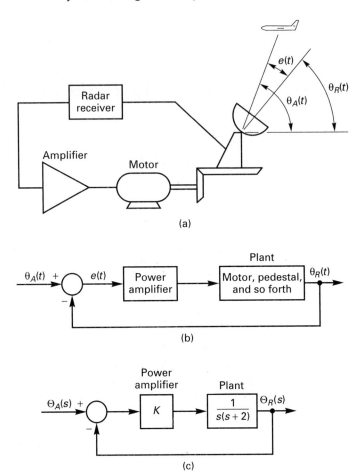

Figure 7.1   Radar tracking system.

It is not evident for this example exactly how the value of K affects the transient response. For example, suppose that a design specification is that the damping ratio ζ should equal 0.707. The characteristic equation for the standard second-order system is, from (4-18),

$$s^2 + 2\zeta\omega_n s + \omega_n^2 = 0 \tag{7-3}$$

Then, from (7-2) and (7-3),

$$2\zeta\omega_n = 2(0.707)\omega_n = 2$$

or $\omega_n = 1.414$ and $\omega_n^2 = 2$. Hence, comparing (7-2) with (7-3), we see that the design requires a gain of $K = 2$ for the power amplifier.

The preceding design was simple, but we have satisfied only one system specification. There might be additional specifications for rise time, settling time, steady-state accuracy, and so forth. If, as shown in Figure 7.1, the only design parameter available is the gain of the power amplifier (that is, if no dynamic compensation is to be used), we can satisfy only one design specification exactly. However, it is usually necessary in practical design to consider trade-offs; that is, a value of K is chosen that satisfies all design specifications to some degree but none exactly.

To investigate some of the effects of choosing different values of K, we plot the roots of the system characteristic equation in the s-plane. From (7-2), these roots are given by

$$s = \frac{-2 \pm \sqrt{4 - 4K}}{2} = -1 \pm \sqrt{1 - K} \tag{7-4}$$

The roots are real and negative for $0 < K \leq 1$, and for $K > 1$,

$$\text{roots} = -1 \pm j\sqrt{K - 1} = -\zeta\omega_n \pm j\omega_n\sqrt{1 - \zeta^2} = \omega_n \big/ 180 \pm \cos^{-1}\zeta$$

[see (4-34)]. These roots are plotted in Figure 7.2 for $0 \leq K \leq \infty$. We can see from the plot that for $0 < K < 1$, the two roots are real with different time constants. For $K = 1$, the roots are real and equal, and the system is critically damped. For $K > 1$, the roots are complex with a time constant of 1s, with the value of ζ decreasing as K increases. Hence, as K increases with the roots complex, the overshoot in the transient response increases.

Note that the plot of the roots of the characteristic equation, as some parameter varies, gives us a great deal of information concerning changes in the natural response of the system as the parameter is varied. The characteristics of the natural response are evident from this plot, but the amplitudes of its various components are not. The amplitudes of the various components of the natural response depend on (1) the poles of the closed-loop transfer function (given on the plot), (2) the zeros of the closed-loop transfer function (cannot be obtained from the plot), and (3) the system input function and the initial conditions (cannot be obtained from the plot).

The plot in Figure 7.2 is called the *root locus* [1] of the system of Figure 7.1.

Root locus   A *root locus* of a system is a plot of the roots of the system characteristic equation (the poles of the closed-loop transfer function) as some parameter of the system is varied.

For an nth-order system, if the parameter is varied in a continuous fashion, the root locus appears as a family of n continuous paths or branches traced out by the n roots of the characteristic equation.

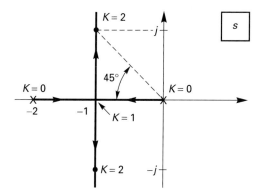

Figure 7.2 Plot of characteristic equation roots.

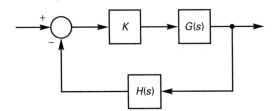

Figure 7.3 System for root locus.

We generally consider the system of Figure 7.3 in discussing the root locus, with $0 \leq K < \infty$. In this figure it is assumed that, in general, $G(s)$ includes both the compensator transfer function $G_c(s)$ and the plant transfer function $G_p(s)$. The characteristic equation for this system is given by

$$1 + KG(s)H(s) = 0 \qquad (7\text{-}5)$$

A value s_1 is a point on the root locus *if and only if s_1 satisfies (7-5) for a real value of K,* with $0 \leq K < \infty$. Plotting the root locus for negative values of K [2] is considered in Section 7.12. As will become evident in the following development, the parameter K must appear linear in the root-locus equation (7-5).

A value of s is on the root locus if (7-5) is satisfied for that value of s. This equation can be written as

$$K = -\frac{1}{G(s)H(s)} \qquad (7\text{-}6)$$

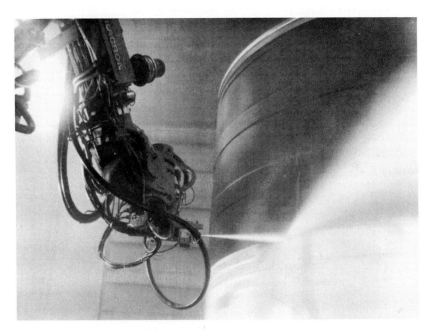

Robotic cleaning system that removes protective coating from recovered solid rocket boosters. (Courtesy of NASA.)

Hence, if s is a point on the root locus, the right side of (7-6) evaluated for that value of s yields a positive real number.

Since, in general, $G(s)$ and $H(s)$ are complex for a given value of s, (7-6) is in fact two equations, which can be expressed as

$$|K| = \frac{1}{|G(s)H(s)|} \tag{7-7}$$

and

$$\underline{/G(s)H(s)} = \arg G(s)H(s) = r(180°) \qquad r = \pm 1, \pm 3, \pm 5, \dots \tag{7-8}$$

where $\underline{/(\cdot)}$ denotes the angle of the function. Also, as shown, $\arg(\cdot)$ is used for the same purpose.

We call (7-7) the *magnitude criterion* of the root locus. Since K can assume any value between zero and infinity, (7-7) is satisfied for arbitrary values of s, and we do not consider this condition further at this time. Hence the condition for a point s to be on the root locus is (7-8), and we call this equation the *angle criterion*.

The angle criterion is illustrated in Figure 7.4 for the function

$$KG(s)H(s) = \frac{K(s - z_1)}{(s - p_1)(s - p_2)} \tag{7-9}$$

Suppose that the point s_1 is to be tested to determine if it is on the root locus. Note that the angle of the factor $(s_1 - z_1)$ is θ_1, that of $(s_1 - p_1)$ is θ_2, and that of $(s_1 - p_2)$ is θ_3. Thus the angle condition (7-8) becomes

$$\theta_1 - \theta_2 - \theta_3 = \pm 180°$$

for the point s_1 to be on the root locus. The locus of all points that satisfies this relationship forms the complete root locus of the system. If a point s_1 is found to be on the root locus, the value of K that places the locus at that point is found from (7-6); that is,

$$K = \frac{-1}{G(s_1)H(s_1)}$$

Recall that this equation must be satisfied with K a positive real value.

For the system of Figure 7.3 we call $KG(s)H(s)$ the *open-loop function.* From the preceding discussion, it is seen that the condition for a point in the s-plane to be on the root locus is that

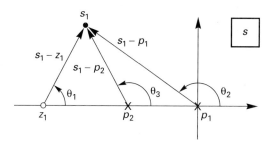

Figure 7.4 Angles of terms.

$$\sum(\text{all angles from the finite zeros}) - \sum(\text{all angles from the finite poles})$$
$$= r(180°) \qquad r = \pm1, \pm3, \pm5, \dots \tag{7-10}$$

where the poles and zeros are those of the open-loop function. Although at one time (7-10) was the basis for the graphical construction of root-locus plots, it is *not* suggested that we use it now. Digital computer programs for calculating a root locus are available that are more convenient and certainly more accurate. However, a good knowledge of the rules and techniques for root-locus construction will offer insight into the effects of both changing parameters and adding poles and zeroes in the design process. This insight is the principal reason for the emphasis on graphical techniques in this chapter. If an accurate root locus is needed, by all means use a digital computer to calculate this locus. If insight into the design process is needed, make rough sketches of the root locus, using the rules to be developed in the following sections.

As a final point, recall that the characteristic equation of a linear, time-invariant system can be expressed as, from Mason's gain formula [see (5-12)],

$$\Delta(s) = 1 + F(s) = 0$$

For example, for the system of Figure 7.3, $F(s)$ is equal to $KG(s)H(s)$. For multi-loop systems, $F(s)$ may be an involved function of the various transfer functions in the system. As stated before, we define $F(s)$ as the *open-loop function*. Hence, the open-loop function is defined such that the system characteristic equation is formed by setting the sum of this function and unity to zero. Many of the analysis and design procedures to be presented are based on the use of the open-loop function. The importance of the open-loop function cannot be overemphasized; *open-loop functions* are used extensively in the analysis and design of the closed-loop systems.

7.2 SOME ROOT-LOCUS TECHNIQUES

In this and the following section we develop several rules applicable to the construction of root loci. Most of these rules are derived using the angle criterion (7-8) or (7-10), which is the basic condition for a point in the s-plane to be on the root locus.

The first rule is as follows:

1. The root locus is symmetrical with respect to the real axis.

This rule applies since we have assumed that the system models used are rational functions (the ratio of two polynomials) with real coefficients. Hence if the characteristic equation has a complex root, the complex conjugate of that root must also be a root; that is, complex roots of real polynomials always appear in complex conjugate pairs.

The second rule is developed next. The characteristic equation for the system of Figure 7.3 may be expressed in the form

$$1 + KG(s)H(s) = 1 + \frac{Kb_m(s-z_1)(s-z_2)\cdots(s-z_m)}{(s-p_1)(s-p_2)\cdots(s-p_n)} = 0 \tag{7-11}$$

This equation can be cleared of fractions to yield

$$(s - p_1)(s - p_2)\cdots(s - p_n) + Kb_m(s - z_1)(s - z_2)\cdots(s - z_m) = 0 \qquad (7\text{-}12)$$

Thus for $K = 0$, the roots of the characteristic equation are simply the poles of the open-loop function $G(s)H(s)$. As K approaches infinity but s remains finite, the branches of the locus approach the zeros of the open-loop function. If the open-loop function has zeros at infinity, that is, if $n > m$ (as is the usual case with the models of physical systems) the locus will also approach these zeros. This aspect of the root locus will be proved in the discussion of one of the additional rules shortly. The second rule is therefore as follows:

2. The root locus originates on the poles of $G(s)H(s)$ (for $K = 0$) and terminates on the zeros of $G(s)H(s)$ (as $K \to \infty$), including those zeros at infinity.

The third rule, which involves the structure of the root locus for very large values of s, will now be developed. With reference to (7-11), let the open-loop function be rewritten in the form

$$KG(s)H(s) = \frac{K(b_m s^m + b_{m-1} s^{m-1} + \cdots)}{s^n + a_{n-1} s^{n-1} + \cdots} = \frac{K(b_m s^m + \cdots)}{s^{m+\alpha} + \cdots} \qquad (7\text{-}13)$$

where

$$\alpha = n - m > 0 \qquad (7\text{-}14)$$

Thus $KG(s)H(s)$ has α zeros at infinity. If we let s approach infinity in (7-13) the polynomials approach their highest-power terms, and

$$\lim_{s \to \infty} KG(s)H(s) = \lim_{s \to \infty} \frac{Kb_m s^m}{s^n} = \lim_{s \to \infty} \frac{Kb_m}{s^\alpha} \qquad \alpha > 0 \qquad (7\text{-}15)$$

The root locus for large values of s then satisfies the relationship

$$\lim_{s \to \infty} [1 + KG(s)H(s)] = \lim_{s \to \infty} \left[1 + \frac{Kb_m}{s^\alpha} \right] = 0 \qquad (7\text{-}16)$$

This equation has roots that are given by

$$s^\alpha + Kb_m = 0 \qquad (7\text{-}17)$$

or

$$s^\alpha = -Kb_m = Kb_m \underline{/r180^\circ} \qquad r = \pm 1, \pm 3, \ldots \qquad (7\text{-}18)$$

The magnitude of these roots approaches infinity because of the assumption that s is approaching infinity. The angles of the roots are the principal values of the angles

$$\theta = \frac{r180^\circ}{\alpha} \qquad r = \pm 1, \pm 3, \ldots \qquad (7\text{-}19)$$

The angles in (7-19) are the angles of *asymptotes* of the root locus, since generally the root loci approach these angles at s (and thus K) approaches infinity. Table 7.1 gives these angles for small values of α. For the case that $\alpha = 0$, there are no asymptotes. Of course, the magnitudes of the roots are $(Kb_m)^{1/\alpha}$. It is seen that for a given α, the initial angle is $180^\circ/\alpha$, with the increment in the angle of $360^\circ/\alpha$. Two examples are now given.

TABLE 7.1 ANGLES OF ASYMPTOTES

α	Angles
0	No asymptotes
1	180°
2	±90°
3	±60°, 180°
4	±45°, ±135°

Example 7.1

Consider again the radar tracking system discussed in Section 7.1, which has the open-loop function

$$KG(s)H(s) = \frac{K}{s(s+2)}$$

Since, in (7-13), m(number of finite zeros) is zero and n(number of finite poles) is two, there are

$$\alpha = n - m = 2$$

zeros at infinity. Hence there are two asymptotes, with angles of ±90°, as seen from Table 7.1. The asymptotes are evident in Figure 7.2, which is a plot of the root locus for this system.

Example 7.2

As a second example, consider the system with the open-loop function

$$KG(s)H(s) = \frac{K}{s^3}$$

Since this function has three zeros at infinity, the root locus has three asymptotes, with angles of ±60° and 180°, from Table 7.1. For this system the roots can be calculated analytically. The characteristic equation is

$$1 + KG(s)H(s) = 0 = s^3 + K$$

The roots are then

$$s = (-K)^{1/3} = (K)^{1/3}\underline{/60°} \qquad (K)^{1/3}\underline{/-60°} \qquad (K)^{1/3}\underline{/180°}$$

Note that the roots of the characteristic equation include any complex number that, when cubed, yields a negative real number. The root locus is plotted in Figure 7.5. This system is unstable for all values of $K > 0$, since the closed-loop system has two poles in the right half-plane for all such values of K. We can obtain the same result concerning the system stability from the Routh–Hurwitz criterion.

$$
\begin{array}{c|cc}
s^3 & 1 & 0 \\
s^2 & \cancel{0}\,\varepsilon & K \\
s^1 & \dfrac{-K}{\varepsilon} & \\
s^0 & K &
\end{array}
$$

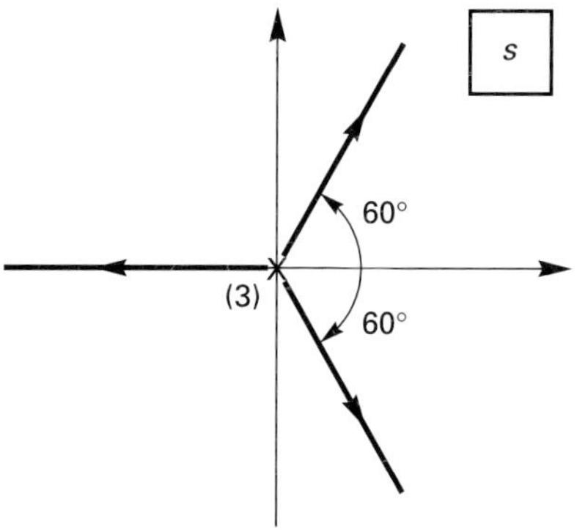

Figure 7.5 Example 7.2.

We see then that the closed-loop system always has two poles in the right half-plane, which agrees with Figure 7.5. A MATLAB program that calculates and plots the root locus by two procedures is given by

```
rlocus ([1], [1 0 0 0])
```

In Example 7.1, the asymptotes intersected the real axis at $s = -1$, whereas in Example 7.2, the asymptotes intersected the real axis at the origin. If we denote the value of s at which the asymptotes intersect the real axis as σ_a, this value is given by

$$\sigma_a = \frac{(\text{sum of finite poles}) - (\text{sum of finite zeros})}{(\text{number of finite poles}) - (\text{number of finite zeros})} \tag{7-20}$$

where the poles and zeros are those of the open-loop function. This relationship is given without proof and is valid for the case that the open-loop function has one or more zeros at infinity ($\alpha \geq 1$). Of course, if $\alpha = 1$, there is only one asymptote (the negative real axis), and (7-20) gives a meaningless result. Hence we use (7-20) for $\alpha \geq 2$. For the case that the open-loop function has no finite zeros, the second term in the numerator of (7-20) is set to zero. We can now state the third rule of root-locus construction:

3. If the open-loop function has α zeros at infinity, $\alpha \geq 1$, the root locus approaches α asymptotes as K approaches infinity. The asymptotes are located at the angles

$$\theta = \frac{r180°}{\alpha} \qquad r = \pm1, \pm3, \dots$$

and these asymptotes intersect the real axis at the point

[eq. (7-20)] $\sigma_a = \dfrac{(\text{sum of finite poles}) - (\text{sum of finite zeros})}{(\text{number of finite poles}) - (\text{number of finite zeros})}$

The root loci of Examples 7.1 and 7.2 illustrate this rule.

Example 7.3

For the radar tracking system of Example 7.1,

$$KG(s)H(s) = \frac{K}{s(s+2)}$$

Since there are no finite zeros and two finite poles, one at $s = 0$ and the other at $s = -2$, (7-20) yields

$$\sigma_a = \frac{[0 + (-2)] - 0}{2 - 0} = -1$$

which checks with Figure 7.2

For the system of Example 7.2,

$$KG(s)H(s) = \frac{K}{s^3}$$

There are no finite zeros, and the three poles all occur at $s = 0$. Thus

$$\sigma_a = \frac{0 - 0}{3 - 0} = 0$$

which checks the results of the example.

In both of the systems in the last two examples the locus followed the asymptotes exactly. In most cases the locus only approaches the asymptotes as s approaches infinity, as will be shown in examples in the next section.

In this section we have developed three rules to aid in sketching the root locus of a system. Recall that the root locus is a plot of the roots of the *system characteristic equation* (poles of the *closed-loop transfer function*) as some parameter is varied. In the following section we develop three additional rules that will allow us to make rough sketches of the root locus for some low-order systems. As we have stated, for more accurate plots of the root locus, we must use a digital computer to calculate the locus (see Example 7.2).

7.3 ADDITIONAL ROOT-LOCUS TECHNIQUES

In this section we develop three additional rules for constructing root loci. First, we develop a rule for the real-axis portion of the root locus, using the angle requirement of the root locus, (7-10). This requirement states that

Σ angles from finite zeros $- \Sigma$ angles from finite poles $= r180°$ $\quad r = \pm1, \pm3, \ldots$

where Σ denotes the sum. We consider first that all finite poles and zeros of the open-loop function are on the real axis, and we test points on the real axis to determine if these points are on the root locus.

Consider the open-loop function, illustrated in Figure 7.6(a), of two poles and one zero such that

$$KG(s)H(s) = \frac{K(s - z_1)}{(s - p_1)(s - p_2)}$$

We first test a point s on the real axis to the right of the pole p_1. Figure 7.6(b) shows the factor $(s - p_1)$, which has an angle of $0°$. It is also seen that the angle of the zero factor $(s - z_1)$ and that of the other pole factor $(s - p_2)$ are also $0°$. Hence, the angle requirement just stated is not satisfied, and we see that any point to the right of the pole p_1 cannot be on the root locus.

Consider now a point s between the pole p_1 and the zero z_1. The pole factor $(s - p_1)$ now has an angle of $180°$, as shown in Figure 7.6(c). However, the angles from the zero z_1

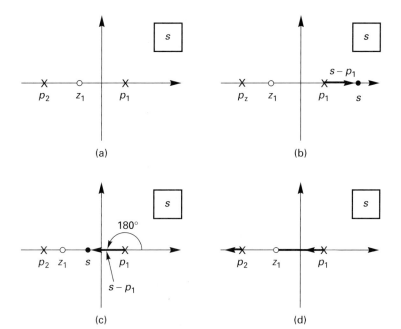

Figure 7.6 Real-axis locus.

and from the pole p_2 are still 0°. Thus the angle requirement is satisfied, and any point between z_1 and p_1 is on the locus.

For a point s between p_2 and z_1, the angle from p_1 is still 180°, as now is the angle from z_1. The angle from p_2 is still 0°; hence the angle requirement is not satisfied and no points between p_2 and z_1 are on the locus. If the point s is to the left of the pole p_2, the angles from p_1, p_2, and z_1 are all 180°, and the angle requirement is satisfied. The resulting locus on the real axis is then as shown in Figure 7.6(d), with arrowheads to indicate the direction of movement of the roots with increasing K.

If we are considering a point on the real axis, the angle from any real *critical frequency* (by definition, a pole or a zero) of the open-loop function to that point will be either 0° or 180°. We now justify not differentiating between poles and zeros for this rule. We obviously get the same result by adding 0° to an angle that we get by subtracting 0° from that angle. It is not as obvious, but we get the same result by adding 180° to an angle that we get by subtracting 180° from that angle. Hence the contribution of either a real pole or a real zero to the angle requirement is the same if the test point is on the real axis, since all angles are either 0° or 180°. Thus it is not necessary to differentiate between poles and zeros, and we speak only of critical frequencies.

If a real critical frequency is to the *left* of a test point, it contributes an angle of 0° to the angle requirement; thus the number of critical frequencies to the left of the test point is not important. If a real critical frequency is to the *right* of the test point, it effectively contributes an angle of 180° to the angle requirement. Thus if the number of real critical frequencies to the right of a test point is odd, the point is on the locus. If this number is even, the test point is not on the locus.

For the case that the open-loop function has complex poles or zeros, the preceding discussion still applies. For example, two complex-conjugate poles are shown in Figure 7.7. Since complex poles (and zeros) must occur in conjugate pairs, the sum of the angles from a pair of poles (or zeros) to a point on the real axis will always be 0° (or 360°), as shown in the figure. Hence complex poles and zeros do not affect the part of the root locus that lies on the real axis. We then have the following rule:

4. The root locus includes all points on the real axis to the left of an odd number of real critical frequencies (poles and zeros).

The fifth rule to be developed concerns points at which two or more branches of the locus come together and then part (break away). Points of this type are called *breakaway points*. The distinguishing feature of a breakaway point is that, for the value of K that places the locus at the breakaway point, the characteristic equation has multiple roots. For example, the open-loop function of Example 7.2 was

$$KG(s)H(s) = \frac{K}{s^3}$$

and, as shown in Figure 7.5, a breakaway point occurs at $s = 0$, where the three branches of the locus break away. At this point $(K = 0)$, the characteristic equation has a root of order three. We now develop rule 5, based on this discussion.

If we express $G(s)H(s)$ as the ratio of two polynomials $N(s)$ and $D(s)$, the system characteristic equation can be written as

$$1 + KG(s)H(s) = 1 + K\frac{N(s)}{D(s)} = 0 \tag{7-21}$$

Then

$$D(s) + KN(s) = Q(s) = 0 \tag{7-22}$$

where $Q(s)$ is the system characteristic polynomial. Also, from (7-21),

$$K = -\frac{1}{G(s)H(s)} = -\frac{D(s)}{N(s)} \tag{7-23}$$

Suppose that $s = s_b$ is a breakaway point. Then, from the discussion above, the characteristic equation has a multiple-order root at s_b; that is, we can express the characteristic polynomial as

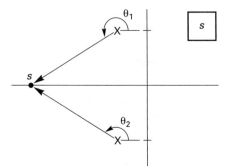

Figure 7.7   Real-axis locus.

$$Q(s) = (s - s_b)^\gamma Q_1(s) \qquad (7\text{-}24)$$

where $\gamma \geq 2$ is the order of the root. The derivative of $Q(s)$ with respect to s is then

$$\frac{dQ(s)}{ds} = Q'(s) = \gamma(s - s_b)^{\gamma - 1} Q_1(s) + (s - s_b)^\gamma Q_1'(s)$$

$$= (s - s_b)^{\gamma - 1} [\gamma Q_1(s) + (s - s_b) Q_1'(s)] \qquad (7\text{-}25)$$

where the prime indicates the derivative with respect to s. Thus $Q'(s)$ also has a root of order $(\gamma - 1)$ at $s = s_b$; that is, $Q'(s_b) = 0$. From (7-22),

$$Q'(s) = D'(s) + KN'(s) \qquad (7\text{-}26)$$

Substituting (7-23) into (7-26) yields

$$Q'(s) = D'(s) - \frac{D(s)}{N(s)} N'(s) \qquad (7\text{-}27)$$

which can be expressed for $s = s_b$ as

$$Q'(s_b) = 0 = D'(s)N(s) - D(s)N'(s)\big|_{s = s_b} \qquad (7\text{-}28)$$

Thus this equation will be satisfied at a breakaway point.

Finally, note that

$$\frac{d[G(s)H(s)]}{ds} = \frac{d}{ds}\left[\frac{N(s)}{D(s)}\right] = \frac{D(s)N'(s) - D'(s)N(s)}{D^2(s)} \qquad (7\text{-}29)$$

and also that

$$\frac{d}{ds}\left[\frac{1}{G(s)H(s)}\right] = \frac{d}{ds}\left[\frac{D(s)}{N(s)}\right] = \frac{N(s)D'(s) - N'(s)D(s)}{N^2(s)} \qquad (7\text{-}30)$$

Either (7-29) or (7-30) set to zero, with $s = s_b$, results the same expression as (7-28). Thus (7-29) or (7-30) or direct use of (7-28) will give a polynomial among whose roots are the breakaway points of the root locus. Rule 5 can then be stated as follows:

5. The breakaway points on a root locus will appear among the roots of the polynomial obtained from either

$$\frac{d[G(s)H(s)]}{ds} = 0$$

or, equivalently,

$$N(s)D'(s) - N'(s)D(s) = 0$$

where $N(s)$ and $D(s)$ are the numerator and denominator polynomials, respectively, of $G(s)H(s)$.

As a final point concerning rule 5, if s is a value on the real axis, $G(s)H(s)$ is real. From (7-30) and using (7-23), since K is a function of s, we can write

$$\frac{dK(s)}{ds} = -\frac{d}{ds}\left[\frac{1}{G(s)H(s)}\right] = 0$$

Hence, at any breakaway point on the real axis, K is either a maximum or a minimum for that part of the locus on the real axis.

An additional rule may be developed concerning the angle at which a branch of the root locus leaves a pole (*angle of departure*) or approaches a zero (*angle of arrival*). To illustrate this rule, consider a system with the open-loop function

$$KG(s)H(s) = \frac{K}{(s+2)(s^2+2s+2)}$$

We are interested in finding the angle θ_{1d} in Figure 7.8, for the case that ε is very small, such that s_1 is a point on the root locus. Thus θ_{1d} is the angle at which the locus leaves the pole p_1, which in this case is the point $-1+j$.

For Figure 7.8, from the angle criterion (7-10),

$$-\theta_{1d} - \theta_2 - \theta_3 = \pm 180°$$

Since $s_1 \approx p_1$, then $\theta_2 \approx 90°$ and $\theta_3 \approx 45°$, and the angle of departure is given by

$$\theta_{1d} \approx 180° - 90° - 45° = 45°$$

The complete root locus is given in Figure 7.8(b). It is seen from this development that the sixth rule may be stated as follows:

6. Loci will depart from a pole p_j (arrive at a zero z_j) at the angle $\theta_d(\theta_a)$, where

$$\theta_d = \sum_i \theta_{zi} - \sum_{\substack{i \\ i \neq j}} \theta_{pi} + r(180°)$$

$$\theta_a = \sum_i \theta_{pi} - \sum_{\substack{i \\ i \neq j}} \theta_{zi} + r(180°)$$

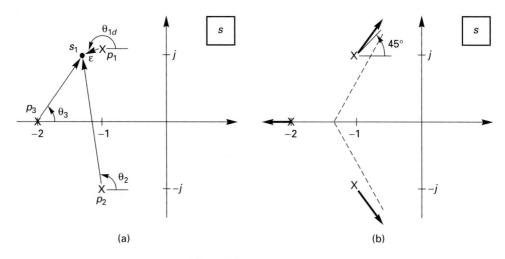

(a)   (b)

Figure 7.8   Angle of departure.

and where $r = \pm 1, \pm 2, \pm 3, \ldots$, and $\theta_{pi}(\theta_{zi})$ represent the angles from pole p_i (zero z_i), respectively, to $p_j(z_j)$.

In summary, the six rules for sketching the root locus of a closed-loop system with the characteristic equation

$$1 + KG(s)H(s) = 0$$

are given in Table 7.2. Recall that the root locus is a plot of the roots of the *system characteristic equation* (poles of the *closed-loop system*) as the parameter K is varied. Some examples will be given now to illustrate the use of these rules.

TABLE 7.2   RULES FOR ROOT LOCUS DEVELOPMENT

1. The root locus is symmetrical with respect to the real axis.
2. The root locus originates on the poles of $G(s)H(s)$ (for $K = 0$) and terminates on the zeros of $G(s)H(s)$ (as $K \to \infty$), including those at infinity.
3. If the open-loop function has α zeros at infinity $\alpha \geq 1$, the root locus will approach α asymptotes as K approaches infinity. The asymptotes are located at the angles

$$\theta = \frac{r180°}{\alpha} \qquad r = \pm 1, \pm 3, \ldots$$

   and those asymptotes intersect the real axis at the point

$$\sigma_a = \frac{\Sigma \text{ finite poles} - \Sigma \text{ finite zeros}}{\# \text{ finite poles} - \# \text{ finite zeros}}$$

   where the symbol # denotes number.
4. The root locus includes all points on the real axis to the left of an odd number of real critical frequencies (poles and zeros).
5. The breakaway points on a root locus will appear among the roots of the polynomial obtained from either

$$\frac{d\,[G(s)H(s)]}{ds} = 0$$

   or, equivalently,

$$N(s)D'(s) - N'(s)D(s) = 0$$

   where $N(s)$ and $D(s)$ are the numerator and denominator polynomials, respectively, of $G(s)H(s)$.
6. Loci will depart from a pole p_j (arrive at a zero z_j) of $G(s)H(s)$ at the angle $\theta_d(\theta_a)$, where

$$\theta_d = \sum_i \theta_{zi} - \sum_{\substack{i \\ i \neq j}} \theta_{pi} + r(180°)$$

$$\theta_a = \sum_i \theta_{pi} - \sum_{\substack{i \\ i \neq j}} \theta_{zi} + r(180°)$$

   and where $r = \pm 1, \pm 3, \ldots$ and $\theta_{pi}(\theta_{zi})$ represent the angles from pole p_i (zero z_i), respectively, to $p_j(z_j)$.

Example 7.4

We consider the system that has the open-loop function

$$KG(s)H(s) = \frac{K}{(s-1)(s+2)(s+3)} = \frac{K}{s^3 + 4s^2 + s - 6}$$

From the rules of Table 7.2:

Rule 2 The locus originates at the poles $s = 1$, $s = -2$, and $s = -3$, and terminates on the three zeros at infinity.

Rule 3 Three asymptotes, at $\pm 60°$ and $180°$ from Table 7.1, intersect the real axis at

$$\sigma_a = \frac{\Sigma \text{ finite poles} - \Sigma \text{ finite zeros}}{\# \text{ finite poles} - \# \text{ finite zeros}} = \frac{(1 - 2 - 3) - 0}{3 - 0} = -\frac{4}{3}$$

as shown in Figure 7.9.

Rule 4 The root locus occurs on that part of the real axis $-2 < s < 1$ (to the left of one critical frequency), and for $s < -3$ (to the left of three critical frequencies). The range $s > 1$ is to the left of zero critical frequencies, and $-3 < s < -2$ is to the left of two critical frequencies.

Rule 5 To determine the breakaway points,

$$N(s)D'(s) - N'(s)D(s) = N(s)D'(s) = (1)\frac{d}{ds}(s^3 + 4s^2 + s - 6)$$

$$= 3s^2 + 8s + 1$$

The roots of this polynomial occur at $s = -0.132$ and at $s = -2.54$. From Rule 4 and Figure 7.9 we see that the point $s = -2.54$ is not on the locus, so we ignore this point. Actually this point is a breakaway point for that part of the locus for $K < 0$, but we do not consider that part in this section. The point $s = -0.132$ is on the locus of Figure 7.9, and this is a breakaway point. The complete locus is given in Figure 7.9. Note that the system is unstable for K small (one root in the right half-plane) and for K larger (two roots in the right half-plane).

The range of K for stability is found most easily by using the Routh–Hurwitz criterion. The characteristic equation is obtained from

$$1 + KG(s)H(s) = 0$$

and substituting the open-loop function into this relationship yields

$$s^3 + 4s^2 + s - 6 + K = 0$$

The Routh array is then

$$
\begin{array}{c|cc}
s^3 & 1 & 1 \\
s^2 & 4 & K - 6 \\
s^1 & \dfrac{10 - K}{4} & \Rightarrow K < 10 \\
s^0 & K - 6 & \Rightarrow K > 6
\end{array}
$$

Hence the system is stable for $6 < K < 10$.

From the array we see that for $K = 6$ the closed-loop system has a pole at $s = 0$. This pole is also shown on the root locus. Note that the value of K required to place the pole at $s = 0$ can also be found from the general criterion (7-6).

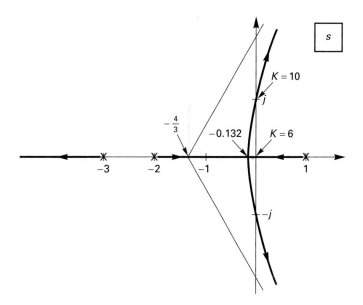

Figure 7.9 Example 7.4.

$$K = -\frac{1}{G(s)H(s)}\bigg|_{s=0} = -(s^3 + 4s^2 + s - 6)\bigg|_{s=0} = 6$$

which is the same result as obtained from the Routh–Hurwitz criterion. For the upper stability limit, $K = 10$, the Routh array gives the auxiliary polynomial

$$4s^2 + 4 = 4(s^2 + 1)$$

Thus the closed-loop transfer function has poles at $\pm j$ for $K = 10$. This result can also be verified from (7-6).

$$K = -\frac{1}{G(s)H(s)}\bigg|_{s=j} = -(s^3 + 4s^2 + s - 6)\bigg|_{s=j} = 10$$

Note that the Routh–Hurwitz criterion is a useful tool in root-locus analysis in determining the gains required for stability and the imaginary-axis crossings. The root locus is verified with the MATLAB program

```
rlocus ([1], [1 4 1 -6])
```

Example 7.5

We now sketch the root locus for the system that has the open-loop function

$$KG(s)H(s) = \frac{K(s+1)}{s^2}$$

Rule 2   The root locus originates at the double pole at $s = 0$, and terminates at the zeros at $s = -1$ and $s = \infty$.

Rule 3   There is one asymptote, at $180°$.

Rule 4 The root locus occurs on the real axis $s < -1$ (to the left of three critical frequencies). The range $s > 0$ is to the left of zero critical frequencies, and the range $-1 < s < 0$ is to the left of two critical frequencies.

Rule 5 To determine the breakaway points,

$$N(s)D'(s) - N'(s)D(s) = (s+1)(2s) - (1)(s^2) = s(s+2) = 0$$

Both roots of this equation are on the locus, and hence each root must be a breakaway point. The point $s = 0$ is obvious, since the real axis in the vicinity of this point is not a part of the root locus. At this point the locus leaves the real axis. The point $s = -2$ is a breakaway point at which the locus returns to the real axis and then approaches the zeros of the open-loop function. The root locus is given in Figure 7.10. This root locus can be verified with the MATLAB program

```
rlocus([0 1 1],[1 0 0])
```

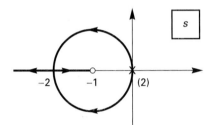

Figure 7.10 Example 7.5.

Example 7.6

In this example we will consider a simplified ship-steering system [3], which is to be a part of an automatic ship-steering system. A block diagram of the components of the ship-steering system is given in Figure 7.11(b). The controlling voltage $u_0(t)$ (to be generated by a compensator) is a low-power signal; hence a power amplifier is required to operate the steering gear, which controls the ship's rudder position angle, $u_r(t)$. The power amplifier and the steering gear are each modeled as first order; the ship dynamics are modeled as second order. The heading of the ship is denoted as $c(t)$. The parameters τ_a, τ_g, and τ_s are the time constants of the various components of the system. Assuming that the ship is large, we expect the time constants of the power amplifier and the steering gear to be small compared to that of the ship; for example, we expect to be able to turn the rudder much faster than the ship will respond to the rudder. The integrator in the ship dynamics is required, since if the rudder is set at a constant nonzero value, the ship will travel in a circle. Hence the heading of the ship, an angle, will grow without limit [$c(t)$ is a ramp function in the steady state].

Suppose that the time constants are $\tau_a = 0.05$ s, $\tau_g = 0.5$ s, and $\tau_s = 10$ s. Since the power amplifier and the steering gear are much faster than the ship, these time constants will be ignored in this example. However, these time constants will be included in the next two examples, to illustrate some of the effects of ignoring poles in systems as well as in the construction of root loci. For this example, the open-loop system has the form shown in Figure 7.11(c). For this system, in standard notation, the open-loop function is given by

$$KG(s)H(s) = \frac{(0.5)(0.1)K_a}{s(10s+1)} = \frac{0.005K_a}{s(s+0.1)} = \frac{K}{s^2 + 0.1s}$$

where $K = 0.005K_a$, and K_a is the gain of the power amplifier. We will now construct the root locus.

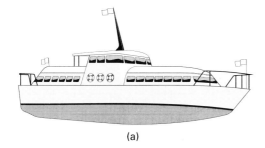

(a)

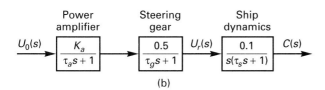

(b)

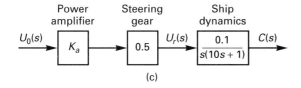

(c)

Figure 7.11   Ship-steering system.

Rule 2   The root locus originates at the poles $s = 0$ and $s = -0.1$, and terminates at the two zeros at infinity.

Rule 3   Two asymptotes occur at $\pm 90°$, and intersect the real axis at

$$\sigma_a = \frac{\Sigma \text{ finite poles} - \Sigma \text{ finite zeros}}{\# \text{ finite poles} - \# \text{ finite zeros}} = \frac{-0.1 - 0}{2 - 0} = -0.05$$

Rule 4   The root locus occurs on that part of the real axis $-0.1 < s < 0$.

Rule 5   To find the breakaway points,

$$N(s)D'(s) - N'(s)D(s) = (1)(2s + 0.1) - 0$$
$$= 2(s + 0.05) = 0$$

and the breakaway point is $s = -0.05$.

The locus is then of the form in Figure 7.12, with only the upper half of the s-plane shown. The lower half of the plane is the mirror image of the upper half. Note that the system is stable for all values of positive gain for the power amplifier. If the system is compensated by varying the gain of the power amplifier K_a, we expect the gain to be chosen large to reduce steady-state errors. The breakaway point occurs for the gain

$$K_a = -\frac{s^2 + 0.1s}{0.005}\bigg|_{s = -0.05} = -\frac{(-0.05)^2 + (0.1)(-0.05)}{0.005}$$

$$= \frac{0.0025}{0.005} = 0.5$$

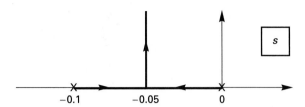

Figure 7.12 Root locus for Example 7.6.

Thus, for $K_a > 0.5$, the poles of the closed-loop transfer function will be complex, with the time constant of $\tau = 1/0.05 = 20$ s. Hence we expect to be able to attain a different direction of travel of the ship in approximately 80 s.

Example 7.7

We will again consider the ship-steering system of the last example, except only the dynamics of the power amplifier will be ignored. The open-loop function is then

$$KG(s)H(s) = \frac{0.05 K_a}{s(0.5s+1)(10s+1)} = \frac{0.01 K_a}{s(s+2)(s+0.1)}$$

$$= \frac{K}{s^3 + 2.1s^2 + 0.2s}$$

The root locus will now be constructed.

Rule 2 The root locus originates at the poles $s = 0$, $s = -0.1$, and $s = -2$, and terminates at the three zeros at infinity.

Rule 3 Three asymptotes occur, at $\pm 60°$ and $180°$, and intersect the real axis at

$$\sigma_a = \frac{\Sigma \text{ poles} - \Sigma \text{ zeros}}{\text{no. poles} - \text{no. zeros}} = \frac{-0.1 - 2 - 0}{3 - 0} = \frac{-2.1}{3} = -0.7$$

Rule 4 The root locus occurs on that part of the real axis $-0.1 < s < 0$ and $s < -2$.

Rule 5 To find the breakaway points,

$$N(s)D'(s) - N'(s)D(s) = (1)(3s^2 + 4.2s + 0.2) - 0$$
$$= 3(s^2 + 1.4s + 0.0667) = 0$$

Thus the breakaway points are among the roots of this equation, which are $s = -0.0494$ and $s = -1.351$. Since the second point is not on the root locus, $s = -0.0494$ is the only breakaway point.

The root locus is plotted in Figure 7.13. Note that the system becomes unstable with increasing gain; we will use the Routh–Hurwitz criterion to determine the range of the gain for stability. The characteristic equation is seen to be

$$1 + KG(s)H(s) = 0 = s^3 + 2.1s^2 + 0.2s + 0.01 K_a$$

The Routh array is then

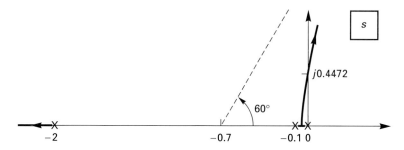

Figure 7.13 Root locus for Example 7.7.

$$
\begin{array}{c|cc}
s^3 & 1 & 0.2 \\
s^2 & 2.1 & 0.01\,K_a \\
s^1 & \dfrac{0.42 - 0.01\,K_a}{2.1} & \Rightarrow \quad K_a < \dfrac{0.42}{0.01} = 42 \\
s^0 & 0.01\,K_a & \Rightarrow \quad K_a > 0
\end{array}
$$

Hence the system is stable for $0 < K_a < 42$. For the marginally stable system, the gain of the power amplifier is $K_a = 42$, and the auxiliary polynomial of the Routh array (from the coefficients of the s^2-row) is

$$ Q_a(s) \;=\; 2.1 s^2 + 0.01\,(42) \;=\; 2.1\,(s^2 + 0.2) $$

Thus, for $K_a = 42$, the system will oscillate with a frequency of $\omega = \sqrt{0.2} = 0.4472$ rad/s. The period of the oscillation is $T = (2\pi/0.4472) = 14.1$ s; we expect this period to be large because of slow dynamics of the ship.

From the last example, if the pole at $s = -2$ in the open-loop function is ignored, the system is stable for all positive gains of the power amplifier. If this pole is included in the model of the system, the system is stable only for $0 < K_a < 42$. We will comment on this point further after the next example.

Example 7.8

We continue the analysis of the ship-steering system of the last two examples and consider the complete fourth-order system model. Hence, from Figure 7.11, the open-loop function is

$$ KG(s)H(s) \;=\; \frac{0.05\,K_a}{s(0.05 s + 1)(0.5 s + 1)(10 s + 1)} $$

$$ \;=\; \frac{0.2\,K_a}{s(s + 20)(s + 2)(s + 0.1)} \;=\; \frac{K}{s^4 + 22.1 s^3 + 42.2 s^2 + 4 s} $$

It is left as an exercise for the reader to apply the root-locus rules. The resulting root locus is given in Figure 7.14. Note that the system becomes unstable with increasing gain; we will use the Routh–Hurwitz criterion to determine the range of the gain for stability. The characteristic equation is seen to be

$$ 1 + KG(s)H(s) \;=\; 0 \;=\; s^4 + 22.1 s^3 + 42.2 s^2 + 4 s + 0.2\,K_a $$

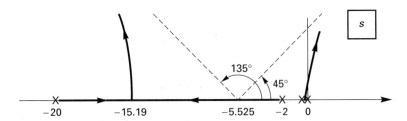

Figure 7.14 Root locus for Example 7.8.

The Routh array results in $0 < K_a < 38.03$ for stability. Hence, $K_a = 38.03$ for the marginally stable system. The Routh array yields the auxiliary polynomial

$$Q_a(s) = 42.02s^2 + 0.2(38.03) = 42.02(s^2 + 0.1810)$$

Thus, for $K_a = 38.03$, the system will oscillate with a frequency of $\omega = \sqrt{0.1810} = 0.4254$ rad/s. The following MATLAB program calculates the breakaway point and verifies the root locus.

```
roots([4 66.3 84.4 4]), pause
rlocus ([1], [1 22.1 42.2 4 0])
```

The last three examples illustrate root-locus construction and also the reduction of the order of systems. The comparisons of these three examples are given in Tables 7.3, 7.4, and 7.5. Table 7.3 indicates that the second-order model (Example 7.6) of the system is

TABLE 7.3 STABILITIES FOR EXAMPLES 7.6, 7.7, AND 7.8

Example	K_a, gain for marginal stability	Frequency of oscillation
7.6	Always stable	—
7.7	42	0.4472 rad/s
7.8	38.03	0.4254 rad/s

TABLE 7.4 POLES FOR EXAMPLES 7.6, 7.7, AND 7.8

Example	$K_a = 1$	$K_a = 20$	$K_a = 40$
7.6	$-0.05 \pm j0.05$	$-0.05 \pm j0.3123$	$-0.05 \pm j0.4444$
7.7	$-0.0486 \pm j0.0515$	$-0.0250 \pm j0.3113$	$-0.00218 \pm j0.4369$
	-2.003	-2.050	-2.0096
7.8	$-0.0486 \pm j0.0513$	$-0.0226 \pm j0.3111$	$0.00236 \pm j0.4358$
	-2.002	-2.055	-2.106
	-19.99997	-19.9994	-19.9990

inadequate if the gain is large (of the order of 40), since the system is unstable for large gains. This instability is not indicated in the second-order model. Also, from Table 7.4, the dominant poles for $K_a = 1$ are approximately equal for all three models; this is not true for

$K_a = 20$ and $K_a = 40$, indicating again that the second-order model is not adequate for higher values of gain.

For gains of approximately 20, the third-order model is adequate; adding the pole at $s = -20$ does not significantly affect the damping ratio ζ and the natural frequency ω_n of the dominant poles for this gain (Table 7.5). For gains of approximately 40 and larger, the fourth-order model is required. The third-order model indicates that the system is stable for $K = 40$, while it is actually unstable.

These three examples indicate that this system may be approximated with the complex dominant poles for low gains, but for higher gains it is necessary to increase the order of the system. This may be viewed by considering the angle contributions of the poles to the angle criterion (7-8) of the root locus. For low gains, the angles contributed by the poles at $s = -2$ and $s = -20$ are negligible. For somewhat larger gains, the angle contributed by the pole at $s = -2$ can no longer be ignored. For even larger gains, neither of the angles can be ignored.

As a final point regarding order reduction for a system model, note that zeros and poles must be eliminated in a manner that does not change the system's low-frequency characteristics. A first-order factor must first be expressed as $(\tau s + 1)$; then the elimination of this factor does not affect the system's low-frequency characteristics. A second-order factor should be expressed as

$$\text{factor} = \left(\frac{s}{\omega_n} \right)^2 + 2\zeta \frac{s}{\omega_n} + 1$$

so as not to change the system's low-frequency characteristic.

TABLE 7.5 ζ AND ω_n FOR EXAMPLES 7.6, 7.7, AND 7.8

Example	$K_a = 1$	$K_a = 20$	$K_a = 40$
7.6	$\zeta = 0.707$	$\zeta = 0.158$	$\zeta = 0.111$
	$\omega_n = 0.0707$	$\omega_n = 0.562$	$\omega_n = 0.669$
7.7	$\zeta = 0.686$	$\zeta = 0.080$	$\zeta = 0.0497$
	$\omega_n = 0.0708$	$\omega_n = 0.559$	$\omega_n = 0.661$
7.8	$\zeta = 0.688$	$\zeta = 0.072$	Unstable
	$\omega_n = 0.0707$	$\omega_n = 0.558$	

Six rules for the construction of a root locus have been developed. Many additional rules are given in the literature, and these rules are valuable to those who wish to construct accurate root loci graphically. However, as has been stated previously, a digital computer can calculate a root locus that is much more accurate than any graphical procedure will yield. For accurate construction of a root locus, a digital computer should be used; for quickly sketching a root locus to give insight into the characteristics of a system, the rules given here should be sufficient.

The rules in Table 7.2 are stated for the case that the system characteristic equation is

$$1 + KG(s)H(s) = 0$$

In the more general case, the characteristic equation may be stated as

$$\Delta(s) = 1 + F(s) = 1 + KF_1(s) = 0$$

where $\Delta(s)$ is Δ from Mason's gain formula and $F(s) = KF_1(s)$ is the open-loop function for the system. For this case, the function $G(s)H(s)$ in the preceding rules should be replaced with the function $F_1(s)$. Of course, $F_1(s)$ cannot be a function of K.

7.4 ADDITIONAL PROPERTIES OF THE ROOT LOCUS

In this section some additional properties of the root locus are given. These properties will aid in the rapid sketching of root loci for low-order systems.

Sketching root loci relies heavily on experience. It is useful to have seen many root loci previously to aid in giving quick, accurate sketches. Table 7.6 gives several loci for low-order systems; these should be studied to familiarize the reader with some of the characteristics of root loci. The root loci of parts (f) and (g) of this table illustrate some of the problems in sketching root loci. Each open-loop function has one real pole and two complex poles. However, note the effect on the root loci of shifting the complex poles relative to the real pole, For an open-loop function of this type, the calculation of the breakaway points will indicate which of the sketches apply, since (f) has two breakaway points to the left of p_1 and (g) has none. An important characteristic of the root locus is now derived, which will extend the usefulness of Table 7.2. As shown in the proof of rule 5, the system characteristic equation can be written as, in (7-22),

$$D(s) + KN(s) = 0$$

where $N(s)$ and $D(s)$ are the numerator and the denominator polynomials, respectively, of $G(s)H(s)$. For a given gain K_1, the roots of the characteristic equation occur on the root locus at the values of s that satisfy the equation

$$D(s) + K_1 N(s) = 0 \tag{7-31}$$

Suppose now that K is increased by an amount K_2 from the value K_1, that is, suppose that $K = K_1 + K_2$. Then the characteristic equation becomes

$$D(s) + (K_1 + K_2) N(s) = 0$$

which can be expressed as

$$1 + K_2 \left[\frac{N(s)}{D(s) + K_1 N(s)} \right] = 1 + K_2 \frac{N(s)}{D_1(s)} = 0 \tag{7-32}$$

With respect to the gain K_2, the locus appears to originate on roots as placed by letting $K = K_1$ but still terminates on the zeros of the original open-loop function. Hence the branches follow the same trajectories as the original root locus, but the starting points are different. The following example illustrates this concept.

Example 7.9

For the system of Example 7.5,

$$G(s)H(s) = \frac{s+1}{s^2}$$

and the characteristic equation is given by

$$s^2 + K(s + 1) = 0$$

For $K = 2$, this equation becomes

$$s^2 + 2s + 2 = (s + 1)^2 + 1$$

with roots at $s = -1 \pm j$. Thus the root locus for the system with the open-loop function

$$KG(s)H(s) = \frac{K(s + 1)}{s^2 + 2s + 2}$$

is that of Example 7.5, as shown in Figure 7.10, except that the locus begins at the points $s = -1 \pm j$. The resultant root locus is shown in Figure 7.15. This example illustrates that the loci given in Table 7.6 can be interpreted to represent many systems in addition to those for which the loci are sketched.

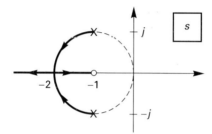

Figure 7.15 Example 7.9.

An additional important property is now derived. The equation for a root locus is given by

$$1 + KG(s)H(s) = 0 \qquad (7\text{-}33)$$

Consider now the root locus that is generated by replacing s with $s - s_1$ in (7-33), that is, the root locus that is generated by

$$1 + KG(s - s_1)H(s - s_1) = 0 \qquad (7\text{-}34)$$

where s_1 is a real number. Suppose that, for a given $K = K_0$, $s = s_0$ satisfies (7-33). Then for the same value K_0, $(s - s_1) = s_0$ satisfies (7-34). That is, for that value K_0, the value

$$s = s_0 + s_1$$

satisfies (7-34). Thus the effect of shifting all poles and zeros of the open-loop function by a constant amount s_1 is to shift the entire root locus by the same amount. This property, which greatly extends the usefulness of Table 7.6, is illustrated by the following example.

Example 7.10

The root locus for the system for which

$$G(s)H(s) = \frac{s + 1}{s^2}$$

TABLE 7.6

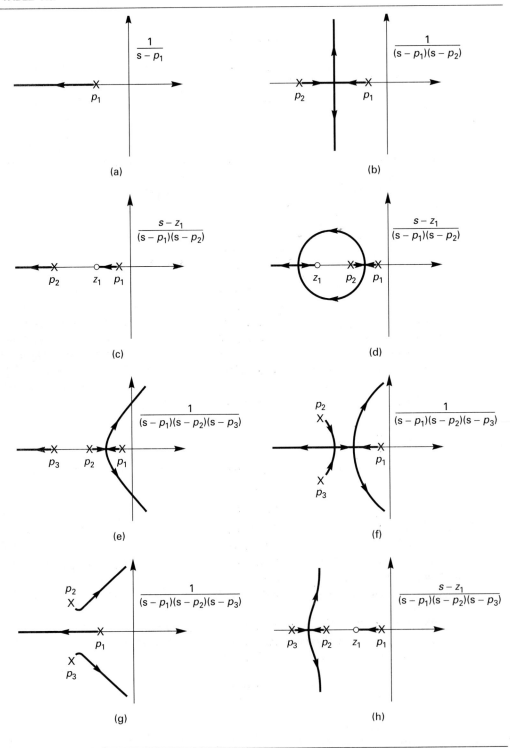

was derived in Example 7.5 and is repeated in Figure 7.16(a). Thus the root locus for the system for which

$$G(s)H(s) = \frac{(s+2)+1}{(s+2)^2} = \frac{s+3}{(s+2)^2}$$

is of the same geometry but is shifted two units to the left in the s-plane. This locus is shown in Figure 7.16(b).

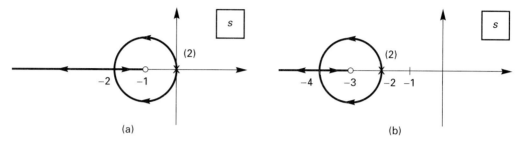

Figure 7.16   Root locus for Example 7.10.

The last property to be discussed is presented without proof. This property is that a root locus possesses symmetry at a breakaway point. For example, if two loci come together and then break away, the angles between the four branches at the breakaway point will always be $360°/4 = 90°$. This property is illustrated in Figures 7.2 and 7.16. If three loci come together and break away, the angles between the six branches at the breakaway point will always be $360°/6 = 60°$. If the locus originates on a third-order pole, as in Figure 7.5, the angles between the three branches will be $120°$. For the locus originating on a second-order pole, the angles between branches will be $180°$, as in Figure 7.16.

7.5 OTHER CONFIGURATIONS

In all the preceding sections, the root locus was presented as a method of plotting the characteristic equation roots for the equation

$$1 + KG(s)H(s) = 0 \tag{7-35}$$

where the system has the open-loop function $KG(s)H(s)$. The gain K appears *linear* in this equation and is varied over the range $0 \le K < \infty$. However, this procedure may be extended to other system configurations and to parameters other than gain. All that is necessary is for the characteristic equation to be manipulated into the form of (7-35), where K is the parameter that varies and $G(s)H(s)$ is a function that satisfies the characteristic equation and is not a function of K.

We now illustrate this procedure with an example. Consider the closed-loop system of Figure 7.17(a). We wish to plot the roots of the system characteristic equation as the parameter α varies. First we write the characteristic equation.

$$1 + G(s) = 1 + \frac{5}{s(s+\alpha)} = 0$$

or

$$s^2 + \alpha s + 5 = 0$$

Next we group terms that are not multiplied by α and divide the equation by these terms.

$$(s^2 + 5) + \alpha s = 0$$

and

$$1 + \alpha \frac{s}{s^2 + 5} = 0$$

The equation is of the form of (7-35), where the equivalent gain is α and the equivalent function $G(s)H(s)$ is $s/(s^2 + 5)$. We then plot the root locus for these functions. This locus originates on the poles at $\pm j\sqrt{5}$ and terminates on the zeros at 0 and $-\infty$, as sketched in Figure 7.15(b).

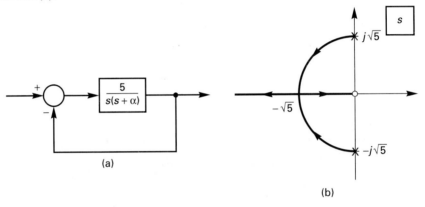

(a)

(b)

Figure 7.17   Root locus as α varies.

For the general case, the system characteristic equation must be expressed as a polynomial equation, and the terms independent of the parameter α are grouped into a function $D_e(s)$. The remaining terms that are multiplied by α are grouped into a function $N_e(s)$, such that the characteristic equation is expressed as

$$D_e(s) + \alpha N_e(s) = 0$$

This equation is then divided by $D_e(s)$, resulting in

$$1 + \alpha \frac{N_e(s)}{D_e(s)} = 1 + K_e G_e(s) H_e(s) = 0 \tag{7-36}$$

with $K_e = \alpha$ and $G_e(s)H_e(s) = N_e(s)/D_e(s)$. The root locus is then plotted for the equivalent gain K_e with the equivalent open-loop function $K_e G_e(s)H_e(s)$. This procedure is illustrated by an example.

Example 7.11

We now use the root locus to design the temperature-control system of Example 5.7, which utilizes a PI controller. The system characteristic equation is given by

$$1 + G_c(s)G_p(s)H_k = 1 + \left(K_P + \frac{K_I}{s}\right)\left[\frac{0.25}{s + 0.1}\right]$$

$$= 1 + \frac{0.25(K_Ps + K_I)}{s(s + 0.1)}$$

Suppose that to meet steady-state error requirements, we specify K_I to be equal to unity, and we wish to plot a root locus as K_P is varied. The characteristic equation can then be expressed as

$$s(s + 0.1) + 0.25(K_Ps + 1) = 0$$

or, in terms of (7-36),

$$s^2 + 0.1s + 0.25 + 0.25K_Ps = 0$$

Then, from (7-37),

$$1 + 0.25K_P\frac{s}{s^2 + 0.1s + 0.25} = 0$$

The equivalent open-loop function has a zero at the origin and two poles at $s = -0.05 \pm j\sqrt{0.99}$ $/2 = -0.05 \pm j0.497$. The root locus is sketched in Figure 7.18. It is left as an exercise to the reader to verify the sketch.

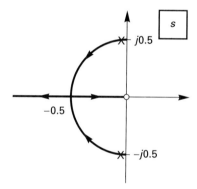

Figure 7.18 Root locus.

Now that the root locus is available as a function of K_P, the characteristic equation roots can be specified rather easily. Suppose, for example, that critical damping is desired. From Figure 7.18, the characteristic equation roots will appear at $s = -0.5$. From the preceding equivalent characteristic equation, the value of K_P that yields these roots is

$$K_P = -\frac{s^2 + 0.1s + 0.25}{0.25s}\bigg|_{s = -0.5} = \frac{0.45}{0.125} = 3.6$$

The required PI compensator has the transfer function

$$G_c(s) = 3.6 + \frac{1}{s}$$

and the design is complete. Note that K_I is chosen to satisfy a steady-state error requirement, and K_P is then chosen to satisfy a transient-response requirement.

7.6 ROOT-LOCUS DESIGN

The design of closed-loop control systems by root-locus techniques is introduced in this section. This technique allows us to control at least some of the closed-loop system pole locations and hence control the transient response to an extent. In addition, we see how the steady state response can also be influenced through root locus design.

We introduce root-locus design with a practical example—the design of an attitude-control system for a rigid satellite. The satellite is assumed to be rigid in order to keep the model second order (see Section 2.6). The assumed model of the satellite is given in Figure 7.19. The input to the model is the torque, $\tau(t)$, as generated by the thrustors, and the output is the attitude angle θ of the satellite. We first assume that the attitude control system is of the form shown in Figure 7.19(b). The modulator is the device that converts the electrical error signal into the thrustor torque that is directly proportional to the error signal. If the only design parameter is the gain K, the root locus is as shown in Figure 7.19(c), since

$$KG(s)H(s) = K\frac{1}{s^2} \tag{7-37}$$

Hence the satellite attitude θ will vary sinusoidally for any value of K, since the system is marginally stable for all $K > 0$. Obviously, the control-system structure of Figure 7.19(b) is unacceptable.

To obtain an acceptable design, the control system shown in Figure 7.20(a) will be employed. The total feedback signal is now the sum of position and of the velocity multiplied by a gain K_v. This control system is said to utilize *rate feedback,* which is also called *velocity* or *derivative feedback.* The open-loop function is now, from Mason's gain formula,

$$KG(s)H(s) = K\left[\frac{K_v}{s} + \frac{1}{s^2}\right] = K\left[\frac{K_v s + 1}{s^2}\right]$$

$$= KK_v\left[\frac{s + 1/K_v}{s^2}\right] \tag{7-38}$$

The effect of adding rate feedback is then to change the gain term to KK_v and to add a zero at $s = -1/K_v$. The root locus for this system is shown in Figure 7.20(b), and the system is seen to be stable for all $K > 0$, $K_v > 0$.

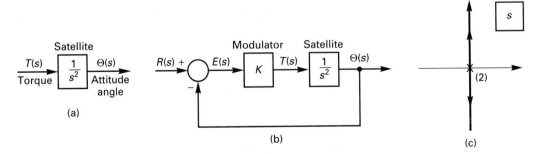

Figure 7.19   Design example.

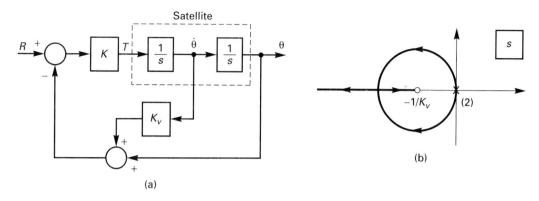

Figure 7.20   Satellite control system with rate feedback.

It can be shown that since there are two design parameters K and K_v, the two roots of the characteristic equation of the attitude-control system can be placed at any point in the s-plane (see Problem 7.22). One implication of this result is that the system can be made to respond arbitrarily fast. This response can be realized in the model; however, remember that no model exactly corresponds to the physical system. For example, for the satellite to rotate with an arbitrarily fast rise time, an unbounded torque must be available from the thrustors. Of course, this type of torque is not available in any physical system, but the model that we have used does not show any physical limits that must exist on the torque. The point of this discussion is that in design, we must not lose sight of the characteristics of the physical system, since the design procedure is normally performed using an imperfect (and linear) model of the system.

7.6.1 Rate Feedback

The limitations just discussed exist using any design procedure. Nevertheless, the preceding development does show the value of the root locus in design. In addition, the design also shows the value of rate feedback, which is implemented in many physical control systems. Because of its importance, rate feedback is discussed further at this point, and this will lead us to some general design considerations. Then root-locus design principles are developed.

We now show that rate feedback is actually a PD (proportional-plus-derivative) compensation. As shown previously, rate feedback adds a zero to the open-loop function, and we can consider this addition to be a compensator of the satellite system with the transfer function

$$G_c(s) = KK_v\left(s + \frac{1}{K_v}\right) \tag{7-39}$$

[compare (7.38) with (7.37)]. The PD compensator is *defined* by the equation

$$m(t) = K_P e(t) + K_D \frac{de(t)}{dt} \tag{7-40}$$

where $e(t)$ is the compensator input and $m(t)$ its output. Hence the transfer function is

$$M(s) = G_c(s)E(s) = (K_P + K_D s)E(s) \tag{7-41}$$

It is seen then, from (7-39) and (7-41), that rate feedback is a form of PD compensation with

$$K_P = K \qquad K_D = KK_v \qquad (7\text{-}42)$$

A problem with PD compensation is evident from the transfer function in (7-41), in that the gain of the compensator continues to increase with increasing frequency; that is

$$G_c(j\omega) = K_P + j\omega K_D \qquad (7\text{-}43)$$

If high-frequency noise is present in the system, this noise will be amplified by the PD compensator, which is generally unacceptable. The usual method for overcoming this problem is to add a pole to the compensator transfer function, which limits the high-frequency gain. Then

$$G_c(s) = K_p + K_D \frac{s}{s - p_o} = \frac{K_c(s - z_o)}{(s - p_o)} \qquad (7\text{-}44)$$

Now $G_c(j\omega)$ approaches the value of K_c as ω approaches infinity. This transfer function is the general form of a first-order compensator, with three parameters (K_c, z_0, p_0) to be determined by the design process. With these three parameters we can expect to be better able to satisfy given design criteria, compared to the case in which we have only one or two design parameters.

7.6.2 General First–Order Compensators

The pole and the zero of the compensator transfer function are usually confined to the left half-plane; that is, z_0 and p_0 are negative real numbers. Thus, with respect to the pole and zero, we have two possibilities: (1) $|z_0| < |p_0|$ and (2) $|z_0| > |p_0|$. The system design is usually classified as being one of these two cases. For the first case, the zero is closer to the origin than the pole, resulting in a contribution to the angle criterion of the root locus that is always positive. This angle contribution is seen from Figure 7.21, where θ_1 is always

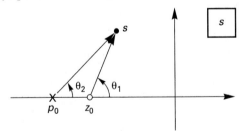

Figure 7.21   Phase of compensator.

greater than θ_2 for the value of s in the upper half-plane. This type of compensator is called a *phase-lead* compensator. The second case, with the pole closer to the origin than the zero, is called *phase-lag,* since the angle contribution is always negative for the case shown. These names, phase-lead and phase-lag, have a clearer meaning with respect to frequency-response design, which is covered in Chapter 9. The design procedures are different for the two compensators, as are the effects of the compensators on the control system. Root-locus design procedures for phase-lead and phase-lag compensators are developed in the next three sections.

7.7 PHASE-LEAD DESIGN

In this section we consider the design of phase-lead compensators by root-locus techniques. The transfer function of the first-order phase-lead compensator is given by

$$G_c(s) = \frac{K_c(s - z_0)}{(s - p_0)} \tag{7-45}$$

where $|z_0| < |p_0|$, and K_c, z_0 and p_0 are to be determined to satisfy certain design criteria. First, before we present the design procedures, consider the characteristic equation of the compensated system.

$$1 + KG_c(s)G_p(s)H(s) = 1 + \frac{KK_c(s - z_0)}{s - p_0}G_p(s)H(s)$$

Obviously the product KK_c can be considered to be a single gain parameter. Quite often we consider K to be unity and solve for K_c. Actually, we are solving for the product KK_c.

In the preceding section, it was shown that a phase-lead compensator contributes a positive angle to the root-locus angle criterion

[eq. (7-10)] $\sum(\text{all angles from finite zeros}) - \sum(\text{all angles from finite poles})$

$$= r(180°) \qquad r = \pm 1, \pm 3, \pm 5, \ldots$$

We now show that the positive angle contributed by the phase-lead compensator will tend to shift the root locus of the plant toward the left in the s-plane, that is, toward the more stable region. Hence a phase-lead compensator will improve the transient response of the system by improving the system's stability, and, as we shall see, the speed of response is also increased.

Consider again the satellite example in the last section. The root locus of the uncompensated system is given again in Figure 7.22(a), and it is noted again that the system is marginally stable. Suppose that the system specifications require that the dominant closed-loop poles be at s_1 and $\bar{s}_1$. The contribution of the plant poles to the angle criterion, (7-10), is $-2\theta_1$, and this contribution is obviously more negative than $-180°$. However, when the phase-lead compensator is added, as shown in Figure 7.20(b), the angle criterion of (7-20) becomes

$$\theta_z - \theta_p - 2\theta_1 = -180°$$

Hence the design problem is to choose z_0 and p_0 such that this equation is satisfied.

The last paragraph illustrates root-locus design for a particular plant. Consider now an uncompensated system. In general, the order of the denominator of the plant is greater than that of the numerator (more finite poles than finite zeros). As a result, a shift of the root locus of the uncompensated system to the left in the s-plane will make the sum of angles in the angle criterion of (7-10) less than 180°. Hence a phase-lead compensator, which adds positive angle to the sum of angles in the angle criterion, is *required* if we desire to shift the root locus to the left.

The design can be accomplished by trial and error. In Figure 7.22(b), given s_1, a value for z_0 can be arbitrarily chosen such that when an appropriate value p_0 is chosen to the left of z_0, the angle criterion can be satisfied. There are obviously many compensators that will satisfy the angle criterion, if at least one can be found. However, an analytical solution exists for this design problem. This analytical procedure will now be presented.

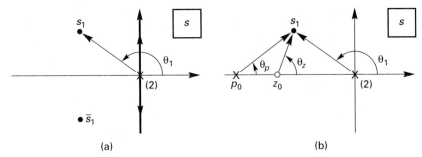

Figure 7.22   Angle effects of compensation.

7.8 ANALYTICAL PHASE-LEAD DESIGN

This section presents an analytical procedure for phase-lead design. The proofs of the equations are not given here [4]. For this procedure it is convenient to express the compensator transfer function as

$$G_c(s) = \frac{a_1 s + a_0}{b_1 s + 1} \tag{7-46}$$

Note that this form of the compensator transfer function is general. The object of the design is to choose a_0, a_1, and b_1 such that given s_1, the equation

$$KG_c(s)G_p(s)H(s)\big|_{s = s_1} = -1 \tag{7-47}$$

is satisfied; that is, we are designing a compensator that places a root of the system characteristic equation at $s = s_1$. In general, s_1 must be considered to be complex.

In equation (7-47) we have four unknowns, K, a_0, a_1, and b_1, and only two relationships (magnitude and angle) that must be satisfied. Hence we can assign values to two of the unknowns. Since we can write

$$KG_c(s) = \frac{Ka_1 s + Ka_0}{b_1 s + 1} \tag{7-48}$$

we see that in fact there are only three independent unknowns, and we let $K = 1$ to simplify the design procedure.

From the equations above we write, with $K = 1$,

$$G_c(s)G_p(s)H(s)\big|_{s = s_1} = \left[\frac{a_1 s_1 + a_0}{b_1 s_1 + 1}\right] G_p(s_1)H(s_1) = -1 \tag{7-49}$$

In general, s_1 is complex and hence each factor in this equation is complex. Equating magnitudes and angles, we get the two equations

$$\left| \frac{a_1 s_1 + a_0}{b_1 s_1 + 1} \right| |G_p(s_1)H(s_1)| = 1$$

(7-50)

$$\arg\left[\frac{a_1 s_1 + a_0}{b_1 s_1 + 1} \right] + \arg[G_p(s_1)H(s_1)] = 180°$$

We have two equations and the three unknowns a_0, a_1, and b_1; hence we can specify any one of the unknowns and solve for the other two. In general, we derive the desired pole location, s_1, from the transient–response specifications for the control system. In addition, we assume that the compensator dc gain a_0 is calculated from the steady state specifications. For this case, in (7-50), only a_1 and b_1 are unknown.

First we express

$$s_1 = |s_1| e^{j\beta}$$

(7-51)

and

$$G_p(s_1)H(s_1) = |G_p(s_1)H(s_1)| e^{j\psi}$$

(7-52)

Solving (7-50) for a_1 and b_1 [4] yields

$$a_1 = \frac{\sin\beta + a_0 |G_p(s_1)H(s_1)| \sin(\beta - \psi)}{|s_1||G_p(s_1)H(s_1)| \sin\psi}$$

$$b_1 = \frac{\sin(\beta + \psi) + a_0 |G_p(s_1)H(s_1)| \sin\beta}{-|s_1| \sin\psi}$$

(7-53)

Given a_0, $G_p(s)H(s)$, and the desired closed-loop pole location s_1, (7-53) gives the remaining compensator coefficients. This procedure places a closed-loop pole at s_1; however, the locations of the remaining closed-loop poles are unknown and may not be satisfactory. In fact, *some may be in the right half-plane*, resulting in an unstable system. Of course, such a design is unsatisfactory.

For the case that ψ is 180° (s_1 on the negative real axis), Equations (7-53) must be modified to give the single equation

$$a_1 |s_1| \cos\beta - \frac{b_1 |s_1|}{|G_p(s_1)H(s_1)|} \cos\beta - \frac{1}{|G_p(s_1)H(s_1)|} + a_0 = 0$$

(7-54)

For this case, the value of either a_1 or b_1 can also be assigned. An example is now given to illustrate this design procedure.

Example 7.12

In this example we design the satellite system of Section 7.6 and Figure 7.19, where

$$G_p(s) = \frac{1}{s^2} \qquad H(s) = 1$$

We choose

$$s_1 = |s_1| e^{j\beta} = -2 + j2 = 2\sqrt{2} e^{j135°}$$

to yield $\zeta = 0.707$ and $\tau = 0.5$ s for these poles. Then

$$G_p(s_1)H(s_1) = |G_p(s_1)H(s_1)| e^{j\psi} = \frac{1}{s^2}\bigg|_{s_1 = -2 + j2} = \frac{1}{8} e^{-j270°}$$

Hence $\beta = 135°$ and $\psi = -270°$. We somewhat arbitrarily choose the dc gain a_0 to be 8/3. Then from (7-53),

$$a_1 = \frac{\sin 135° + (\frac{8}{3})(\frac{1}{8})\sin 45°}{(2\sqrt{2})(\frac{1}{8})\sin(-270°)} = \frac{(1/\sqrt{2})(1 + 1/3)}{\sqrt{2}/4} = \frac{8}{3}$$

$$b_1 = \frac{\sin(-135°) + (\frac{8}{3})(\frac{1}{8})\sin(135°)}{-(2\sqrt{2})\sin(-270°)} = \frac{-(1/\sqrt{2})(1 - \frac{1}{3})}{-2\sqrt{2}} = \frac{1}{6}$$

The compensator is then

$$KG_c(s) = \frac{(\frac{8}{3})s + \frac{8}{3}}{(\frac{1}{6})s + 1}$$

This design is verified with the MATLAB program

```
a0 = 8/3;
sl = -2+2*j;
Gp = tf([1], [1 0 0]);
Gpsl = evalfr(Gp, sl);
beta = angle(sl);
psi=angle (Gpsl);
al = (sin(beta)+a0*abs(Gpsl)*sin(beta-psi))/...
(abs(sl) * abs(Gpsl)*sin(psi))
bl = (sin(beta+psi)+a0*abs(Gpsl)*sin(beta))/...
(-abs(sl)*sin(psi))
```

In Example 7.12 the value of a_0 (the compensator dc gain) was chosen in an arbitrary manner. This coefficient could have been chosen to satisfy some design specification such as steady-state errors, steady-state disturbance rejection, sensitivity specifications, and so forth. In a more practical sense, compensators could be designed for several different values of a_0. Since the model of the plant is never exact, the physical system will never respond exactly as the model does, and in fact the response may be quite different. Then the many different designs may be tested in the physical system or in a more accurate simulation of the systems (including nonlinearities), and perhaps one of the designs will be satisfactory. If none of the designs result in a satisfactory response, the experience gained in testing the system may suggest another approach to try. Generally design is an *iterative procedure;* usually we are not successful on the first try with the physical system, even though we may be quite successful with the model. The next example illustrates some of the topics mentioned in this paragraph.

Example 7.13

The satellite control system of Example 7.12 is considered further. To investigate the effects of the choice of the compensator dc gain on the system response, the design of Example 7.12 was repeated for several different values of a_0. The resulting compensator transfer functions are given in Table 7.7. The compensated system is third order, with two of the closed-loop transfer function poles placed at $s = -2 \pm j2$. The location of the third pole, obtained by computer calculation, is also given in Table 7.7 for each of the compensators. Compensators were designed for compensator dc gains in the range of 0.1 to 6.0. For dc gains above a value of approximately 7, the design procedure yielded an unstable closed-loop system, that is, the third pole was in the right half-plane.

Note from Table 7.7 that, as the compensator dc gain increases, the time constant of the third closed-loop pole decreases from approximately 20 s to 0.083 s. Unit step responses for the compensated systems are given in Figure 7.23; note that the overshoot in these step responses increases as the compensator dc gain increases. For the closed-loop poles at $s = -2 \pm j2$, $\zeta = 0.707$, and the standard second-order system has an overshoot of approximately 5 percent, from Figure 4.8. The overshoot for the designed third-order systems of this example varies from approximately

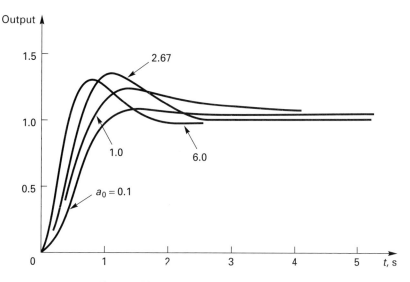

Figure 7.23   Step responses for Example 7.13.

TABLE 7.7   RESULTS FOR EXAMPLE 7.13

Compensator dc gain	$G_c(s)$	Location of third pole	Percent overshoot
0.1	$\dfrac{8.203s + 0.4051}{s + 4.051}$	-0.0506	7
1.0	$\dfrac{10.29s + 4.571}{s + 4.571}$	-0.571	23
2.667	$\dfrac{16.0s + 16.0}{s + 6.0}$	-2.0	33
6.0	$\dfrac{56.0s + 96.0}{s + 16.0}$	-12.0	30

7 to 33 percent. For the system with the compensator dc gain of 0.1, the overshoot is approximately 7 percent, but the time constant of the third pole is approximately 20 s. For the system with the compensator dc gain of 6, the overshoot is approximately 30 percent, but the time constant of the third pole is 0.083 s. Hence we see a trade-off between overshoot and settling time.

One additional point can be made concerning this example. For the compensator with the dc gain of 0.1, the compensator high-frequency gain is given by

$$\lim_{\omega \to \infty} G_c(j\omega) = \lim_{\omega \to \infty} \frac{8.203(j\omega) + 0.4051}{j\omega + 4.051} = 8.203$$

Hence the ratio of high-frequency gain to dc gain for the compensator is 82.03, which is *very* high. By the same procedure, we can calculate the ratio for the compensator with the dc gain of 6 to be 56/6, or 9.33, which is much more satisfactory. High-frequency noise problems and nonlinear operation would probably occur in the physical system employing the compensator with the dc gain of 0.1.

For this design, achieving a reasonable settling time results in the system having a large overshoot. However, reducing the overshoot increases the settling time and possibly introduces high-frequency problems. This example illustrates some of the aspects of trade-off in design.

An important practical point is now discussed. The compensator has the transfer function

[eq. (7-45)] $$G_c(s) = \frac{K_c(s - z_0)}{s - p_0}$$

and its frequency response function is given by

$$G_c(j\omega) = \frac{K_c(j\omega - z_0)}{j\omega - p_0} \qquad (7\text{-}55)$$

The dc gain is $G_c(0) = K_c z_0 / p_o$, and the high-frequency gain is

$$\lim_{\omega \to \infty} G_c(j\omega) = K_c$$

Hence the ratio of high-frequency gain to dc gain is p_0/z_0, and this ratio is greater than 1 for the phase-lead compensator. If we choose this ratio to be too large, high-frequency noise problems can, and probably will, become significant. A rule of thumb that is sometimes used is to limit this ratio to 10, but this depends on the physical system under consideration and particularly on the noise present.

Example 7.14

This example is a continuation of the design in Example 7.13. We will consider the computer calculation of Table 7.7 and Figure 7.23, using MATLAB. A program that calculates the data of Table 7.7 is given by

```
a0 = [0.1 1 2.667 6];
Gp = tf([1],  [1 0 0]);
s1 = -2 + i*2;  s1mag = abs(s1);  beta = angle(s1);
Gps1 = evalfr(Gp,s1);
Gps1mag = abs(Gps1);  psi = angle(Gps1);
t = 0:0.05:5;       %  for simulation
for k = 1:4
   a1 = (sin(beta) + a0(k)*Gps1mag*sin(beta - psi))/...
            (s1mag*Gps1mag*sin(psi));
   b1 = (sin(beta + psi)  + a0(k)*Gps1mag*sin(beta))/...
            (-s1mag*sin(psi));
   Gc = tf([a1 a0(k)], [b1 1]);
   T = minreal(Gc*Gp/(1+Gc*Gc*Gp));
      pole (T)
%      step(T,t)
%      hold on
end
%hold off
```

In this program, a0 = a_0, a1 = a_1, and b1 = b_1, the compensator parameters in (7-46). The design equations of (7-53) are realized directly. The variable names beginning with Gp are plant parameters, those beginning with Gc are compensator parameters, and those with T are the closed-loop system parameters.

The time responses of Figure 7.23 can be generated by the MATLAB program above, by deleting the three % symbols, and placing a % symbol on the *pole(T)* statement.

The MATLAB programs illustrate one procedure for designing a control system using a digital computer. We can easily increase the number of values for the dc gain a_0 and determine the system time constants from the resultant pole locations. We can then investigate the overshoot in the step responses for those values of a_0. For further investigations, we can vary the chosen locations of the two dominant poles, s_1 and $\bar{s}_1$.

Once we have decided on several compensator transfer functions, we can then test these compensators in a more complete system simulation, which would possibly include a higher-ordered plant transfer function for more accuracy, system nonlinearities, and typical input signals and disturbances that may appear in the physical system.

7.9 PHASE-LAG DESIGN

In this section we consider the design of phase-lag compensators. As in the preceding sections, we assume that the compensator transfer function is first order and is given by

[eq. (7-45)]
$$G_c(s) = \frac{K_c(s - z_0)}{s - p_0}$$

It was shown that the effect of the addition of phase-lead compensation is to shift the root locus to the left in the s-plane, since the compensation adds a positive angle to the angle criterion. The compensator in (7-45) is a *phase-lag compensator* if $|z_0| > |p_0|$ and will always add a negative angle to the angle criterion. Hence the phase-lag compensator will tend to shift the root locus to the right in the s-plane, that is, toward the unstable region. Thus, in general, the angle contribution of the phase lag compensator *must be small*, to minimize the destabilizing effects. This small contribution is assured by placing the pole and the zero of the compensator very close to each other. The design requirements of the phase-lag compensator are now developed.

As was shown earlier, the phase-lead compensator can be used to shift the root locus a significant amount in the s-plane. However, the phase-lag compensator as normally used shifts the root locus an insignificant amount. Hence the transient-response characteristics from the dominant poles of a system are not changed appreciably by the phase-lag compensator. Instead the phase-lag compensator is used to improve the system steady-state response, as we now show.

First, for convenience in the design, we assume that the compensator of (7-45) has unity dc gain, that is,

$$G_c(s)\big|_{s=0} = \frac{K_c z_0}{p_0} = 1 \tag{7-56}$$

Thus

$$K_c = \frac{p_0}{z_0} < 1 \tag{7-57}$$

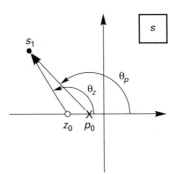

Figure 7.24 Angles for phase-lag compensation.

Suppose that the root locus of the uncompensated system passes through the value s_1 for the gain K_0 and that this point on the root locus gives a satisfactory transient response. Since the characteristic equation of the uncompensated system is

$$1 + KG_p(s)H(s) = 0,$$

then

$$K_0 = \frac{-1}{G_p(s_1)H(s_1)} \qquad (7\text{-}58)$$

As stated earlier, we choose the values of z_0 and p_0 to be approximately equal, and for reasons that will become evident later, we choose the magnitudes of z_0 and p_0 small compared to $|s_1|$ (see Figure 7.24). Then

$$(7\text{-}59)$$

$$G_c(s_1) = \frac{K_c(s_1 - z_0)}{s_1 - p_0} \simeq K_c$$

Now the gain required to place a root of the locus at approximately s_1 for the compensated system is given by

$$K = \frac{-1}{G_c(s_1)G_p(s_1)H(s_1)} \simeq \frac{-1}{K_cG_p(s_1)H(s_1)} = \frac{K_0}{K_c} \qquad (7\text{-}60)$$

from (7-58) and (7-59). From (7-57), since K_c is less than one, K is greater than K_0. The compensator has been chosen to have a unity dc gain; thus the loop dc gain has been increased, but the transient response appears to remain unaffected. Hence the steady-state response of the system has been improved, and this is the principal use of the phase-lag compensator.

Note that, by choosing z_0 and p_0 small in magnitude compared to $|s_1|$, we can satisfy (7-59) and still make K_c in (7-57) much smaller than 1. Hence we can get a significant improvement in the steady-state response. If the magnitudes of z_0 and p_0 are large compared to $|s_1|$, then z_0 and p_0 must be approximately equal to satisfy (7-59), and K_c in (7-57) is approximately unity. For this case (7-60) indicates no significant improvement in the steady-state response.

The steps in designing a phase-lag compensator may be summarized as follows:

1. Choose K_0 in (7-58) to yield the value of the desired closed-loop pole s_1 in the uncompensated system.

2. Calculate the value of K required to yield the desired steady-state response, assuming that the compensator dc gain is unity. Then solve (7-60) for $K_c = K_0/K$.

3. Choose the magnitude of the compensator zero, $|z_0|$, small compared to $|s_1|$.

4. Solve (7-56) for the compensator pole, $p_0 = K_c z_0$.

The compensator transfer function is then given by (7-45), and the gain K is increased to the value calculated in step 2.

We now illustrate phase-lag design with an example. It seems reasonable to use the satellite example that was employed to illustrate phase-lead design; however, recall that phase-lag compensation shifts the uncompensated locus to the right in the s-plane. This shift will destabilize the satellite attitude control system [see Figure 7.19(c)]. A phase-lag compensation cannot be used with the satellite control system; phase-lead compensation must be employed. Instead of the satellite example, a servo system is used as an example.

Example 7.15

We design the radar tracking system of Figure 7.1 in this example. The uncompensated open-loop function is given by

$$KG_p(s)H(s) = \frac{K}{s(s+2)}$$

and the root locus is plotted in Figure 7.25. Suppose that the system design requirements are such that a time constant of 1 s and $\zeta = 0.707$ are satisfactory. Thus the root location $s = -1 + j$ is acceptable; this point is already on the locus, without the addition of compensation, for a value of $K = 2$. Then, in step 1 of the design procedure, $s_1 = -1 + j$ and $K_0 = 2$.

Suppose that this system is required to track aircraft that have essentially constant velocity, which will appear to the control system as a ramp input; that is, the antenna must rotate at a constant velocity to remain pointed directly at the aircraft. Also, suppose that a design specification is a steady-state error of $0.2°$ with a unit ramp input, to keep the aircraft well within the beam width of the antenna. From (5-38), the steady-state error for the compensated system is

$$e_{ss} = \frac{1}{K_v} = \lim_{s \to 0} \frac{1}{sKG_c(s)G_p(s)H(s)}$$

We have specified that the dc gain of the compensator will be unity; hence

$$\lim_{s \to 0} sKG_c(s)G_p(s)H(s) = \lim_{s \to 0} \frac{K}{s+2} = \frac{K}{2}$$

and thus

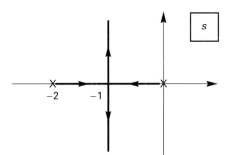

Figure 7.25   Root locus for Example 7.15.

$$e_{ss} = \frac{2}{K}$$

A value of $K = 10$ is then required to satisfy the steady-state error specifications in step 2. From (7-57) and (7-60),

$$K_c = \frac{p_0}{z_0} = \frac{K_0}{K} = \frac{2}{10}$$

since, from the preceding discussion, $K_0 = 2$. In step 3, we choose z_0 to be -0.1, which is small in magnitude compared to $|s_1| = \sqrt{2}$, and thus $p_0 = K_c z_0 = -0.02$ in step 4. Hence the compensator transfer function is

$$G_c(s) = \frac{0.2(s + 0.1)}{s + 0.02}$$

and a gain of $K = 10$ is required. This design will be continued after some further discussion.

In the last example, the closed-loop system has two poles at $s \approx -1 \pm j$. However, the addition of the compensator increases the system order to three; hence the system has a third pole. The approximate position of this third pole can be seen from the system root locus, which is constructed in the next example.

Example 7.16

The open-loop function for the radar tracking system designed in Example 7.15 is

$$KG_c(s)G_p(s)H(s) = \frac{0.2K(s + 0.1)}{s(s + 0.02)(s + 2)}$$

The root locus for this system is shown in Figure 7.26, with the region in the vicinity of the origin greatly expanded. The pole positions for the designed value of $K = 10$ occur at

$$s = -0.1092, -0.9554 \pm j0.9584$$

These values were calculated by computer. Note that real pole has a time constant of $\tau = (1/0.1092)$ s $= 9.16$ s, which would appear to make the system so slow as to be unusable; that is, the system would be so slow that the aircraft would fly out of the beam width of the antenna before steady-state conditions are reached.

To investigate this problem, consider the closed-loop transfer function, which can be expressed as

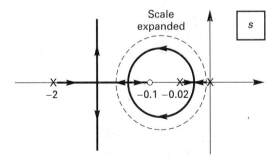

Figure 7.26   Root locus for Example 7.16.

$$T(s) = \left. \frac{KG_c(s)G_p(s)H(s)}{1 + KG_c(s)G_p(s)H(s)} \right|_{K = 10}$$

$$= \frac{2(s + 0.1)}{s^3 + 2.02s^2 + 2.04s + 0.2}$$

$$= \frac{2(s + 0.1)}{(s + 0.1092)(s + 0.9554 + j0.9584)(s + 0.9554 - j0.9584)}$$

We see that the poles at $s = -1 \pm j$ have been shifted to $s = -0.9554 \pm j0.9584$. In addition, a pole at $s = -0.1092$ has been added, and this pole has the very slow time constant of 9.16 s. This time constant appears to make the system unacceptable, and this may be the case. However, consider the unit-step response.

$$C(s) = \frac{2(s + 0.1)}{s(s + 0.1092)(s^2 + 1.9108s + 1.8371)}$$

$$= \frac{1}{s} + \frac{-0.0113}{s + 0.1092} + \frac{k_1 s + k_2}{s_2 + 1.9108s + 1.8371}$$

Even though one term in the response has a time constant of 9.16 s, the amplitude of this term is small and may be insignificant. With other excitations, for example, a ramp function, this term may be more significant. A MATLAB program that performs all calculations in this example is given by

```
K=10; Gc = tf([0.2 0.02], [1 0.02]);
Gp = tf([1],[1 2 0]); H = 1;
T = minreal(K*Gc*Gp/(1+K*Gc*Gp*H)), pause
pole (T), pause
[n,d] = tfdata(T, 'v'), [r,p,k] = residue (n,d)
```

As a final point, note that the analytical design procedure presented in Section 7.8 may also be used to design phase-lag controllers. To use this procedure, assume that the desired pole position is slightly to the right of the uncompensated pole position, and the design proceeds as in the phase-lead case.

We now summarize the preceding design concepts. The addition of a compensator zero adds positive angle to the angle criterion of the root locus and shifts the locus toward the left in the s-plane, in the direction of increased stability and decreased time constants. The addition of a pole to the compensator adds negative angle to the angle criterion, which shifts the locus to the right in the s-plane. This shift increases time constants and decreases stability. The first-order compensator has both a pole and a zero. For the phase-lead compensator, the zero dominates. For the phase-lag compensator, the pole dominates. Of course, for the phase-lag compensator, the shift of the locus to the right must be minimal to maintain the stability of the system.

7.10 PID DESIGN

The design of proportional-plus-integral-plus-derivative (PID) compensators is introduced in this section. The PID compensator is probably the most commonly used compensator in feedback control systems. With $e(t)$ the compensator input and $m(t)$ the output, the PID compensator is *defined* by the equation

$$m(t) = K_P e(t) + K_I \int_0^t e(\tau)d\tau + K_D \frac{de(t)}{dt} \tag{7-61}$$

The Laplace transform of this equation yields the transfer function:

$$M(s) = \left(K_P + \frac{K_I}{s} + K_D s \right) E(s)$$

or

$$G_c(s) = \frac{M(s)}{E(s)} = K_P + \frac{K_I}{s} + K_D s \tag{7-62}$$

A block diagram representation of (7-61) is given in Figure 7.27(a), and the transfer functions are shown in Figure 7.27(b). Quite often it is not necessary to implement all three terms (7-61) to meet the design specifications for a particular control system. For example, a PI compensator is described by (7-61) with $K_D = 0$, and so on. The design of PID compensators by root-locus techniques is described next.

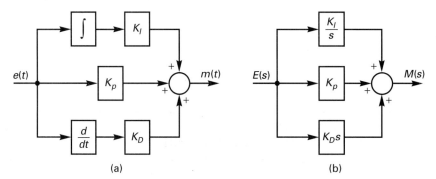

(a)   (b)

Figure 7.27   PID controller.

7.10.1 P Controllers

The P controller is a pure gain (no dynamics) of value K_P. Thus we see that the gain K that we have been varying to generate the root locus is a P controller; that is, the system characteristic equation is given by

$$1 + K_P G_p(s)H(s) = 0 \tag{7-63}$$

This compensator is used in situations in which satisfactory transient and steady-state responses can be obtained simply by setting a gain in the system, with no dynamic compensation required. Hence we can consider the construction of a root locus for an uncompensated system to also be the design of that system, using a P compensator.

7.10.2 PI Controllers

The PI controller has the transfer function

$$G_c(s) = K_P + \frac{K_I}{s} \tag{7-64}$$

This controller increases the system type by 1 (see Section 5.5) and is used to improve the steady-state response. The transfer function of the PI controller can be expressed as

$$G_c(s) = \frac{K_P s + K_I}{s} = \frac{K_P(s + K_I/K_P)}{s} \tag{7-65}$$

The controller has a pole at the origin and at zero at $-K_I/K_p$. Since the pole is nearer to the origin than is the zero, the controller is phase-lag (see Figure 7.24), and the controller adds a negative angle to the angle criterion of the root locus. Hence this controller is used to improve the steady-state response of the system, as stated earlier.

The compensated open-loop function can be expressed as

$$KG_c(s)G_p(s)H(s) = \left(KK_P + \frac{KK_I}{s} \right)G_p(s)H(s) \tag{7-66}$$

and we see that there are only two independent parameters to be determined in the design process. We can arbitrarily set $K = 1$ without affecting the generality of the design. The problem is then to determine K_P and K_I to meet certain steady-state design criteria.

7.10.3 PD Controllers

The transfer function of the PD controller is

$$G_c(s) = K_P + K_D s = K_D\left(s + \frac{K_P}{K_D} \right) \tag{7-67}$$

Thus the PD controller introduces a single zero at $s = -K_P/K_D$, as discussed in Section 7.6, and it is seen that this controller adds a positive angle to the angle criterion of the root locus. Therefore, the PD controller is a type of phase-lead controller and improves the system transient response. Since only a single zero is introduced, the trial-and-error procedure of Section 7.6 is easily applied to the PD controller design. As in the PI controller case, we can write

$$KG_c(s) = KK_P + KK_D s$$

and we see that there are only two independent parameters to be determined. Hence we arbitrarily set $K = 1$ and determine K_P and K_D by the design process.

The frequency response of the PD controller is given by

$$G_c(j\omega) = K_P + j\omega K_D \tag{7-68}$$

and a practical problem is seen from this expression, as was discussed in Section 7.6. The gain of the PD controller continues to increase as frequency increases. This problem occurs in the differentiation section of the controller. If a signal is changing rapidly with respect to time, the signal will have large slopes. Hence its derivative is large. High-frequency noise then will be amplified by a PD controller, and the higher the frequency, the more amplification. To reduce problems with high-frequency noise, it is usually necessary to add a pole to the PD controller transfer function, such that the transfer function becomes

$$G_c(s) = \frac{K_D(s + K_P/K_D)}{s - p_0} \tag{7-69}$$

The high-frequency gain is now limited to the value of K_D, as can be seen by replacing s with $j\omega$ in (7-69) and letting ω approach infinity. The pole in this compensator is chosen larger in magnitude than the zero, such that the compensator is still phase-lead. This compensator is now simply a phase-lead compensator of the type of Section 7.7, and the design procedure given there should be used for a PD compensator that requires a pole to limit the high-frequency gain.

7.10.4 PID Controllers

The PID controller is employed in control systems in which improvements in both the transient response and the steady-state response are required. The transfer function of the PID controller is given by

$$G_c(s) = K_P + \frac{K_I}{s} + K_D s = \frac{K_D s^2 + K_P s + K_I}{s} \tag{7-70}$$

Thus the PID controller has two zeros and one pole. Generally an additional pole is required to limit the high-frequency gain. One method that can be used to design the PID controller is first to design the PI part to give a satisfactory steady-state response. Then the PI controller is considered to be a part of the plant, and the PD portion is designed to improve the transient response. A different procedure for designing PID controllers is given in the next section.

7.11 ANALYTICAL PID DESIGN

In this section an analytical procedure is given for designing PID controllers. As in the analytical procedure given in Section 7.8, we assume that the design specifications require a closed-loop pole location at s_1, which, in general, is complex. We then define the angles β and ψ by the relationships

$$s_1 = |s_1| e^{j\beta} \tag{7-71}$$

and

$$G_p(s_1)H(s_1) = |G_p(s_1)H(s_1)| e^{j\psi} \tag{7-72}$$

The design equations for K_P, K_I, and K_D are [4]

$$K_P = \frac{-\sin(\beta + \psi)}{|G_p(s_1)H(s_1)| \sin\beta} - \frac{2K_I \cos\beta}{|s_1|} \tag{7-73}$$

$$K_D = \frac{\sin\psi}{|s_1||G_p(s_1)H(s_1)| \sin\beta} + \frac{K_I}{|s_1|^2} \tag{7-74}$$

Since there are three unknowns and only two equations to be satisfied, one of the gains may be chosen to satisfy a different design specification, such as choosing K_I to achieve a certain steady-state response. These equations can also be used to design PI and PD controllers, by setting the appropriate gain to zero. This design procedure is now illustrated with examples.

Example 7.17

For this example we first use a pole–zero approach to design a PD compensator for the radar tracking system of Example 7.15. Then we repeat the design using the analytical design approach. For the radar tracking system, the open-loop function is given by

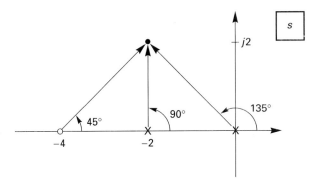

Figure 7.28   Example 7.17.

$$KG_p(s)H(s) = \frac{K}{s(s+2)}$$

Suppose that the design specifications require a time constant of $\tau = 0.5$ s and $\zeta = 0.707$. Thus the closed-loop pole locations of $s = -2 \pm j2$ are needed. From Figure 7.28, we see that the plant poles contribute an angle of $-225°$ to the angle criterion, and thus the controller must contribute an angle of $45°$ to make the total angle of $180°$. Hence the zero of the PD controller must be placed at $s = -4$, as shown in Figure 7.28. Then, from (7-67), $K_P/K_D = 4$. To place the closed-loop poles in the correct location, with $K = 1$, the magnitude requirement gives

$$G_c(s)G_p(s)H(s)\big|_{s=-2+j2} = \frac{K_D(s+4)}{s(s+2)}\bigg|_{s=-2+j2} = -1$$

or

$$\frac{K_D(2+j2)}{(-2+j2)(j2)} = \frac{K_D 2\sqrt{2}\ \underline{/45°}}{2\sqrt{2}\ \underline{/135°}\,2\ \underline{/90°}} = \frac{-K_D}{2} = -1$$

Thus $K_D = 2$, and since $K_P/K_D = 4$, then $K_P = 8.0$. Hence

$$G_c(s) = 8 + 2s$$

The complete root locus is given in Figure 7.29. Note that the PD compensator does not increase the system order; hence, for this example there is no third closed-loop pole.

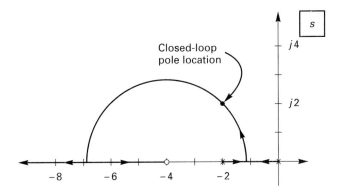

Closed-loop
pole location

Figure 7.29   Root locus for
Example 7.17.

We now repeat the design using the analytical procedure. From (7-71) and (7-72),

Thus, from (7-73) and (7-74),

$$s_1 = -2 + j2 = 2\sqrt{2}\angle 135°$$

$$G_p(s_1)H(s_1) = \frac{1}{(2\sqrt{2}\angle 135°)(-2+j2+2)} = \frac{1}{4\sqrt{2}}\angle 225°$$

$$K_P = \frac{-\sin 270°}{(1/4\sqrt{2})\sin 135°} - \frac{1}{(1/4\sqrt{2})(1/\sqrt{2})} = 8$$

$$K_D = \frac{\sin(-225°)}{(2\sqrt{2})(1/4\sqrt{2})\sin 135°} + \frac{1/\sqrt{2}}{(1/2)(1/\sqrt{2})} = 2$$

A MATLAB program for these coefficient calculations is given in Example 7.21.

Example 7.18

For this example we design a voltage control system for an electrical generator with a resistive load, as described in Section 2.7. The open-loop transfer function is given by

$$KG_p(s)H(s) = \frac{K}{(s+1)(s+2)}$$

The input to the plant is the voltage applied to the generator's field winding, and the output is the voltage out of the generator. Since the uncompensated system is type 0, there will be a non-zero steady-state error for a constant input, that is, for the case that we are attempting to maintain a constant voltage output. In addition, a constant disturbance, such as a change in speed of the prime mover (the energy source driving the generator), will cause a reduction in the output voltage in the steady state. In each of these cases the addition of an integrator will cause the steady-state errors to go to zero, as described in Examples 5.4 and 5.5.

The root locus of the uncompensated system is given in Figure 7.30. If the roots of the uncompensated system are chosen to be complex, the time constant is $\tau = 1/1.5$, or 0.67 s. Suppose that the design specifications require a time constant of 0.25 s and no steady-state errors for a constant input. The first specification can be satisfied by using a PD controller and the second, by a PI controller, hence a PID controller will be designed. The time-constant specification requires that the real part of s_1 be equal to -4, and we choose $\zeta = 0.707$. Then

$$s_1 = |s_1|e^{j\beta} = -4 + j4 = 4\sqrt{2}e^{j135°}$$

and

$$G_p(s_1)H(s_1) = |G_p(s_1)H(s_1)|e^{j\psi} = \frac{1}{(-3+j4)(-2+j4)}$$

$$= 0.04472 \angle -243.4°$$

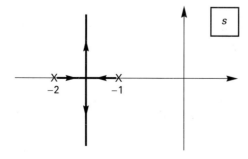

Figure 7.30 Root locus for Example 7.18.

We assume a value of $K_I = 0.1$. Then, since $\beta = 135°$ and $\psi = -243.4°$, from (7-73) and (7-74),

$$K_P = \frac{-\sin(-108.4°)}{0.04472 \sin(135°)} - \frac{2K_I \cos(135°)}{4\sqrt{2}}$$

$$= 30.01 + 0.25K_I = 30.03$$

$$K_D = \frac{\sin(-243.4°)}{4\sqrt{2}\,(0.04472)\sin(135°)} + \frac{K_I}{(4\sqrt{2})^2}$$

$$= 4.999 + 0.0313K_I = 5.002$$

Thus the PID controller has the gains $K_P = 30.03$, $K_I = 0.1$, and $K_D = 5.002$. The step response of the compensated system, obtained by simulation, is shown in Figure 7.31. While the steady-state error is guaranteed to be zero (the system is type 1), the error after 10 s is still 0.06, or 6 percent of the final value. This design is considered further in the next example, and a MAT-LAB program for this design is given in Example 7.20.

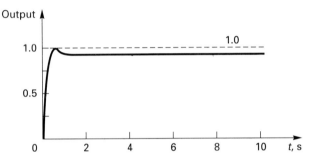

Figure 7.31   Time response for Example 7.18.

Example 7.19

In this example, we consider further the design of the last example. In that design, even though the steady-state error is zero, a very long time is required to achieve steady state. To investigate this problem further, different values of K_I will be assumed in the design, with the desired dominant roots remaining at $s = -4 \pm j4$. The values of the additional integrator gains to be considered are 1.0 and 10.0. As in the last example, the design equations (7-73) are solved for each value of K_I. The resulting PID gains are given in Table 7.8. The addition of the PID controller results in a third-order system. The location of two of the closed-loop poles are, from the design, $s = -4 \pm j4$. The location of the third closed-loop pole was calculated by computer and is also given in Table 7.8. The reason for the slow response for $K_I = 0.1$ of Example 7.18 is seen from the location of the third closed-loop pole. Note that this pole has a time constant of over 300 s. With $K_I = 10.0$, the time constant of the third pole has decreased to approximately 3 s. For $K_I = 0.1$, the overshoot in the step response is approximately 3 percent; this overshoot increases to approximately 7 percent for $K_I = 10.0$. We can then see some of the effects of different gains in the integrator path of the PID compensator.

Example 7.20

This example is a continuation of the design of Example 7.18. We will consider the computer calculation of Table 7.8 and Figure 7.31, using MATLAB. In addition, step responses for the designs yielding $K_I = 1$ and 10 are plotted. A program that calculates the data of Table 7.8 is given by

```
KI = [0.1 1 10];
Gp = tf([0 0 1],[ 1 3 2]);
s1 = -4 + i*4; beta = angle(s1); slmag = abs (s1);
Gps1 = evalfr(Gp,s1);psi = angle(Gps1);
Gps1mag = abs(Gps1);
t = 0:0.05:5;
for k = 1:3
   KP = -sin(beta + psi)/(Gps1mag*sin(beta)) - ...
        2*KI(k)*cos(beta)/slmag;
   KD = sin(psi)/(slmag*Gps1mag*sin(beta)) + KI(k)/slmag^2;
   [KP KI(k) KD]
     Gc = tf([KD KP KI(k)], [0 1 0]);
     T = Gc*Gp/(1 + Gc*Gp);
%      P = pole(T);
     step (T,t)
     hold on
end
hold off
```

In this program, the % symbol must be removed from the statement shown and placed at the beginnings of the next two statements. The program as shown calculates the step responses as discussed below. The compensator gains are KP, KI, and KD. The design equations of (7-73) and (7-74) are realized directly. The variable names beginning with Gp are plant parameters, those beginning with Gc are compensator parameters, and those beginning with T are closed-loop parameters.

TABLE 7.8   PID CONTROLLERS FOR EXAMPLE 7.19

K_P	K_I	K_D	Location of third pole
30.03	0.1	5.003	−0.00313
30.25	1.0	5.031	−0.03125
32.50	10.0	5.312	−0.31250

We now discuss the results from this program. Note the location of the third closed-loop pole in Table 7.8. For $K_I = 0.1$, the time constant is $\tau = 1/0.00313 = 319$ s. This very long time is evident in the step response of Figure 7.32. The error at $t = 10$ s is approximately 6 percent. Since the system is type 1, the steady-state error is zero, but approximately $4\tau = 1278$ s, or 21 minutes is required to reach steady state. For $K_I = 1$, the settling time is approximately $4\tau = 4/0.03125 = 128$ s. For $K_I = 10$, the settling time is approximately $4\tau = 4/0.3125 = 12.8$ s.

For the step responses, the response for $K_I = 0.1$ and $K_I = 1$ are approximately the same, and the response for $K_I = 10$ shows the reduced settling time. For $K_I = 0.1$, the error at t = 10 s is 6 percent, while for $K_I = 10$, the error at the same time is 0.2 percent. The percent overshoot figures in the three step responses are approximately equal.

The MATLAB program in the last example again illustrates one procedure for designing a control system, as discussed after Example 7.14.

The PID controller can be considered from a different viewpoint. The proportional term gives the controller output a component that is a function of the present state of the system. Since the integrator output depends upon the input for all previous time, that component of the compensator output is determined by the past state of the system. This output cannot change instantaneously and gives inertia to the system. The output of the differentiator is a function of the slope of its input and thus can be considered to be a predictor of the future state of the system. Hence the derivative part of the compensation can speed up the system response by anticipating the future. Of course, if the input information is faulty (noisy), unsatisfactory results can occur from this prediction. The PID controller then can be viewed as giving control that is a function of the past, the present, and the predicted future.

7.12 COMPLEMENTARY ROOT LOCUS

In this section we present the rules for the construction of the root locus for the case that K is negative, that is, for $-\infty < K \leq 0$. This locus is sometimes called the *complementary locus.*

As in the case for K positive, we assume that the system characteristic equation is given by

$$1 + KG(s)H(s) = 0 \qquad (7\text{-}75)$$

Hence, independent of the sign of K, the condition for a point s_1 to be on the root locus is that (7-75) be satisfied with K a real number, that is,

$$K = \frac{-1}{G(s_1)H(s_1)}$$

For the case that K is negative, this equation can be expressed as

$$|K| = \frac{1}{G(s_1)H(s_1)} \qquad (7\text{-}76)$$

Hence the magnitude requirement for K negative is the same as that for K positive, which is given in (7-7). However, the angle requirement is given by

$$\arg G(s)H(s) = r(360°) \qquad r = 0, \pm1, \pm2, \ldots \qquad (7\text{-}77)$$

We see, then, from (7-8), that the difference in the angle requirement for K negative is that the principal value of the angle of $G(s_1)H(s_1)$ must be $0°$, rather than $180°$ as required for K positive. Noting this difference, we can generate the rules for the root locus for K negative by replacing the $180°$ requirement in the rules with a $0°$ requirement.

The six rules for root-locus construction for K negative are as follows:

1. The root locus is symmetrical with respect to the real axis.
2. The root locus originates on the poles of $G(s)H(s)$ (for $K = 0$) and terminates on the zeros of $G(s)H(s)$ (as $K \to -\infty$), including those at infinity.
3. If the open-loop function has α zeros at infinity, $\alpha \geq 1$, the root locus will approach α asymptotes as K approaches infinity. The asymptotes are located at the angles

$$\theta = \frac{r360°}{\alpha} \qquad r = 0, 1, 2, \ldots$$

and those asymptotes intersect the real axis at the point

[eq. (7-20)] $\sigma_a = \dfrac{(\text{sum of finite poles}) - (\text{sum of finite zeros})}{(\text{number of finite poles}) - (\text{number of finite zeros})}$

The root locus includes all points on the real axis to the left of an even number of real critical frequencies (poles and zeros).

4. The breakaway points on a root locus will appear among the roots of the polynomial obtained from either

$$\frac{d\,[G(s)H(s)]}{ds} = 0$$

or, equivalently,

$$N(s)D'(s) - N'(s)D(s) = 0$$

where $N(s)$ and $D(s)$ are the numerator and denominator polynomials, respectively, of $G(s)H(s)$.

5. Loci will depart from a pole p_j (arrive at a zero z_j) of $G(s)H(s)$ at the angle $\theta_d(\theta_a)$, where

$$\theta_d = \sum_i \theta_{zi} - \sum_{\substack{i \\ i \ne j}} \theta_{pi} + r(360°)$$

$$\theta_a = \sum_i \theta_{pi} - \sum_{\substack{i \\ i \ne j}} \theta_{zi} + r(360°)$$

where $r = 0, \pm1, \pm2, \pm3, \dots$, and where θ_{pi} (θ_{zi}) represent the angles from pole p_i (zero z_i), respectively, to $p_j(z_j)$.

Note that rules 3, 4, and 6 have changed from those for positive K. These are the three rules based on the angle requirement. The proofs of these rules can be seen easily from the proofs of the same rules for positive K, which are given in Sections 7.2 and 7.3. Hence the proofs are not presented here. An example is now given.

Example 7.21

The system of Example 7.4 is used in this example. In Example 7.4, the open-loop function is given by

$$KG(s)H(s) = \frac{K}{(s-1)(s+2)(s+3)}$$

and the root locus is plotted for positive K. This locus is given in Figure 7.9 and is repeated as a part of Figure 7.32. The intersection of the asymptotes was found to be at $s = -\frac{4}{3}$; from rule 3, this value is also the intersection of the asymptotes for negative K. This value does not depend on the sign of K. Since the open-loop function has three zeros at infinity, the three asymptotes for positive K occur at the angles of $\pm60°$ and $180°$. From rule 3, the angles of the asymptotes for negative K are given by

$$\theta = \frac{r360°}{\alpha} \qquad r = 0, 1, 2, \dots$$

where α, the number of asymptotes, is equal to 3. Hence the angles of the asymptotes for negative K can be expressed as $0°$ and $\pm120°$.

Note, from rule 4, that for negative K, the root locus is on those parts of the real axis not covered by the locus for positive K. Hence all parts of the real axis will be on the roots locus for any system, if the locus is plotted for both positive and negative K.

In Example 7.4, the candidate points for the breakaway points were found to be –0.132 and –2.54. It is seen then that the point $s = -0.132$ is a breakaway point for positive K, and $s = -2.54$ is a breakaway point for negative K.

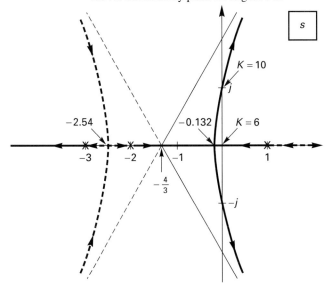

Figure 7.32 Example 7.21.

The total root locus, for both positive and negative K, is given in Figure 7.32. The locus for positive K is shown with solid lines, and the locus for negative K is shown with dashed lines. The arrowheads on the locus indicates increasing K for both cases. Note that the individual loci form continuous curves for this system, as K increases from $-\infty$ through zero to ∞. A MATLAB program that plots the root locus of this example for K negative is given by

```
GH = zpk ([], [1 -2 -3], -1);
rlocus(GH);
```

In the last example, the root locus was plotted with the arrowheads indicating K increasing from $-\infty$ to ∞. This procedure has the effect of changing rule 2 to the following:

2. The root locus originates on the zeros of $G(s)H(s)$ [$K = -\infty$], including those at infinity, and terminates on the poles of $G(s)H(s)$ [$K = 0$].

This change, which only reverses the direction of the arrowheads for K negative, has the effect of showing the root locus passing through the poles of the open-loop function in a continuous manner, as shown in Figure 7.32.

7.13 COMPENSATOR REALIZATION

In this section we consider the realization of compensator transfer functions using electronic circuits. In particular, operational amplifiers and resistor-capacitor networks are utilized to realize the various types of compensators described earlier in this chapter.

The basic circuit used to realize compensators, or any type of analog filters, is given in Figure 2.9 and is repeated in Figure 7.33(a). As shown in Section 2.2, equation (2-9), the transfer function of this circuit is

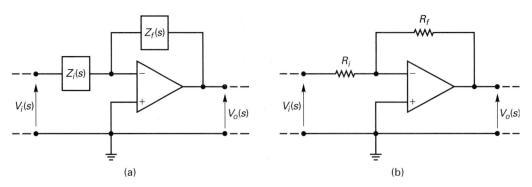

Figure 7.33 Op-amp circuits for compensators.

$$\frac{V_o(s)}{V_i(s)} = -\frac{Z_f(s)}{Z_i(s)} \tag{7-78}$$

By the proper choice of the impedances $Z_i(s)$ and $Z_f(s)$, we can realize phase-lead compensators, phase lag compensators, and PI, PD, and PID compensators. However, note that the transfer function in (7-78) contains a negative sign. If this sign is not acceptable in the physical system, the circuit of Figure 7.33(b), which is an inverting amplifier, must be connected to either the input or the output of the circuit of Figure 7.33(a). From (7-78), we see that the gain of this circuit is $-R_f/R_i$ and thus the results of cascading the two circuits is the transfer function

$$\frac{V_2(s)}{V_1(s)} = \frac{R_f Z_f(s)}{R_i Z_i(s)}$$

where $V_1(s)$ is the input voltage and $V_2(s)$ is the output voltage. The circuit of Figure 7.33(b) is also useful in changing the loop gain in a closed-loop control system and in amplifying weak sensor signals.

We first consider the circuit of Figure 7.34. The impedance of a parallel RC circuit of the type shown in the figure is given by

$$Z(s) = \frac{R(1/sC)}{R + 1/sC} = \frac{R}{RCs + 1} \tag{7-79}$$

The transfer function of the circuit of Figure 7.35 is then, from (7-78) and (7-79),

$$\frac{V_o(s)}{V_i(s)} = -\frac{Z_f(s)}{Z_i(s)} = -\frac{R_f/(R_f C_f s + 1)}{R_i/(R_i C_i s + 1)} = -\frac{R_f(R_i C_i s + 1)}{R_i(R_f C_f s + 1)}$$

or,

$$\frac{V_o(s)}{V_i(s)} = -\frac{C_i(s + 1/R_i C_i)}{C_f(s + 1/R_f C_f)} = -\frac{K_c(s - z_0)}{s - p_0} \tag{7-80}$$

Hence this circuit will realize the following compensators by the proper choice of circuit elements:

1. *Phase-lead.* The choice of $R_i C_i > R_f C_f$ will make the compensator zero magnitude smaller than that of its pole; thus the compensator is phase-lead.

2. *Phase-lag.* The choice of $R_i C_i < R_f C_f$ will make the compensator zero magnitude larger than that of its pole; thus the compensator is phase-lag.

3. *Proportional-plus-integral.* Letting R_f approach infinity (removing R_f from the circuit) yields the transfer function,

$$\frac{V_o(s)}{V_i(s)} = -\frac{C_i}{C_f} - \frac{1/R_iC_f}{s} = -\left(K_P + \frac{K_I}{s}\right) \tag{7-81}$$

which results in a PI compensator.

4. *Proportional-plus-derivative.* Removing C_f (letting $C_f = 0$) results in the transfer function

$$\frac{V_o(s)}{V_i(s)} = -\frac{R_f}{R_i} - R_fC_is = -(K_P + K_Ds) \tag{7-82}$$

However, as a practical matter, the value of C_f is usually chosen to be a small nonzero value, in order to limit the high-frequency gain. In this case, the PD compensator becomes phase-lead, as described in 1.

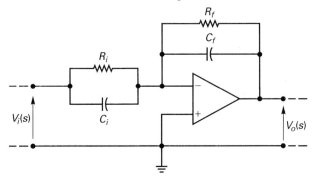

Figure 7.34   General compensator circuit.

5. *Proportional-plus-integral-plus-derivative.* The PID compensator cannot be realized with the circuit of Figure 7.34. Instead, one circuit for the realization of the PID compensator is given in Figure 7.35. This circuit also introduces an additional pole to the PID compensator, in order to limit the high-frequency gain. It is left as an exercise to the reader to show that this circuit realizes a practical PID compensator.

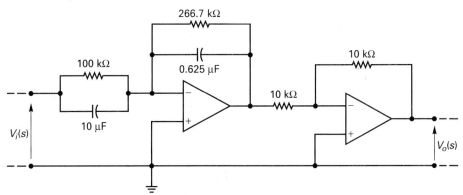

Figure 7.35   PID compensator circuit.

The circuits given above are not the only ones that can be used to realize compensators. Furthermore, compensators can also be realized with either hydraulic systems or pneumatic systems [5,6] in addition to the electronic circuits. An example of an electronic compensator will now be given.

Example 7.22

In this example, we design a circuit to realize the phase-lead compensator for the satellite control system designed in Example 7.12. The compensator transfer function was calculated as

$$G_c(s) = \frac{16(s+1)}{s+6}$$

Comparing this transfer function with (7-80), we see that the circuit elements of Figure 7.34 must satisfy the equations

$$R_i C_i = 1$$

$$R_f C_f = \frac{1}{6}$$

$$\frac{C_i}{C_f} = 16$$

Since there are four unknowns and only three equations to be satisfied, we choose $R_i = 100\text{k}\Omega$. The other circuit elements are then found to be $R_f = 266.7 \text{ k}\Omega$, $C_i = 10\mu \text{ F}$, and $C_f = 0.625 \text{ μF}$.

The required circuit is shown in Figure 7.36, where the unity-gain inverting amplifier has been added to the circuit to give the correct sign to the transfer function.

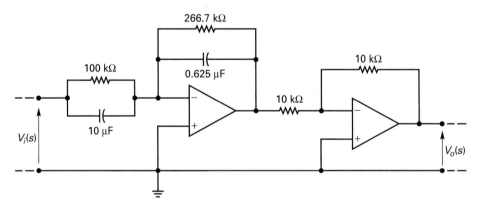

Figure 7.36   Phase-lead compensator for Example 7.22.

7.14 SUMMARY

This chapter presented analysis and design techniques based on the root locus of a closed-loop system. The design is broken down into two cases. The first case is the one in which the design specifications require that the system transient response be faster, and a phase-lead compensator is employed here. The second case is the one in which the design specifications require that the steady-state response be improved, and a phase-lag compensator is employed here. The phase-lag compensator can also be used for the case that the system response is to be made slower. Many of these effects become more evident when frequency-response techniques are covered in Chapters 8 and 9.

Two design procedures were covered in this chapter. One is based on trial and error and the other is an analytical technique. However, generally the analytical technique requires that certain parameters be assigned values in a somewhat arbitrary manner, making this method a modified trial-and-error procedure.

Generally the design techniques of this chapter place some of the poles of the closed-loop transfer function at certain locations in the *s*-plane. Since other poles of the closed-loop transfer function are not placed, the complete root locus of the compensated system must be examined to determine if these pole locations are satisfactory.

Frequency-response analysis and design techniques are presented in the next two chapters. The design techniques result in phase-lead and phase-lag controllers as do the root-locus procedures of this chapter. However, the frequency-response techniques give information not available from the root-locus methods, and the root-locus methods give information not available from the frequency-response techniques. Hence the two methods complement each other, and both can be used to arrive at a satisfactory design in a given design situation.

REFERENCES

1. W. R. Evans. "Graphical Analysis of Control Systems," *Trans. AIEE,* 67 (1948): 547–551.

2. B. C. Kuo. *Automatic Control Systems,* 7th ed. Upper Saddle River, NJ: Prentice Hall, 1996.

3. T. Arie, M. Itoh, A. Senoh, N. Takahashi, S. Fujii, and N. Mizuno. "An Adaptive Steering System for a Ship," *IEEE Control Sys.* (October 1986): 3–8.

4. C. L. Phillips and R. D. Harbor. *Feedback Control Systems*, 3rd ed. Upper Saddle River, NJ: Prentice Hall, 1996.

5. J. Van de Vegte. *Feedback Control Systems,* 3rd ed. Upper Saddle River, NJ: Prentice Hall, 1994.

6. W. J. Palm. *Control System Engineering.* New York: Wiley, 1986.

PROBLEMS

7.1. For the system with the open-loop function

$$KG(s) = \frac{K}{s(s+1)(s+3)} \qquad 0 \le K \le \infty$$

determine if the following points are on the root locus.

(a) $s = -0.5$

(b) $s = 0.5$ (unstable)

(c) $s = j1.732$

(d) $s = -1 + j$

(e) Use the *roots* statement of MATLAB to verify all results.

7.2. Given the system with the open-loop function

$$KG(s) = \frac{K(s+2)}{s^2(s+4)}$$

Determine if a value of $K > 0$ can be chosen such that the closed–loop transfer function has a pole at

(a) $s = -1.0$

(b) $s = -3.0$

(c) $s = -5.0$

(d) $s = 1.0$ (unstable)

(e) Use the *roots* statement of MATLAB to verify all results.

7.3. Given the system shown in Figure P7.3.

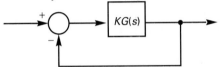

Figure P7.3

(a) Sketch the root locus for the transfer function

$$G(s) = \frac{s + 2}{s + 1}$$

(b) Verify the results of (a) by calculating the locations of the roots as a function of K.

(c) Verify the root locus with the *rlocus* statement in MATLAB.

7.4. Given the system shown in Figure P7.4.

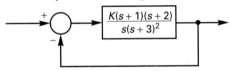

Figure P7.4   System for Problem 7.4.

(a) Accurately sketch the root locus of the system.

(b) Verify the root locus of this sytem.

7.5. (a) Let the transfer function in the block of Figure P7.4 be

$$G(s) = \frac{K(s^2 + 1)}{s(s + 2)}$$

Accurately sketch the root locus for this system.

(b) Verify the root locus with the *rlocus* statement in MATLAB.

7.6. Consider the system of Figure P7.6. Note that the sensor gain is not unity.

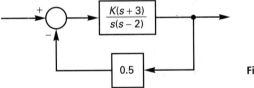

Figure P7.6

(a) Accurately sketch the root locus for the system.

(b) Find any points at which the locus crosses the $j\omega$-axis. Use the Routh-Hurwitz criterion as required.

(c) From (a) and (b), find the range of K for which the system is stable.

(d) From (a) and (b), find the range of K for which the system is stable and the closed-loop transfer function poles are real.

(e) From the results above, find all values of gain for which the system is critically damped.

(f) Verify the root locus in (a) using MATLAB.

(g) Use SIMULINK to verify the results in (b).

7.7. Sketch the root locus of the single-loop systems having the open-loop functions $KG(s)H(s)$ given by the following functions. Solve for the values of s at any crossings of the imaginary axis.

(a) $\dfrac{K(s+1)}{s^2}$

(b) $\dfrac{K}{s(s+2)^2}$

(c) $\dfrac{K}{s[(s+10)^2+1]}$

(d) $\dfrac{K}{s[(s+5)^2+25]}$

(e) Verify each root locus with MATLAB.

(f) Use SIMLINK to check all crossings of the $j\omega$-axis..

7.8. The PD compensator in the system in Figure P7.8 adds a zero to the system open-loop function, since

$$KG_c(s)G_p(s) = \frac{K(K_P + K_D s)}{s(s+2)} = \frac{KK_D(s+a)}{s(s+2)} \qquad a = \frac{K_P}{K_D}$$

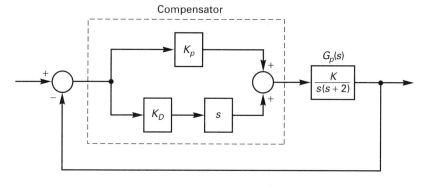

Figure P7.8

(a) To show the effects of the compensator, give a sketch of the root locus for the zero in the following ranges.

   (i) $-a > 0$

   (ii) $-2 < -a < 0$

   (iii) $-a < -2$

(b) Which of the three cases in (a) will result in the system with the fastest settling time?

(c) Which of the three cases in (a) can result in an unstable system?

7.9. A model of a temperature-control system of a large test chamber is shown in Figure P7.9. This control system is discussed in Problem 6.12. For this problem, assume that $D(s) = 0$ (the door of the chamber remains closed).

(a) Find time constant of the closed-loop system for $G_c(s) = 1$.

(b) The chamber temperature is to be commanded to the constant temperature of 40°C. Find the constant input $r(t)$, in volts, required.

(c) Draw the unity feedback model for this system (see Figure 5.2).

(d) If $G_c(s) = 1$, find the steady-state error, in degrees Celsius, for (b), using the unity feedback model.

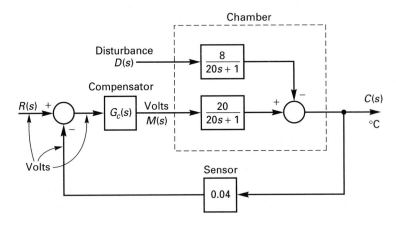

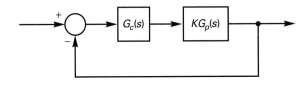

Figure P7.9

(e) Design a phase-lag compensator such that the pole in (a) does not move significantly, and the steady-state error is less than 5 percent of the desired temperature.

(f) Design a PI compensator that satisfies the specifications of (e) with the closed-loop pole location at s = -0.085.

(g) Verify the results in (d) with SIMULINK. Why is the long simulation time required?

(h) Verify the results in (e) with SIMULINK.

(i) Verify the results in (f) with SIMULINK.

7.10. For the closed-loop system of Figure P7.10, the plant transfer function is given by

$$KG_p(s) = \frac{50K}{(s+1)(s+2)(s+10)}$$

Figure P7.10

(a) Sketch the root locus with $G_c(s) = 1$.

(b) With $G_c(s) = 1$ and $K = 0.377$, the closed-loop system poles occur at $s = -1.377 \pm j1.377$ and -10.25. Hence, for the complex poles, $\zeta = 0.707$. Find the steady-state error for a unit step input.

(c) A phase-lag compensator with a dc gain of unity is to be designed, and K is to be increased to a value of 3. Find the steady-state error for a unit step input for this case.

(d) Design a phase-lag compensator such that, with $K = 3$, the closed–loop poles in (b) are shifted by only a small amount.

(e) Use MATLAB to verify the closed-loop pole in (d).

(f) Use SIMULINK to verify the results in (b) and (d).

7.11. For the system of Figure P7.10, suppose that

$$KG_p(s) = \frac{K}{s(s+4)}$$

(a) With $G_c(s) = 1$, sketch the root locus.

(b) With $G_c(s) = 1$, find the closed-loop system time constant if K is chosen such that the closed-loop system poles are complex.

(c) Suppose that the system specifications require a time constant of 0.333 s. What types of compensators can be designed to satisfy this condition?

(d) Design a PD compensator to yield closed-loop poles with $\zeta = 0.707$, $\tau = 0.333$ s, and $K = 1$.

(e) Use MATLAB to verify the closed-loop poles in (d).

(f) Use the MATLAB statement *step* to verify the system time constant.

7.12. The attitude-control system of a space booster is shown in Figure P7.12(a). The attitude angle θ is controlled by controlling the engine angle δ, which is then the angle of the applied thrust, F_T. The vehicle velocity is denoted by v. These control systems are sometimes open-loop unstable, which occurs if the center of aerodynamic pressure is forward of the booster center of gravity. For example, the rigid–body transfer function of the Saturn V booster was

$$G_p(s) = \frac{0.9407}{(s^2 - 0.0297)}$$

This transfer function does not include vehicle bending dynamics, 'iquid fuel slosh dynamics, and the dynamics of the hydraulic motor that positioned the engine. These dynamics added 25 orders to the transfer function. The rigid-body vehicle was stabilized by the addition of rate feedback, as shown in Figure P7.12(b). (Rate feedback, in addition to other types of compensation, was used on the actual vehicle.)

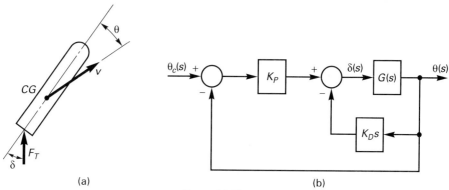

(a)   (b)

Figure P7.12

(a) With $K_D = 0$ (the rate feedback removed), plot the root locus and state the different types of (nonstable) responses possible.

(b) Design the compensator shown, which is PD, to place a closed-loop pole at $s = -0.25 + j0.25$. Note that the time constant of the pole is 4 s, which is not unreasonable for a large space booster.

(c) Plot the root locus of the compensated system, with K_p variable and K_D set to the value found in (b).

(d) In this problem, the magnitude of the poles of the plant are very small, compared to other systems that we have considered. Recall that this transfer function is that of the Saturn V booster. What does the small magnitude signify, in terms of the transient response? Is this small magnitude reasonable?

(e) Use MATLAB to verify the closed-loop poles in (b).

(f) Use SIMULINK to verify the system time constant in (b).

(g) Use MATLAB to verify the root locus in (c).

7.13. For the Saturn V booster attitude control system of Figure P7.12, let $K_D = 0$.

(a) Plot the root locus and state the different types of (nonstable) responses possible.

(b) Replace K_p with a compensator with the transfer function $G_c(s)$. Based on the root locus of (a), which types of compensators can be employed to stabilize the closed-loop system?

(c) Design a phase-lead compensator with a dc gain of 0.1 that places a closed-loop pole at $s = -0.25 + j0.25$.

(d) Use MATLAB to verify the closed-loop poles in (c).

(e) Use SIMULINK to verify the system time constant in (c).

7.14. Shown in Figure P7.14 is the block diagram of the servo control system for one of the axes of a digital plotter. This system is described in detail in Problem 6.13.

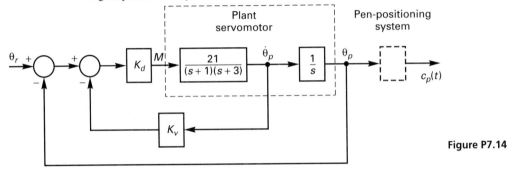

Figure P7.14

(a) Let $K_d = K$ and $K_v = 0$. Sketch the root locus for this system.

(b) Remove the rate feedback path and replace K_d with a PD compensator, with the transfer function $G_c(s) = K_p + K_D s$. Calculate the compensator gains K_p and K_D to give a characteristic equation root at $s = -1 + j$.

(c) Find the equivalent gains K_d and K_v in the rate feedback system that yield the same root. These values can be found by equating the characteristic equation of the PD compensated system to that of the rate-compensated system.

(d) Sketch the root locus of the compensated system to find the approximate location of the third root of the system characteristic equation. For this part, assume that a gain has been added in the error path (at the output of the first summing junction) and that this gain is varied to generate the root locus.

(e) Use MATLAB to verify the results in (a), (b), (c), and (d).

(f) Use SIMULINK to verify the system time constants in (b) and (c).

7.15. For the digital plotter servo of Problem 7.14 shown in Figure P7.14, the rate feedback path is removed and K_d is replaced with the phase-lead compensator of (7-46).

(a) Design the compensator such that the compensator dc gain is 0.15 and the closed-loop system has a pole at $s = -1 + j$.

(b) Sketch the root locus of the compensated system to indicate the approximate locations of the other two roots of the fouth-order characteristic equation.

(c) Use MATLAB to verify the results in (a) and (b).

(d) Use SIMULINK to verify the system time constant in (a).

7.16. Shown in Figure P7.16 is the block diagram of the servo-control system for one of the joints of a robot. This system is described in Section 2.12.

(a) Show that the plant transfer function is given by

$$\frac{\theta_L(s)}{E_a(s)} = \frac{0.15}{s(s+1)(s+5)}$$

and sketch the root locus of the uncompensated system (with the compensator replaced with a gain K).

(b) Calculate the compensator gains K_p and K_D to give a characteristic equation root at $s = -1 + j$.

(c) Design a phase-lead compensator to give a characteristic equation root at $s = -1 + j$. The dc gain of the phase-lead compensator is to be unity.

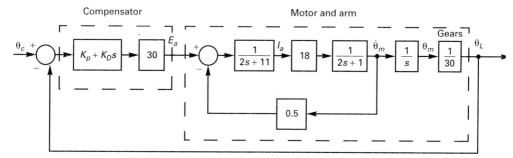

Figure P7.16

(d) Sketch the root locus of the compensated system of (b) to find the approximate locations of the third root of the system characteristic equation. For this part, assume that a gain has been added in the error path (at the output of the first summing junction) and that this gain is varied to generate the root locus.

(e) Repeat (d) for the compensated system of (c).

(f) Use MATLAB to verify the results in (b), (c), (d), and (e).

(g) Use SIMULINK to verify the system time constants in (b) and (c).

7.17. Consider the automobile cruise-control system of Figure P7.17. This system is described in Problem 3.12. For this problem, ignore the disturbance torque.

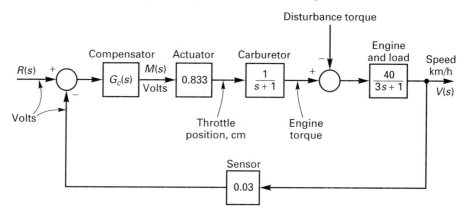

Figure P7.17

(a) Sketch the uncompensated-system root locus, and give the time constant of the complex roots.

(b) Design a phase-lead compensator with a dc gain of 8 that places the system complex poles such that $\zeta = 0.707$ and the time constant is one-half that found in (a).

(c) Sketch the root locus for the compensated system, showing the closed-loop transfer function poles.

(d) Design a PD compensator such that the criteria of (b) are satisfied. The gain specification must be ignored. Why?

(e) Sketch the root locus for the compensated system of (d), showing the closed-loop transfer function poles.

(f) Use MATLAB to verify the closed-loop pole location in (b) and (d).

(g) Use SIMULINK to verify the system constants in (b) and (d).

7.18. Consider the radar-tracking system of Example 7.15. Suppose that the steady-state error is to be 0.1° with a unit-ramp input, with all other specifications the same.

   (a) Design a phase-lag controller that satisfies these specifications.
   (b) Compare the closed-loop time constants for the system in (a) with those of the system in Example 7.15.
   (c) Simulate the system of (a) with a unit-ramp input, and note the approximate errors at 20 s and at 50 s.
   (d) Repeat (c) for the system of Example 7.15, and compare errors.

7.19. Consider the satellite control system of Section 7.6 and Figure 7.19. Let the transfer function of the satellite be given by

$$G_p(s) = \frac{0.5}{s^2}$$

   (a) Sketch the root locus for the system for both positive and negative K.
   (b) Locate all crossings of the $j\omega$-axis by the root locus, and find the value of K at each of these crossings.
   (c) From the results in (b), state the complete ranges of K for which the system is stable, unstable, or marginally stable.

7.20. Consider the system with the open-loop function

$$KG(s)H(s) = \frac{50K}{(s+1)(s+2)(s+10)}$$

   (a) Sketch the root locus for this system for both positive and negative K.
   (b) Locate all crossings of the $j\omega$-axis by the root locus, and find the value of K at each of these crossings.
   (c) From the results in (b), state the complete range of K for which the system is stable.
   (d) Verify the root locus by computer. Given a computer program that calculates a root locus only for $K > 0$, how can this program be used to calculate a root locus for $K < 0$?

7.21. In the following designs, use resistors in the range of 1 kΩ to 10 MΩ and capacitors less than or equal to 10 μF, for practical considerations.

   (a) Design an op-amp circuit to realize the PI compensator of Example 7.11: $G_c(s) = 3.6 + 1/s$.
   (b) Design an op-amp circuit to realize the phase-lag compensator of Example 7.14: $G_c(s) = 0.2(s + 0.1)/(s + 0.02)$.
   (c) Design an op-amp circuit to realize the phase-lag compensator of Example 7.13: $G_c(s) = (10.29s + 4.571)/(s + 4.571)$.
   (d) Design an op-amp circuit to realize the PD compensator of Example 7.17: $G_c(s) = 8.0 + 2.0s$.

7.22. Consider the rigid-satellite control system with the rate feedback depicted in Figure 7.20. The open-loop function is given in (7-38). Prove that the roots of the system characteristic equation may be placed at any point in the s-plane by proper choice of the gains K and K_v.

8

Frequency-Response Analysis

The use of frequency-response analysis has been illustrated several times in the previous chapters. Frequency-response methods are among the most useful techniques available for systems analysis and design. As has been mentioned previously, the various analysis and design techniques supplement each other, rather than one method excluding others. There are obvious advantages to the root-locus techniques, in that very good indicators of the transient response of a system are available. However, we do need a relatively accurate model of the system to obtain these benefits from the root locus. As we shall see, the benefits of the frequency-response methods may be obtained from measurements on the physical system without deriving the system transfer function. In fact, we can design a control system by frequency-response methods without developing a transfer function, provided that we can measure the system frequency response.

Also, one of the most common methods used to verify a derived transfer function is to compare the frequency response as calculated from the transfer function with that obtained from measurements on the physical system. For example, the U.S. Navy has obtained frequency responses for aircraft by applying sinusoidal inputs to the autopilots and measuring the resulting position of the aircraft while the aircraft were in flight [1].

In addition, one of the best methods of verifying a simulation is to obtain a frequency response of the simulation by inputting sinusoids and comparing this response with the calculated frequency response of the transfer function of the system being simulated.

In this chapter we consider frequency-response analysis methods, and in the following chapter we extend these methods to the design of control systems.

8.1 FREQUENCY RESPONSES

In Section 4.4 we considered the response of stable linear time-invariant systems to sinusoidal inputs. The gain of a system to a sinusoidal input of frequency ω_1 in both amplitude and phase, is given by the transfer function evaluated at $s = j\omega_1$. We illustrate this using the system of Figure 8.1. For this system,

$$C(s) = G(s)E(s) \tag{8-1}$$

We designate $G(j\omega)$, which is a complex number for a given ω, as

$$G(j\omega) = |G(j\omega)|e^{j\theta(\omega)} = |G(j\omega)| \underline{/\theta(\omega)} \tag{8-2}$$

Then, for the case that the input signal $e(t)$ is sinusoidal, that is,

$$e(t) = A \sin \omega_1 t \tag{8-3}$$

the output signal $c_{ss}(t)$ in the steady state is given by

$$c_{ss}(t) = A|G(j\omega_1)| \sin [\omega_1 t + \theta(\omega_1)] \tag{8-4}$$

(See Section 4.4.) As an example, suppose that $e(t) = 8 \sin 2t$ and that

$$G(s) = \frac{2}{s^2 + 3s + 2}$$

for the system of Figure 8.1. Then in (8-2),

$$G(j\omega)|_{s = j2} = \frac{2}{(j2)^2 + 3(j2) + 2} = \frac{2}{-2 + j6} = 0.316 \underline{/-108.4°}$$

From (8-4), the steady-state output is then given by

$$c_{ss}(t) = 8(0.316) \sin(2t - 108.4°) = 2.528 \sin(2t - 108.4°) \tag{8-5}$$

This calculation is verified with the MATLAB program

```
G = tf([2], [1 3 2]);
Gj2 = evalfr(G, j*2)
Gmag = abs(Gj2), argGj2 = angle(Gj2)*180/pi
```

If the transfer function is written as

$$G(s) = \frac{2}{(s + 1)(s + 2)} = \frac{1}{(s + 1)(0.5s + 1)} = \frac{1}{(\tau_1 s + 1)(\tau_2 s + 1)}$$

we see that the system time constants are $\tau_1 = 1$ s and $\tau_2 = 0.5$ s. Thus after approximately 4 s, or four times the longer time constant, the system is in the steady-state condition, and the response is given by (8-5).

With respect to the preceding example, we wish to make one additional point. For the system,

$$G(s) = \frac{C(s)}{E(s)} = \frac{2}{s^2 + 3s + 2}$$

Thus

$$(s^2 + 3s + 2)C(s) = 2E(s)$$

and taking the inverse Laplace transform yields the differential equation of the system.

$$\ddot{c}(t) + 3\dot{c}(t) + 2c(t) = 2e(t)$$

By the preceding calculations we have found the steady-state solution of this differential equation for the forcing function $e(t) = 8 \sin 2t$. This solution applies for time greater than approximately 4 s after the application of the forcing function. For time less than 4 s, the natural response, which is of the form

$$c_c(t) = k_1 e^{-t/\tau_1} + k_2 e^{-t/\tau_2} = k_1 e^{-t} = k_2 e^{-2t}$$

is significant and cannot be neglected.

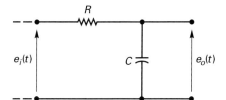

Figure 8.1   System.

We see then that, from the complex function $G(j\omega)$, we can obtain the steady-state response for any sinusoidal input, provided that the system is stable. We call $G(j\omega)$, $0 \le \omega \le \infty$, the *frequency-response function*. For a given value of ω, $G(j\omega)$ is a complex number. Hence, two numbers are required to specify $G(j\omega)$ for a given value of ω. These two numbers may be magnitude and angle, or real part and imaginary part, or some of the other forms that we consider later.

We usually plot $G(j\omega)$ versus ω in some form to characterize the frequency response. We illustrate two forms by a simple example. For the *RC* circuit of Figure 8.2, using the impedance approach of Section 2.2 yields the voltage transfer function of

$$G(s) = \frac{E_o(s)}{E_i(s)} = \frac{(1/sC)}{R + (1/sC)} = \frac{1}{RCs + 1}$$

Suppose that the circuit has a time constant of 1 s; then

$$G(s) = \frac{1}{s + 1}$$

The frequency-response function of this system is given by

$$G(j\omega) = \frac{1}{1 + j\omega} = (1 + \omega^2)^{-1/2} \underline{/-\tan^{-1}(\omega)} \qquad (8\text{-}6)$$

One common method of displaying this frequency response is in the form of a *polar plot*. In such a plot, the magnitude and angle of the frequency-response function (or its real and imaginary parts) are plotted in the complex plane as the frequency, ω, is varied. For the function of (8-6), to construct a polar plot we first evaluate the function for values of ω. As an example, a table of these values is given in Table 8.1. Next these values are plotted in the complex plane, as shown in Figure 8.3. Note that, mathematically, the frequency response is a mapping from the *s*-plane to the $G(j\omega)$-plane. The upper half of the $j\omega$-axis, which is a straight line, is mapped into the complex plane via the mapping $G(j\omega)$.

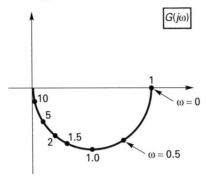

Figure 8.3   Frequency response.

TABLE 8.1   FREQUENCY RESPONSE

ω	$G(j\omega)$
0	$1\underline{/0°}$
0.5	$0.894\underline{/-26.6°}$
1.0	$0.707\underline{/-45°}$
1.5	$0.555\underline{/-56.3°}$
2.0	$0.447\underline{/-63.4°}$
3.0	$0.316\underline{/-71.6°}$
5.0	$0.196\underline{/-78.7°}$
10.0	$0.100\underline{/-84.3°}$
∞	$0\underline{/-90°}$

In the polar plot of Figure 8.3, the frequency ω appears as a parameter. A second form for displaying the frequency response is to plot the magnitude, or gain, of $G(j\omega)$ versus ω and to plot the angle of $G(j\omega)$ versus ω. These plots for the example above are shown in Figure 8.4. Note that the gain of the system as frequency increases is somewhat clearer in this plot than in the polar plot. This plot, or a variation of it, is normally used to display the frequency response of filters. From the plot of Figure 8.4, the circuit is seen to be a lowpass filter; that is, the circuit will pass frequencies that are low relative to the cutoff frequency (the upper edge of the bandwidth) and will attenuate the frequencies that are high relative to the cutoff frequency. Bandwidth is defined in Section 4.4. For this example the cutoff frequency is $\omega = 1$ rad/s. A practical use of this circuit (filter) is to reduce the variations in the voltage output of a dc power supply [2].

Consider now a general first-order transfer function:

$$G(s) = \frac{K}{\tau s + 1} \tag{8-7}$$

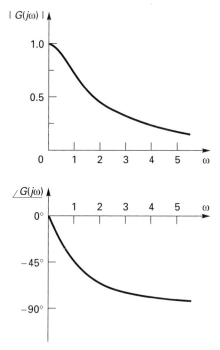

Figure 8.4 Frequency response.

The frequency-response function is given by

$$G(j\omega) = \frac{K}{1 + j\omega\tau} = \frac{K/\tau}{1/\tau + j\omega} \tag{8-8}$$

The term in the denominator can be represented by the vector, shown in Figure 8.5 (a), that originates on the pole at $-1/\tau$ and terminates at the point $j\omega$ in the s-plane. As ω varies from zero to infinity, the length of the vector increases without limit. Thus, from (8-8), the magnitude of $G(j\omega)$ decreases from a value of K to zero, since the magnitude of the denominator increases from $1/\tau$ to infinity. Also, in Figure 8.4(a), the angle of the vector increases from $0°$ to $90°$. Thus the angle of $G(j\omega)$ decreases from $0°$ to $-90°$. We can then see the frequency-response characteristics of (8-7) from the vector in Figure 8.5(a).

This concept of vectors as shown in Figure 8.4(a) can be extended to higher-order systems. Consider now the second-order transfer function with real poles

$$G(j\omega) = \frac{K}{(1 + j\omega\tau_1)(1 + j\omega\tau_2)} = \frac{K/\tau_1\tau_2}{(1/\tau_1 + j\omega)(1/\tau_2 + j\omega)} \tag{8-9}$$

The two factors in the denominator are equal to the two vectors shown in Figure 8.5(b). Hence, as ω increases from zero to infinity, the magnitude of $G(j\omega)$ decreases to zero, since the lengths of the vectors increase without limit. In addition, the angle of $G(j\omega)$ decreases from $0°$ to $-180°$, since the angle of each vector increases from $0°$ to $90°$.

The case of the transfer function having two complex poles is shown in Figure 8.4(c), with the angle result the same as in the case of the two real poles. However, the magnitude of $G(j\omega)$ may be quite different. Suppose that the pole $-p_1$ is close to the $j\omega$-axis. The vector $(j\omega + p_1)$ decreases to a minimum and then increases without limit, as ω increases from

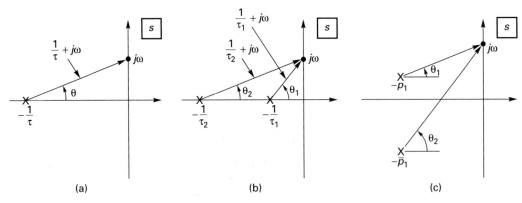

Figure 8.5   Characteristics of the frequency response.

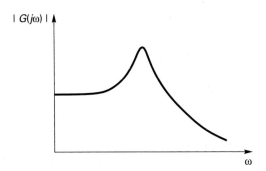

Figure 8.6   Frequency response indicating resonance.

zero. Hence, as ω increases, the magnitude $G(j\omega)$ will have a maximum and then decrease to zero, as shown in Figure 8.6. This same effect (called a *resonance*) is shown in Figure 4.14 for a second-order system with complex poles.

Obviously, if a transfer function has a real zero, this zero will contribute a numerator vector of the type described above, which will increase without limit as ω increases. In addition, the angle of this vector will increase from 0° to 90°, giving the same increase in angle to $G(j\omega)$.

We see from the preceding discussion that for low-order transfer functions we can quickly make an approximate sketch of the frequency response by considering vectors in the *s*-plane. These sketches are quite useful in the preliminary analysis of a control system. If more than a rough sketch is required, the frequency response should be calculated by computer. An example of sketching a frequency response is given next.

Example 8.1

Consider the system described by the transfer function

$$G(s) = \frac{K(\tau_3 s + 1)}{(\tau_1 s + 1)(\tau_2 s + 1)} \qquad \tau_1 > \tau_2 > \tau_3$$

The frequency response is then given by

$$G(j\omega) = \frac{(K\tau_3/\tau_1\tau_2)(1/\tau_3 + j\omega)}{(1/\tau_1 + j\omega)(1/\tau_2 + j\omega)}$$

As ω varies from zero to infinity, each pole contributes a phase shift to the frequency response from $0°$ to $-90°$, and the zero contributes a phase shift from $0°$ to $90°$. Hence the total phase shift of the frequency response begins at $0°$ and ends at $-90°$. Note that the two limits are given by

$$G(j0) = K\,\underline{/0°} \qquad \lim_{\omega \to \infty} G(j\omega) = \lim_{\omega \to \infty} \frac{K\tau_3/\tau_1\tau_2}{j\omega}$$

Hence $G(j\omega)$ begins at a value of K on the real axis for $\omega = 0$, and approaches the origin at an angle of $-90°$ as ω approaches infinity. The angles that $G(j\omega)$ assumes for other values of ω are dependent on the actual values of τ_1, τ_2, and τ_3. Because of the magnitude of the zero relative to the poles, the frequency response extends into the third quadrant. A typical frequency response for this transfer function is shown in Figure 8.7.

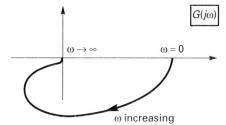

Figure 8.7 Frequency response.

8.2 BODE DIAGRAMS

This section presents a method for plotting a frequency response that is different from the two methods given in the first section of this chapter. This method results in a plot of magnitude versus frequency and phase versus frequency, but the frequency scale is logarithmic. In addition, the magnitude is also plotted on a logarithmic scale. The plot that is presented here is called a *Bode plot*, or a *Bode diagram*. The main advantage of the Bode plot over other types of plots for frequency response is that the effects of adding a real pole or a real zero to a transfer function can be seen rather easily. For this reason, Bode diagrams are very useful in designing control systems.

We develop the Bode diagram by using as an example the second-order transfer function

$$G(s) = \frac{K(1 + \tau_3 s)}{(1 + \tau_1 s)(1 + \tau_2 s)} = \frac{K(1 + s/\omega_3)}{(1 + s/\omega_1)(1 + s/\omega_2)} \tag{8-10}$$

where it is assumed that both poles and the zero are real. Note that we have defined a constant ω_i equal to the reciprocal of τ_i, which is a time constant for the denominator terms. The reason for using the symbol for frequency will become evident later. Also, we call the value ω_i a *break frequency*, for a reason to be explained later.

First we form the magnitude of $G(j\omega)$.

$$|G(j\omega)| = \frac{|K||1 + j\omega/\omega_3|}{|1 + j\omega/\omega_1||1 + j\omega/\omega_2|} \tag{8-11}$$

In general, a gain K can be negative; we show this possibility by giving K magnitude symbols. For a given value of frequency, (8-11) will be evaluated as a positive real number, since each factor is a positive real number.

Next we used the property of logarithms given by

$$\log\left(\frac{ab}{cd}\right) = \log ab - \log cd = \log a + \log b - \log c - \log d$$

Also we define the unit decibel (dB) as

$$dB = 20 \log a$$

where a is a gain. We plot the magnitude of the frequency response in decibels; that is, we plot $20 \log|G(j\omega)|$. For the transfer function of (8-11),

$$\begin{aligned} 20 \log|G(j\omega)| = {}& 20 \log|K| + 20 \log\left|1 + \frac{j\omega}{\omega_3}\right| \\ & -20 \log\left|1 + \frac{j\omega}{\omega_1}\right| - 20 \log\left|1 + \frac{j\omega}{\omega_2}\right| \end{aligned} \tag{8-12}$$

The effect of plotting in decibels is then to cause the individual factors in the numerator to add to the total magnitude and the individual factors in the denominator to subtract from the total magnitude. If the transfer function has additional numerator or denominator factors, these factors add or subtract, respectively, to the total magnitude function, as shown in (8-12).

Consider now the general frequency-dependent term in (8-12).

$$dB_i = 20 \log\left|1 + \frac{j\omega}{\omega_i}\right| = 20 \log\left[1 + \left(\frac{\omega}{\omega_i}\right)^2\right]^{1/2} \tag{8-13}$$

This term is plotted versus $\log \omega$ in Figure 8.8. Note that the value of the term at the frequency ω_i (called the break frequency) is equal to $20 \log (2)^{1/2}$, or 3.0103. We usually approximate this value as 3 dB and say that, for a general first-order numerator term, the value of the magnitude is equal to 3 dB at the break frequency of that term. For a first-order denominator term, the value is equal to -3 dB at its break frequency. Note also that the first-order term has a value of $20 \log (101)^{1/2} = 20.04$, or approximately 20 dB, at the frequency $10 \omega_i$.

If we need an accurate Bode diagram, we usually calculate it using a digital computer program. However, there are situations in which approximate sketches of a Bode diagram

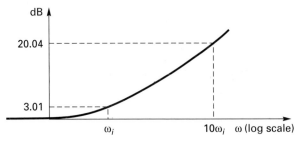

Figure 8.8 First-order term.

are adequate. These sketches are especially useful in gaining insight into both the type of compensation required for a particular control system design and the effects of adding compensator poles and zeros. These approximations are usually made only for real poles and zeros, since the approximations can be very inaccurate for a complex pole or zero depending on the damping ratio, ζ, of the term. Thus we emphasize the approximations for the first order terms.

We now develop the approximations for the first-order terms. Consider the first-order term of (8-13):

$$dB_i = 20 \log [1 + (\omega/\omega_i)^2]^{1/2}$$

For frequencies very small compared to the break frequency ω_i,

$$dB_i \simeq 20 \log(1) = 0 \qquad \omega \ll \omega_i \tag{8-14}$$

and for frequencies very large compared to ω_i,

$$dB_i \simeq 20 \log\left(\frac{\omega}{\omega_i}\right) = 20 \log \omega - 20 \log \omega_i \qquad \omega \gg \omega_i \tag{8-15}$$

For low frequencies, as given in (8-14), the term is approximated by a straight line (the ω-axis). For high frequencies, as given in (8-15), if viewed as a function of $\log \omega$, the term is again approximated by a straight line. Differentiation of (8-15) with respect to $\log \omega$ reveals that this line has a slope of 20 dB per decade of frequency. Equating the high-frequency and low-frequency expressions of (8-14) and (8-15) shows that the two straight lines intersect at $\omega = \omega_i$. These two terms are plotted in Figure 8.9(a). Comparing this figure with the exact curve of Figure 8.8, we see that the exact curve approaches the straight lines asymptotically, as is shown in Figure 8.9(b). As an approximation in sketching, we quite often extend the straight lines to the intersection at $\omega = \omega_i$ and use this straight-line approximation instead of the exact curve. The maximum error in using the straight-line approximations instead of the exact curve occurs at the break frequency $\omega = \omega_i$ and is 3 dB, as was shown above. Also, ω_i is called the *break frequency* because of the break in the slope at that frequency, as shown in Figure 8.9(b).

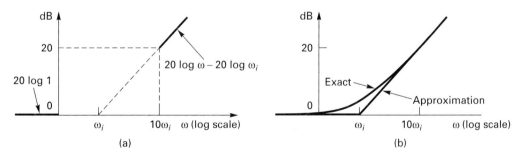

Figure 8.9 First-order approximation.

In constructing frequency responses, we consider the following types of transfer-function factors:

1. Constant gain
2. Poles and zeros at the origin

3. Real poles and zeros not at the origin
4. Complex poles and zeros
5. Ideal time delays

We now consider each of these factors in order. First we develop the magnitude plots, and then we develop the phase plots.

8.2.1 Constant Gain

For the case of a constant gain.

$$dB = 20 \log|K|$$

and the magnitude of this term does not vary with frequency. The two possible cases are shown in Figure 8.10. If $|K|$ is greater than unity, dB is positive; if $|K|$ is less than unity, dB is negative. In either case, the magnitude plot is a straight line with a slope of zero.

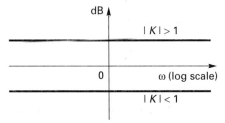

Figure 8.10   Constant-gain term.

8.2.2 Poles and Zeros at the Origin

For the case that a transfer function has a zero at the origin, the magnitude of this term is given by

$$dB = 20 \log|j\omega| = 20 \log \omega$$

Hence a plot of this term is a straight line, with a slope of 20 dB per decade of frequency, that intersects the ω-axis at $\omega = 1$. The plot of this term is shown in Figure 8.11(a). For this case, the straight line is the exact plot of the magnitude. In this figure the slope, in dB per decade of frequency, is shown directly above the Bode diagram. We usually include the slope in this manner on straight-line diagrams for clarity.

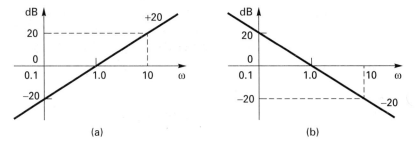

(a)   (b)

Figure 8.11   Zero and pole at $s = 0$.

For the case that a transfer function has a pole at the origin, the magnitude of the term is given by

$$dB = -20 \log \omega$$

and the curve is the negative of that for a zero at the origin, described earlier. Thus the curve is a straight line with a slope of -20 dB per decade that intersects the log ω axis at $\omega = 1$. This curve is shown in Figure 8.11(b). Once again, this curve is exact.

For the case of an Nth-order zero at the origin, the magnitude is

$$dB = 20 \log |(j\omega)^N| = 20 \log \omega^N = 20N \log \omega$$

Thus the curve is still a straight line that intersects the ω-axis at $\omega = 1$, but the slope is now $20N$ dB per decade. For the case of an Nth-order pole at $s = 0$, it is seen that the curve is the negative of that of the Nth-order zero.

8.2.3 Nonzero Real Poles and Zeros

The case of real poles and zeros was considered previously. For a term of this type,

$$20 \log \left| 1 + \frac{j\omega}{\omega_i} \right| = 20 \log \left[1 + \left(\frac{\omega}{\omega_i} \right)^2 \right]^{1/2}$$

$$\simeq \begin{cases} 0 & \omega \leq \omega_i \\ 20 \log \omega - 20 \log \omega_i & \omega > \omega_i \end{cases} \tag{8-16}$$

This straight-line approximation is shown in Figure 8.12(a) for a zero and in Figure 8.12(b) for a pole. Note that the terms have been normalized to have a dc gain of unity, or 0 dB. This normalization is not necessary but is usually done for convenience. If the dc gain is not unity, each term will have a different low-frequency gain, and the Bode diagram is somewhat more difficult to plot. It is recommended that the terms of the transfer function always be normalized to have unity dc gain (0-dB gain).

Suppose that a first-order term is repeated, that is, suppose that we have an Nth-order term of the form $(1 + s/\omega_i)^N$. The magnitude term is then given by

$$dB_i = 20 \log \left[1 + \left(\frac{\omega}{\omega_i} \right)^2 \right]^{N/2} \simeq \begin{cases} 0 & \omega \ll \omega_i \\ 20N \log \omega/\omega_i & \omega \gg \omega_i \end{cases} \tag{8-17}$$

The straight-line approximation for this term is shown in Figure 8.13 for the case of a numerator term. It is seen that for $\omega > \omega_i$, the line has a slope of $20N$. For a given denominator term, the slope is $-20N$. Three examples of the Bode diagram are given next.

Example 8.2

Consider the transfer function

$$G(s) = \frac{10(s + 1)}{(s + 10)} = \frac{(s + 1)}{(s/10 + 1)}$$

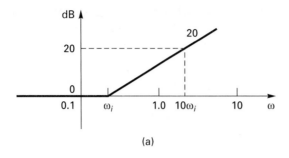

(a)

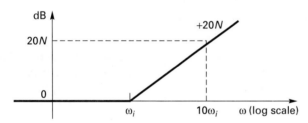

(b)

Figure 8.12   First-order terms.

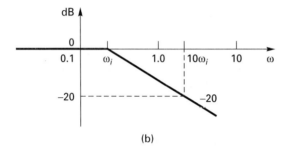

Figure 8.13   Bode diagram for repeated zeros.

The break frequency of the numerator is $\omega = 1$, and the break frequency of the denominator is $\omega = 10$. The numerator term, the denominator term, and the total magnitude [which, from (8-12), is the sum of the two terms] are shown in Figure 8.14. Note how easily the approximate magnitude of the frequency response is sketched for this simple transfer function.

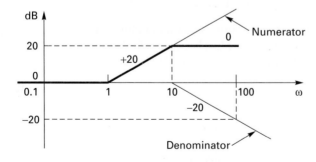

Figure 8.14   Example 8.2.

We can always easily check the magnitudes at the extremes of the Bode diagram, that is, the magnitude at the low-frequency end and the magnitude at the high-frequency end. We illustrate this with the system of Example 8.2. For this example,

$$G(j\omega) = \frac{1 + j\omega}{1 + j\omega/10}$$

Hence,

$$\lim_{\omega \to 0} G(j\omega) = 1 \Rightarrow 0 \text{ dB}$$

and

$$\lim_{\omega \to \infty} G(j\omega) = \lim_{\omega \to \infty} \frac{j\omega}{j\omega/10} = 10 \Rightarrow 20 \text{ dB}$$

Note that these two calculations check the values sketched in Figure 8.14. These two evaluations are very useful in checking the construction of Bode diagrams and should always be employed, even if digital computer calculations are used (programs can be in error, data can be entered incorrectly, or numerical errors can occur).

Example 8.3

As a second example, consider the transfer function of the system of the preceding example with a different constant gain and with an added pole at $s = 10$. Then

$$G(s) = \frac{200(s + 1)}{(s + 10)^2} = \frac{2(1 + s)}{(1 + s/10)^2}$$

The Bode diagram has three terms. The first term is the constant gain, which adds a term of value 20 log 2 = 6 dB at all frequencies. The second term is the zero term with the break frequency at $\omega = 1$, and the third term is the second-order pole at $s = 10$. The three terms are plotted in Figure 8.15 along with the complete magnitude diagram (the sum of the three terms).

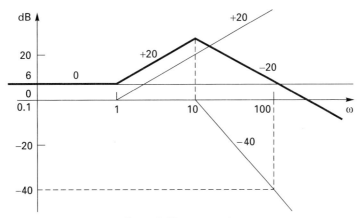

Figure 8.15   Example 8.3.

The low-frequency gain is $G(0) = 2$, which is 6 dB. At high frequencies,

$$\lim_{\omega \to \infty} |G(j\omega)| = \frac{200}{\omega}$$

Hence the Bode diagram has a slope of −20 dB at high frequencies from the $1/\omega$ factor. Note that the diagram in Figure 8.15 agrees with these calculations at the two extremes in frequency. The Bode diagram for this example can be plotted with the MATLAB program

```
bode(tf([200 200], [1 20 100]));
```

Example 8.4

As a third example of the Bode diagram, consider the transfer function

$$G(s) = \frac{1000(s + 3)}{s(s + 12)(s + 50)} = \frac{5(1 + s/3)}{s(1 + s/12)(1 + s/50)}$$

The constant-gain term is obtained from

$$\frac{(1000)(3)}{(12)(50)} = 5$$

Now 20 log 5 = 14.0 dB, and the five terms of the Bode diagram are as shown in Figure 8.16. Also given in this figure is the complete magnitude diagram.

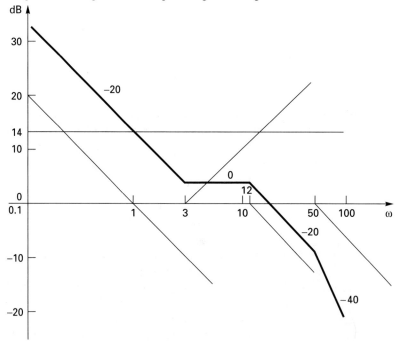

Figure 8.16 Example 8.4.

At low frequencies,

$$\lim_{\omega \to 0} |G(j\omega)| = \frac{5}{\omega}$$

and the magnitude diagram has a slope of -20 dB per decade at the low-frequency extreme. At high frequencies,

$$\lim_{\omega \to \infty} |G(j\omega)| = \frac{1000}{\omega^2}$$

Hence the magnitude diagram has a slope of -40 dB per decade at the high-frequency extreme. Both these calculations agree with Figure 8.16. Of course, numerical values of ω may also be used in these relationships to check the diagram further. The low–frequency slope intersects the ω-axis at $\omega = 5$, and the high-frequency slope intersects at $\omega = (1000)^{1/2} = 31.6$.

8.2.4 Phase Diagrams

Before we consider the final two types of terms that can appear in a Bode diagram, we construct the phase diagrams for the three types of terms already considered. First, for the constant-gain term, the phase angle is either $0°$ or $\pm180°$. If the gain term is positive, the phase angle is $0°$; if the gain term is negative, the phase angle can be plotted as either $180°$ or $-180°$.

For a zero of the transfer function at the origin, the phase angle is $90°$, since

$$s\big|_{s = j\omega} = j\omega = \omega \underline{/90°} \tag{8-18}$$

In a like manner, a pole at the origin gives a phase angle of $-90°$, since

$$\frac{1}{s}\bigg|_{s = j\omega} = \frac{1}{j\omega} = \frac{1}{\omega} \underline{/-90°} \tag{8-19}$$

Note that for both these terms, the angle is independent of frequency.

For a real zero of the transfer function, with the zero not at the origin, the term is given by

$$\left(1 + \frac{s}{\omega_i}\right)\bigg|_{s = j\omega} = 1 + \frac{j\omega}{\omega_i} = \left[1 + \left(\frac{\omega}{\omega_i}\right)^2\right]^{1/2} \underline{/\theta(\omega)} \tag{8-20}$$

where

$$\theta(\omega) = \tan^{-1}\left(\frac{\omega}{\omega_i}\right) \tag{8-21}$$

The values of θ for various values of the ration ω/ω_i are given in Table 8.2, and these values are plotted in Figure 8.17. We can approximate this exact curve with the straight-line construction shown in Figure 8.17. The straight-line approximation for the phase characteristic breaks from $0°$ at the frequency $0.1\omega_i$ and breaks back to the constant value of $90°$ at $10\omega_i$. Although the errors in this straight-line approximation can be appreciable, we use it in this book because of its convenience. If a more-accurate phase curve is required, computer calculations should be used. Note that the phase characteristic for a pole is the negative of that for a zero, since, for a pole,

$$\frac{1}{1 + s/\omega_i}\bigg|_{s = j\omega} = \frac{1}{1 + j\omega/\omega_i} = \frac{1}{[1 + (\omega/\omega_i)^2]^{1/2}} \underline{/\theta(\omega)} \tag{8-22}$$

where

$$\theta(\omega) = -\tan^{-1}\left(\frac{\omega}{\omega_i}\right)$$

Three examples of the construction of the phase characteristic are now given.

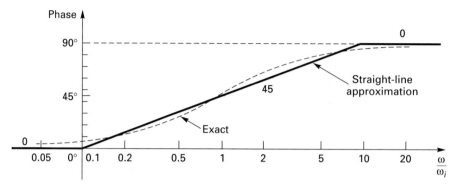

Figure 8.17   Phase characteristic of a real zero.

Example 8.5

Consider again the system of Example 8.2:

$$G(s) = \frac{s+1}{s/10+1}$$

The straight-line approximation to the phase characteristic of the zero breaks at $\omega = 0.1(1) = 0.1$ and breaks back at $\omega = 10(1) = 10$. The pole characteristic breaks at $\omega = 0.1(10) = 1$ and breaks

TABLE 8.2

ω/ω_i	Exact value (deg)	Straight-line approximation (deg)
0.05	2.9	0
0.1	5.7	0
0.2	11.3	13.5
0.5	26.6	31.5
0.8	38.7	40.6
1.0	45.0	45.0
2	63.4	58.5
5	78.7	76.5
8	82.9	85.6
10	84.3	90
20	87.1	90

back at $\omega = 10(10) = 100$. These characteristics, along with the total phase diagram (which is the sum of the two characteristics), are plotted in Figure 8.18. The slopes of the phase characteristics, in degrees per decade, are also shown in the figure.

As in the case for plots of the magnitude characteristics, we should check the extremes of the phase diagram. For low frequencies,

$$\lim_{\omega \to 0} G(j\omega) = 1$$

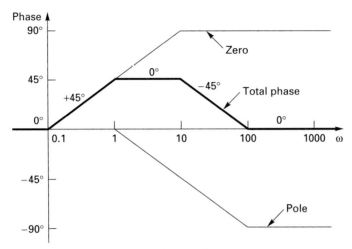

Figure 8.18   Example 8.5.

which has an angle of 0°. For high frequencies,

$$\lim_{\omega \to \infty} G(j\omega) = 10$$

which also has an angle of 0°, as indicated in Figure 8.18. Note that the extremes in both magnitude and phase are evident from the preceding calculations, and separate calculations are not required.

Example 8.6

As a second example illustrating the phase characteristic of the Bode diagram, consider the system of Example 8.4, with the transfer function

$$G(s) = \frac{5(1 + s/3)}{s(1 + s/12)(1 + s/50)}$$

The phase characteristics of the various terms, along with the total phase characteristic of the system, are given in Figure 8.19. The low-frequency characteristic is given by

$$\lim_{\omega \to 0} G(j\omega) = \frac{5}{j\omega} = \frac{5}{\omega}\underline{/-90°}$$

and the phase at low frequencies is −90°. For high frequencies,

$$\lim_{\omega \to \infty} G(j\omega) = \frac{1000}{(j\omega)^2} = \frac{1000}{\omega^2}\underline{/-180°}$$

and the phase at high frequencies is −180°. These values are the same as those obtained graphically in Figure 8.19. Note that the extremes on the magnitude characteristic are also obtained by these calculations.

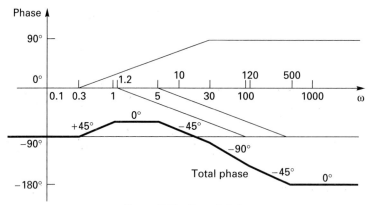

Figure 8.19   Example 8.6.

Example 8.7

As a final example, the complete Bode diagram will be constructed for the transfer function

$$G(s) = \frac{(1-s)}{(1+s/10)}$$

This transfer function is the same as that in Examples 8.2 and 8.5, except that the zero has been moved from $s = -1$ to $s = 1$. Hence the transfer function is *non-minimum phase* (some poles and/or zeros are in the right half-plane). First, note that the magnitude characteristic is unchanged from that of Example 8.2, since the magnitude of a complex number is independent of the signs of either the real part or the imaginary part, that is,

$$|1 + j\omega| = |1 - j\omega| = (1 + \omega^2)^{1/2}$$

Thus the magnitude characteristic is as shown in Figure 8.20 (see also Figure 8.14).

The phase of the zero term varies from $0°$ to $-90°$, because of the minus sign on the imaginary part.

$$1 - j\omega = (1 + \omega^2)^{1/2} \underline{/\,\theta(\omega)} \qquad \theta(\omega) = \tan^{-1}(-\omega)$$

The total phase characteristic is then as shown in Figure 8.20. The characteristics at the extremes in frequency are verified through the calculations

$$\lim_{\omega \to 0} G(j\omega) = 1 \underline{/0°}$$

$$\lim_{\omega \to 0} G(j\omega) = \frac{-j\omega}{j\omega/10} = -10 = 10\underline{/-180°}$$

To illustrate typical errors that occur when using straight-line approximations for simple transfer functions, the exact Bode diagram for this system is also given in Figure 8.20. The Bode diagram of Figure 8.20 can be plotted with the MATLAB program

```
bode([-10 10], [1 10])
```

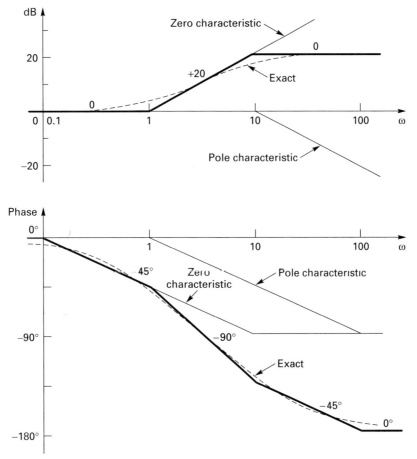

Figure 8.20   Bode diagram for Example 8.7.

As a final point, the polar plot (for example, Figure 8.3) is not changed by a rotation of either 360° or −360°. The equivalent operation on the Bode diagram is to shift the phase characteristic either up by 360° or down by 360°.

In this section the Bode diagrams of transfer functions that contain only real poles and zeros have been covered extensively. In the next section the difficult topic of complex poles and zeros is considered.

8.3 ADDITIONAL TERMS

In this section we consider two additional types of terms that can be encountered in constructing a Bode diagram. First we consider poles or zeros of the form

$$s^2 + 2\zeta\omega_n s + \omega_n^2 \qquad 0 \le \zeta < 1 \qquad (8\text{-}23)$$

Then we consider the transfer function of the phenomenon known as ideal time delay, or transport lag.

8.3.1 Complex Poles and Zeros

First we consider complex poles and zeros resulting from factors of the type given in (8-23). As in the last section, for convenience in plotting we normalize this factor, (8-23), to have a dc gain of unity; this is accomplished by factoring out ω_n^2. Hence the factor that we consider is

$$1 + 2\zeta\frac{s}{\omega_n} + \left(\frac{s}{\omega_n}\right)^2 \tag{8-24}$$

The magnitude and phase of this expression for $s = j\omega$ is an involved function of the damping ratio ζ, and in general it does not lend itself to approximations by straight lines. However, we illustrate some special cases that do and then consider the general case that does not.

Consider first the case that $\zeta = 1$. For this case, (8-24) has two real equal zeros.

$$\left. 1 + 2\zeta\frac{s}{\omega_n} + \left(\frac{s}{\omega_n}\right)^2 \right|_{\zeta = 1} = \left(1 + \frac{s}{\omega_n}\right)^2$$

Since the zeros are real, this case is covered by the methods given in the preceding section. The straight-line approximations for this case are given in Figure 8.21, along with the exact curves. The maximum error in the magnitude plot, which occurs at $s = j\omega = j\omega_n$, is 6 dB. This is an appreciable error, since it represents a factor of 2. At $\omega = \omega_n$, the exact value of the magnitude is 2, whereas the value of the approximation is 1.

For cases in which ζ is less than unity, the straight-line curves are sometimes used, but the errors can be large. Figure 8.22 illustrates some exact curves for cases that the maximum error in the magnitude characteristic is less than approximately 6 dB. This value for error was chosen arbitrarily; in any particular situation the maximum allowable error may be different. Allowing this value of error, the straight-line approximations may be used then for complex zeros and poles for $\zeta \geq 0.3$. But note that, even with this limitation, the phase errors can be quite large. However, remember that in general we use the approximate Bode diagram only to obtain an idea of the characteristics of a system; for accurate analysis and design we calculate the exact Bode diagram using a digital computer.

The Bode plots of Figure 8.22 are for complex zeros. Of course, the figures should be inverted for complex poles. For the case that $\zeta < 0.3$, the straight-line approximations are very inaccurate and are seldom used. Instead, exact curves are used. Exact curves are given in Figure 8.23 for several values of ζ between zero and unity for complex zeros. Once again, the curves for complex poles are obtained by inverting these curves.

The limiting case for $\zeta = 1$ was discussed earlier. The other limiting case, for $\zeta = 0$, is also of interest. For this case, the factor is given by

$$1 + \left(\frac{s}{\omega_n}\right)^2$$

which has purely imaginary zeros at $s = j\omega_n$. Hence, at the frequency $\omega = \omega_n$, the term has a magnitude of zero, or $-\infty$ in decibels. The phase of this term is $0°$ for $\omega < \omega_n$ and $180°$ for $\omega > \omega_n$. Thus the phase characteristic has a discontinuity of $180°$ at $\omega = \omega_n$. This characteristic is shown in Figure 8.23. An example is now given to illustrate complex terms in a Bode diagram.

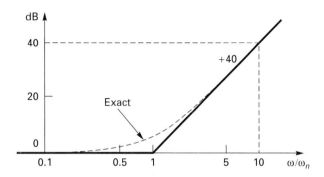

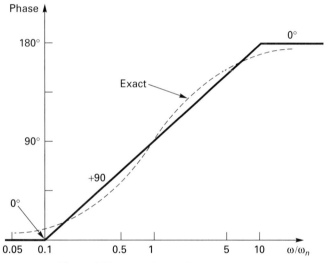

Figure 8.21   Bode diagram for a repeated zero.

Example 8.8

As an example, consider the transfer function

$$G(s) = \frac{200(s+1)}{s^2 + 4s + 100} = \frac{2(s+1)}{(s/10)^2 + 2(0.2)(s/10) + 1}$$

For the complex poles, $\zeta = 0.2$ and we do not expect the straight-line approximation to be very accurate in the vicinity of $\omega_n = 10$. Both the straight-line approximation and the exact Bode diagram are given in Figure 8.24. The maximum error in the magnitude diagram for the straight-line approximation is seen to be approximately 8 dB. Note also the very large errors in the straight-line approximations for the phase. The exact curves in Figure 8.24 are plotted with the MATLAB statement

```
bode([0 200 200],[1 4 100])
```

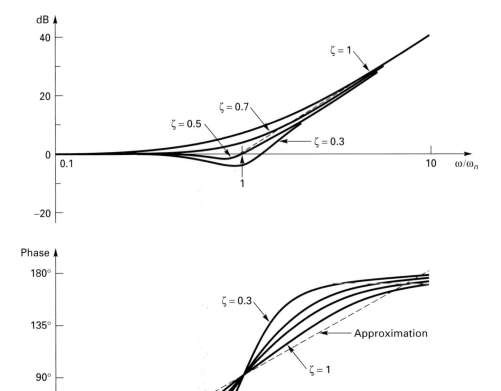

Figure 8.22 Bode diagram for complex zero.

8.3.2 Ideal Time Delay

We first define *ideal time delay* mathematically and then give an example of a system that contains ideal time delay. Suppose that a signal $e(t)$ is applied to an ideal time delay, as shown by the block diagram in Figure 8.25(a). Then, by definition, the output of the ideal time delay is $e(t - t_0) u(t - t_0)$, where t_0 is the amount of the time delay. With the Laplace transform of $e(t)$ equal to $E(s)$ and that of $c(t)$ equal to $C(s)$, the output of the ideal time delay, $C(s)$, is

$$C(s) = \mathcal{L}\,[e(t-t_0)u(t-t_0)] = e^{-t_0 s}E(s) \tag{8-25}$$

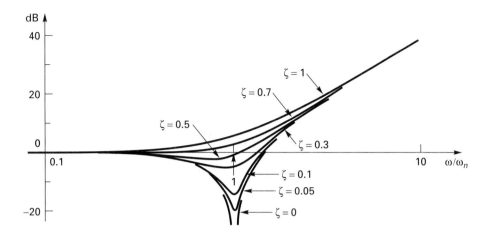

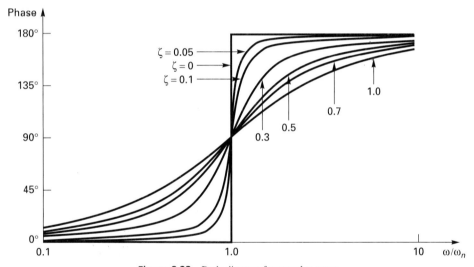

Figure 8.23   Bode diagram for complex zeros.

Therefore, the transfer function of the ideal time delay is

$$G(s) = e^{-t_0 s} \qquad (8\text{-}26)$$

as shown in Figure 8.25(b). Note that this transfer function is different from all others considered thus far, in that the transfer function is not a ratio of polynomials. This fact complicates the analysis of systems that contain ideal time delays. For example, the Routh–Hurwitz criterion cannot be used with systems of this type, since the system characteristic equation is not a polynomial in s. We consider this problem further when we introduce the Nyquist criterion.

Before we construct the Bode diagram of the transfer function (8-26), we consider a system that exhibits ideal time delay (also called *transport lag*) in order to better understand

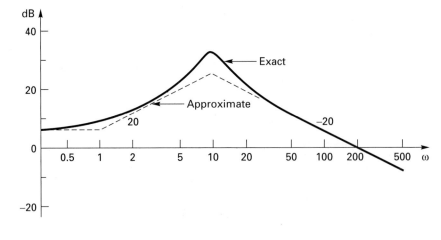

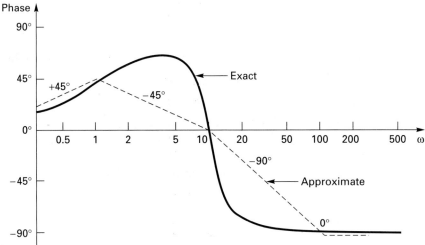

Figure 8.24 Bode diagram for Example 8.8.

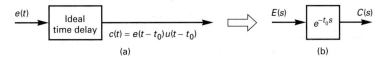

Figure 8.25 Ideal time delay.

the phenomenon. Figure 8.26 depicts the process of steel rolling. The purpose of this system is to control accurately the thickness, w, to which the steel is rolled. Let $w(t)$ be the distance between the outer surfaces of the rollers. This distance between the rollers is usually controlled hydraulically and is difficult to measure accurately. Instead the thickness of the steel is measured downstream from the rollers. Hence the measured thickness of the steel is the distance between the rollers that occurred some t_0 seconds earlier, where t_0 is determined by the distance of the sensor form the rollers and by the speed of the steel. If this measurement is used in a closed-loop system to control the thickness of the steel, the control is based on knowing the distance between rollers not at the present time, but at a time some t_0 seconds earlier. This time delay can cause some severe problems in the design of the control system, as discussed later.

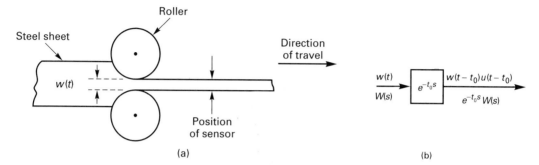

Figure 8.26   Example of ideal time delay.

The frequency response of the transfer function is obtained from (8-26) by replacing s with $j\omega$. Thus

$$G(j\omega) = e^{-j\omega t_0} = 1\,\underline{/-\omega t_0} \tag{8-27}$$

Therefore, the magnitude of the transfer function is unity, and the phase is a linear function of frequency. The phase lag introduced by the ideal time delay increases without limit as frequency becomes large. The unity magnitude is evident, since a signal is not distorted by transmission through an ideal time delay; it is only delayed. For example, suppose that the signal $e(t) = A \cos \omega t\, u(t)$ is applied to an ideal time delay of t_0 seconds. The delay output, by definition, is

$$A \cos\,[\omega(t - t_0)]\,u(t - t_0) = A \cos(\omega t - \omega t_0)u(t - t_0)$$

$$= A \cos(\omega t - \theta)u(t - t_0) \qquad \theta = \omega t_0 \tag{8-28}$$

For $t > t_0$, the output is equal to $A \cos(\omega t - \theta)$. Thus it is seen that the amplitude of the sinusoid is unchanged, and the phase now includes a phase lag of ωt_0 radians.

The Bode diagram of the transfer function of an ideal time delay is then as shown in Figure 8.27. The magnitude is, of course, 0 dB, and the phase is given by

$$\text{phase} = -57.3\omega t_0 \quad \text{in degrees} \tag{8-29}$$

The Bode diagram of a system containing an ideal time delay is now illustrated by an example.

Example 8.9

Consider the system of Figure 8.28, which contains an ideal time delay of 0.2 s. The system transfer function is given by

$$G(s) = \frac{e^{-0.2s}}{s(s + 1)}$$

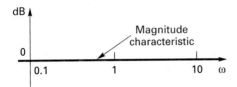

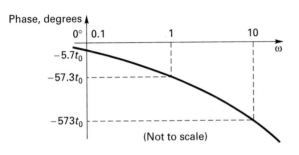

Figure 8.27   Frequency response of an ideal time delay.

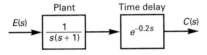

Figure 8.28   System for Example 8.9.

The two poles contribute terms to the Bode diagram, as described before. The ideal time delay of 0.2 s contributes 0 dB to the magnitude term and a phase of $-57.3\omega t_0 = -11.46\omega$, in degrees. The Bode diagram is then as shown in Figure 8.29. The Bode diagram for this example is plotted with the MATLAB program

```
nn = 30; a = -1;  b = 1;
w = logspace(a,b,nn);
[mag,phase,w] = bode([1], [1 1 0],w);
   db=20*log10(mag);
      phased1 = (-0.2)*57.296*w;
   phase = phase + phased1;
      subplot(211), semilogx(w,db)
      title('Bode diagram'); xlabel('frequency');
      ylabel('dB'); grid
      subplot(212), semilogx(w,phase)
      xlabel('frequency'); ylabel('phase'); grid
```

8.4 NYQUIST CRITERION

In this section we consider closed-loop systems of the type shown in Figure 8.30. The characteristic equation of this system is given by

$$1 + G(s)H(s) = 0 \tag{8-30}$$

This equation determines the stability of the system and, if the system is stable, certain useful characteristics of the transient response, as discussed earlier. As indicated in the previous sections, the Bode diagram of the open-loop function can be plotted using the transfer function $G(j\omega)H(j\omega)$. In this section we investigate the determination of stability of a *closed-loop* system from the Bode diagram (or frequency response) of the *open-loop* func-

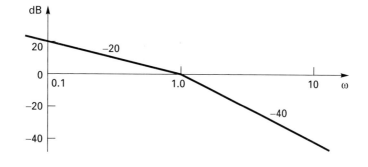

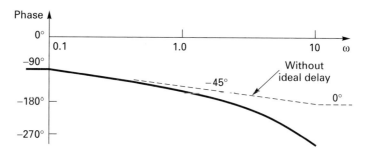

Figure 8.29   Bode diagram for Example 8.9.

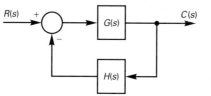

Figure 8.30   Closed-loop system.

tion $G(j\omega)H(j\omega)$. The method that allows us to do this is based on the Nyquist criterion, which is the topic of this section.

In order to introduce the Nyquist criterion, we consider some mappings (functions) from the complex s-plane to the $F(s)$-plane. For example, consider the case that the function (mapping) $F(s)$ is given by

$$F(s) = s - s_0 \tag{8-31}$$

where s_0 is a given value that is possibly complex. Suppose that we want to map a circle centered at s_0 in the s-plane into the $F(s)$-plane, as shown in Figure 8.31. The curve C in the s-plane in Figure 8.31(a) is to be mapped into the curve Γ in the $F(s)$-plane by evaluating $F(s)$ for points on the curve C and plotting the resulting complex values in the $F(s)$-plane. For the simple function (8-31), $F(s)$ is the vector shown in Figure 8.31(a), and the resulting curve in the $F(s)$-plane is seen to be a circle of the same radius as that of C but centered at the origin. Note that the curve C encircles the zero of $F(s)$ in the s-plane in the clockwise direction, and the curve Γ encircles the origin in the $F(s)$-plane in the clockwise direction. This point is important, as is shown shortly.

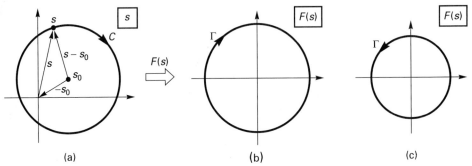

Figure 8.31 Mappings to the complex plane.

Consider now the function

$$F(s) = \frac{1}{s - s_0} \tag{8-32}$$

which is the reciprocal of the function in (8-31). If the curve C in Figure 8.31(a) is mapped into the $F(s)$-plane through (8-32), the vector $s - s_0$ remains as shown in Figure 8.31(a). Hence $F(s)$ is the reciprocal of this vector. The magnitude of $F(s)$ is the reciprocal of that shown in Figure 8.31(b), and the angle is the negative of that in the figure. Thus in this case, the curve Γ in the $F(s)$-plane is also a circle, as shown in Figure 8.31(c), except that the direction of travel is now counterclockwise. For this function, a clockwise encirclement of the pole in the s-plane leads to a counterclockwise encirclement of the origin in the $F(s)$-plane.

As a third example of mappings, suppose that the mapping $F(s)$ is given by

$$F(s) = (s - s_0)(s - s_1) \tag{8-33}$$

and suppose that the curve C in the s-plane encircles both zeros s_0 and s_1, as shown in Figure 8.32(a). In this case the curve C is not necessarily a circle. The two vectors that comprise $F(s)$ are shown in Figure 8.32(a). As the point s travels around the curve C, the angle of the vector $(s - s_0)$ changes by an amount of $-360°$, as does also the angle of the vector $(s - s_1)$. Hence the angle of the function $F(s)$ changes by the amount of $-720°$. At the same time the magnitudes of the two vectors are bounded and nonzero. Therefore, the curve Γ, the mapping of C into the $F(s)$-plane, must encircle the origin twice, as shown in Figure 8.32(b). Note that the curve C in the s-plane encircles the two zeros of $F(s)$, in the clockwise direction. In the $F(s)$-plane, the curve Γ encircles the origin twice in the clockwise direction.

If the mapping $F(s)$ is the reciprocal of (8-33), that is, if

$$F(s) = \frac{1}{(s - s_0)(s - s_1)} \tag{8-34}$$

the vectors for the curve C are still as shown in Figure 8.32. Since the angle of the product of the two vectors rotates through $-720°$, the angle of $F(s)$ in (8-34) rotates through $720°$. Hence the mapping into the curve Γ will result in two counterclockwise encirclements of the origin. The clockwise encirclement of the two poles in the s-plane results in two counterclockwise encirclements of the origin in the $F(s)$-plane.

We see then a relationship between the number of poles and zeros encircled by C in the s-plane and the number and direction of encirclements of the origin in the $F(s)$-plane.

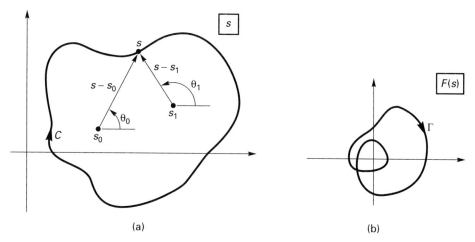

Figure 8.32   Path encircling two poles.

Many additional examples of mappings of these types could be constructed to indicate further this relationship, but the results would confirm the theorem that is known as *Cauchy's principle of argument* [3].

Theorem.   Let $F(s)$ be the ratio of two polynomials in s. Let the closed curve C in the s-plane be mapped into the complex plane through the mapping $F(s)$. If $F(s)$ is analytic within and on C, except at a finite number of poles, and if $F(s)$ has neither poles nor zeros on C, then

$$N = Z - P \qquad (8\text{-}35)$$

where Z is the number of zeros of $F(s)$ in C, P is the number of poles of $F(s)$ in C, and N is the number of encirclements of the origin, taken in the same sense as C.

Cauchy's principle of argument is the mathematical basis for the Nyquist criterion. We will not define the term *analytic,* but the functions that we consider in this text do satisfy the conditions of the theorem. Those interested in pursuing the mathematical aspects of this topic are referred to any book on complex-variable theory (see [3], for example).

We now develop the Nyquist criterion. Suppose that we let the mapping of $F(s)$ be the characteristic function of the closed-loop system of Figure 8.30; that is,

$$F(s) = 1 + G(s)H(s)$$

Furthermore, let the curve C be as shown in Figure 8.33(a). This curve, which is composed of the imaginary ($j\omega$)-axis and an arc of infinite radius, completely encloses the right half of the s-plane. Then, in Cauchy's principle of argument, (8-35), Z is the number of zeros of the system characteristic function in the right half of the s-plane. *Therefore, Z must be zero for the system to be stable.*

P in (8-35) is the number of poles of the characteristic function in the right half of the s-plane and thus is the number of poles of the open-loop function $G(s)H(s)$ in the right half-plane, since the poles of $[1 + G(s)H(s)]$ are also those of $G(s)H(s)$. Hence, P indicates

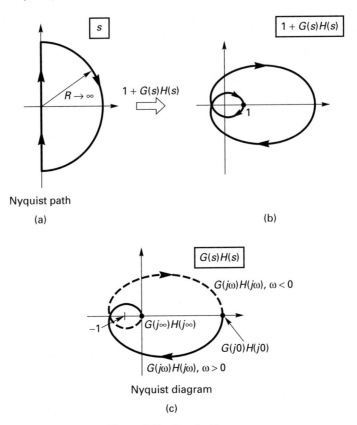

Nyquist path

(a)

(b)

Nyquist diagram

(c)

Figure 8.33   Nyquist diagram.

the stability of the system if operated open-loop. If P is not zero, the open-loop system is unstable. As is shown later, to apply the Nyquist criterion we must first determine P.

The curve in Figure 8.33(a) is called the *Nyquist path,* and a typical mapping is shown in Figure 8.33(b). The mapping encircles the origin two times in the clockwise direction, and from (8-35),

$$N = 2 = Z - P$$

or

$$Z = 2 + P$$

Since P is the number of poles of a function inside the Nyquist path, it cannot be a negative number. Thus in this example, Z is greater than or equal to 2, and the closed-loop system is unstable.

Although the preceding example illustrates a correct application of the Nyquist criterion, a modification is usually made to simplify the application. Suppose that instead of plotting $1 + G(s)H(s)$, as in Figure 8.33(b), we plot just $G(s)H(s)$. The resulting plot has the same shape but is shifted 1 unit to the left, as shown in Figure 8.33(c). Hence, rather than plotting $1 + G(s)H(s)$ and counting encirclements of the origin, we get the same result by plotting

$G(s)H(s)$ and counting encirclements of the point $-1 + j0$, hereafter referred to as the -1 point. The Nyquist criterion is usually applied in this fashion, and the resultant plot of the open-loop function $G(s)H(s)$ is called the *Nyquist diagram*. Note that we are plotting the *open-loop function* to determine the *closed-loop stability*.

The Nyquist criterion may be stated with respect to Figure 8.34.

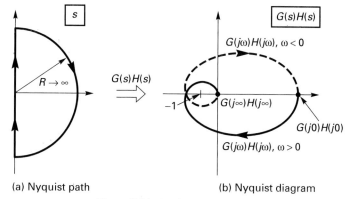

(a) Nyquist path   (b) Nyquist diagram

Figure 8.34   Typical Nyquist diagram.

The Nyquist path is shown in Figure 8.34(a). This path is mapped through the open-loop function $G(s)H(s)$ into the Nyquist diagram, as illustrated in Figure 8-34(b). Then

$$Z = N + P \tag{8-36}$$

where Z is the number of roots of the system characteristic equation in the right half-plane, N is the number of clockwise encirclements of the -1 point, and P is the number of poles of the open-loop function $G(s)H(s)$ in the right half-plane.

A simple example is given to illustrate the Nyquist criterion.

Example 8.10

Consider the system with the open-loop function

$$G(s)H(s) = \frac{5}{(s+1)^3}$$

Then

$$G(j\omega)H(j\omega) = \frac{5}{(1+j\omega)^3}$$

An evaluation of this function is given in Table 8.3 for certain values of ω, and a plot of these values is shown in Figure 8.35. The dc gain, $G(0)H(0)$, is equal to 5 and is shown as part I. The solid curve, part II, is obtained by directly plotting the values of Table 8.3. Since

$$\lim_{s \to \infty} G(s)H(s) = 0$$

the infinite arc of the Nyquist path maps into the origin in the $G(s)H(s)$-plane, part III of the Nyquist diagram. For negative ω, we use the complex conjugate of the values given in Table 8.3. This part of the Nyquist diagram, part IV, is shown as a dashed curve in Figure 8.35.

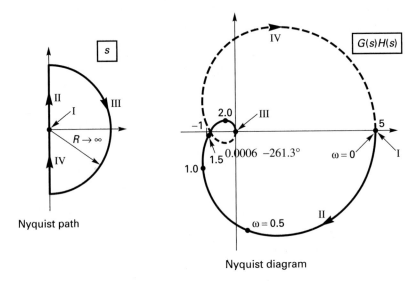

Nyquist path

Nyquist diagram

Figure 8.35 Nyquist criterion for Example 8.10.

The Nyquist equation is

$$Z = N + P$$

from (8-36). The value of P is zero, since $G(s)H(s)$ has no poles in the right half-plane. Note that P is the number of poles of the open-loop function $G(s)H(s)$ in the *right half-plane,* not simply the number of poles of the open-loop function. From the Nyquist diagram of Figure 8.35, we see that the number of encirclements of the -1 point in the $G(s)H(s)$-plane, N, is zero. Then the number of zeros of the characteristic function in the right half-plane is given by

$$Z = N + P = 0 + 0 = 0$$

and the closed-loop system is stable. The Nyquist diagram of Figure 8.35 can be plotted with the MATLAB program

```
nyquist([0 0 0 5],[1 3 3 1])
```

The stability of this system can be verified through the use of the Routh–Hurwitz criterion. Suppose that we determine the stability of the system for the case that a gain K is added to the open-loop function. The characteristic equation is then given by

$$1 + KG(s)H(s) = 1 + \frac{5K}{(s+1)^3} = 1 + \frac{5K}{s^3 + 3s^2 + 3s + 1} = 0$$

or

$$s^3 + 3s^2 + 3s + 1 + 5K = 0$$

The Routh array is then (see Section 6.1)

TABLE 8.3 FREQUENCY RESPONSE

ω	$1 + j\omega$	$G(j\omega)$
0	1	5
0.5	$1.118\underline{/26.6°}$	$3.58\underline{/-79.8°}$
1.0	$1.414\underline{/45°}$	$1.77\underline{/-135°}$
1.5	$1.80\underline{/56.3°}$	$0.85\underline{/-169°}$
2.0	$2.24\underline{/63.4°}$	$0.45\underline{/-190.3°}$
5.0	$5.10\underline{/78.7°}$	$0.038\underline{/-236.1°}$
20.0	$20.0\underline{/87.1°}$	$0.0006\underline{/-261.3°}$

$$
\begin{array}{c|cc}
s^3 & 1 & 3 \\
s^2 & 3 & 1 + 5K \\
s^1 & \dfrac{8 - 5K}{3} & \Rightarrow K < 5/8 \\
s^0 & 1 + 5K & \Rightarrow K > -0.2
\end{array}
$$

The system is stable then for $-0.2 < K < \frac{8}{5}$. Thus the system is stable for $K = 1$, which is the case for the Nyquist diagram of Figure 8.35.

The question arises as to why we would use the Nyquist criterion for stability analysis if the Routh–Hurwitz criterion is simpler to apply, as indicated in the previous example. One answer is that the Nyquist criterion not only indicates stability, but it also gives us information useful in the design of compensators, as is shown in the next chapter.

A second point illustrating the usefulness of the Nyquist criterion is evident from the Nyquist diagram of Figure 8.35. *Part II of the Nyquist diagram is a plot of the frequency response of the open-loop function.* In fact, the complete Nyquist diagram can be constructed from the frequency response $G(j\omega)H(j\omega)$, since parts I and III are also parts of the frequency response and part IV is the complex conjugate of the frequency response. Since the frequency response can be obtained experimentally, the Nyquist diagram can be constructed from measurements on the physical system, without knowing the open-loop function $G(s)H(s)$.

An additional point can be made with respect to stability analysis using the Nyquist criterion. Suppose that the Nyquist diagram intersects the -1 point for some value $\omega = \omega_1$. Then

$$G(j\omega_1)H(j\omega_1) = -1$$

or

$$1 + G(j\omega_1)H(j\omega_1) = 0 \tag{8-37}$$

This indicates that the closed-loop system has a pole at $s = j\omega_1$, since (8-37) is the denominator of the closed-loop transfer function. Thus, the system is marginally stable and oscillates with the frequency ω_1, provided that all other poles of the closed-loop system are in the left half-plane. In the preceding example, the system was shown to be marginally stable for

$K = \frac{8}{5}$. For this value of gain, the auxiliary equation from the Routh–Hurwitz criterion (see Section 6.1) is, from the above array,

$$3s^2 + 1 + 5K\big|_{K = 8/5} = 3s^2 + 9 = 3(s^2 + 3) = 0$$

The roots of this equation occur at $s = \pm j\sqrt{3}$. Therefore, the open-loop function evaluated at this frequency yields

$$G(j\omega)H(j\omega)\big|_{\omega = \sqrt{3}} = \frac{5}{(1 + j\sqrt{3})^3} = \frac{5}{(2\underline{/60°})^3} = -\frac{5}{8}$$

The Nyquist diagram intersects the negative real axis at the value −5/8. Thus an increase in gain of 8/5 will cause the Nyquist diagram to intersect the −1 point, and the system will be marginally stable. The factor by which the open-loop gain must be changed to make a stable system marginally stable is called the system *gain margin*. The gain margin is one measure that is used to indicate the *relative stability* of a system. The very important topic of relative stability will be discussed later.

8.5 APPLICATION OF THE NYQUIST CRITERION

In the preceding section, the Nyquist criterion was introduced. This criterion is very important, since it allows us to determine the stability of the *closed-loop* system from a knowledge of the frequency-response of the *open-loop* function. If the physical system is open-loop stable, the frequency response can be determined experimentally. Thus the stability of the closed-loop system can be determined from measurements on the open-loop system. Furthermore, the Nyquist criterion is important in that the Nyquist diagram gives us important information concerning the type of compensation required to stabilize certain types of systems.

Before additional examples of the Nyquist diagram are presented, the Nyquist criterion is reviewed. First, the Nyquist path in the *s*-plane and a typical Nyquist diagram are shown in Figure 8.36. The path is shown broken down into four parts. These parts are as follows:

I. This part of the Nyquist path is the point $s = 0$, or the origin of the *s*-plane. The open-loop function evaluated at this point, $G(0)H(0)$, yields the dc gain of the open-loop

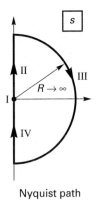

Nyquist path

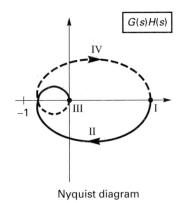

Nyquist diagram

Figure 8.36   Nyquist diagram.

function and may be measured experimentally. How would you perform this measurement on a physical system?

II. This part of the Nyquist path is the imaginary ($j\omega$)-axis, and the open-loop function evaluated along this axis is the frequency response of the open-loop function, $G(j\omega)H(j\omega)$. How would you experimentally evaluate the function $G(j\omega)H(j\omega)$?

III. This part of the Nyquist path is the infinite arc, along which s approaches infinity. Since physical systems are low-pass in nature, the open-loop function evaluated along this arc is zero. Hence this arc will generally map into the origin of the $G(s)H(s)$-plane.

IV. This section of the Nyquist path is the lower half of the imaginary axis. The open-loop function evaluated along this axis, $G(j\omega)H(j\omega)$, is the complex conjugate of the open-loop function evaluated along part II of the Nyquist path. Part IV of the Nyquist diagram is sometimes omitted, since its shape is obvious from part II.

From this discussion we see that the complete Nyquist diagram can be evaluated either mathematically if the open-loop function is available or experimentally if the physical system is available for experimentation. In general, a digital computer program is required to mathematically evaluate $G(j\omega)H(j\omega)$, $0 \leq \omega < \infty$, except for some very simple functions.

The Nyquist diagram is evaluated by mapping the Nyquist path into the complex plane via the mapping $G(s)H(s)$, as described above. A typical Nyquist diagram is shown in Figure 8.36. For this diagram,

$$Z = N + P \qquad (8\text{-}38)$$

where

- Z is the number of roots of the system characteristic equation in the right half-plane, or, equivalently, the number of poles of the closed-loop transfer function in the right half-plane. For the closed-loop system to be stable, Z must be zero. Note that Z cannot be negative.

- N is the number of clockwise encirclements of the -1 point by the Nyquist diagram. If the encirclements are in a counterclockwise direction, N is the negative of the number of encirclements. Thus N may be either positive or negative. If the Nyquist diagram intersects the -1 point, the closed-loop system has poles on the $j\omega$-axis.

- P is the number of poles of the open-loop function $G(s)H(s)$ in the right half-plane. Thus P cannot be a negative number. Note that the closed-loop system may be stable ($Z = 0$) with the open-loop system unstable ($P > 0$), provided that $N = -P$. An example of this case was the attitude control system of the Saturn V booster stage (see Problem 8.24). For this system, $P = 1$, and a counterclockwise encirclement of the -1 point on the Nyquist diagram, obtained by adding proper compensation, was required to stabilize the closed-loop control system.

Some simple but valuable examples are given next. In these examples, for convenience of notation it is assumed that $H(s) = 1$, and the open-loop function is then $G(s)$.

Example 8.11

Consider the system with the open-loop function

$$G(s) = \frac{5}{(s+1)^2}$$

The dc gain is $G(0) = 5$, and part I of the Nyquist path of Figure 8.36 maps into this point. For part II of the Nyquist path,

$$G(j\omega) = \frac{5}{(1+j\omega)(1+j\omega)}$$

Each factor in the denominator, $(1 + j\omega)$, has a magnitude that increases from a value of unity to infinity as ω increases from zero to infinity. Therefore, the magnitude of $G(j\omega)$ decreases from a value of 5 to 0. In addition, the angle of each term in the denominator increases from $0°$ to $90°$ as ω increases from zero to infinity. Thus the angle of $G(j\omega)$ decreases from $0°$ to $-180°$ as its magnitude decreases from 5 to 0. The Nyquist diagram of this system is then as shown in Figure 8.37.

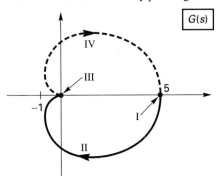

Figure 8.37   Nyquist diagram for Example 8.11.

Note that the angle of $G(j\omega)$ cannot be more negative than $-180°$, and thus the Nyquist diagram in Figure 8.37 cannot cross the negative real axis. If a gain factor K is added to the system such that the system characteristic equation is

$$1 + KG(s) = 1 + \frac{5K}{(1+s)^2} = 0$$

the Nyquist diagram cannot be forced to encircle the -1 point by increasing K, $K > 0$. By the Nyquist criterion,

$$Z = N + P = 0 + 0 = 0$$

N is zero since the Nyquist diagram does not encircle the -1 point, and P is zero since $G(s)$ has no poles in the right half-plane (both poles occur at $s = -1$). Hence the system is stable for all positive K and has an infinite gain margin.

This result can be checked using the Routh–Hurwitz criterion. From the above, the characteristic equation can be expressed as

$$1 + KG(s) = s^2 + 2s + 1 + 5K = 0$$

The Routh array is

$$
\begin{array}{c|cc}
s^2 & 1 & 1 + 5K \\
s^1 & 2 & \\
s^0 & 1 + 5K &
\end{array}
$$

It is seen that the closed-loop system is stable for all $K > 0$.

In the preceding example, we are considering the *model* of a physical system, not the physical system itself. This model does not go unstable as the gain K increases, but in all probability the physical system represented by the model would. This point is illustrated in the next example, in which a pole is added to the transfer function.

Example 8.12

Consider now the system with the open-loop function

$$G(s) = \frac{50}{(s+1)^2(s+10)}$$

Note that this is the open-loop function of Example 8.11, with a pole added. For part I of the Nyquist path, the dc gain $G(0)$ is equal to 5. The frequency response, $G(j\omega)$, for part II of the Nyquist path is obtained from

$$G(j\omega) = \frac{50}{(1+j\omega)^2(10+j\omega)}$$

Since each factor in the denominator increases without limit with increasing frequency, the magnitude of $G(j\omega)$ decreases to zero as ω approaches infinity. The angle of each factor in the denominator increases from $0°$ to $90°$ with increasing frequency; therefore, the angle of $G(j\omega)$ decreases from $0°$ to $-270°$ as ω increases. A sketch of the Nyquist diagram is given in Figure 8.38. It is common practice not to show the complex-conjugate portion of the Nyquist diagram so as not to clutter the diagram. For diagrams of this type, if the reader encounters difficulty in counting the encirclements of the -1 point, the remaining part of the diagram can be sketched in.

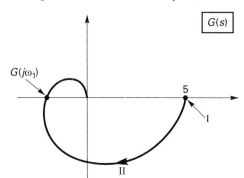

Figure 8.38 Nyquist diagram for third-order system.

For this system we have some difficulty in counting the encirclements of the -1 point, since the magnitude and frequency of $G(j\omega)$ at the point that the Nyquist diagram crosses the $180°$-axis is not evident. The value of $G(j\omega)$ at this point is indicated on the figure as $\mathrm{G}(j\omega_1)$. We can solve for $\mathrm{G}(j\omega_1)$ using the Routh–Hurwitz criterion. First we add a gain K to $G(s)$ and find the system characteristic equation.

$$1 + KG(s) = 1 + \frac{50K}{(s+1)^2(s+10)} = 0$$

or

$$s^3 + 12s^2 + 21s + 10 + 50K = 0$$

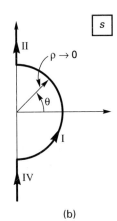

(a)

(b)

Figure 8.39 Detour around poles at the origin.

that is, on the detour with $\theta = 0°$, and let θ increase. On part I of the Nyquist path,

$$G(s)\big|_{s\,=\,\rho e^{j\theta}} = \frac{K}{\rho e^{j\theta}(\tau\rho e^{j\theta} + 1)}, \qquad 0° \leq \theta \leq 90°$$

and

$$\lim_{\rho \to 0} G(\rho e^{j\theta}) = \lim_{\rho \to 0}\frac{K}{\rho e^{j\theta}} = \lim_{\rho \to 0}\frac{K}{\rho}\ \underline{/-\theta}$$

Hence part I of the Nyquist path generates a very large arc on the Nyquist diagram, as shown in Figure 8.40. This arc swings past the $-90°$-axis somewhat, since the pole at $s = -1/\tau$ contributes a very small negative angle to the Nyquist diagram. Recall that this Nyquist diagram cannot be drawn to scale.

For part II of the Nyquist path, s is equal to $j\omega$, and this portion of the Nyquist diagram is a plot of the function

$$G(j\omega) = \frac{K}{j\omega(1 + j\omega\tau)}$$

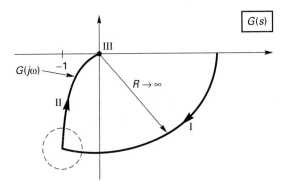

Figure 8.40 Nyquist diagram.

The magnitude and angle of $G(j\omega)$ are then given by

$$|G(j\omega)| = \frac{K}{\omega(1 + \omega^2\tau^2)^{1/2}}$$

and

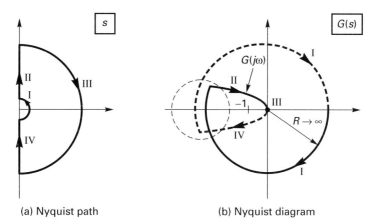

(a) Nyquist path   (b) Nyquist diagram

Figure 8.41   Nyquist diagram for Example 8.14.

$$\underline{/G(j\omega)} = -90° - \tan^{-1}(\omega\tau)$$

for $\rho \le \omega \le \infty$, with ρ very small. Therefore, the magnitude function decreases from a very large value to zero, and the angle function decreases from a value slightly more negative than $-90°$ to $-180°$. The resulting Nyquist plot is shown as part II of Figure 8.40.

Note again that the Nyquist diagram cannot be drawn to scale for this example, since we must show very large magnitudes and on the same figure show the point -1. The shape of the Nyquist diagram in the region shown enclosed by a dashed circle is not important, since this region is at a very large magnitude; the shape of the Nyquist diagram in the region cannot affect the number of encirclements of the -1 point. Note then that for this example

$$Z = N + P = 0 + 0 = 0$$

and the system is stable for all gains K, $K > 0$. The reader may verify these results using the Routh–Hurwitz criterion.

Example 8.14

As a second example of an open-loop function with poles at the origin, consider the transfer function

$$G(s) = \frac{K}{s^2(s+1)}$$

For certain aircraft, the reduced-order transfer function of the aircraft in an automatic landing system is of this form, with two poles at the origin and one pole on the negative real axis. Since the transfer function has two poles at the origin, the Nyquist path must detour around the origin as shown in Figure 8.41(a). On this detour (part I),

$$G(s)\Big|_{s=\rho e^{j\theta}} = \frac{K}{\rho^2 e^{j2\theta}(1 + \rho e^{j\theta})}$$

and

$$\lim_{\rho \to 0} G(\rho e^{j\theta}) = \lim_{\rho \to 0} \frac{K}{\rho^2 e^{j2\theta}} = \lim_{\rho \to 0} \frac{K}{\rho^2} \underline{/-2\theta} \qquad 0° \le \theta \le 90°$$

Thus the magnitude of the function is very large, and its angle rotates from $0°$ to slightly past $-180°$, as shown in Figure 8.41(b). This rotation past $-180°$ occurs because the pole at $s = -1$ contributes a very small negative angle to the function.

For part II of the Nyquist path, $s = j\omega$ and

$$G(j\omega) = \frac{K}{(j\omega)^2(1 + j\omega)} = \frac{K}{-\omega^2(1 + j\omega)}$$

Thus

$$|G(j\omega)| = \frac{K}{\omega^2(1 + \omega^2)^{1/2}}$$

and

$$\underline{/G(j\omega)} = -180° - \tan^{-1}\omega$$

As ω increases from a very small value to a very large value, the magnitude function decreases from a very large value to zero, and the angle decreases from $-180°$ to $-270°$. The Nyquist diagram is then as shown in Figure 8.41(b). As in the preceding example, the shape that is assigned to the Nyquist diagram in the encircled area is not important, since this shape does not affect the number of encirclements.

From this diagram we see that there are two clockwise encirclements of the -1 point, and hence $N = 2$. Since the open-loop function has no poles inside the Nyquist path, $P = 0$ and

$$Z = N + P = 2 + 0 = 2$$

The closed-loop transfer function has two poles in the right half-plane. Furthermore, the number of encirclements is not affected by the value of the gain K, provided that K is positive. Thus the system is unstable for all gains $K > 0$. This same result can be obtained by either the Routh–Hurwitz criterion or a root locus of the system. Of course, these three methods of stability analysis always give the same result for a given system. The frequency response of the Nyquist diagram is plotted by the MATLAB program

```
nyquist([0 0 0 1],[1 1 0 0])
```

It is instructive to run this program and compare the resulting plot with Figure 8.41(b).

In the preceding two examples, we considered the case first of one pole at the origin and then of two poles at the origin. For a single pole at the origin, the first-quadrant segment of part I of the Nyquist path resulted in a $-90°$ rotation in the Nyquist diagram at a very large magnitude; two poles at the origin resulted in a $-180°$ rotation in the Nyquist diagram. In the same manner it can be shown that three poles at the origin cause a $-270°$ rotation, four poles at the origin cause a $-360°$ rotation, and so forth. Note that any negative angle added to the Nyquist diagram tends to rotate the diagram towards the -1 point and increases the probability of encirclements of this point. Stability can be adversely affected by the addition of negative angles (phase angle) to the diagram, and thus the addition of poles at the origin can be destabilizing. Very few physical systems have more than two poles at the origin. In general, these systems are very difficult, but not impossible to stabilize. An example of a system with three poles at the origin is the aircraft-carrier automatic landing system used by the U.S. Navy [4].

Given in Table 8.4 is a comparison of the root loci and the Nyquist diagrams for several open-loop functions. Note, in (c) and (d) of this table, the effect of moving the zero relative to the pole locations. Some of the effects of adding a zero to a transfer function are shown in (e) and (f).

TABLE 8.4   COMPARISON OF ROOT LOCI AND NYQUIST DIAGRAMS

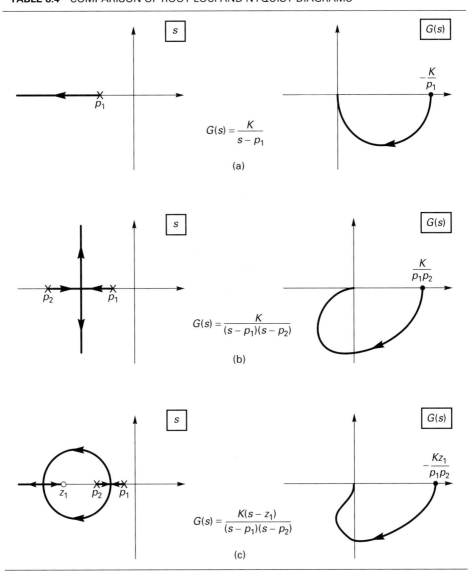

We consider one further example in this section. This example is unusual in that the open-loop function is nonminimum phase, resulting in an unusual Nyquist diagram.

TABLE 8.4   (CONTINUED)

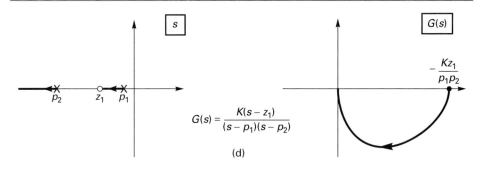

$$G(s) = \frac{K(s - z_1)}{(s - p_1)(s - p_2)}$$

(d)

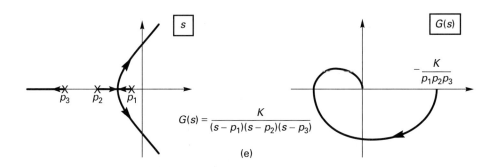

$$G(s) = \frac{K}{(s - p_1)(s - p_2)(s - p_3)}$$

(e)

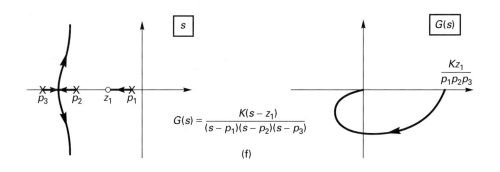

$$G(s) = \frac{K(s - z_1)}{(s - p_1)(s - p_2)(s - p_3)}$$

(f)

Example 8.15

Consider a system with the open-loop function

$$G(s) = \frac{K}{s(s - 1)}$$

Since $G(s)$ has a pole at the origin, the Nyquist path must detour around the origin, as shown in Figure 8.42(a). Thus

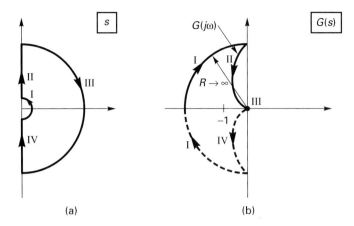

Figure 8.42 Nyquist construction for Example 8.15.

$$\lim_{\rho \to 0} G(\rho e^{j\theta}) = \lim_{\rho \to 0} \frac{K}{\rho e^{j\theta}(-1)} = \lim_{\rho \to 0} \frac{K}{\rho} \underline{/-180° - \theta} \qquad 0° \leq \theta \leq 90°$$

The magnitude of the open-loop function on part I of the Nyquist path is then very large, and the angle rotates from −180° to −270°. The infinite arc appears as shown in Figure 8.42(b). Note that the usual infinite arc generated by a pole at the origin has been rotated by −180° because of the negative sign in the preceding limit.

On part II of the Nyquist path, the frequency response is given by

$$G(j\omega) = \frac{K}{j\omega(j\omega - 1)}$$

The magnitude of this term decreases to zero as ω increases. The angle of the first term in the denominator is 90°, and the angle of the second term varies from 180° to 90°. The angle of $G(j\omega)$ then varies from −270° to −180° as the magnitude decreases from a very large value to zero. The complete Nyquist diagram is as shown in Figure 8.42(b). From this figure, $N = 1$ since there is one clockwise encirclement of the −1 point. Also, $P = 1$, since the open-loop function has one pole inside the Nyquist path, at $s = 1$. Thus

$$Z = N + P = 1 + 1 = 2$$

The system characteristic equation has two zeros in the right half-plane and the system is unstable. If we form the system characteristic equation,

$$1 + G(s) = 0 = s^2 - s + K$$

the roots occur at

$$s = \frac{1 \pm \sqrt{1 - 4K}}{2}$$

and the two roots in the right half-plane are evident. The MATLAB program for this example is

```
nyquist([1], [1 -1 0])
```

It is worthwhile to run this program and note the scaling problem in computer-generated Nyquist diagrams.

We make one final point in this section. Difficulties can occur in counting the number of encirclements of the −1 point for complex Nyquist diagrams. For example, how many encirclements of the −1 point occur for the Nyquist diagram in Figure 8.43? One procedure for counting encirclements is given here. First a ray is drawn from the −1 point in any convenient direction. One such ray is shown in Figure 8.43. The number of clockwise encirclements of the −1 point is then equal to the number of crossings of this ray by the Nyquist diagram, in the clockwise direction, minus the number of crossings of the ray in the counterclockwise direction. For Figure 8.43, there is one crossing in each direction, and thus the number of encirclements is zero. If the −1 point were at point x in this figure rather than at the point shown, then there are two clockwise crossings and no counterclockwise crossings. For this case, the Nyquist diagram has two clockwise encirclements of the −1 point.

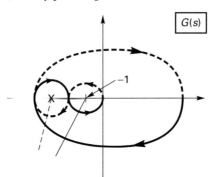

Figure 8.43   Counting encirclements on Nyquist diagram.

8.6 RELATIVE STABILITY AND THE BODE DIAGRAM

In the preceding sections, the Nyquist diagram was used to determine if a system was stable or unstable. Of course, in general a usable system must be stable. However, there are concerns beyond simple stability, for two reasons. First, a stable system must also have, among other characteristics, an acceptable transient response. Also, the model that is used in the analysis and design of a control system is never exact. Hence, the model may indicate that a system is stable, whereas in fact the physical system is unstable. Generally we require not only that a system be stable but also that it be stable by some margin of safety.

For these two reasons we define the *relative stability* of a system. For this book we define the relative stability of a system in terms of the closeness of the Nyquist diagram to the −1 point in the complex plane. This concept is illustrated in Figure 8.44(a). Note that the vector from the −1 point to the Nyquist diagram has a value of $[1 + G(j\omega)]$, where it is assumed that $G(s)$ is the open-loop function. The minimum magnitude of this vector is a good measure of the relative stability of a closed-loop system; however, this measure is seldom used, as explained below. Instead, the two measures indicated in Figure 8.44(b) are commonly used.

The first of these measures is the *gain margin,* which was defined earlier in this chapter. The gain margin is the factor by which the open-loop gain of a stable system must be changed to make the system marginally stable. In Figure 8.44(b), we denote the value of the Nyquist diagram at the −180° crossover as −α. If the open-loop function is multiplied by the

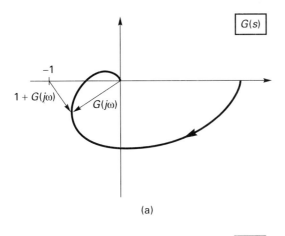

(a)

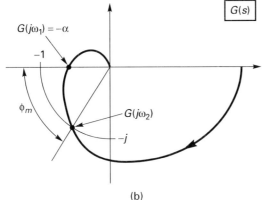

(b)

Figure 8.44   Relative stability margins.

gain of $K = 1/\alpha$, the Nyquist diagram intersects the -1 point, and the closed-loop system is marginally stable. Thus we make the following definition:

Gain margin   If the magnitude of the open-loop function of a stable closed-loop system at the $-180°$ crossover on the Nyquist diagram is the value α, the *gain margin* is $1/\alpha$. This margin is usually given in decibels. If the Nyquist diagram has multiple $-180°$ crossovers, the gain margin is determined by that point which results in the gain margin with the smallest magnitude, in decibels.

The second commonly used measure of relative stability is the *phase margin:*

Phase margin   The *phase margin* is the magnitude of the minimum angle by which the Nyquist diagram must be rotated in order to intersect the -1 point for a stable closed-loop system.

The phase margin is indicated by the angle ϕ_m in Figure 8.44(b). The magnitude of the Nyquist diagram, $G(j\omega)$, is unity at the frequency that the phase margin occurs. This frequency is indicated as ω_2 in Figure 8.44(b), and thus $|G(j\omega_2)| = 1$. The phase margin is then

$$\phi_m = \underline{/G\,(j\omega_2)} - 180° \tag{8-44}$$

A third stability margin will now be defined. This stability indicator, used in modern control theory, is the *return difference* [5], and is the function $[1 + G(s)]$, the denominator of the closed-loop transfer function. For multiple-input multiple-output systems, the return difference is a matrix; however, we will consider only single-input single-output systems. We now define the third stability margin.

Minimum return difference   The *minimum return difference* is the minimum value of $|1 + G(j\omega)|$, $0 < \omega < \infty$, where $1 + G(s) = 0$ is the system characteristic equation [5].

If the minimum return difference occurs at the value of ω for which $G(j\omega)$ is real, it is directly related to the gain margin. If the minimum return difference occurs for the value of ω for which $|G(j\omega)| = 1$, it is directly related to the phase margin. It is seen in Figure 8.44(a) that the minimum return difference is the minimum distance from the Nyquist diagram to the -1 point. We increase the stability of a system by increasing the minimum return difference. The minimum return difference is seldom used in classical control analysis and design.

Although the gain and phase margins may be obtained directly from a Nyquist diagram, they are more often determined from a Bode diagram. As shown in Figure 8.44, the Nyquist diagram is a plot of $G(j\omega)$ in polar coordinates. Of course, a Bode diagram of the open-loop function is also a plot of the same function but with different coordinates. Hence we are able to determine the gain and phase margins directly from a Bode diagram. For example, the Bode diagram for the system whose Nyquist diagram is shown in Figure 8.44 is given in Figure 8.45. This diagram is evident since the magnitude at low frequencies is greater than unity (positive dB) and decreases towards zero (negative dB for a magnitude less than unity) for high frequencies. As the magnitude characteristic decreases from a positive value towards minus infinity, in decibels, the angle varies in a negative direction from $0°$ toward $-270°$.

Refer again to Figure 8.44(b), in which the stability margins are defined. The gain margin occurs at the frequency at which the phase angle of $G(j\omega)$ is $-180°$. This frequency is evident on the Bode diagram of Figure 8.45 and is labeled as ω_1. The gain margin is the reciprocal of the magnitude of $G(j\omega_1)$. Since the effect of taking the logarithm of a reciprocal of a number is

$$\log\left(\frac{1}{\alpha}\right) = -\log \alpha$$

the gain margin can also be expressed in decibels as the value dB_α in Figure 8.45. It is common to express the gain margin in decibels, because of the popularity of Bode diagrams.

In Figure 8.44(b) we see that the phase margin occurs at the frequency ω_2 at which the magnitude of the open-loop gain is unity, or 0 dB. The phase margin ϕ_m is the difference between the angle of $G(j\omega_2)$ and $-180°$, as shown in Figure 8.45. Thus we see that both the gain margin and the phase margin can be determined directly from the Bode diagram for the open-loop function of a closed-loop system.

For a system in which multiple $-180°$ crossovers occur, the simplest approach is probably to sketch the Nyquist diagram from the Bode diagram in order first to determine stability and then to determine the frequencies at which the stability margins occur. Then the stability margins may be determined from the Bode diagram.

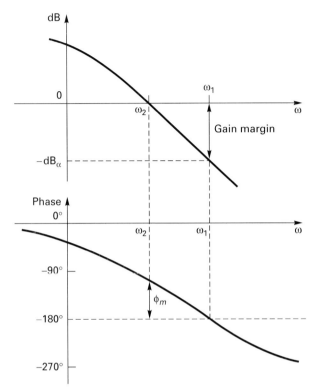

Figure 8.45   Relative stability margins on a Bode diagram.

In practical control-system design, the straight-line approximations for the Bode diagram are usually inadequate to determine the stability margins. For example, controls engineers have found that an 8-dB gain margin (a factor of 2.51) is usually adequate, whereas a 5-dB gain margin (a factor of 1.78) is usually not adequate. In addition, a 50° phase margin is often adequate, whereas a 30° phase margin seldom is. The values given here are rules of thumb obtained from experience and are very difficult to justify mathematically. They are used here to indicate that the usual approximation errors associated with the straight-line Bode diagrams are not acceptable in control-system design. These approximation errors can be large compared to the 3-dB difference and the 20° difference just given. The straight-line Bode diagrams are generally useful in indicating the characteristics of a system and in understanding the basics of compensation. We usually require digital computer-calculated data for both analysis and design of practical control systems.

An example is now given to illustrate the use of the Bode diagram to determine stability margins.

Example 8.16

Consider the system with the open-loop function

$$G(s) = \frac{1}{s(s+1)^2}$$

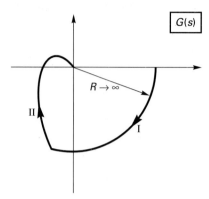

Figure 8.46 Nyquist diagram.

The Nyquist diagram is sketched in Figure 8.46. It is left as an exercise for the reader to justify that the sketch is correct. The frequency response, $G(j\omega)$, is listed in Table 8.5, and the exact Bode diagram is given in Figure 8.47. The gain margin and the phase margin are shown in this figure, but more exact values can be obtained from the data in Table 8.5. First we see that the $-180°$ crossover occurs at $\omega = 1$ and that the magnitude of the transfer function is -6 dB; thus

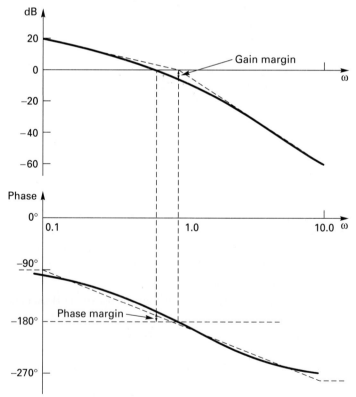

Figure 8.47 Bode diagram for Example 8.16.

TABLE 8.5 FREQUENCY RESPONSE FOR EXAMPLE 8.16

ω	$|G(j\omega)|$	$|G(j\omega)|_{dB}$	$\angle G(j\omega)$
0.0100	99.98999	39.99913	−91.14587
0.0200	49.98000	33.97593	−92.29153
0.0300	33.30336	30.44976	−93.43671
0.0400	24.96006	27.94491	−94.58122
0.0500	19.95012	25.99891	−95.72482
0.0600	16.60688	24.40576	−96.86726
0.0700	14.21605	23.05558	−98.00835
0.0800	12.42051	21.88279	−99.14785
0.0900	11.02183	20.84508	−100.28550
0.1000	9.90099	19.91357	−101.42120
0.2000	4.80769	13.63873	−112.61990
0.3000	3.05810	9.70904	−123.39850
0.4000	2.15517	6.66964	−133.60280
0.5000	1.60000	4.08240	−143.13010
0.6000	1.22549	1.76620	−151.92750
0.7000	0.95877	−0.36569	−159.98410
0.8000	0.76219	−2.35868	−167.31960
0.9000	0.61387	−4.23842	−173.97440
1.0000	0.50000	−6.02060	−180.00000
2.0000	0.10000	−20.00000	−216.86990
3.0000	0.03333	−29.54243	−233.13010
4.0000	0.01471	−36.65018	−241.92750
5.0000	0.00769	−42.27887	−247.38010
6.0000	0.00450	−46.92706	−251.07530
7.0000	0.00286	−50.88136	−253.73980
8.0000	0.00192	−54.32007	−255.75000
9.0000	0.00136	−57.36113	−257.31960
10.0000	0.00099	−60.08643	−258.57880

the gain margin is 6 dB. The transfer function has a magnitude of unity (0 dB) at $\omega \approx 0.7$, and the phase of the transfer function is approximately −160° at this frequency. Hence the phase margin is approximately

$$\phi_m \simeq -160° - 180° = -340° = 20°$$

The straight-line approximations are also included in Figure 8.47 to give an indication of the errors encountered when using these approximations. The straight-line approximation indicates a marginally stable system, while the exact Bode diagram indicates a stable system with a 6-dB gain margin. A table of the form of Table 8.5 can be calculated with the MATLAB program

```
[mag,phase,w] = bode ([1], [1 2 1 0]);
magdb = 20*log10(mag);
[w,mag,magdb,phase], pause
[Gm,Gp,wcg,wcp] = margin(mag,phase,w)
```

The last statement yields the gain margin of 2.013 and at $\omega = 0.680$ rad/s, the phase margin of 21.43°.

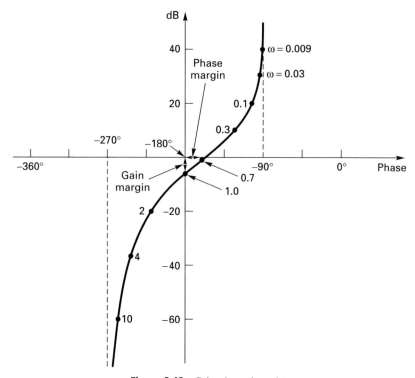

Figure 8.48 Gain–phase plane plot.

In the preceding discussions, two methods were given for plotting the frequency response of the open-loop function. A third method that is often used is shown in Figure 8.48 and presents exactly the same information as the Bode diagram. This plot is called the *gain–phase* plot, with the frequency response plotted as gain, in decibel, versus phase, with frequency as a parameter. As an example, the frequency response of the system of Example 8.16 is shown plotted in Figure 8.48. The gain and phase margins are shown in the figure. This figure is verified with the MATLAB program

```
nichols([1], [1 2 1 0])
```

The basis for the term *nichols* is given in Section 8.7.

8.6.1 Second–Order Systems

One final topic is covered in this section. For the standard second-order system, the transfer function is

$$T(s) = \frac{\omega_n^2}{s^2 + 2\zeta\omega_n s + \omega_n^2} \tag{8-45}$$

We assume that, for this derivation, this transfer function applies to the closed-loop system of Figure 8.49, where $G(s)$ is given by

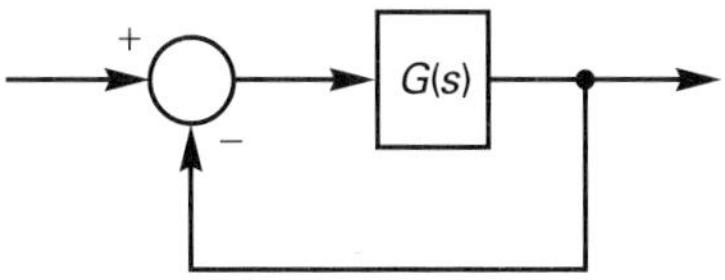

Figure 8.49 System for deriving phase margin.

$$G(s) = \frac{\omega_n^2}{s(s + 2\zeta\omega_n)} \tag{8-46}$$

Most position control systems that employ a servomotor are of this general configuration (see Section 2.7). We now relate the phase margin of this system to the damping ratio ζ.

The open-loop function can be expressed as

$$G(j\omega) = \frac{\omega_n^2}{-\omega^2 + j2\zeta\omega_n\omega} = \frac{1}{-\omega_v^2 + j2\zeta\omega_v}\bigg|_{\omega_v = \omega/\omega_n} \tag{8-47}$$

where the change of variables is made for convenience. The phase margin occurs at the frequency at which $|G(j\omega_v)|^2 = 1$, or

$$\frac{1}{\omega_v^4 + 4\zeta^2\omega_v^2} = 1$$

This equation can be expressed as

$$\omega_v^4 + 4\zeta^2\omega_v^2 - 1 = 0$$

and the roots of the equation are given by

$$\omega_v^2 = -2\zeta^2 \pm \sqrt{4\zeta^4 + 1}$$

Since the root that indicates the phase margin must be real,

$$\omega_v = \left[\sqrt{4\zeta^4 + 1} - 2\zeta^2\right]^{1/2} \tag{8-48}$$

With ω_v satisfying (8-48), the phase margin is defined [with use of (8-47)] as

$$\phi_m = 180° + \underline{/G(j\omega_v)} = 180° - 90° - \tan^{-1}\left(\frac{\omega_v}{2\zeta}\right)$$

$$= 90° - \tan^{-1}\left(\frac{\omega_v}{2\zeta}\right) = \tan^{-1}\left(\frac{2\zeta}{\omega_v}\right)$$

or, using (8-48),

$$\phi_m = \tan^{-1}\left[\frac{2\zeta}{\left(\sqrt{4\zeta^4 + 1} - 2\zeta^2\right)^{1/2}}\right] \tag{8-49}$$

Equation (8-49) relates the damping ratio of the standard second-order system to the system's phase margin, based on Figure 8.49. This equation is plotted in Figure 8.50. Also plotted are the other characteristics of the standard second-order system as derived in Chapter 4 and plotted in Figure 4.16. In this figure, M_{pt} is the peak value of the unit step response, $M_{p\omega}$ is the peak gain at the resonance in the frequency response, and ω_r is the frequency at which the resonance occurs. Note that except for ω_r, which is also a function of the natural frequency ω_n, these characteristics are determined by the damping ratio ζ. We can see from this figure why controls engineers are interested in the damping ratio associated with any complex poles of the closed-loop transfer function. If a physical system is accurately modeled as a second-order system with the open-loop transfer function of (8-46), Figure 8.50 applies directly. If the second-order model is less accurate, the figure gives indications of the characteristics of the system. If a higher-order model is required but the closed-loop transfer function has complex poles of importance, the figure gives some indication of the nature of that part of the response that depends on those poles.

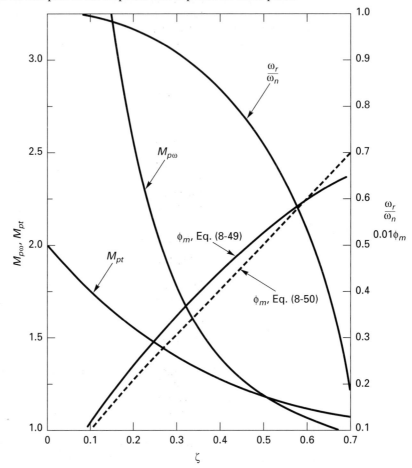

Figure 8.50   Second-order system characteristics.

An approximation that is sometimes used in (8-49) is the linear relationship

$$\phi_m = 100\zeta \tag{8-50}$$

This equation is much simpler than that of (8-49), and is reasonably accurate for many applications. Equation (8-50) is also plotted on Figure 8.50 to indicate the accuracy. An example is now given to illustrate the use of Figure 8.50.

Example 8.17

We again consider the system of Example 8.16, which has the open-loop function

$$G(s) = \frac{1}{s(s+1)^2}$$

In Example 8.16, the phase margin was found to be approximately 20°. *If the system were the standard second-order system,* from Figure 8.50 we see that ζ would be approximately 0.17 and the peak value in the system's unit step response would be approximately 1.6, based on the phase margin of 20°. A simulation of the third-order closed-loop system yielded a peak value in the system's unit step response of approximately 1.55. For this system we see that Figure 8.50 accurately predicts one of the system's characteristics, even though the system is third order. This system is considered further in the next section.

Example 8.18

The system of Example 7.4 will be considered in this example. This system has the nonminimum phase open-loop function

$$G(s) = \frac{K}{(s-1)(s+2)(s+3)} = \frac{K}{s^3 + 4s^2 + s - 6}$$

The root locus for this system was found in Example 7.4 and is repeated in Figure 8.51(a). This system is stable for $6 < K < 10$. A system that can be made unstable by increasing the gain and also by decreasing the gain is called a *conditionally stable system.* It will now be shown that both the Nyquist diagram and the root locus indicate the same stability characteristics, but in a different form.

The Nyquist diagram for this system was calculated by the MATLAB program

```
nyquist([8],[1 4 1 -6])
```

and is sketched (not to scale) in Figure 8.51 (b). For $K = 8$, the -1 point is located as shown in Figure 8.51(b), and the Nyquist criterion shows that the system is stable, since

$$Z = N + P = -1 + 1 = 0$$

The encirclement is counterclockwise ($N = -1$) and the open-loop function has one pole in the right-half plane ($P = 1$), at $s = 1$.

For $K = 8$, the gain margin for the gain increasing is $|20 \log(1/0.8)| = 1.94$ dB, and for the gain decreasing is $|20 \log(1.33)| = 2.48$ dB. Each of the margins is too small for a practical control system. The phase margin, from computer calculations, is $\phi_m = 2.6°$, which is also much too small. For this case, a simulation of the system shows that the step response has a 75 percent overshoot.

For $K < 6$, the -1 point is outside the Nyquist diagram. Therefore, there are no encirclements and $N = 0$; hence

$$Z = N + P = 0 + 1$$

We see then that the closed-loop transfer function has one pole in the right half-plane, which is also indicated by the root locus.

For $K > 10$, $N = 1$ and

$$Z = N + P = 1 + 1 = 2$$

The closed-loop transfer function has two poles in the right half-plane, and this is also indicated by the root locus.

Example 8.19

In this example, we will consider the ship-steering system of Examples 7.6, 7.7, and 7.8. This system was used in these examples to illustrate model-order reduction based on root loci. In this example, model-order reduction based on Bode diagrams will be illustrated. We will consider only the reduction of the third-order model to a second-order model. The third-order model of the ship-steering system is shown in Figure 8.52. From Figure 8.52, the third-order model is given by

$$G_3(s) = \frac{0.05 K_a}{s\left(1 + \frac{s}{0.1}\right)\left(1 + \frac{s}{2}\right)}$$

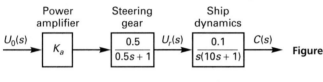

Figure 8.52   Ship-steering system.

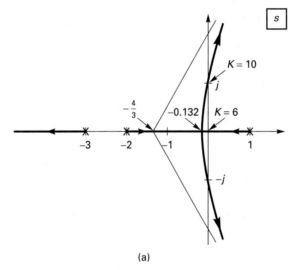

(a)

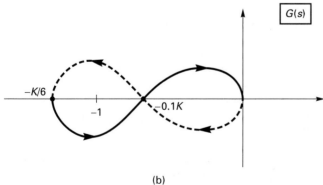

(b)

Figure 8.51   Root locus and Nyquist diagram for Example 8.18.

where K_a is the gain of the power amplifier. The Bode diagram for this transfer function is given in Figure 8.53, for $K_a = 20$. Also shown in Figure 8.53 is the Bode diagram with the pole at $s = -2$ ignored; hence this transfer function is given by

$$G_2(s) = \frac{0.05 K_a}{s\left(1 + \dfrac{s}{0.1}\right)}$$

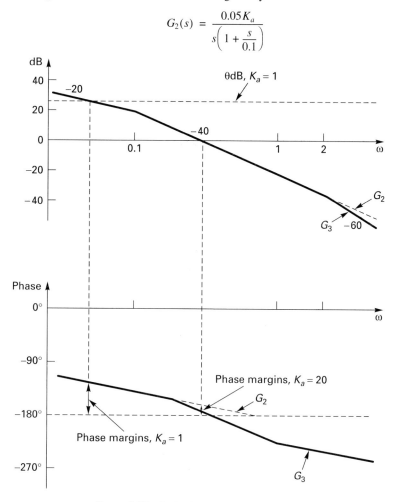

Figure 8.53 Bode diagrams for Example 8.19.

This Bode diagram is shown by dashed lines. Note that the differences in the two Bode diagrams appear only at the higher frequencies for the straight-line approximations.

The phase margins for the two transfer functions occur at the same frequency for $K_a = 20$, within the approximations of the straight-line diagrams. Note that the two phase margins are different; computer calculations give a phase margin $19°$ for $G_2(s)$ and $9°$ for $G_3(s)$. A $10°$ difference is significant (see Figure 8.50); hence the pole at $s = 2$ cannot be ignored.

Consider next the case that $K_a = 1$. This reduction in K_a has the effect of shifting the 0-dB axis higher, by the amount $20 \log(20) = 26$ dB, as indicated on Figure 8.53. Note that the phase margins occur at the frequency indicated, and that the phase margins are approximately equal for the two transfer functions, at 66°. Hence the pole at $s = 2$ does not significantly affect the systems for $K_a = 1$, since the gain of the open-loop systems is very small at frequencies above $s = j\omega = j2$; any changes that occur in the frequency response above this frequency can be ignored. Thus, for $K_a = 1$, the pole at $s = 2$ can be ignored. This result was also illustrated in Example 7.7, using a root-locus approach. The exact Bode diagrams in Figure 8.53 are plotted with the MATLAB program

```
bode([0.2], [1 2.1 0.22 0]), hold on
bode([0.1], [1 0.1 0]), hold off
```

As noted in Section 7.3, in model-order reduction of a system model, a first-order term must be expressed in the form $(\tau s + 1)$, and a second-order term as

$$\text{term} = \left(\frac{s}{\omega_n}\right)^2 + 2\zeta\frac{s}{\omega_n} + 1$$

The elimination of terms in these forms does not change the low-frequency characteristics of a system, as illustrated in the MATLAB program in Example 8.19.

Example 8.20

Consider the system shown in Figure 8.54. This system contains the same plant as in Example 8.16, but an ideal time delay has been added (see Section 8.3). The characteristic equation is given by

$$1 + G(s)e^{-t_0 s} = 1 + \frac{e^{-t_0 s}}{s(s+1)^2} = 0$$

or

$$s^3 + 2s^2 + s + e^{-t_0 s} = 0$$

We wish to determine if any values of s in the right half-plane satisfy this equation. Note that the characteristic equation is not a polynomial; thus the Routh–Hurwitz criterion does not apply. However, we can plot the Nyquist diagram for the open-loop function, $G(s)e^{-t_0 s}$. The open-loop frequency response is given by $G(j\omega)e^{-jt_0\omega}$, and hence

$$\left|G(j\omega)e^{-jt_0\omega}\right| = |G(j\omega)| \qquad \underline{/G(j\omega)e^{-jt_0\omega}} = \underline{/G(j\omega)} - t_0\omega$$

The frequency response for $G(j\omega)$ is given in Table 8.5, Example 8.16, and a sketch of the Nyquist diagram for $G(s)$ is given in Figure 8.46. Thus the frequency response for the system of this example is the same as given in Table 8.5, except the angle $-t_0\omega$ must be added to that in the table. Note that the units of the angle $t_0\omega$ is radians and that the magnitude of this angle increases with increasing frequency. The resulting Nyquist diagram is sketched in Figure 8.55, along with the Nyquist diagram for $G(s)$, which is shown with a dashed line. We see that the

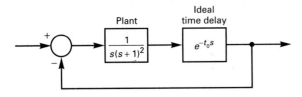

Figure 8.54 System for Example 8.20.

only effect of ideal delay is to add phase lag to the system. Of course, this phase lag is destabilizing, since the Nyquist diagram is rotated toward the −1 point.

For this example we wish to determine the maximum amount of time delay t_0 allowable such that the closed-loop system remains stable. Thus we find the maximum angle through which the Nyquist diagram can be rotated without an encirclement of the −1 point. This angle is, of course, the phase margin. From the last example, the phase margin is 21.43° at $\omega = 0.680$ rad/s. Thus,

$$\omega t_0 \big|_{\omega = 0.680} = 0.680 t_0 \approx \frac{21.43}{57.3} = 0.374 s$$

Hence the system is stable for $t_0 \leq 0.374/0.680 = 0.550$ s. The results in this example are verified with the MATLAB program

```
t0 = [0.75 0.550 0]; w = 0.3:0.1:4;
for k = 1:3
    [re,im,w] = nyquist([1], [1 2 1 0], w);
    re,im
      D1 = exp(-j*t0(k)*w)
      Gw = (re+j*im) .* D1
        Greal = real(Gw); Gimag = imag (Gw);
        plot(Greal,Gimag)
        title('Nyquist Diagram'); xlabel('Re G(jw)');
        ylabel('Im G(jw)'); grid; hold on
end
hold off
```

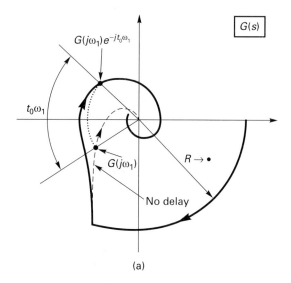

(a)

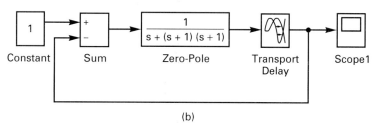

(b)

Figure 8.55 (a) Nyquist diagram and (b) SIMULINK simulation for Example 8.20.

where the Nyquist diagram is plotted for three different values of t_0. A SIMULINK example is depicted in Figure 8.55(b). The ideal time delay is denoted as transport delay. This simulation yielded a steady–state oscillation of $\omega \simeq 0.68$ rad/s at $t_0 = 0.550$ s.

The general effect of adding time delay to a stable closed-loop system is destabilizing, since the Nyquist diagram is rotated toward the -1 point, decreasing the phase margin. Note that if the system open-loop gain is greater than unity at any frequency, a sufficient amount of time delay can be added to force the Nyquist diagram to intersect the -1 point. Hence, in general, almost any closed-loop system can be destabilized by the addition of time delay in the control system loop.

An example of the effects of the addition of time delay is the driving of a car, which is a closed-loop system. As a person ages, the person's responses become slower; this slowing of the responses is essentially time delay. Hence, for an older person operating a car, stability problems occur. This is the primary reason for older persons reducing the speed of the car, so that larger responses for required changes are not needed. A larger response must be modeled as a larger gain.

As a final point in this section, we will relate the phase margin ϕ_m to the settling time T_s for a standard second-order system (see Section 4.3). From the derivation below (8-48),

$$\phi_m = \tan^{-1}\left(\frac{2\zeta}{\omega_v}\right) = \tan^{-1}\left(\frac{2\zeta\omega_n}{\omega_1}\right) \tag{8-51}$$

where ω_1 is the frequency at which the phase margin occurs; that is, ω_1 is defined by $G(j\omega_1)$ $= 1\underline{/-180°} + \phi_m$ (see Figure 8.44). From (4-27), the settling time is given by $T_s = 4/(\zeta\omega_n)$. Hence (8-51) can be expressed as

$$\tan \phi_m = \frac{2\zeta\omega_n}{\omega_1} = \frac{8}{\omega_1 T_s} \tag{8-52}$$

This equation relates the phase margin ϕ_m, the frequency ω_1 at which the phase margin occurs, and the settling time T_s, for the standard second-order system. For a given phase margin, increasing ω_1 reduces the settling time T_s. This equation will prove to be useful in the design procedures of Chapter 9.

8.7 CLOSED-LOOP FREQUENCY RESPONSE

In the preceding sections, we were concerned with the frequency response of an *open-loop system*. From this frequency response we were able to determine the stability and the relative stability of the *closed-loop system*. We stated that the relative stability margins are related, in some undefined way, to the response characteristics of the closed-loop system. However, the response characteristics of the closed-loop system are determined exactly by the *closed-loop frequency response*. We consider the frequency response of closed-loop systems in this section.

First we derive a graphical transformation from the open-loop frequency response to the closed-loop frequency response. The usefulness of this transformation becomes evident later. Consider the system of Figure 8.56(a). Note that we are limiting this derivation to unity feedback systems. For this system the closed-loop transfer function $T(s)$ is given by

$$T(s) = \frac{C(s)}{R(s)} = \frac{G(s)}{1 + G(s)}$$

The closed-loop frequency-response function is given by

$$T(j\omega) = \frac{C(j\omega)}{R(j\omega)} = \frac{G(j\omega)}{1 + G(j\omega)} \qquad (8\text{-}53)$$

Suppose that $G(j\omega)$ is as shown in the polar plot in Figure 8.56(b). Note that this plot is actually a portion of the system's Nyquist diagram. The numerator of the closed-loop transfer function $T(j\omega)$ is the vector $G(j\omega)$ shown in the figure, and the denominator of $T(j\omega)$ is the vector $1 + G(j\omega)$ shown, for the frequency ω_a. The frequency response is the ratio of these vectors, and we denote this ratio as

$$T(j\omega) = \frac{|G(j\omega)|e^{j\theta}}{|1 + G(j\omega)|e^{j\beta}} = Me^{j(\theta - \beta)} = Me^{j\phi} \qquad (8\text{-}54)$$

Thus M is the magnitude of the closed-loop frequency response, and ϕ is the phase.

The locus of points in the $G(j\omega)$-plane for which the magnitude of the closed-loop frequency response, M, is constant is called a *constant magnitude locus,* or a constant M circle. To see that these loci are in fact circles, consider the following derivation. Let

$$G(j\omega) = X + jY \qquad (8\text{-}55)$$

Then, from (8-54),

$$M^2 = \frac{X^2 + Y^2}{(1 + X)^2 + Y^2}$$

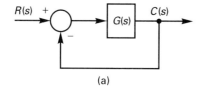

(a)

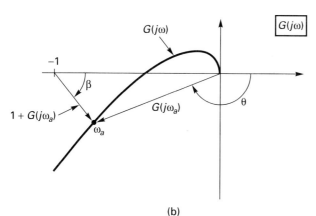

(b)

Figure 8.56 Closed-loop frequency response.

which can be rearranged to yield

$$X^2(M^2 - 1) + 2M^2X + M^2 + (M^2 - 1)Y^2 = 0 \tag{8-56}$$

For $M \neq 1$ we can express this relationship as

$$\left[X + \frac{M^2}{M^2 - 1}\right]^2 + Y^2 = \frac{M^2}{(M^2 - 1)^2}$$

This relationship is the equation of a circle of radius $|M/(M^2 - 1)|$ with the center at $X = -M^2/(M^2 - 1)$ and $Y = 0$. For $M = 1$, (8-56) yields $X = -\frac{1}{2}$, which is a straight line. Figure 8.57 illustrates the constant M circles.

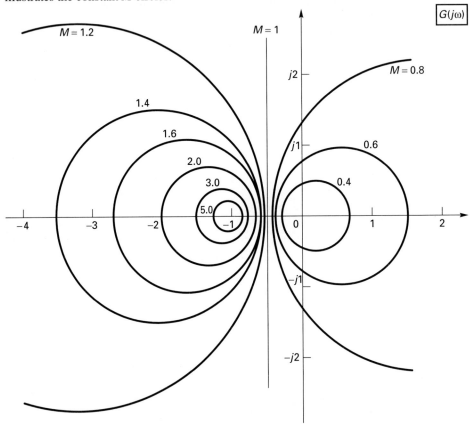

Figure 8.57 Constant M circles. (From C. L. Phillips and H. T. Nagle, *Digital Control System Analysis and Design.* © 1995, 3rd ed., p. 268. Adapted by permission of Prentice Hall, Upper Saddle River, NJ.)

The loci of points of constant phase are also circles. It is seen from (8-54) and (8-55) that

$$\phi = \theta - \beta = \tan^{-1}\left(\frac{Y}{X}\right) - \tan^{-1}\left(\frac{Y}{1 + X}\right) \tag{8-57}$$

Then, letting $N = \tan \phi$,

$$N = \tan(\theta - \beta) = \frac{\tan \theta - \tan \beta}{1 + \tan \theta \tan \beta} \qquad (8\text{-}58)$$

Equations (8-57) and (8-58) can be solved to yield

$$\left(X + \frac{1}{2}\right)^2 + \left(Y - \frac{1}{2N}\right)^2 = \frac{1}{4} + \left(\frac{1}{2N}\right)^2 \qquad (8\text{-}59)$$

(see Problem 8.34). This is the equation of a circle of radius $\sqrt{\frac{1}{4} + (1/2N)^2}$ with center at $X = -\frac{1}{2}$ and $Y = 1/(2N)$. Figure 8.58 illustrates the constant N (phase) circles.

It is common to represent the constant magnitude curves and the constant phase curves derived above on different sets of axes. One form is the Nichols chart. The gain-phase plane was introduced in Section 8.6 and was shown to be useful in presenting the open-loop frequency response. The circles of Figures 8.57 and 8.58 can also be plotted in this plane. If this is done, the resulting diagram is called a *Nichols chart* [6]. The Nichols chart is shown in Figure 8.59. This chart contains the same information as do the constant M and N circles; the information is plotted on different axes.

The constant M and N circles or the Nichols chart can be used to find graphically the closed-loop frequency response, given the open-loop frequency response. However, the frequency response $G(j\omega)$ is usually computed via computer programs. Thus it is logical to add statements to the program that will, at the same time, calculate the closed-loop frequency response rather than using the figures above in some type of graphical procedure. The computer procedure is much more accurate, saves time, and *is not limited to the unity feedback system of Figure* 8.56(a).

The importance of the open-loop to closed-loop frequency response graphs just given is in illustrating the general effects of the open-loop frequency response on the closed-loop frequency response. As described relative to Figure 8.56(b), the denominator of the closed-loop transfer function, $[1 + G(j\omega)]$, is the vector from the -1 point to the Nyquist diagram $G(j\omega)$. If the magnitude of this vector is large compared to unity, the vector can be approximated as $G(j\omega)$, and the closed-loop transfer function can be approximated by

$$T(j\omega) = \frac{G(j\omega)}{1 + G(j\omega)} \simeq 1 \qquad |G(j\omega)| \gg 1 \qquad (8\text{-}60)$$

In the frequency range to which this condition applies, the output is approximately equal to the input.

If the magnitude of $G(j\omega)$ is small compared to unity, the vector $[1 + G(j\omega)]$ is approximately equal to unity, and

$$T(j\omega) \simeq G(j\omega) \qquad |G(j\omega)| \ll 1 \qquad (8\text{-}61)$$

Thus the closed-loop system greatly attenuates the frequencies for which (8-61) is satisfied.

The remaining frequency band is that band of frequencies for which the magnitude of $G(j\omega)$ is of the order of unity. Of course, if the Nyquist diagram passes close to the -1 point, that part of the Nyquist diagram falls within this band. If, within this band, $|1 + G(j\omega)|$ is small, then the closed-loop transfer function has a large magnitude, as can be seen from (8-53) and Figures 8.56(b), 8.57 and 8.59, and *the closed-loop system is resonant.*

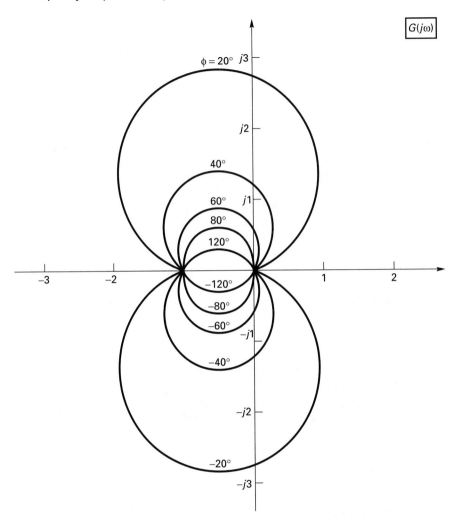

Figure 8.58 Constant phase circles. (From C. L. Phillips and H. T. Nagle, *Digital Control System Analysis and Design.* © 1995, 3rd ed., p. 269. Adapted by permission of Prentice Hall, Upper Saddle River, NJ.)

From this discussion we see that the frequency axis can be partitioned into three regions:

1. Those frequencies for which $|G(j\omega)| \gg 1$. For these frequencies, the closed-loop gain is approximately unity, and the output tracks the input very well.

2. Those frequencies for which $|G(j\omega)| \ll 1$. The gain of the closed-loop system is approximately equal to $|G(j\omega)|$, and the closed-loop system greatly attenuates frequencies in this band. Hence the actual magnitude and phase of $G(j\omega)$ are not important.

3. All the remaining frequencies, that is, those frequencies for which $|G(j\omega)|$ is of the order of unity. If $|1 + G(j\omega)|$ is small (if the Nyquist diagram passes close to the -1

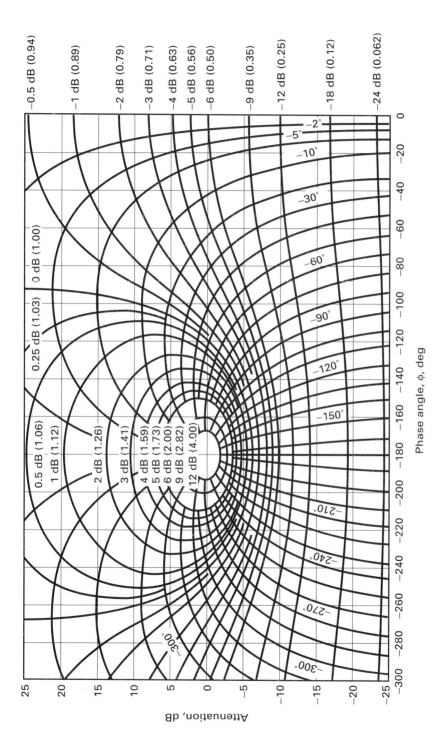

Figure 8.59 Constant *M* circles. (From C. L. Phillips and H. T. Nagle, *Digital Control System Analysis and Design*. © 1995, 3rd ed., p. 270. Adapted by permission of Prentice Hall, Upper Saddle River, NJ.)

point, indicating small stability margins), the closed-loop system gain is very large (see Figures 8.57 and 8.59), indicating a resonance in the closed-loop system. We see then that *the closed-loop system of Figure* 8.56(a) *can have a resonance only if the stability margins are small.* It is assumed that the gain and phase margins are true indicators of the minimum distance between the Nyquist diagram and the −1 point. This point is also evident from the constant *M* circles in Figure 8.57. *M* has large values only in the vicinity of the −1 point.

An example is given to illustrate the preceding.

Example 8.21

We consider the system of Example 8.16, for which the open-loop transfer function is given by, with $K = 1$,

$$G(s) = \frac{K}{s(s + 1)^2}$$

It was seen in Example 8.16 that for the case that $K = 1$, the phase margin is approximately 20°. The frequency response of the open-loop function is given in Table 8.5 and Figure 8.47.

For the frequency $\omega = 0.42$, $G(j0.42) \simeq 2\angle{-136°}$. Thus, if the gain K is given a value of 0.5, the phase margin will be 180° −136°, or 44°. Hence, for $K = 1$, the phase margin is 20°, and for $K = 0.5$, the phase margin is 44°.

From Figure 8.50, for the standard second-order system with a phase margin of 20°, the unit step response has a peak value of approximately 1.6, and the closed-loop frequency response has a peak value of approximately 3.0 or 9.5 dB. The unit-step response for this system for $K = 1$, obtained from a simulation, is given in Figure 8.60, and the peak value is approximately 1.55. The closed-loop frequency response is plotted in Figure 8.61, and the peak value is approximately 9 dB. Note that the resonance is evident in this response.

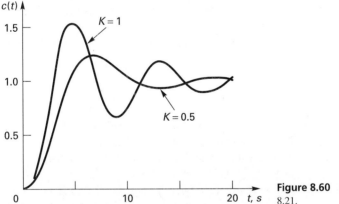

Figure 8.60 Step response for Example 8.21.

For $K = 0.5$, the phase margin is 44°, and for the equivalent second-order system, the peak value in the unit-step response is approximately 1.25. The peak value in the frequency response is approximately 1.35, or 2.6 dB. For this third-order system, the unit-step response is plotted in Figure 8.60 for $K = 0.5$, and the peak value is 1.24. The peak value of the frequency response, shown in Figure 8.61, is seen to be approximately 2.5 dB. The closed-loop frequency responses and step responses are calculated with the MATLAB program

```
G = tf([1], [1 2 1 0]);
T1 = G/(1+G);  T2 = 0.5*G/(1+0.5*G);
    bode(T1),  hold on
```

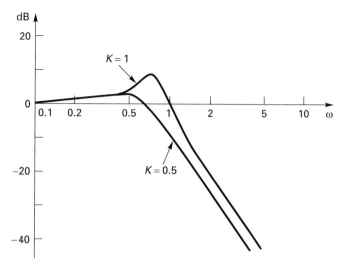

Figure 8.61   Frequency response for Example 8.21.

```
    bode(T2),   hold off, pause
t = 0:0.5:20
    step (T1,t), hold on
    step (T2,t), hold off
```

Two points are illustrated in the last example. First, small stability margins lead to resonance in the closed-loop frequency response and resulting overshoot in the time response. When these stability margins are increased, the resonance decreases, which decreases the overshoot in the time response. The second point is that the second-order characteristics plotted in Figure 8.50 are reasonably accurate indicators of this third-order system's characteristics.

8.8 SUMMARY

Although the frequency response of systems has been considered in the preceding chapters, this chapter presented frequency-response analysis in detail. The two important techniques were developed by Bode and Nyquist. The Bode diagram illustrates the frequency characteristics of a system, and the straight-line approximations present a quick and convenient procedure of presenting the approximate frequency-response characteristics. As is shown in the next chapter, the straight-line approximations are especially useful in giving insight into the design of control systems by frequency-response techniques.

The Nyquist criterion relates the closed-loop stability of a system to the open-loop frequency response. This relationship is important for two reasons. First it allows us to use the open-loop frequency response to design the closed-loop system. Also, it gives us a procedure to design systems based directly on data obtained experimentally, that is, on the frequency response.

The closed-loop frequency response determines the closed-loop system characteristics. Important indicators of the system response are peaking in the closed-loop frequency response and the system bandwidth. The peaking in the frequency response, called resonance, indicates overshoot, or looseness, in the time response. The bandwidth indicates the speed of response of the system; for example, to decrease the rise time in the step response by a factor of 2, the bandwidth must be increased by approximately a factor of 2.

While the system frequency response indicates system time response, the important use of the frequency response is design. Generally, frequency-response techniques are based on the use of the Bode diagram or some variation of it. However, the Nyquist criterion forms the foundations of frequency-response design. The following chapter presents frequency-response design techniques for closed-loop control systems.

REFERENCES

1. A. P. Schust, Jr. "Determination of Aircraft Response Characteristics in Approach/Landing Configuration for Microwave Landing System Program," Report FT-61R-73, Naval Air Test Center, Patuxent River, MD, 1973.

2. R. C. Jaeger. *Microelectronic Circuit Design (Introduction to Electronics).* New York: McGraw-Hill, 1997.

3. C. R. Wylie and L. C. Barrett. *Advanced Engineering Mathematics,* 6th ed. New York: McGraw-Hill, 1995.

4. R. F. Wigginton. "Evaluation of OPS-II Operational Program for the Automatic Carrier Landing System," Naval Electronics Systems Test and Evaluation Facility, Saint Inigoes, MD, 1971.

5. B. Freidland. *Control System Design.* New York: McGraw-Hill, 1986.

6. H. M. James, N. B. Nichols, and R. S. Phillips. *Theory of Servomechanisms.* New York: McGraw-Hill,1947.

PROBLEMS

8.1. Consider the third-order system with the transfer function

$$G(s) = \frac{C(s)}{R(s)} = \frac{50}{(s+1)(s+2)(s+10)}$$

The frequency response $G(j\omega)$ was calculated by computer, with the results given in Table P8.1.

(a) Verify the entries for ω equal to 2 and 10.

(b) We define the system bandwidth as that frequency at which the system gain $|G(j\omega)| = 0.707G(0)$. Find the system bandwidth.

(c) Give the time constants of the system.

(d) The system input is the sinusoid $r(t) = A\cos(\omega t)\,u(t)$. The system is in steady state for $t > T_s$. Find T_s.

(e) Use MATLAB to plot a Bode diagram for $G(s)$. Compare the diagram and Table P8.1 for $\omega = 10$.

8.2. Consider the system of Problem 8.1 and Table P8.1.

(a) The system input is the sinusoid $r(t) = 10\cos \omega t$. Write the steady-state response for each input frequency (no calculations are required)

  (i) $\omega = 0.2$

  (ii) $\omega = 2$

  (iii) $\omega = 20$

Note the differences in the amplitudes of the responses.

TABLE P8.1   FREQUENCY RESPONSE FOR PROBLEM 8.1

ω	Magnitude	dB	Phase
0.1000	2.48436	7.90431	−9.14594
0.2000	2.43880	7.74352	−18.16629
0.3000	2.36701	7.48399	−26.94837
0.4000	2.27430	7.13694	−35.40195
0.5000	2.16660	6.71557	−43.46370
0.6000	2.04964	6.23354	−51.09664
0.7000	1.92838	5.70385	−58.28625
0.8000	1.80677	5.13807	−65.03515
0.9000	1.68774	4.54613	−71.35773
1.0000	1.57329	3.93619	−77.27565
2.0000	0.77522	−2.21153	−119.74490
3.0000	0.42003	−7.53430	−144.57420
4.0000	0.25177	−11.97997	−161.20010
5.0000	0.16287	−15.76342	−173.45370
6.0000	0.11145	−19.05861	−183.06650
7.0000	0.07957	−21.98492	−190.91650
8.0000	0.05873	−24.62326	−197.49860
9.0000	0.04452	−27.02972	−203.11820
10.0000	0.03450	29.24445	−207.97950
20.0000	0.00556	−45.10556	−234.86200
30.0000	0.00175	−55.12954	−245.84180
40.0000	0.00076	−62.42105	−251.66930
50.0000	0.00039	−68.13782	−255.25370
60.0000	0.00023	−72.83470	−257.67370
70.0000	0.00014	−76.81866	−259.41490
80.0000	0.00010	−80.27673	−260.72670
90.0000	0.00007	−83.33112	−261.75020
100.0000	0.00005	−86.06599	−262.57070

(b) The results in (a) apply for $t > T_s$. Find T_s.

(c) The amplitude of the sinusoid in (a) is changed such that $r(t) = 100 \cos \omega t$. Repeat (a).

(d) Verify the results of (a) and (b) for $\omega = 2$ using a SIMULINK simulation.

8.3. Consider the system with the transfer function

$$G(s) = \frac{2(s + 10)}{(s + 1)(s + 2)}$$

(a) Plot the straight-line amplitude and phase response of the Bode diagram.

(b) Find the steady-state response for the input signal $\cos 2t$. Use the results in (a) and do no calculations.

(c) Calculate the result in (b).

(d) Verify $G(j2)$ in (c) with MATLAB.

(e) Use MATLAB to plot the Bode diagram. Transfer the results in (a) to this plot, and indicate the approximate maximum errors.

8.4. (a) Sketch the Bode diagram for the transfer function

$$G(s) = \frac{50}{(s + 1)(s + 2)(s + 10)}$$

(b) Computer-calculated values for $G(j\omega)$ are given in Table P8.1. Plot these values on the Bode diagram of (a), and estimate the maximum errors in decibels and angle for the straight-line diagrams. Also, give the frequencies at which the maximum errors occur.

8.5. Sketch the Bode diagrams for the given transfer functions.

(a) $G(s) = \dfrac{20}{s\,(s+1)^2}$

(b) $G(s) = \dfrac{8s}{(s+1)^2}$

(c) $G(s) = \dfrac{s+2}{s^2}$

(d) $G(s) = \dfrac{2}{s^2(s+2)}$

(e) Verify all Bode diagrams using MATLAB.

8.6. Sketch the Bode diagrams for the following transfer functions.

(a) $G(s) = \dfrac{1-s}{s(s+1)}$

(b) $G(s) = \dfrac{s-1}{s(s+1)}$

(c) $G(s) = \dfrac{-s}{(s+1)(s-1)}$

(d) $G(s) = \dfrac{s}{(s+1)\,(s-1)}$

(e) What is the difference in the Bode diagrams for $G(s)$ and for $-G(s)$?

(f) Verify all Bode diagrams using MATLAB

8.7. Consider the magnitude plot of the straight–line Bode diagram for a minimum–phase transfer function $G(s)$ in Figure P8.7.

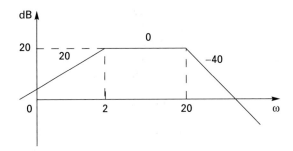

Figure P8.7

(a) Find $G(s)$.

(b) Suppose that G(s) is not specified to be minimum phase. Give three additional transfer functions that have the magnitude characteristic of Figure P8.7.

(c) Suppose that the section of the diagram of Figure P8.7 for $\omega < 2$ is extended for all ω. Find the transfer function that has this modified magnitude characteristic.

(d) Suppose that the section of the diagram of Figure P8.7 for $2 < \omega < 20$ is extended for all ω. Find the transfer function that has this modified magnitude characteristic.

(e) Suppose that the section of the diagram of Figure P8.7 for $\omega > 20$ is extended for all ω. Find the transfer function that has this modified magnitude characteristic.

(f) Use MATLAB to verify the results in (a).

8.8. For the system of Figure P8.8, let

$$G(s) = \frac{150}{s(s+4)^2}$$

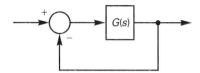

Figure P8.8

(a) Sketch the Nyquist diagram for this system.
(b) Using the Nyquist criterion only, find the number of poles of the closed-loop transfer function in the right half-plane. Note that the frequency at which arg $G(j\omega) = -180°$ can be easily determined.
(c) Check the results of (b) via the Routh–Hurwitz criterion.
(d) Suppose that the gain K is cascaded with $G(s)$ such that the open-loop function is now $KG(s)$. Find the value of K at which the system is marginally stable and the frequency at which the marginally stable system will oscillate.
(e) Verify the results of (a), (b), and (d) with MATLAB. Note the difficulty in plotting and interpreting the Nyquist diagram when drawn to scale.

8.9. The system of Problem 8.8(d) has the characteristic equation

$$1 + \frac{128}{s^3 + 8s^2 + 16s} = 0$$

and oscillates at the frequency $\omega = 4$ rad/s.
(a) Find the period T of this sinusoidal term in the system response.
(b) Find the time T_s required for this system to reach steady-state response. Do not use a computer or calculator for this part.
(c) How long should a simulation of this system run such that at least one cycle of the sinusoid in steady state is clearly visible? The system is to be excited by initial conditions.
(d) Verify the results of (a) and (c) with SIMULINK.

8.10. For this problem, let each $G(s)$ in Problem 8.5 be the open-loop function in the system of Figure 8.8.
(a) Sketch each Nyquist diagram. If available, the Bode plots from Problem 8.5 are useful.
(b) Determine the stability of each system, using only the Nyquist criterion.
(c) Verify each Nyquist diagram in (a) with MATLAB. Show clearly the Nyquist diagram in the vicinity of the -1 point.
(d) Verify the results in (b) using the *pole* statement of MATLAB.

8.11. For this problem, let each $G(s)$ in Problem 8.6 be the open-loop function in the system of Figure 8.8.
(a) Sketch each Nyquist diagram. If available, the Bode plots from Problem 8.6 are very useful.
(b) Determine the stability of each system, using only the Nyquist criterion.
(c) Verify each Nyquist diagram in (a) with MATLAB. Show clearly the Nyquist diagram in the vicinity of the -1 point.
(d) Verify the results in (b) using the *pole* statement of MATLAB.

8.12. For this problem, let each $G(s)$ in Problem 8.5 be the open-loop function in the system of Figure 8.8.
(a) Sketch each Nichols chart. If available, the Bode plots from Problem 8.5 are useful.
(b) Label the gain and phase margins for the stable systems.
(c) Verify each Nichols chart in (a) with MATLAB.
(d) Use MATLAB to find the gain and phase margins for the stable systems.

8.13. Consider the system of Figure P8.8 with

$$KG(s) = \frac{50K}{(s+1)(s+2)(s+10)}$$

The open-loop frequency response G(jw) is given in Table P8.1.

(a) Let $K = 1$. Find the approximate system gain margin and phase margin from Table P8.1.

(b) Repeat (a) for $K = 2.5$.

(c) The closed-loop system input is $r(t) = 5\cos(t + 30°)$, with $K = 1$. Find the steady–state output, using Table P8.1.

(d) Use MATLAB to find the gain and phase margins for (a) and (b).

(e) Verify the result in (c) with SIMULINK.

8.14. For the system of Figure P8.14, let

$$G(s) = \sqrt{2}/(s+1)$$

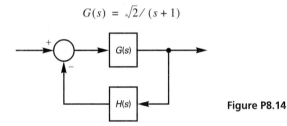

Figure P8.14

(a) With $H(s) = 1$, sketch the Bode diagram, and find the values of the gain margin and the phase margin.

(b) What are the effects on the diagram of (a) if $H(s) = e^{-0.3s}$ (an ideal time delay of 0.3 s)?

(c) Sketch the Bode diagram for (b) on the graph of (a).

8.15. This problem extends Problem 8.14. For the system of Figure P8.14, let

$$G(s) = \frac{\sqrt{2}}{1+s}$$

(a) With $H(s) = 1$, sketch the Nyquist diagram, and find the values of the gain margin and the phase margin.

(b) What are the effects on the diagram of (a) if $H(s) = e^{-0.3s}$ (an ideal time delay of 0.3 s)?

(c) Sketch the Nyquist diagram for (b) on the graph of (a).

(d) With $H(s) = e^{-t_0 s}$ (an ideal time delay of t_0 seconds), find the range of $t_0 >$ zero for which the system is stable.

(e) Modify the MATLAB program in Example 8.20 to plot the Nyquist diagram with t_0 equal to the value found in (d).

8.16. Consider the rigid satellite control system of Figure P8.16. This system is discussed in Problem 4.5. This control system utilized rate feedback to stabilize the system.

(a) The open-loop function $G_{0L}(s)$ is defined by the system characteristic equation

$$1 + G_{0L}(s) = 0$$

Find the open-loop function.

(b) Sketch the Bode diagram for the open-loop function, with $K = 1$ and $K_v = 0$ (the rate feedback is removed).

(c) Use the results of (b) to sketch the Nyquist diagram.

(d) Find the system gain and phase margins, from (b) or (c).

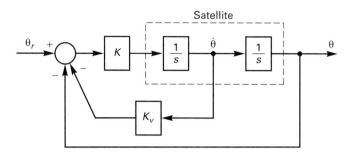

Figure P8.16

8.17. This problem extends Problem 8.16. Consider the rigid satellite control system of Figure P8.16.

(a) The open-loop function $G_{OL}(s)$ defined by the system characteristic equation

$$1 + G_{OL}(s) = 0$$

Find the open-loop function.

(b) Use any procedure to find the range of K_v for which the control system is stable.

(c) Sketch the Bode diagram for the open-loop function, with $K = 1$ and $K_v = 0.5$.

(d) Use the results of (c) to sketch the Nyquist diagram.

(e) Verify the Bode diagram of (c) using MATLAB.

(f) From the results in (e), estimate the gain and phase margins.

(g) Use MATLAB to calculate the gain and phase margins.

8.18.   Shown in Figure P8.18 is the block diagram of the servo control system for one of the axes of a digital plotter. The input θ_r is the output of a digital computer, and the output θ_p is the position of the servomotor shaft. This system is described in Problem 6.13.

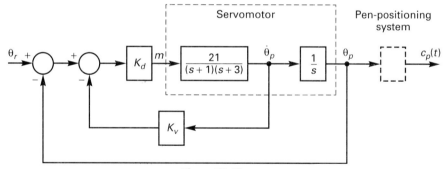

Figure P8.18

(a) With $K_d = 1$ and $K_v = 0$ (no rate feedback), sketch the Nyquist diagram.

(b) Determine the stability of the system of (a).

(c) Repeat (a) and (b) for the case that $K_d = 0.1414$ and $K_v = 1.414$.

(d) Use MATLAB to verify the Nyquist diagrams of (a) and (c).

(e) Use MATLAB to find the gain and phase margins in (c).

8.19. (a) For the digital plotter servo of Figure P8.18, the transfer function of the motor from voltage input to rotational velocity output is

$$\frac{\Omega_p(s)}{M(s)} = \frac{21}{(s+1)\ (s+3)}$$

The dc gain of this transfer function is 7. Given the physical system with the velocity sensor, how would you experimentally verify this dc gain?

(b) In (a) the measurement of the dc gain of a physical system was considered. The model given in this problem, evaluated at $s = j2$, is

$$\frac{\Omega_p(j2)}{M(j2)} = \frac{21}{(1 + j2)(3 + j2)} = 2.60\underline{/-97.12^o}$$

Given the physical system with the velocity sensor, how would you experimentally verify this complex number?

(c) In both (a) and (b), the motor speed had to reach steady state. How many seconds are required for the motor to reach steady state.

(d) Use SIMULINK to demonstrate the results expected in (a) and (b).

8.20. Consider the satellite attitude-control system shown is Figure P8.16.

(a) Plot the Nyquist diagram of this system, with $K_v = 0$ (no rate feedback).

(b) Determine system stability as a function of K, from the Nyquist diagram of (a).

(c) Verify the results of (b) using the Routh-Hurwitz criterion.

(d) Verify the results of (b) using the root locus.

(e) Use the Nyquist diagram to show that the system is stable for all $K > 0$, $K_v > 0$.

(f) Use MATLAB to verify the Nyquist diagram in (a).

8.21. The Nyquist diagram of a system of the configuration of Figure P8.8 is shown in Figure P8.21. Assume that $G(s)$ has no poles in the right half-plane. For this system, the stability may be changed by either increasing or decreasing gain, with the gain positive. Systems of this type are called *conditionally stable*.

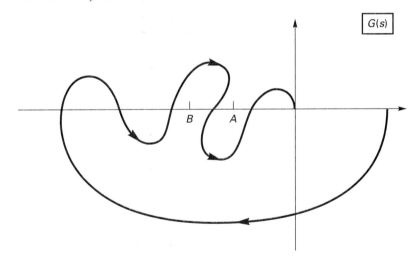

Figure P8.21

(a) Find the number of zeros of $1 + G(s)$ in the right half-plane if the point A is the -1 point.

(b) Repeat (a) if point B is the -1 point.

8.22. The Nyquist diagram of a system of the configuration of Figure P8.8 is shown in Figure P8.22. Assume that $G(s)$ has no poles in the right half-plane. A gain K is cascaded with $G(s)$. Find the ranges of positive K for which the system is stable.

8.23. Derive the equation for the constant N circles of the closed-loop response, Equation (8-59).

8.24. Shown in Figure P8.24 is the attitude control system of a large space booster. The attitude angle θ is controlled by controlling the engine angle δ, which is the angle of the applied thrust. The

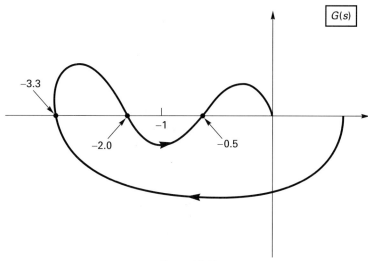

Figure P8.22

vehicle velocity is denoted by the v. These control systems are sometimes open-loop unstable, which occurs if the booster center of aerodynamic pressure is forward of its center of gravity. We will assign a transfer function to the booster:

$$G(s) = \frac{1}{s^2 - 0.04}$$

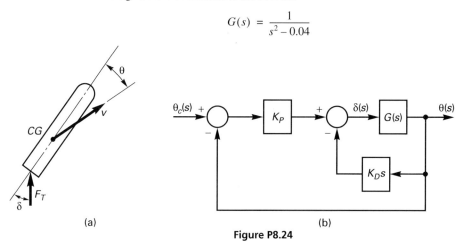

(a)   (b)

Figure P8.24

This transfer function is approximately that of the *Saturn V* booster stage, with bending, fuel slosh, and the dynamics of the mechanism that controlled the angle of the thrust ignored. If these effects are not ignored, the transfer function is 27th order. In this problem, let $K_P = 1$ and $K_D = 0$ (the rate feedback is ignored).

(a) Plot the Bode diagram of the open-loop function $K_P G(s)$.
(b) From the Bode diagram, sketch the Nyquist diagram.
(c) Determine stability from the Nyquist diagram.
(d) Verify the results in (c) by sketching the root locus.
(e) Verify the results in (c) by finding the closed-loop poles.

9

Frequency-Response Design

In Chapter 7, root-locus analysis and design procedures were developed. The root-locus design was presented as a method for influencing the transient response of a control system in a limited manner; a control system was designed to have a pair of complex poles in a desired location. However, since the system was generally of higher order than two, the other poles of the transfer function were not placed. If these poles were not in acceptable locations, it would be necessary to change the design procedure in some manner in order to shift these locations.

In this chapter we develop a different design procedure. This procedure is a frequency-response method and is based on the Nyquist criterion. As shown in the last chapter, we can determine the stability of a closed-loop system from a mapping of the open-loop function. If the open-loop function has no poles on the $j\omega$-axis, this mapping is the frequency response of the open-loop function. If the open-loop function has poles on the $j\omega$-axis, the mapping includes curves in addition to the frequency response.

While the root-locus design procedure gives us direct information on the closed-loop system's transient response, frequency-response design gives us this information indirectly. The closed-loop transient-response information is contained in the open-loop frequency response, but in not so visible a form as in the root locus. However, the frequency response does give us information on the steady-state response (low-frequency response), on stability margins, and on the system bandwidth. *We can see then the usefulness of using both the root-locus and the frequency response in any practical design project.*

9.1 CONTROL SYSTEM SPECIFICATIONS

In the previous chapters we have discussed some of the general characteristics that a well-designed control system should have. All the characteristics discussed may not be required for a particular control system. For example, the servo systems of a digital plotter should be as fast as possible, since even then the plotter cannot stay current with the computer. However, the attitude-control systems of space boosters are purposely designed to respond slowly, since the large torques required for the booster to respond quickly would break the booster apart. Remember, then, when we list desired characteristics that a control system may have, these characteristics do not apply to all control systems; in fact, in some cases the required characteristics may be quite different.

These required characteristics are called the system specifications. The specifications of a control system must be developed before the design begins. We can state general specifications, such as the system must be stable (of course), but we also need particular specifications, such as steady-state errors allowable, disturbance rejection required, overshoot allowed to step changes in the input, the types of inputs that the system must track accurately, and so forth. Once the specifications are developed, we are in a position to begin the design.

We begin this chapter by reviewing some common control-system specifications and relate these to the system open-loop and closed-loop frequency responses. We consider a control system of the form shown in Figure 9.1. We have assumed that the system has been converted to the equivalent unity feedback model (see Section 5.1), with the system characteristic equation

$$1 + G_c(s)G_p(s) = 0 \tag{9-1}$$

Given the plant transfer function $G_p(s)$, we wish to determine the compensator transfer function $G_c(s)$ such that the closed-loop transfer function has at least some of the characteristics reviewed below.

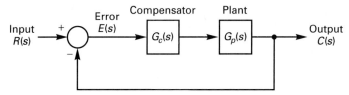

Figure 9.1   Closed-loop control system.

In working with control systems, we must be very careful not to confuse the *open-loop function* [$G_c(s)G_p(s)$ in Figure 9.1] with the *closed-loop transfer function*

$$T(s) = \frac{G_c(s)G_p(s)}{1 + G_c(s)G_p(s)}$$

In analysis and design, we are always ultimately concerned with the closed-loop system. However, some of our analysis and design procedures require that we work with the open-loop function. We must always be careful to state which function (open-loop or closed-loop) is being considered in analysis and design. We now review some important design specifications.

9.1.1 Steady-State Accuracy

We define the system error to be the error signal $E(s)$ shown in Figure 9.1. Steady-state accuracy is discussed in Section 5.5. In that section it is shown that, provided that the system is stable, a type 1 system has no steady-state error for a constant input, a type 2 system has no steady-state error for a constant or ramp input, a type 3 system has no steady-state error for a constant, a ramp, or a parabolic input, and so forth. The system type is defined as the number of poles of the open-loop function at $s = 0$. It appears that in order to improve the steady-state performance, the compensator should add poles at $s = 0$. However, it was shown in Section 8.5 that a system with poles at $s = 0$ is more difficult to stabilize. In addition, since the addition of poles at $s = 0$ is a type of phase-lag compensation, the transient response generally contains some terms with long time constants.

The frequency response of the closed-loop system of Figure 9.1 is given by

$$T(j\omega) = \frac{G_c(j\omega)G_p(j\omega)}{1 + G_c(j\omega)G_p(j\omega)} \tag{9-2}$$

For a given frequency ω_1, if the open loop gain $G_c(j\omega_1)G_p(j\omega_1)$ is large, the closed-loop gain is approximately unity. Thus for small errors at any frequency, the loop gain must be large. If we can then design the system to have a high open-loop gain over a wide band of frequencies, the system will track input signals composed of these frequencies very well.

9.1.2 Transient Response

The step response of a system with dominant second-order poles is given in Figure 9.2. Characteristics of this response that might be important for a particular control system are the rise time T_r, the peak value of the signal M_{pt} (or the percent overshoot), and the settling time T_s. The rise time is closely related to the closed-loop system bandwidth, as given in (4-52). In general, if we desire to reduce the rise time by a factor of 2, it is necessary to increase the closed-loop system bandwidth by a factor of approximately two. If we want to decrease the settling time, the closed-loop transfer function poles must be shifted to the left in the s-plane, thereby reducing the time constants. Shifting the closed-loop poles to the left also generally increases the system bandwidth. Hence to decrease the system's response time, the closed-loop system's bandwidth must be increased (see Section 4.4).

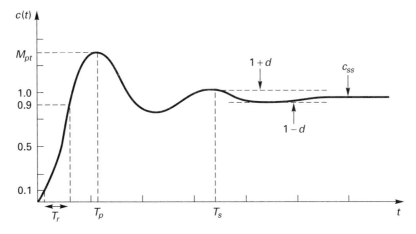

Figure 9.2   Typical step response.

As shown in Section 8.7, overshoot in the transient response is related to resonance in the closed-loop system. Resonance appears as a peak in the closed-loop frequency response, as shown in Figure 9.3. Resonances in the closed-loop system can occur only if the relative stability margins of the system are small (see Section 8.7). Hence to reduce overshoot in the system's transient response, the relative stability margins must be increased. The overshoot is generally more sensitive to the phase margin than to the gain margin, since the open-loop gain is higher at the phase-margin frequency than at the gain-margin frequency. The specification to limit resonance effects may be in terms of the percent overshoot to a step input, in terms of the maximum allowable peak in the closed-loop frequency response, shown as $M_{p\omega}$ in Figure 9.3, and in terms of the phase margin.

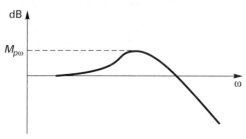

Figure 9.3   Closed-loop frequency response.

9.1.3 Relative Stability

Some of the effects of small stability margins are given in the preceding paragraph. For a reasonable transient response we require an adequate phase margin and an adequate gain margin. A second reason for large stability margins originates in the inaccuracies of modeling. Since the models upon which our designs are based are inaccurate to some degree, stability margins may also be viewed as safety margins. It is difficult to give numbers to the margins for the general case, but usually control systems with adequate stability margins have phase margins of 45° or higher and gain margins of 8 dB or higher. These values are not absolute but are approximate. However, note from (9-2), that if the stability margins are increased by reducing the gain of $G_c(j\omega)G(j\omega)$ in a frequency band, the system bandwidth is reduced.

9.1.4 Sensitivity

Generally the plant has parameters that vary with temperature, humidity, age, and so on. Sensitivity as defined in Section 5.3 is a measure of the change in system characteristics to small (incremental) changes in parameters. For example, in the system of Figure 9.1, the sensitivity of the closed-loop transfer function to changes in the plant transfer function, as a function of frequency, is given by

$$S_{G_p}^T = \left.\frac{\partial T}{\partial G_p}\frac{G_p}{T}\right|_{s=j\omega} = \frac{1}{1 + G_c(j\omega)G_p(j\omega)} \tag{9-3}$$

as derived in (5-19). We see, then, for the sensitivity to be small over a given frequency band, the gain of the open-loop function over that band must be large. Generally, for good sensitivity in the forward path of the control system, the loop gain is made large over as wide a band of frequencies as possible. But, in general, increasing the loop gain degrades the stability margins. Hence we usually have a trade-off between low sensitivity and adequate stability margins.

Flexible fabrication system, comprised of nine computer-controlled machining centers, with automatic movement of parts between machines and automatic inspection. (Courtesy of Hughes Aircraft Company.)

9.1.5 Disturbance Rejection

A closed-loop system with a disturbance input to the plant may be modeled as shown in Figure 9.4 (see Section 5.4). In this figure, $D(s)$ is a disturbance (unwanted input), such as wind turbulence on a commercial airliner. The system of Figure 9.4 could be the model of the altitude control system of such an aircraft, and $G_c(s)$ would then be the vertical autopilot. All who have flown in these aircraft are well aware of the disturbance effects on the aircraft and the problems with designing the control system to reject the wind-turbulence disturbance. If this were not such a difficult problem, it would have been solved long ago.

By superposition, the output expression for the system of Figure 9.4 is given by

$$C(s) = \frac{G_c(s)G_p(s)}{1 + G_c(s)G_p(s)}R(s) + \frac{G_d(s)}{1 + G_c(s)G_p(s)}D(s) \qquad (9\text{-}4)$$

In order to reduce the effects of the disturbance of the output, we must make the gain of the open-loop function, $G_c(j\omega)G_p(j\omega)$, large over the frequency band of the disturbance $d(t)$.

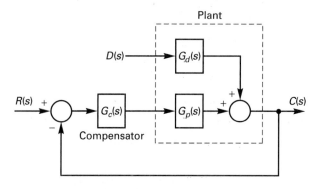

Figure 9.4   System with disturbance.

If this cannot be done, the open-loop gain is made large over as wide a frequency band as possible. The rule for disturbance rejection by the method presented here is to make the open-loop gain large over as wide a frequency band as possible without increasing the gain between the disturbance input and the system output. However, recall that increasing the loop-gain can degrade the stability margins, and again we have trade-offs.

9.1.6 Summary

From the preceding discussion we see that a fast transient response, good tracking accuracy, good sensitivity, and good disturbance rejection require a high loop gain over a wide band of frequencies. However, efforts to increase the loop gain can degrade the system stability margins and the transient response. These reduced stability margins can lead to resonance in the system and to the possibility of instability in the physical system due both to modeling inaccuracies and to nonlinearities in the physical system that are ignored in the linear model. Hence a trade-off is required between these conflicting requirements, as is generally required in all design. Control system design is an iterative process in which, after the design of each controller, the controller is tested in the physical system. The inadequacies are noted, and the controller is redesigned in an effort to overcome these problems. With each iteration, the designer gains more insight into the system; the value of this insight cannot be overrated.

9.2 COMPENSATION

In this chapter we discuss, for the most part, the compensation of single-input, single-output systems of the form shown in Figure 9.5(a). This system has the characteristic equation

$$1 + G_c(s)G_p(s)H(s) = 0 \qquad (9\text{-}5)$$

We wish to design $G_c(s)$ such that the system has certain specified characteristics. The compensation of the type shown in Figure 9.5(a) is called cascade, or series, compensation. Many of the effects of the compensation on system characteristics are indicated by the locations of the roots of the characteristic equation, (9-5).

It is sometimes more feasible to place the controller within a loop internal to the system; an example is shown in Figure 9.5(b). The characteristic equation for this system, from Mason's gain formula, is

$$\Delta(s) = 1 + G_1(s)G_2(s)H_1(s) + G_c(s)G_2(s)H_2(s) = 0 \qquad (9\text{-}6)$$

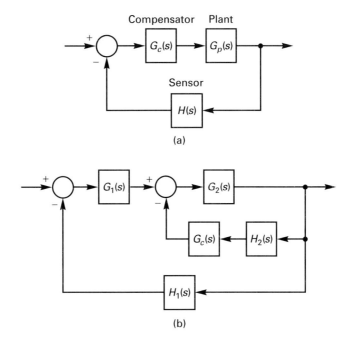

Figure 9.5　Closed-loop control system.

The effects of the controller $G_c(s)$ on the system characteristics are difficult to see from this equation. None of the conclusions expressed above apply directly to (9-6). However, if (9-6) is divided by the first two terms of the equation, the characteristic equation can be expressed as

$$1 + \frac{G_c(s)G_2(s)H_2(s)}{1 + G_1(s)G_2(s)H_1(s)} = 0 \tag{9-7}$$

This equation can then be expressed as

$$1 + G_c(s)G_e(s) = 0 \tag{9-8}$$

where

$$G_e(s) = \frac{G_2(s)H_2(s)}{1 + G_1(s)G_2(s)H_1(s)} \tag{9-9}$$

The characteristic equation of (9-8) is of the same form as (9-5), and thus many of the results given earlier can be applied to the design of the system of Figure 9.5(b). However, others, such as those for disturbance rejection, steady-state accuracy, and so forth, must be derived again for this system.

　　The characteristic equation of a multiloop system in the form of (9-8) can be derived directly by the following procedure. First the system input is ignored, and the system is opened at the input of the controller, as shown in Figure 9.6. The signal from the open into the controller is labeled as $E_i(s)$, and that coming into the open from the system is labeled as $E_o(s)$. Next $E_o(s)$ is expressed as a function of the input $E_i(s)$. In Figure 9.6,

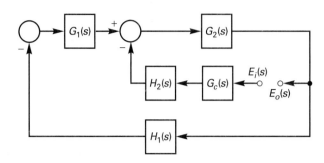

Figure 9.6   System with open loop.

$$E_o(s) = \frac{-G_c(s)G_2(s)H_2(s)}{1 + G_1(s)G_2(s)H_1(s)}E_i(s) = G_{ol}(s)E_i(s) \tag{9-10}$$

where $G_{ol}(s)$ is the transfer function from the open back to the open. We call this transfer function the *open-loop transfer function*. For the closed-loop system, $E_i(s) = E_o(s)$, and (9-10) can then be expressed as

$$[1 - G_{ol}(s)]\, E_o(s) = 0$$

Even though the closed-loop system input has been specified to be zero, $E_o(s)$ is not necessarily zero, since we can place initial conditions on the system. Hence the first factor must be zero for the initial condition response, or

$$1 - G_{ol}(s) = 0 \tag{9-11}$$

This equation must then be the system characteristic equation.

The preceding procedure is a general method for calculating the system characteristic equation. If the open is placed at the compensator, as in Figure 9.6, the characteristic equation will be of the form of (9-8), which is the form that is suitable for design. The equivalent transfer function, $G_e(s)$, is then seen to be the negative of the transfer function from the controller output to the controller input. As one final point, note that the *open-loop transfer function*, defined earlier as the transfer function from an open back to that open, is not the *open-loop function* defined with respect to the root locus, the frequency response, the Nyquist criterion, and so on. In the characteristic equation

$$1 + G(s)H(s) = 0$$

$G(s)H(s)$ is defined as the open-loop function, and this function is useful in many methods of analysis. However, if we open the loop on a system that has this characteristic equation, the open-loop transfer function is $-G(s)H(s)$. We see then that the open-loop function is the negative of the open-loop transfer function. This result is general.

9.3 GAIN COMPENSATION

First we consider compensation by varying a gain K. We consider the closed-loop system of Figure 9.7(a), in which the compensator $G_c(s)$ is simply a gain K. This, of course, is the proportional or P controller introduced in Section 7.10, and we are very limited in the specifications that we can satisfy with this type of compensation.

Suppose, for example, that the uncompensated system has the Nyquist diagram shown in Figure 9.7(b). The only effect that the gain K can have on this Nyquist diagram is

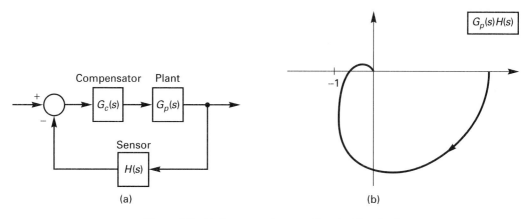

Figure 9.7   (a) Compensated system; (b) typical Nyquist diagram.

to enlarge or reduce it; the shape of the Nyquist diagram cannot be changed. Hence, for example, we can specify the phase margin but must accept both the frequency at which the phase margin occurs and the resulting gain margin.

The same effect can be seen from the system Bode diagram, which is sketched in Figure 9.8 for the Nyquist diagram of Figure 9.7(b). The phase margin of the uncompensated system is shown on the figure. The effect of adding a gain compensator having a gain value less than unity is shown. The magnitude characteristic is shifted down with no change in shape; the phase characteristic is unchanged. Hence the phase margin is increased. Since the frequency at which the phase margin occurs is reduced, the system bandwidth is generally reduced. Hence the system rise time increases; but since the phase margin is increased, the overshoot in the transient response should be less. An example is given next.

Example 9.1

We will design a P compensator for a radar tracking system (see Section 2.7), as depicted in Figure 9.9. The inductance of the servomotor cannot be ignored [see (2-51)], and the open-loop function of the uncompensated system is given by

$$G_p(s)H(s) \; = \; \frac{4}{s(s+1)(s+2)}$$

with $H(s) = 1$. The open-loop frequency response, $G_p(j\omega)H(j\omega)$, is given in Table 9.1. Note that the uncompensated system is stable, but the phase margin is approximately $12°$ and the gain margin is approximately 3.5 dB; these values are much too small for good transient response.

Suppose that the phase margin is specified to be $50°$. Hence we require that $|G(j\omega)H(j\omega)| = 1$ $\angle{-130°}$ for some value of ω. Since we cannot change the angle of the open-loop frequency response with gain compensation, the frequency at which the phase margin occurs must be $\omega \approx 0.5$ rad/s, from Table 9.1. Since the magnitude of the open-loop function at this frequency is approximately 3.5, we choose K to be equal to 1/3.5, or 0.286. The design of the control system is now completed.

In order to determine some of the characteristics of the compensated system, the unit step response was obtained by simulation and is plotted in Figure 9.10. The overshoot is approximately 19 percent.

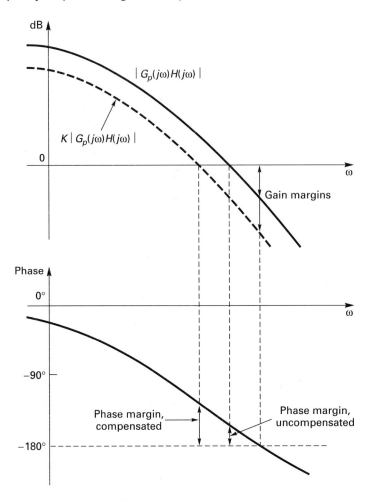

Figure 9.8   Gain compensation.

To investigate further the effects of gain compensation in the example, compensations were also calculated to produce phase margins of 35° and of 65°. From Table 9.1, the gains of $1/2.2 = 0.455$ and $1/6.31 = 0.158$, respectively will result approximately in these margins. The unit step responses for these systems are also given in Figure 9.10. Given in Figure 9.11 are the closed-loop frequency responses for these three compensated systems. The step responses are calculated by the MATLAB simulation

```
K = [0.455 0.286 0.158]; t = 0:0.2:16;
for k = 1:3
  G = tf([4], [1 3 2 0]);
  T = K(k)*G/(1+K(k)*G);
  step (T,t);
  hold on
end, hold off, pause, margin(K(1)*G)
```

Also, the gain and phase margins are calculated for $K = 0.455$. From these two figures we see that the effect of increasing the phase margin by reducing the open-loop gain is to (1)

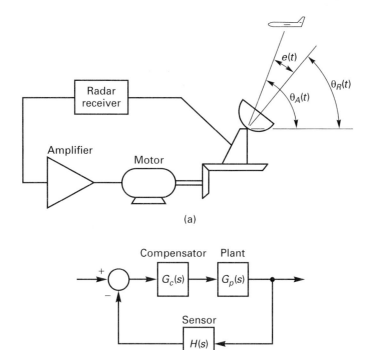

(a)

(b)

Figure 9.9   Radar tracking system.

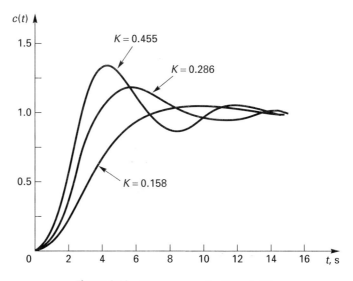

Figure 9.10   Time responses for Example 9.1.

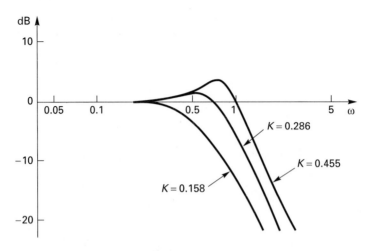

Figure 9.11   Closed-loop frequency responses for Example 9.1.

TABLE 9.1   FREQUENCY RESPONSE FOR EXAMPLE 9.1

ω	Magnitude	dB	Phase
0.1000	19.87591	25.96654	−98.57300
0.2000	9.75714	19.78645	−107.02050
0.3000	6.31486	16.00728	−115.23000
0.4000	4.55223	13.16449	−123.11130
0.5000	3.47089	10.80881	−130.60130
0.6000	2.73776	8.74792	−137.66300
0.7000	2.20925	6.88491	−144.28210
0.8000	1.81255	5.16578	−150.46120
0.9000	1.50628	3.55811	−156.21500
1.0000	1.26491	2.04120	−161.56500
1.1432	1.00001	.00005	−168.57490
1.4142	0.66667	−3.52182	−180.00000
2.0000	0.31623	−10.00000	−198.43500
3.0000	0.11694	−18.64066	−217.87500
4.0000	0.05423	−25.31479	−229.39870
5.0000	0.02913	−30.71192	−236.88870
6.0000	0.01733	−35.22445	−242.10270
7.0000	0.01110	−39.09322	−245.92450
8.0000	0.00752	−42.47483	−248.83870
9.0000	0.00532	−45.47598	−251.13100
10.0000	0.00390	−48.17235	−252.97950

reduce the overshoot in the transient response, (2) reduce the bandwidth, and (3) thus increase the rise time. The comparisons of the three systems are given in Table 9.2. Note that the product of rise time and bandwidth is also included in this table, further illustrating the statement of (4-52) that for a given system, this product is approximately constant.

TABLE 9.2   RESULTS FOR EXAMPLE 9.1

K	Phase margin (deg)	Overshoot (%)	Rise time (s)	Bandwidth (rad/s)	$T_r \times BW$
0.158	65	3.7	4.5	0.5	2.25
0.286	50	18	2.4	0.9	2.16
0.455	35	35	1.7	1.2	2.04

9.4 PHASE-LAG COMPENSATION

In this section we assume that the compensator is first order and has the transfer function

$$G_c(s) = \frac{1 + s/\omega_0}{1 + s/\omega_p} \qquad (9\text{-}12)$$

Thus the compensator has unity dc gain [$G_c(0)$]; we consider the case of nonunity dc gain later. The design problem is then to determine the compensator zero, $-\omega_0$, and the compensator pole, $-\omega_p$, such that the closed-loop control system will have certain specified characteristics.

As in root-locus design, we consider two cases. First, if $\omega_0 > \omega_p$, the compensator is called a phase-lag compensator; if $\omega_0 < \omega_p$, it is a phase-lead compensator (we are assuming that the compensator is minimum phase). Generally, if we design a phase-lag compensator, we are attempting to satisfy different specifications than if we design a phase-lead compensator. In this section we consider the design of phase-lag compensators by frequency-response, or Bode, techniques.

Since the compensator of (9-12) is of unity dc gain and we have stated that $\omega_0 > \omega_p$, the Bode diagram of the phase-lag compensator is as shown in Figure 9.12. The straight-line approximation is given for the magnitude characteristic. On the phase characteristic, the angle must be negative at every frequency, and the maximum phase lag, θ_m, is less than 90° in magnitude. The maximum phase lag occurs at the frequency ω_m, which can be shown to be the geometric mean of ω_0 and ω_p, that is, $\omega_m = \sqrt{\omega_p \omega_0}$.

Note that the effect of the phase-lag compensator of Figure 9.12 is to reduce the gain at higher frequencies and to introduce phase lag. The high-frequency gain is reduced by the factor

$$\lim_{\omega \to \infty} G_c(j\omega) = \lim_{\omega \to \infty} \frac{1 + j\omega/\omega_0}{1 + j\omega/\omega_p} = \frac{\omega_p}{\omega_0} \qquad (9\text{-}13)$$

Note that the phase-lag compensator is, in fact, a form of a low-pass filter; that is, the high frequencies are attenuated relative to the low frequencies. We know from the developments in the last chapter that generally the reduced gain will tend to stabilize a system and the phase lag will tend to destabilize a system. Hence we must be careful to place the phase lag in a frequency range such that stability is not unduly affected.

We again consider the design of the closed-loop system of Figure 9.7, which has characteristic equation

$$1 + G_c(s)G_p(s)H(s) = 0 \qquad (9\text{-}14)$$

The Nyquist diagram for this system, for example, might appear as shown in Figure 9.13. In this figure the Nyquist diagram is divided into three frequency regions, denoted by A, B, and

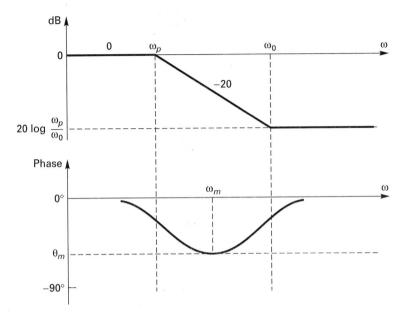

Figure 9.12   Bode diagram phase-lag compensation.

C. We now discuss the placement of the pole and zero of the phase-lag compensator in each of the three regions.

The gain of the system in region *C* is quite small; placing the compensator pole and zero in this region further reduces the gain and thus has very little effect on the closed-loop system. Thus there is no obvious reason to place the compensator pole and zero in *C*, the system's high-frequency region.

In the midfrequency region denoted by *B* in Figure 9.13, the Nyquist diagram is in the vicinity of the −1 point. Hence placing the compensator's pole and zero in this region tends to destabilize the system because of the added phase lag. We do not place the pole and zero in this region.

Consider that the compensator pole and zero are placed in the low-frequency region *A*. The added phase lag has little effect on the system stability, since the Nyquist diagram is not in the vicinity of the −1 point. As is seen from the compensator Bode diagram of Figure 9.12, the phase decreases toward 0° as frequency increases well above the value of the compensator zero. However, the reduced gain of the compensator in the higher-frequency range increases the system's stability margins. Hence the compensator pole and zero are placed in the frequency range of region *A*, the low-frequency region.

From the preceding discussion we see that the high-frequency effect of the phase-lag compensator is the same as the gain-reduction compensation described in the last section. However, there is little effect from the compensator in the low-frequency range. Hence the low-frequency characteristics of the closed-loop system are not degraded by the phase-lag compensator, whereas the stability margins are improved.

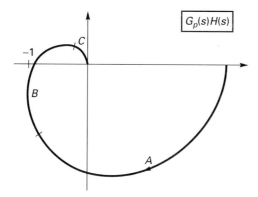

Figure 9.13   Typical Nyquist diagram.

The effect of phase-lag compensation is seen from the Bode diagram of Figure 9.14. The magnitudes of the compensator zero and pole, ω_0 and ω_p, are chosen to be small compared to the frequency at which the magnitude of $G_p(j\omega)H(j\omega)$ is unity. Then the phase lag introduced by the compensator has little effect on the Nyquist diagram in the vicinity of the -1 point. The high-frequency gain is reduced by the factor ω_p/ω_0, thereby increasing the phase margin.

We assumed that the compensator dc gain is unity. However, the dc gain of the open-loop function, $G_p(0)H(0)$, is normally not of the value required to meet system specifications; hence we must be able to design a nonunity dc gain compensator. We solve this problem by the following steps:

1. Find the value of open-loop dc gain required to satisfy system requirements, such as for steady-state errors, low-frequency disturbance rejection, and so on.
2. Adjust the dc gain of $G_p(s)H(s)$ to be equal to this required value.
3. Design a unity dc gain compensator using this adjusted open-loop function.
4. Adjust the dc gain of the compensator such that the dc gain of $G_c(s)G_p(s)H(s)$ meets system specifications.

A technique for determining ω_p and ω_0 to yield the desired phase margin is given next. It is assumed that the dc gain of $G_p(s)H(s)$ has been adjusted by the factor K_c to meet the low-frequency specifications. In Figure 9.14 the frequency ω_1, at which the phase of $K_c G_p(j\omega)H(j\omega)$ is equal to $(180° + \phi_m + 5°)$, is noted, where ϕ_m is the required phase margin. The 5° term is explained shortly. We force the phase margin to occur at the frequency ω_1 by adjusting the open-loop gain of the compensated system to be unity at this frequency. Assuming that the magnitude of the zero ω_0 (and thus the pole ω_p) is small compared to ω_1, the gain of the compensator at this frequency is then

$$G_c(j\omega_1) = \frac{1 + j\omega_1/\omega_0}{1 + j\omega_1/\omega_p} \approx \frac{\omega_p}{\omega_0} \qquad (9\text{-}15)$$

Since

$$\left| K_c G_c(j\omega_1) G_p(j\omega_1) H(j\omega_1) \right| = 1 \qquad (9\text{-}16)$$

then

$$\frac{\omega_p}{\omega_0} \approx \frac{1}{|K_c G_p(j\omega_1)H(j\omega_1)|} \qquad (9\text{-}17)$$

Thus we have one equation and the two unknowns ω_0 and ω_p. Next we choose

$$\omega_0 = 0.1\omega_1 \qquad (9\text{-}18)$$

to ensure that the compensator introduces very little phase lag at ω_1. In fact, this choice will cause the phase of the compensator to be approximately $-5°$ at ω_1, accounting for the $5°$ added earlier. Thus (9-17) and (9-18) are the two equations required to be solved for the pole and zero of the compensator.

The phase-lag design procedure is then as follows:

1. Adjust the dc gain of $G_p(s)H(s)$ by the factor K_c to satisfy low-frequency specifications.
2. Find the frequency ω_1 at which the angle of $K_c G_p(j\omega)H(j\omega)$ is equal to $(-180° + \phi_m + 5°)$, where ϕ_m is the specified phase margin.
3. The magnitude of the zero is given by

[eq. (9-18)] $\omega_0 = 0.1\omega_1$

4. From (9.15) and (9.16), the ratio of the compensator pole and zero is then given by

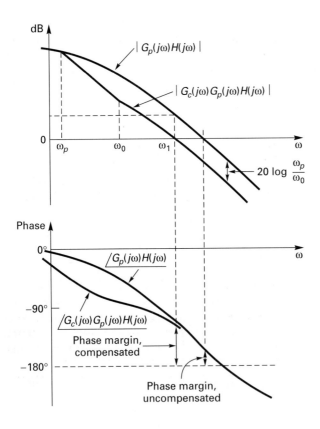

Figure 9.14 Phase-lag design.

$$\frac{\omega_p}{\omega_0} = \frac{1}{|K_c G_p(j\omega_1)H(j\omega_1)|}$$

and thus

$$\omega_p = \frac{0.1\omega_1}{|K_c G_p(j\omega_1)H(j\omega_1)|} \tag{9-19}$$

5. The compensator transfer function is then

$$G_c(s) = \frac{K_c(1 + s/\omega_0)}{1 + s/\omega_p}$$

An example is now given to illustrate phase-lag compensation.

Example 9.2

We consider the radar tracking system as in Example 9.1, with the open-loop transfer function

$$G_p(s)H(s) = \frac{4}{s(s + 1)(s + 2)}$$

We assume that the low-frequency gain meets the system specifications; thus $K_c = 1$. Suppose that a phase margin of 50° is required, as in Example 9.1. Hence the frequency ω_1 is determined by the equation

$$\angle G_p(j\omega_1)H(j\omega_1) = -180° + 50° + 5° = -125°$$

The frequency response, $G_p(j\omega)H(j\omega)$, is given in Table 9.1, and from this table we see that $\omega_1 \approx 0.4$. We then let $\omega_1 = 0.4$. The magnitude of the open-loop function at this frequency is 4.55. From (9-18), $\omega_0 = 0.1$, and from (9-19),

$$\omega_p = \frac{0.1\omega_1}{|K_c G_p(j\omega_1)H(j\omega_1)|} = \frac{0.04}{4.55} = 0.0088$$

The compensator transfer function is then

$$G_c(s) = \frac{1 + s/0.04}{1 + s/0.0088} = \frac{1 + 25s}{1 + 113.6s} = \frac{0.22s + 0.00880}{s + 0.00880}$$

Note that the low-frequency gain is unity, and the high-frequency gain is $1/4.55 = 0.220$, as required. The phase margin was calculated to be 52.3° and the gain margin to be 6.5, using the MATLAB program

```
Gp = tf([4], [1 3 2 0]); Gc = tf([25 1], [113.6 1]);
[Gm, Pm, Wcg, Wcp] = margin(Gp*Gc)
```

The unit step response of the compensated system is plotted in Figure 9.15, along with that of the system of Example 9.1 that was designed to have a 50° phase margin. Note that the response of the phase-lag compensated system lags that of the gain-compensated system. However, the low-frequency response of the phase-lag compensated system is superior in terms of steady-state errors, disturbance rejection, and so on. The characteristics of these two systems will be discussed further after a phase-lead compensator is designed for this system.

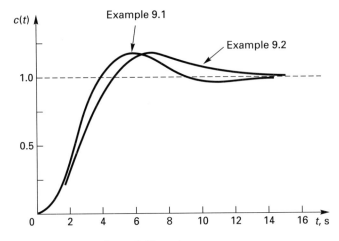

Figure 9.15   Unit step response.

9.5 PHASE-LEAD COMPENSATION

For our development of phase-lead compensator design we assume, as in the phase-lag case, a unity dc gain compensator of the form

$$G_c(s) = \frac{1 + s/\omega_0}{1 + s/\omega_p} \tag{9-20}$$

For the phase-lead compensator, $\omega_0 < \omega_p$. We wish to determine the values of ω_0 and ω_p such that certain design criteria will be satisfied for the closed-loop system of Figure 9.7, which has the characteristic equation

$$1 + G_c(s)G_p(s)H(s) = 0 \tag{9-21}$$

The Bode diagram for the phase-lead controller of (9-20) has the general form given in Figure 9.16, since $\omega_0 < \omega_p$. Thus we see that the phase-lead controller is a form of a high-pass filter, in that the high frequencies are amplified relative to the low frequencies. The controller introduces gain at high frequencies, which in general is destabilizing. However, the positive phase angle of the controller tends to rotate the Nyquist diagram away from the −1 point and thus is stabilizing. Hence we must carefully choose the pole and zero locations so that the stabilizing effect of the positive phase angle dominates.

Note that the Bode diagram of the phase-lead controller is the mirror image, about the ω-axis, of that of the phase-lag controller of Figure 9.12. The value of θ_m, the maximum phase lag introduced by the phase-lag controller, is generally of no particular significance, and we did not calculate it. However, it is important in the phase-lead case, since the phase enhances stability. The relationship between θ_m and the controller pole and zero is now calculated.

The controller transfer function is given by

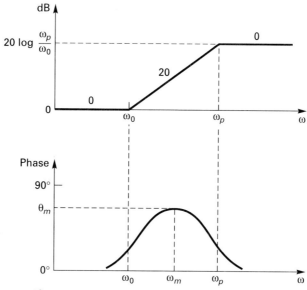

Figure 9.16 Bode diagram for phase-lead compensation.

$$G_c(j\omega) = |G_c(j\omega)|e^{j\theta} = \frac{1 + j(\omega/\omega_0)}{1 + j(\omega/\omega_p)} \tag{9-22}$$

Then

$$\tan\theta = \tan\left(\tan^{-1}\frac{\omega}{\omega_0} - \tan^{-1}\frac{\omega}{\omega_p}\right) = \tan(\alpha - \beta)$$

Thus

$$\tan\theta = \frac{\tan\alpha - \tan\beta}{1 + \tan\alpha\tan\beta} = \frac{\omega/\omega_0 - \omega/\omega_p}{1 + \omega^2/\omega_0\omega_p} \tag{9-23}$$

As stated in the preceding section, the maximum phase shift occurs at the frequency

$$\omega_m = \sqrt{\omega_0\omega_p} \tag{9-24}$$

Thus, from (9-23) and (9-24), the value of θ_m is calculated from

$$\tan\theta_m = \frac{1}{2}\left(\sqrt{\frac{\omega_p}{\omega_0}} - \sqrt{\frac{\omega_0}{\omega_p}}\right) \tag{9-25}$$

The maximum value of the phase shift θ_m is then a function only of the ratio ω_p/ω_0. This function, from (9-25), is plotted in Figure 9.17. From (9-20) and (9-24), the gain of the controller at ω_m is

$$|G_c(j\omega_m)| = \left(\frac{1 + (\omega_m/\omega_0)^2}{1 + (\omega_m/\omega_p)^2}\right)^{1/2} = \left(\frac{1 + (\omega_p/\omega_0)}{1 + (\omega_0/\omega_p)}\right)^{1/2}$$

$$= \left(\frac{\omega_p}{\omega_0}\right)^{1/2} \tag{9-26}$$

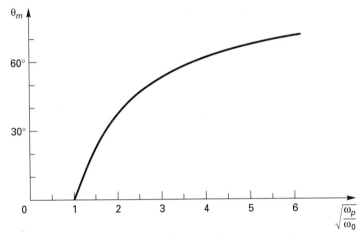

Figure 9.17   Maximum phase shift for phase-lead filter.

Phase-lead compensation is discussed with respect to the typical Nyquist diagram in Figure 9.18. As in the phase-lag controller discussion above, the Nyquist diagram is divided into three frequency ranges. First consider range C, in which the open-loop gain is quite small. Since the gain is small, the introduction of the controller pole and zero into this range has little effect.

In range A, since the Nyquist diagram is not in the vicinity of the point -1, the phase lead of the controller does not increase the stability margins of the closed-loop system. However, the increased gain at high frequencies decreases these margins. Hence the pole and zero of the phase-lead controller cannot be placed in the low-frequency range of the open-loop function.

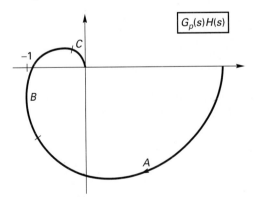

Figure 9.18   Typical Nyquist diagram.

For the system to benefit from the phase-lead controller, the controller pole and zero must be placed in frequency range B, which represents the vicinity of the -1 point on the Nyquist diagram. In contrast, recall that the pole and zero of the phase-lag controller must be placed in frequency range A to enhance system stability.

Phase-lead design is illustrated in Figure 9.19. The controller pole and zero are placed in the vicinity of the 0-dB, or unity gain, frequency of the open-loop function $G_p(j\omega)H(j\omega)$. Note that both the increased gain and the positive angle of the controller have an effect in determining the stability margins of the closed-loop system. Hence the placement of the pole and zero is much more critical than in the phase-lag case.

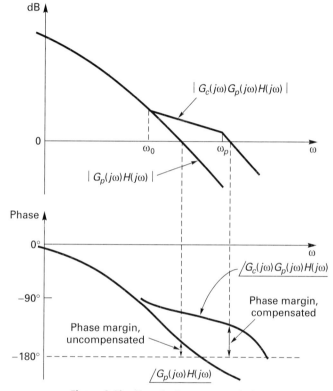

Figure 9.19 Phase-lead compensator design.

Since both gain and phase of the controller affect the phase margin in Figure 9.19, phase-lead design tends to be a trial-and-error process. One possible trial-and-error procedure is as follows:

1. Choose a zero location in the vicinity of the 0-dB crossover of the open-loop function $G_p(j\omega)H(j\omega)$, as shown in Figure 9.19.

2. Note the phase margin of the uncompensated system. Then, from Figure 9.17, choose a ratio of pole to zero that gives a value of θ_m larger than the additional phase lead needed to give the required phase margin. This value of θ_m must be larger, since the compensator also adds gain to the open-loop frequency response.

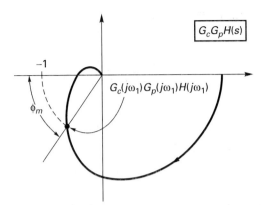

Figure 9.20   Compensated Nyquist diagram.

3. From the zero in part 1 and the ratio in part 2, calculate the pole magnitude ω_p. Next calculate the compensated Bode diagram, and determine if the phase margin is adequate. If not, move the pole in the direction that will adjust the phase margin to the desired value. If moving the pole does not give the desired results, try moving the zero.

There are many variations on this procedure, and most if not all are unsatisfactory. An analytical procedure is given in the next section.

9.6 ANALYTICAL DESIGN

In the previous section a trial-and-error design procedure was given for phase-lead design. While almost all design procedures for physical systems contain some trial and error (we sometimes call it "tuning"), we can develop an analytical design procedure for models of control systems [1,2]. This procedure allows us to set the phase margin of the system and the dc gain of the compensator, if a first-order compensator exists that satisfies these specifications.

We begin the development with the system characteristic equation (see Figure 9.7)

$$1 + G_c(s)G_p(s)H(s) = 0 \qquad (9\text{-}27)$$

and the compensator transfer function

$$G_c(s) = \frac{a_1 s + a_0}{b_1 s + 1} \qquad (9\text{-}28)$$

Note that the assumed form of the compensator is different from that in the preceding sections, since we do not limit the compensator to having unity dc gain. The compensator is given the form of (9-28) so that its dc gain is represented by the single coefficient a_0.

To develop the design equations, it is assumed that the *compensated* Nyquist diagram is as shown in Figure 9.20. Note that, in this figure, the specified phase margin ϕ_m has been realized at the frequency ω_1. From this figure

$$G_c(j\omega_1)G_p(j\omega_1)H(j\omega_1) = 1\ \underline{/-180° + \phi_m} \qquad (9\text{-}29)$$

We then wish to determine the compensator coefficients a_1 and b_1 such that (9-29) is satisfied for a specified compensator dc gain a_0. We now give the required equations.

From (9-28) and (9-29) we write

$$G_c(j\omega_1)G_p(j\omega_1)H(j\omega_1) = \left[\frac{a_0 + j\omega_1 a_1}{1 + j\omega_1 b_1}\right]G_p(j\omega_1)H(j\omega_1)$$

$$= 1 \ \underline{/-180° + \phi_m} \tag{9-30}$$

Since the factors in this equation are complex, we can expand (9-30) into two equations by equating real to real and imaginary to imaginary. There are four unknowns in (9-30): the compensator coefficients a_0, a_1, and b_1, and the frequency ω_1 at which the phase margin ϕ_m occurs. We choose to solve (9-30) for a_1 and b_1 in terms of a_0 and ω_1. It is convenient to define θ as the angle of $G_c(j\omega_1)$, such that, from (9-30),

$$\theta = \arg G_c(j\omega_1) = -180° + \phi_m - \arg G_p(j\omega_1)H(j\omega_1) \tag{9-31}$$

where $\arg(\cdot)$ denotes the angle of $(\cdot)$.

From Ref. 3, the design equations are

$$a_1 = \frac{1 - a_0|G_p(j\omega_1)H(j\omega_1)|\cos\theta}{\omega_1|G_p(j\omega_1)H(j\omega_1)|\sin\theta}$$

$$b_1 = \frac{\cos\theta - a_0|G_p(j\omega_1)H(j\omega_1)|}{\omega_1\sin\theta} \tag{9-32}$$

In a given design, steady-state error specifications may be used to determine the compensator dc gain a_0 (see Table 5.1). In addition, if the settling time T_s is specified, (8-52) may be expressed as

$$\omega_1 = \frac{8}{T_s\tan\phi_m} \tag{9-33}$$

Recall that this equation is exact only for the standard second-order system. For higher-order systems it is only approximate and may be quite inaccurate.

We see then that phase-lead design using this procedure has the following steps:

1. From the design specifications, determine the phase margin ϕ_m, the compensator dc gain a_0, and the settling time T_s.
2. From T_s and ϕ_m, use (9-33) to calculate ω_1, the frequency at which the phase margin occurs.
3. From (9-31), calculate θ, the angle of the compensator at ω_1.
4. All parameters on the right side of (9-32) are now known; solve equations (9-32) for a_1 and b_1.

The design procedure results in the compensator coefficients such that the Nyquist diagram passes through the point shown in Figure 9.20 as $G_c(j\omega_1)G_p(j\omega_1)H(j\omega_1)$. *If the resulting system is stable*, the system will have the required steady-state accuracy, the required phase margin, and approximately the required settling time. But nothing in the derivation guarantees stability, since we have satisfied only (9-29). Thus once the compensator coefficients are calculated, the Bode diagram (or Nyquist diagram) must be calculated to determine if the closed-loop system is stable. A second procedure for determining stability is to calculate the poles of the closed-loop transfer function.

We have assumed that the design specifications for a particular control system can be translated into a unique phase margin ϕ_m, a unique steady-state accuracy requirement (which determines a_0), and a unique settling time T_s. If this assumption is valid, the resulting design may still produce an unstable system. Then some trade-offs must be made between a_0, ϕ_m, and T_s to result in a designed system that has adequate stability characteristics.

If the design specifications cannot be translated into the unique parameters a_0, ϕ_m, and T_s, usually the design does not result in a unique compensator. It may then be desirable to design several compensators that meet the design specifications. The compensators can then be tested with either an accurate system simulation or the physical system itself to determine which one gives the best results. Furthermore, since the model upon which the design is based is not exact, it may be desirable to design several compensators in practical cases, even if the specifications a_0, ϕ_m, and T_s are known. For this case a_0, ϕ_m, and T_s are varied in some logical manner.

9.6.1 Phase-Lead Requirements

In the preceding derivation, there were no requirements placed on the compensator to make it phase-lead. The equations for the compensator coefficients, (9-32), are applicable to both phase-lead and phase-lag compensator design. However, certain requirements must be met for each case. These requirements are now given.

For a phase-lead controller, the phase angle of the compensator must be positive. Hence, from (9-30),

$$\theta = -180° + \phi_m - \arg G_p(j\omega_1)H(j\omega_1) > 0°$$

or

$$\arg G_p(j\omega_1)H(j\omega_1) < -180° + \phi_m \tag{9-34}$$

Also, since the gain of a phase-lead compensator for $\omega > 0$ is always greater than its dc gain, then $|G_c(j\omega)| > a_0$, and from (9-30),

$$|G_p(j\omega_1)H(j\omega_1)| < \frac{1}{a_0} \tag{9-35}$$

An additional constraint is that the controller must be stable. If we design an unstable controller, then in the Nyquist condition (see Section 8.4),

$$Z = N + P \tag{9-36}$$

the number of poles of the open-loop function in the right half-plane, P, increases by one. The system is no longer stable, since Z is now not zero. In addition, even if the closed-loop system were stable, an unstable controller is unsatisfactory in that if a failure occurred that caused the loop to open, the plant would have a very large signal on its input from the unstable controller. Probably the system would be damaged, and possibly personnel would be injured. Thus the controller coefficient b_1 must be greater than zero. Therefore, from (9-32),

$$\cos\theta > a_0|G_p(j\omega_1)H(j\omega_1)| \tag{9-37}$$

The choice of the phase-margin frequency ω_1 must then satisfy the constraints (9-34), (9-35), and (9-37) for a phase-lead controller. However, recall that even then the closed-loop system may not be stable. An example is given next to illustrate this design procedure. Then the constraints for phase-lag design using (9-32) are given.

Example 9.3

We wish to design a unity dc gain phase-lead compensator for the radar tracking system of Examples 9.1 and 9.2. Recall then that

$$G_p(s)H(s) = \frac{4}{s(s+1)(s+2)}$$

The phase margin is again specified to be 50°, with the additional specification of settling time T_s of less than 4 seconds. Note in the designs of Example 9.1 and 9.2 that the settling time is greater than 10 seconds (see Figure 9.15). The frequency response of this open-loop function is given in Table 9.1. From constraint (9-34),

$$\arg G_p(j\omega_1)H(j\omega_1) < -180° + \phi_m = -130°$$

From Table 9.1, we see then that ω_1 must be greater than 0.5 rad/s. From constraint (9-35), $|G_p(j\omega_1)H(j\omega_1)| < 1$, and hence ω_1 must be greater than approximately 1.15. From (9-33), the settling-time constraint yields

$$\omega_1 > \frac{8}{T_s \tan\phi_m} = \frac{8}{4(\tan 50°)} = 1.68 \text{ rad/s}$$

We choose $\omega_1 = 1.7$ rad/s. Thus

$$\begin{aligned}
G_p(j\omega_1)H(j\omega_1)\big|_{\omega 1 = 1.7} &= \frac{4}{j1.7(1+j1.7)(2+j1.7)} \\
&= \frac{4}{(1.7\,\underline{/90°})(1.972\,\underline{/59.5°})(2.625\,\underline{/40.4°})} \\
&= 0.454\,\underline{/-189.9°}
\end{aligned}$$

This calculation is made with the MATLAB program

```
Gp = tf([4],[1 3 2 0]);
Gpjw1 = evalfr(Gp,j*1.7);
mag = abs(Gpjw1), phase = angle(Gpjw1)*180/pi
```

We specify a dc gain of unity in order to be able to compare results with Example 9.2. From (9-31),

$$\theta = \arg G_c(j\omega_1) = -180° + \phi_m - \arg G_p(j\omega_1)H(j\omega_1) = 59.9°$$

and from (9-32),

$$a_1 = \frac{1 - a_0|G_p(j\omega_1)H(j\omega_1)|\cos\theta}{\omega_1|G_p(j\omega_1)H(j\omega_1)|\sin\theta} = \frac{1-(1)(0.454)(0.502)}{(1.7)(0.454)(0.865)} = 1.155$$

$$b_1 = \frac{\cos\theta - a_0|G_p(j\omega_1)H(j\omega_1)|}{\omega_1\sin\theta} = \frac{(0.502)-(1)(0.454)}{(1.7)(0.865)} = 0.0320$$

The compensator transfer function is then

$$G_c(s) = \frac{1.155s + 1}{0.0320s + 1} = \frac{36.1s + 31.2}{s + 31.2}$$

The open-loop frequency response of the compensated system was calculated and the gain margin was found to be 24 dB. The unit step response was calculated by simulation and is plotted in Figure 9.21. Also plotted in this figure are the step responses from Examples 9.1 (gain compensation) and 9.2 (phase-lag compensation). Note the increased speed of response for the phase-lead system because of the increased high-frequency gain. A MATLAB program that performs the design calculation for this example is given by

```
phim = 50; w1 = 1.7; a0 = 1;   Gp = tf([4],[1 3 2 0]);
Gpjw1 = evalfr (Gp,j*w1);
Gpjw1mag = abs(Gpjw1);
theta = -pi + phim/57.296 - angle(Gpjw1);
   a1 = (1 - a0*Gpjw1mag*cos(theta))/...
            (w1*Gpjw1mag*sin(theta))
   b1 = (cos(theta) - a0*Gpjw1mag)/(w1*sin(theta))
Gc = tf([a1/b1 a0/b1], [1 1/b1]);
T = minreal(Gc*Gp/(1+Gc*Gp));
   pole(T), pause, margin(Gc*Gp)
```

This program also gives the closed-loop poles for the design.

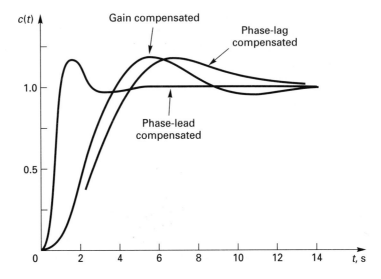

Figure 9.21   Step response of examples.

To compare further the three design techniques presented thus far, the *closed-loop fre-quency responses* of Examples 9.1, 9.2, and 9.3 are plotted in Figure 9.22. Recall that each of these systems has a phase margin of approximately 50°. Note the increased bandwidth of the phase-lead compensated system and the correlation of the time responses and the frequency responses of the three systems. Table 9.3 lists characteristics of the three systems. It is seen once again that the product of rise time and bandwidth is approximately constant. As a further comparison, Table 9.4 gives the poles of the closed-loop transfer function, that is, the zeros of the characteristic equation. Note the very slow pole and the resulting large settling time for the phase-lag system; this is a characteristic of phase-lag compensated systems. See Section 7.9 for a discussion of this slow pole.

The reason for the increase in the speed of response in Example 9.3 is seen clearly in Figure 9.23, which is a plot of the plant input (compensator output). Even though the system input is equal to unity, the plant input at $t = 0$ is 36.1. This value can be verified by expressing the controller transfer function as

$$G_c(s) = \frac{1.155s + 1}{0.0320s + 1} = 36.1 + \frac{-35.1}{0.0321s + 1}$$

We can then consider the controller to be a gain of 36.1 in parallel with a first-order lag transfer function. Since the lag cannot respond instantaneously, a step change of unity on the controller input results in a step change on the controller output of 36.1. This very large input to the plant causes the plant to respond very fast.

TABLE 9.3 RESULTS OF EXAMPLES

Compensation	Overshoot (%)	Rise time (s)	Band-width (rad/s)	$T_r \times BW$	T_s (s)
Gain	18	2.4	0.9	2.16	12.5
Phase-lag	19	2.8	0.7	1.96	36.0
Phase-lead	17	0.75	2.7	2.03	3.9

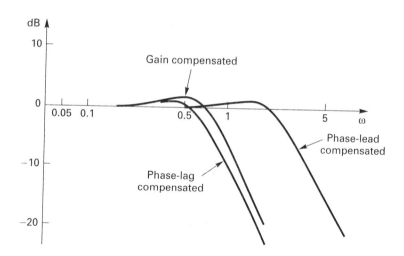

Figure 9.22 Closed-loop frequency responses of examples.

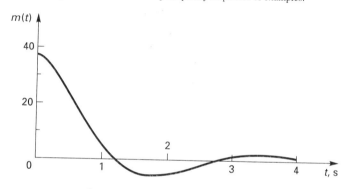

Figure 9.23 Plant input for Example 9.3.

design. For a phase-lag compensator, the angle of the compensator, θ, must be negative; hence criterion (9-34) becomes

$$\arg G_p(j\omega_1)H(j\omega_1) > -180° + \phi_m \tag{9-38}$$

Also, since the gain of a phase-lag compensator for $\omega > 0$ is always less than its dc value, then from (9-30),

$$\left| G_p(j\omega_1)H(j\omega_1) \right| > \frac{1}{a_0} \tag{9-39}$$

Also, consider the equation for the denominator coefficient of the compensator

[eq. (9-32)] $$b_1 = \frac{\cos\theta - a_0 \left| G_p(j\omega_1)H(j\omega_1) \right|}{\omega_1 \sin\theta}$$

Since this coefficient must be positive and $\sin\theta$ is negative,

$$\cos\theta < a_0 \left| G_p(j\omega_1)H(j\omega_1) \right| \tag{9-40}$$

Thus the three constraints that the choice of ω_1 must satisfy for the phase-lag compensator are (9-38), (9-39), and (9-40).

9.6.3 Summary

To summarize the results developed thus far, the advantages of the phase-lag controller are as follows:

1. The low-frequency characteristics are improved, as compared to gain compensation.
2. Stability margins are maintained or improved.
3. Bandwidth is reduced, which is an advantage if high-frequency noise is a problem. Also, for other reasons, reduced bandwidth may be an advantage.

Some disadvantages of phase-lag controllers are the following:

1. Reduced bandwidth is a disadvantage in some systems.
2. The system transient response will have one very slow term.

Some advantages of phase-lead compensation are as follows:

1. Improved stability margins.
2. Improved high-frequency performance, such as speed of response.
3. Required to stabilize certain types of systems.

Some disadvantages are these:

1. Accentuated high-frequency noise problems.
2. May generate large signals, which may damage the system or at least result in non-linear operation of the system. Since the design assumed linearity, the results of the nonlinear operation will not be immediately evident.

It is emphasized that the designs produced in this book are necessarily academic in nature. We have assumed that all transfer functions are accurate and that the design specifications are of a certain type. For the practical case the plant transfer function may be quite inaccurate. Also, it may not be possible to translate the specifications into the parameters required. For such cases, several iterations may be required between the design of a compensator and the testing of the design using the physical system. Each iteration will usually result in more accurate specifications for the system design procedure. In addition, each iteration increases the designer's knowledge of the system.

9.7 LAG-LEAD COMPENSATION

In Example 9.3 a unity dc gain phase-lead compensator was designed to give a system phase margin of 50°. However, the compensator pole-to-zero ratio of 36.1 was required to achieve this phase margin. As stated earlier, the gain to a step change in the compensator input is always equal to the compensator dc gain times the ratio of the pole to the zero. If the compensator were implemented, this high gain would almost certainly lead to nonlinear operation of the system because of the large signal magnitudes, in addition to causing noise-response problems. Since the design was performed on a linear model of the physical system, the results of the nonlinear operation would be unknown without additional nonlinear analysis and tests on the physical system.

Because of this large high-frequency gain, the design would in all probability be unsatisfactory. However, suppose that a fast system response is required. The gain compensator of Example 9.1 and the phase-lag compensator of Example 9.2 would then also be unsatisfactory.

If the design procedure of the last section is used to design different phase-lead compensators, it is found that this high-frequency gain cannot be reduced appreciably while maintaining the 50° phase margin (see Example 9.4). In cases such as this, a compromise is necessary. If either the specified phase margin or the compensator dc gain can be reduced, the high-frequency gain of the compensator can also be reduced. If these specifications cannot be reduced, it may be necessary to employ a section of phase-lag compensation cascaded with a section of phase-lead compensation, as shown in Figure 9.25, to achieve the design requirements. We call this compensator a *lag-lead compensator.*

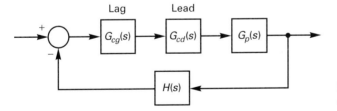

Figure 9.25   System with lag-lead compensator.

The transfer function of the lag-lead compensator is given by

$$G_c(s) = G_{cg}(s)G_{cd}(s) = \frac{1 + s/\omega_{0g}}{1 + s/\omega_{pg}} \frac{a_1 s + a_0}{b_1 s + 1} \qquad (9\text{-}41)$$

where $G_{cg}(s)$ is the phase-lag compensator and $G_{cd}(s)$ is the phase-lead compensator, with the transfer functions in the forms for which the design methods were developed. The typical lag-lead compensator has the Bode diagram shown in Figure 9.26. Note from the figure that the compensator dc gain is a_0. The low-frequency gain in the figure is greater than the high-frequency gain; in a particular design, either gain could be greater.

The design procedure for lag-lead compensators to be used in this section is as follows. The phase-lag section of the lag-lead compensator can be designed to maintain the low-frequency gain while realizing a part of the gain margin. The phase-lead section of the compensator then realizes the remainder of the phase-margin, while increasing the system bandwidth to achieve the faster system response. An example of lag-lead design is now presented.

Example 9.5

The radar tracking system of preceding examples is used again in this example. Recall that the open-loop function is

$$G_p(s)H(s) = \frac{4}{s(s+1)(s+2)}$$

The frequency response of this function is given in Table 9.1. The phase-lag section is designed first. We rather arbitrarily choose the phase margin to be achieved by the phase-lag controller to be 20°. Thus, in step 2 of the design procedure of Section 9.4, we determine the frequency at which the phase of $G_p(j\omega)H(j\omega)$ is equal to $(-180° + 20° + 5°)$, or $-155°$. Thus we choose ω_1 to be 0.9 rad/s and $\omega_{0g} = 0.1\ (\omega_1) = 0.09$. The gain of the open-loop function at $\omega = 0.9$ is 1.51; thus

$$\omega_{pg} = \frac{1}{1.51}\omega_{0g}$$

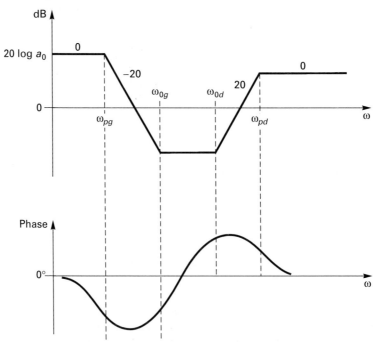

Figure 9.26   Bode diagram of lag-lead compensator.

and $\omega_{pg} = 0.0596$, from step 4 of the procedure. Thus the compensator transfer function is

$$G_{cg}(s) = \frac{1 + s/0.09}{1 + s/0.0596} = \frac{0.662s + 0.0596}{s + 0.0596}$$

To design the phase-lead compensator, we must utilize the frequency response of the open-loop function

$$G_{ole}(s) = G_{cg}(s)G_p(s)H(s) = \frac{4(0.662s + 0.0596)}{s(s + 0.0596)(s + 1)(s + 2)}$$

Note that we consider this function to be the equivalent uncompensated open-loop function and design a phase-lead compensator to give a 50° phase margin.

The phase-margin frequency, which is denoted by ω_1 in (9-31), is chosen to be the same as in Example 9.3, or $\omega_1 = 1.7$ rad/s. At this frequency, the equivalent open-loop function has a value of $G_{ole}(j1.7) = 0.3011 \underline{/-190.9°}$. Then, from (9-31), the required angle of the compensator at $\omega_1 = 1.7$ rad/s is

$$\theta = -180° + 50° - (-190.9°) = 60.9°$$

From (9-32),

$$a_1 = \frac{1 - a_0|G_{ole}(j\omega_1)|\cos\theta}{\omega_1|G_{ole}(j\omega_1)|\sin\theta} = \frac{1 - (1)(0.3011)(0.4863)}{(1.7)(0.3011)(0.8738)} - 1.908$$

$$b_1 = \frac{\cos\theta - a_0|G_{ole}(j\omega_1)|}{\omega_1\sin\theta} = \frac{0.4863 - (1)(0.3011)}{(1.7)(0.8738)} = 0.1247$$

The controller transfer function is then

$$G_{cd}(s) = \frac{1.908s + 1}{0.1247s + 1} = \frac{15.3s + 8.02}{s + 8.02}$$

The lag-lead controller transfer function is

$$G_{cg}(s)G_{cd}(s) = \frac{0.662s + 0.0596}{s + 0.0596}\frac{15.3s + 8.02}{s + 8.02}$$

The unit step response of this system is plotted in Figure 9.27(a), along with the response of the phase-lead compensated system of Example 9.3. Note that very little has been lost in terms of the speed of response. However, the unit step on the input results in a step change of only $(0.662)(15.3) = 10.1$, as compared to a value of 36.1 for the phase-lead compensated system. The plant input of the lag-lead system remains high for a longer period of time to give the fast response. The time responses for the system of this example were calculated with the SIMULINK simulation depicted in Figure 9.27(b).

9.8 PID CONTROLLER DESIGN

In the last section the lag-lead controller was developed. This controller is second order, with a phase-lag controller followed by a phase-lead controller. As we saw in that section, the lag-lead controller offers much more flexibility than does either the phase-lag or the phase-lead controller separately. In this section we consider a different form of the lag-lead controller. This form is probably the most commonly employed controller in closed-loop control systems. In addition, most commercially available controllers are of this type.

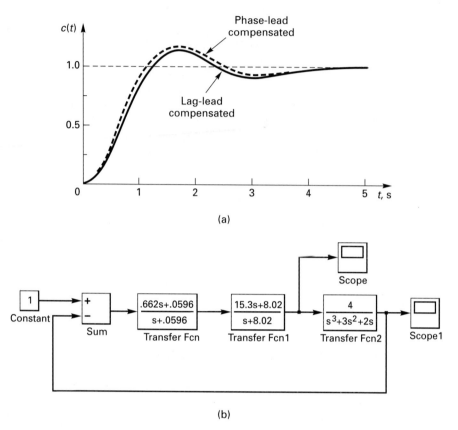

(a)

(b)

Figure 9.27 Step responses and SIMULINK simulation for Example 9.5.

For this form of lag-lead controller we consider again the *PID controller*, which was introduced in Section 7.10. The block diagram of the PID controller is shown again in Figure 9.28. The equation for the controller is given by

$$m(t) = K_P e(t) + K_I \int_0^t e(\tau)\, d\tau + K_D \frac{de(t)}{dt} \tag{9-42}$$

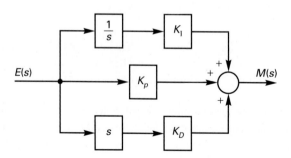

Figure 9.28 PID controller.

and the transfer function is given by

$$G_c(s) = K_P + \frac{K_I}{s} + K_D s \tag{9-43}$$

Three parameters of the controller are to be determined by the design process: the proportional gain K_P, the integral K_I, and the derivative gain K_D. As stated in Chapter 7, for a particular design one or more of the controller gains may be chosen to be zero.

The PID controller in (9-42) is second order, since differentiating (9-42) yields a second-order differential equation. The design procedures are developed by considering lower-order controllers, with certain gains set to zero. We begin by considering the proportional controller.

9.8.1 Proportional Controller

For the proportional controller, only the proportional gain is nonzero. Hence the controller transfer function is

$$G_c(s) = K_P$$

and the controller is a pure gain. This case is developed in Section 9.3 and is not considered further here.

9.8.2 PI Controller

The transfer function of the PI controller is given by

$$G_c(s) = K_P + \frac{K_I}{s} = \frac{K_P s + K_I}{s} \tag{9-44}$$

The controller has a pole at the origin and a zero on the negative real axis. If we write the transfer function as

$$G_c(s) = \frac{K_I(1 + s/\omega_0)}{s} \tag{9-45}$$

we see that the zero is located at $s = -\omega_0 = -K_I/K_P$. The Bode diagram of the PI controller is given in Figure 9.29, and the controller is seen to be a form of the phase-lag controller of Section 9.4 and Figure 9.12. If the high-frequency gain and the zero of the phase-lag controller of Figure 9.12 are held constant and the pole is moved to the origin of the s-plane ($-\infty$ of the log ω axis of the Bode diagram), the PI characteristic of Figure 9.29 results. Note that the high-frequency gain of the PI controller is K_P in magnitude and $20 \log K_P$ in decibels. Note also that the controller pole at the origin increas
Section 5.5).

The design procedure of Section 9.4 can be applie
minor modifications. These modifications occur since the h
controller is given by a different equation. This procedure, a
is as follows:

1. Adjust the dc gain of $G_p(s)H(s)$ by the factor K_c to s
   ifications. Remember that the PI controller increas
2. Find the frequency ω_1 at which the angle $G_p(j\omega_1)H(j$
3. The gain K_p is then given by

$$K_P = \frac{1}{|K_c G_p(j\omega_1) H(j\omega_1)|} \tag{9-46}$$

4. The magnitude of the zero is given by

$$\omega_0 = \frac{K_I}{K_P} = 0.1\omega_1 \tag{9-47}$$

and thus

$$K_I = 0.1\omega_1 K_P = \omega_0 K_P$$

5. The controller transfer function is then

$$G_c(s) = K_c(K_P + K_I/s) \tag{9-48}$$

Hence the implemented proportional gain is $K_c K_P$, and the implemented integral gain is $K_c K_I$.

This design procedure is now illustrated with an example.

Example 9.6

The radar tracking system used in all the examples of this chapter is again employed. For this system,

$$G_p(s)H(s) = \frac{4}{s(s+1)(s+2)}$$

The frequency response of this open-loop function is given in Table 9.1. As in the preceding examples, we specify the phase margin to be 50°. Hence the frequency ω_1 of step 2 is that frequency at which the angle of the open-loop function is equal to approximately $(-180° + 50° + 5°) = -125°$. From Table 9.1, $\omega_1 = 0.4$, and the magnitude of the open-loop function is 4.55. From step 3, $K_p = 1/4.55 = 0.220$.

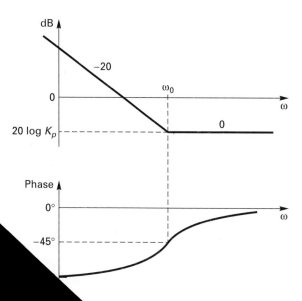

Figure 9.29 Bode diagram for PI controller.

From step 4, $\omega_0 = 0.1(\omega_1) = 0.04$, and

$$K_I = K_P\omega_0 = (0.220)(0.04) = 0.0088$$

The controller transfer function is then given by

$$G_c(s) = 0.220 + \frac{0.0088}{s}$$

The PI-compensated control system has the unit step response shown in Figure 9.30. Also given in this figure is the unit step response of the phase-lag compensated system of Example 9.2. Note that the responses are almost identical; the PI compensator pole at the origin does add some additional lag. The settling time increases from 36 s for the phase-lag compensated system to 41 s for the PI-compensated system. However, the PI-compensated system is a type 2 system and has somewhat better low-frequency characteristics.

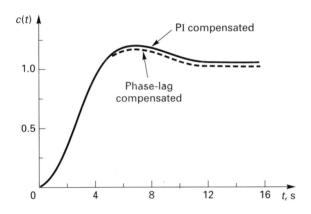

Figure 9.30 Step responses for Example 9.6.

9.8.3 PD Controller

The transfer function of the PD controller is given by

$$G_c(s) = K_P + K_D s \tag{9-49}$$

which may be written as

$$G_c(s) = K_P\left(1 + \frac{s}{\omega_0}\right) \tag{9-50}$$

where $\omega_0 = K_P/K_D$. Hence $G_c(s)$ is seen to have a zero at $s = -\omega_0$ and a pole at infinity. The Bode diagram of the PD controller is shown in Figure 9.31; note that the compensator is phase-lead. The low-frequency gain is K_P, and the high-frequency gain is unbounded. The PD controller is obtained from the phase-lead controller of Figure 9.16 by holding the low-frequency gain and the zero ω_0 constant, and moving the pole ω_p to infinity.

Since the PD compensator is phase-lead, the design procedure given in Section 9.5 can be employed. However, an analytical procedure is developed in the next section, and in general this procedure is used in this book.

9.8.4 PID Controller

The transfer function of the PID controller is given by

$$G_c(s) = K_P + \frac{K_I}{s} + K_D s \tag{9-51}$$

As described earlier, the integral term is phase-lag and the derivative term is phase-lead. Hence the integral term contributes a low-frequency effect and the derivative term contributes a high-frequency effect. The Bode diagram of the PID compensator is then the combination of Figures 9.29 and 9.31, as illustrated in Figure 9.32.

Note the similarity of the PID compensator Bode diagram of Figure 9.32 and that of the lag-lead compensator of Figure 9.26. Hence we see that the PID compensator is a form

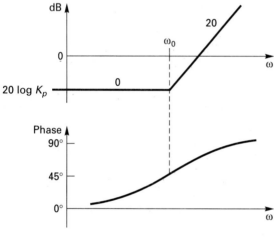

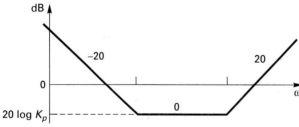

Figure 9.31 Bode diagram of PD controller.

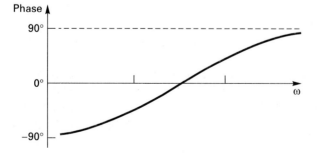

Figure 9.32 Bode diagram of PID controller.

of lag-lead compensation. However, for the PID compensator, the three gains are the only parameters to be chosen. The unbounded high-frequency gain of the differentiator can lead to problems, as discussed before, and a pole is usually added to the derivative path to limit this gain. For this case, the transfer function of the compensator is

$$G_c(s) = K_P + \frac{K_I}{s} + \frac{K_D s}{1 + s/\omega_{pd}} \tag{9-52}$$

and four parameters must be determined by the design process.

The PID compensator can be designed by the procedure given in Section 9.7 for the design of lag-lead controllers. In that design procedure, the PI section is designed first, in order to realize a part of the specified gain margin. In this case, both K_P and K_I are calculated. The remaining parameter, K_D, is then calculated to realize the total phase margin. In the next section an analytical design procedure is developed that will allow us to design a controller to realize a specified phase margin for all variations of the PID controller. Examples of the design of these controllers are given then.

9.9 ANALYTICAL PID CONTROLLER DESIGN

In Section 9.6 an analytical design method was developed for first-order phase-lead and phase-lag controllers. The same type of development leads to design equations for the PID controller. We first assume that the controller transfer function is given by

$$G_c(s) = K_P + \frac{K_I}{s} + K_D s \tag{9-53}$$

As in Figure 9.20, it is assumed that the compensated Nyquist diagram is to pass through the point $1 \, \underline{/-180° + \phi_m}$, for the frequency ω_1, to achieve the phase margin ϕ_m. Or,

$$G_c(j\omega_1)G_p(j\omega_1)H(j\omega_1) = 1 \, \underline{/-180° + \phi_m} \tag{9-54}$$

If the angle of $G_c(j\omega_1)$ is denoted by θ, then from (9-54),

$$\theta = \arg G_c(j\omega_1) = -180° + \phi_m - \arg G_p(j\omega_1)H(j\omega_1) \tag{9-55}$$

From (9-53) and (9-54),

$$K_P + j\left(K_D\omega_1 - \frac{K_I}{\omega_1} \right) = |G_c(j\omega_1)|(\cos\theta + j\sin\theta) \tag{9-56}$$

where, from (9-54),

$$|G_c(j\omega_1)| = \frac{1}{|G_p(j\omega_1)H(j\omega_1)|} \tag{9-57}$$

Assuming that we know ω_1 (we do not), we then know θ and $|G_c(j\omega_1)|$ from (9-55) and (9-57). Thus in (9-56), equating real part to real part and imaginary part to imaginary part, we have two equations and three unknowns. Hence we can assume a value for one of the gains and solve for the other two. Of course, since we do not know ω_1, (9-56) actually is two equations with four unknowns. The phase-margin frequency ω_1 may be calculated to yield a specified settling time T_s, using (9-33). In addition, the gain K_I may be chosen to satisfy low-frequency specifications, since, at low frequencies, the PID compensator is dominated by the integral term [see (9-53)].

First, from (9-56), equating real part to real part yields

$$K_P = \frac{\cos\theta}{|G_p(j\omega_1)H(j\omega_1)|} \tag{9-58}$$

and equating imaginary part to imaginary part yields

$$K_D\omega_1 - \frac{K_I}{\omega_1} = \frac{\sin\theta}{|G_p(j\omega_1)H(j\omega_1)|} \tag{9-59}$$

Given ω_1 and ϕ_m, the gain K_P, is calculated from (9-58). Then, for example, K_I is determined from steady-state specifications and K_D can be calculated from (9-59).

The phase angle of the compensator at ω_1, θ, can be either positive or negative, as indicated in Figure 9.32. From this figure we see also that the magnitude of the compensator transfer function, $|G_c(j\omega)|$, can be either greater than unity or less than unity. Thus the only requirement on the choice of ω_1 is that the magnitude of θ in (9-55) be less than 90°.

The design equations (9-58) and (9-59) are general in that these equations apply for the PID controller or for any variation of it. For example, the equations for the design of the PD controller are obtained by setting K_I to zero in (9-59). Then the equation for K_D is

$$K_D = \frac{\sin\theta}{\omega_1|G_p(j\omega_1)H(j\omega_1)|} \tag{9-60}$$

Examples of PID controller design are now given.

Example 9.7

For this example, the open-loop function of the radar tracking system of the previous examples of this chapter is used:

$$G_p(s)H(s) = \frac{4}{s(s+1)(s+2)}$$

The frequency response of this function is given in Table 9.1. As a first example, a PD controller is designed to yield a phase margin of 50°. We choose ω_1 to be 1.7 rad/s, since this is the value used in the phase-lead design of Example 9.3. From Example 9.3, $G_p(j1.7)H(j1.7) = 0.454\underline{/-189.9°}$. From (9-55),

$$\theta = -180° + \phi_m - \arg G_p(j\omega_1)H(j\omega_1) = -180° + 50° + 189.9°$$

or $\theta = 59.9°$. From (9-58),

$$K_P = \frac{\cos\theta}{|G_p(j\omega_1)H(j\omega_1)|} = \frac{\cos 59.9°}{0.454} = 1.10$$

Since K_P is the dc gain of the compensator, this gain is somewhat larger than the value of unity employed in Example 9.3. Note that the choice of a different value for ω_1 will result in a different dc gain for the PD compensator.

The value of K_D is calculated from (9-59) [or (9-60)]:

$$K_D = \frac{\sin\theta}{\omega_1|G_p(j\omega_1)H(j\omega_1)|} = \frac{\sin 59.9°}{(1.7)(0.454)} = 1.12$$

The controller transfer function is then

$$G_c(s) = 1.10 + 1.12s$$

The unit step response of the compensated system is plotted in Figure 9.33. As a comparison, the unit step response of the phase-lead compensated system of Example 9.3 is so nearly the same that it cannot be plotted as a different curve. For each system the settling time is 3.9 s.

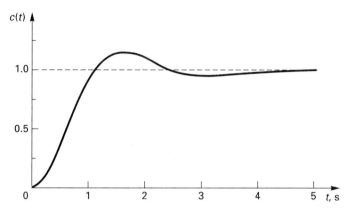

Figure 9.33   Step responses for Example 9.7.

As far as the input–output characteristics of the systems are concerned, the PD-compensated system of Example 9.7 and the phase-lead-compensated system of Example 9.3 are approximately the same. However, the PD-compensated system is somewhat more sensitive to high-frequency noise because of the unlimited gain of the differentiator. This problem is discussed in greater detail in the next section.

Example 9.8

A PID compensator is to be designed for the system of Example 9.7. Since the PID compensator is a type of lag-lead compensator, the results are compared to the lag-lead design of Example 9.5. In Example 9.5, the phase-margin frequency was chosen to be 1.7 rad/s, and the same choice is made in this example. From Example 9.7, which also used $\omega_1 = 1.7$ rad/s,

$$G_p(j1.7)H(j1.7) = 0.454 \angle{-189.9°}$$

and from (9-55),

$$\theta = \arg G_c(j\omega_1) = -180° + \phi_m - \arg G_p(j\omega_1)H(j\omega_1)$$
$$= -180° + 50° + 189.9° = 59.9°$$

Thus, from (9-58),

$$K_P = \frac{\cos\theta}{|G_p(j\omega_1)H(j\omega_1)|} = \frac{\cos 59.9°}{0.454} = 1.10$$

Note that this value must be the same as for a PD controller, as in Example 9.7.

The second design equation is (9-59):

$$K_D\omega_1 - \frac{K_I}{\omega_1} = \frac{\sin\theta}{|G_p(j\omega_1)H(j\omega_1)|}$$

The right side of this equation is determined by the choice of ϕ_m and ω_1. The stabilizing action of the controller can then be divided between the integral term and the derivative term. If K_I is chosen larger, K_D must also be chosen larger. This point is further discussed later.

We choose K_I to have a value of 0.005. Thus from (9-59),

$$K_D = \frac{\sin\theta}{\omega_1|G_p(j\omega_1)H(j\omega_1)|} + \frac{K_I}{\omega_1^2} = \frac{\sin 59.9°}{(1.7)(0.454)} + \frac{0.005}{(1.7)^2}$$

$$= 1.121 + 0.00173 = 1.123$$

The unit step response was calculated and was found to be essentially the same as that of both the PD-compensated system of Example 9.7 and the lag-lead system of Example 9.5. The settling time remained at 3.9 s. Note that K_P and K_D of the PID compensator are approximately the same as those of the PD compensator of Example 9.7. A MATLAB program that performs the design of this example for three different values of K_I is given by

```
KI = [0.005 0.05 0.5]; phim = 50; w1 = 1.7;
Gp = tf([0 0 0 4], [1 3 2 0]);
for k = 1:3
   Gpjw1 = evalfr(Gp,j*w1);
   Gpjw1mag = abs(Gpjw1);
   theta = -pi + phim/57.296 - angle(Gpjw1);
   KP = cos(theta)/Gpjw1mag;
   KD = sin(theta)/(w1*Gpjw1mag) + KI(k)/w1^2;
   [KP,KI(k),KD]
      Gc = tf([0 KD KP KI(k)], [0 0 1 0]);
      T = minreal (Gc*Gp/(1+Gc*Gp));
      pole(T),pause,[Gm,Pm,Wcg,Wcp]=margin(Gc*G),pause
end
end
```

This program also gives the closed-loop poles for each design. The three systems can be simulated by adding statements as in the MATLAB program in Example 7.20.

We noted that the values of K_D and K_I are dependent. If K_I is made larger, the value of K_D must also be larger. Two effects can be seen from making K_I larger. First, more lag is introduced into the system: thus the time response will also exhibit more overshoot and will have a longer settling time. The second effect is that the system may become *conditionally stable*, as is shown in the Nyquist diagram of Figure 9.34. In this Nyquist diagram, the added phase lag at low frequencies is sufficient to cause the diagram to cross the −180° axis at two points where the magnitude of the compensated open-loop function is greater than unity. Note, however, that the system is still stable but that either a reduced loop gain or an increased loop gain will cause the system to become unstable. This type of system is called a *conditionally stable system*.

Some commonly encountered nonlinearities effectively reduce gain for larger signals. One of the most common nonlinearities in physical systems is the limiter, which has the characteristic given in Figure 9.35. For a small input signal, the device is a linear gain of K. However, the device saturates for large signals, and the effective gain of the device is reduced. If the limiter appears in a conditionally stable system, the system may become unstable for large signals. This topic is discussed in detail in Chapter 14.

9.10 PID CONTROLLER IMPLEMENTATION

In this section we discuss some of the problems that occur in the implementation of PID controllers. We have assumed that the PID controller has the structure shown in the control system of Figure 9.36. However, this form of the controller is seldom implemented. Instead, different but mathematically equivalent realizations are used. We now consider some structures that are implemented.

The characteristic equation of the control system of Figure 9.36 is

$$1 + G_c(s)G_p(s) = 0 \qquad (9\text{-}61)$$

This equation determines the stability characteristics and the nature of the transient response of the control system. Also, the design methods given earlier are based on this equation. For these reasons, if we implement the PID controller using different structures, the system characteristic equation must not change.

One problem of the PID controller occurs in the derivative path. The gain of this path increases with increasing frequency. A step change cannot occur in a physical signal, but the rate of change of a physical signal that approximates a step change can be very high. The derivative of a mathematical step function is infinite; the derivative of a physical step function can be very large. For that reason, the structure shown in Figure 9.37 is often used to implement a PID controller. In this structure only the feedback signal is differentiated. Since this signal usually changes slowly, the signal applied to the plant will not be as large

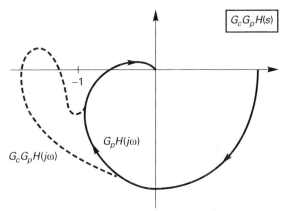

Figure 9.34   Nyquist diagram of a conditionally stable system.

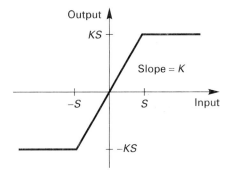

Figure 9.35   Limiter characteristic.

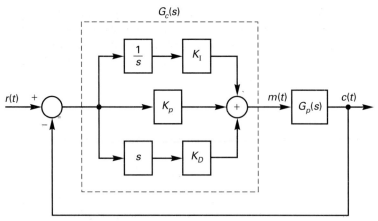

Figure 9.36   PID control system.

as that in the system of Figure 9.36. However, note that the characteristic equation for each of the two systems is the same.

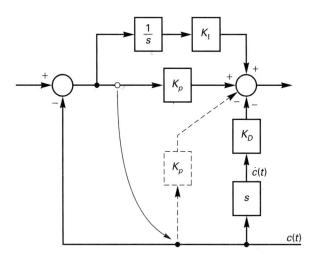

Figure 9.37   PID controller with output differentiation.

A further change in structure that is sometimes employed is illustrated by the dashed lines in Figure 9.37. For this structure the input signal is transmitted only through the integrator path; thus no signals with high rates of change appear on the plant input, assuming that the sensor output has no high-frequency components of significant amplitudes. The plant will not be "bumped" by a step change on the system input.

If the rate of change of the system output can be measured, in addition to the output, the implementation of Figure 9.38 can be employed. In this implementation no differentiation occurs, even though the system characteristic equation is unchanged. This system utilizes rate feedback and emphasizes the point that rate feedback is a form of PD compensation. However, rate feedback does not employ the differentiation of a signal and thus does not create high-frequency noise problems. The rate signal contains some low level of measurement noise, as does the position signal. However, if the position signal is differentiated to give a derived rate signal, as in Figure 9.37, the differentiation amplifies the high-frequency noise in the position signal. Hence a *derived* rate signal will in general be noisier than a *measured* rate signal.

If it is necessary to differentiate the position signal in a system, a low-pass filter is often added to the derivative path. As was mentioned in Section 9.8, the controller transfer function is then

$$G_c(s) \ = \ K_P + \frac{K_I}{s} + \frac{K_D s}{1 + s/\omega_{pd}} \tag{9-62}$$

The filter added to the derivative path has the transfer function

$$G_f(s) \ = \ \frac{1}{1 + s/\omega_{pd}} \tag{9-63}$$

This filter has the Bode diagram of Figure 9.39 and attenuates the higher frequencies of the filter input signal. Any high-frequency noise above the filter cutoff frequency, ω_{pd}, will be reduced in amplitude. The cutoff frequency, ω_{pd}, cannot be chosen arbitrarily low, since decreasing ω_{pd} reduces the effect of the derivative path on system stability and on the speed of response.

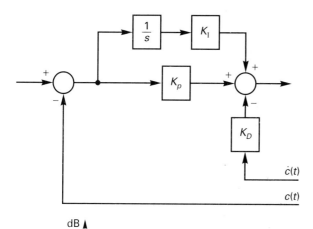

Figure 9.38 PID controller with rate feedback.

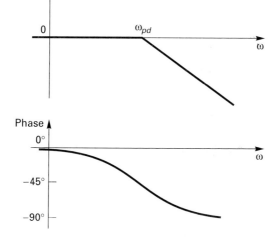

Figure 9.39 Bode diagram of lowpass filter.

In (9-61), if the PD terms are combined, the controller transfer function can be expressed as

$$
G_c(s) = \frac{K_I}{s} + \frac{K_P + K_D s + K_P s / \omega_{pd}}{1 + s / \omega_{pd}}
$$

$$
= \frac{K_I}{s} + \frac{K_P(1 + s / \omega_{0d})}{1 + s / \omega_{pd}}, \qquad \omega_{0d} = \frac{K_P \omega_{pd}}{K_P + K_D \omega_{pd}}
$$

(9-64)

We see then that the zero is smaller in magnitude than the pole and that the PD section of the controller is a phase-lead controller. Hence adding low-pass filtering to the PD controller converts the controller to a phase-lead controller of the type considered in Section 9.5. Of course, the opposite is also true. The phase-lead compensator of Section 9.5 is also a PD controller with low-pass filtering added.

9.11 FREQUENCY-RESPONSE SOFTWARE

Commercial software is available that designs compensators for closed-loop control systems. The software may use various criteria for design; the user's manual for a particular software package will give the algorithms used for design with that package. In this book, we have implemented design procedures using MATLAB.

A problem that occurs in simulation is that of verifying the simulation; that is, how do we know that we have entered the correct system model into the computer? In Chapter 7, we suggested the approach of increasing the gain by the gain-margin factor. The simulation should then oscillate at a known frequency, provided that the numerical integration is accurate (see Section 6.3 for a discussion of this case). If the numerical integration is accurate, this procedure checks one point on the system frequency response. Of course, this same procedure can be applied to the physical system, to verify the system model. However, one must be careful, since forcing the physical system to go unstable may present a danger to personnel or to equipment.

Since the frequency responses can be calculated, the complete frequency response of the model is known. For example, suppose that the closed-loop frequency response, for a given system for $\omega = 2$ rad/s, is $T(j2) = 0.7 \underline{/-50°}$. Then, in the simulation, if the input to the system is $\cos 2t$, the output *in the steady state* should be $0.7 \cos(2t - 50°)$. Several points on the frequency response may be verified in this manner, which increases confidence in the simulation.

In many cases it is desirable to obtain a frequency response of the physical system in order to verify the derived transfer function and the simulation. The procedure in the last paragraph is then applied to the physical system; that is, sinusoids are applied to the input of the physical system, and the steady-state response is recorded. The ratio of the output phasor and the input phasor is the system frequency response. This measured frequency response is then compared with the calculated frequency response. This procedure is probably the best method for the verification of a transfer function and a simulation.

9.12 SUMMARY

In this chapter the design of compensators by frequency-response methods was presented. We classified the compensators as being either phase-lag or phase-lead. The PI compensator then falls under the classification of phase-lag, and the PD compensator under the classification of phase-lead. The second-order filter considered is a section of phase-lag followed by a section of phase-lead and is called a lag-lead compensator. The PID controller is then also a lag-lead filter.

In the design methods presented, the design criteria were considered to be a required open-loop dc gain and a required phase margin, with an adequate gain margin. Since the techniques allow many controllers to be designed to satisfy these criteria, the choice is then among these controllers to satisfy other design criteria, such as bandwidth, disturbance rejection, settling time, and so on. In a practical design, usually a compromise among the design specifications is necessary.

To summarize, some possible advantages of phase-lag compensation are as follows:

1. System low-frequency characteristics maintained or improved
2. Stability margins maintained or improved
3. High-frequency noise response reduced
4. System type increased by one by a PI or PID compensator

Some possible disadvantages of phase-lag compensation are these:

1. Slower response, longer settling time
2. Does not stabilize some systems (see Section 7.6)

Some possible advantages of phase-lead compensation are these:

1. Improved stability margins
2. Improved high-frequency (speed of response) performance
3. Required for certain systems
4. Rate feedback is easy to implement in some systems

Some possible disadvantages of phase-lead compensation are as follows:

1. May accentuate high-frequency noise problems
2. May generate large signals at the plant input

As was stated previously, we are considering the *models* of physical systems in all the analysis and design procedures presented in this book. We are not working with the physical systems themselves. For a given model, the model may or may not accurately describe the physical system's characteristics. If the model is accurate, it will be accurate only over a limited range of operation of the physical system. Hence we should be very careful in attaching undue significance to the analysis and design techniques presented here or elsewhere. We can design a model to have 10 percent overshoot for a step input, to have very low steady-state errors, and so forth. However, all this is meaningless until it is tested using the physical system; generally the response of the physical system will not be as good as that of the model. In fact, many iterations on the design and on improving the system model are often needed before there is even a resemblance between the model response and the physical system response. The point to be made here is that all analytical work is with the *model* of a system; the major part of a controls engineer's work is in applying the results of analysis and design to the physical system (see Figure 1.4).

REFERENCES

1. W. R. Wakeland. "Bode Compensator Design," *IEEE Trans. Autom. Control*, AC-21 (October 1976): 771.

2. J. R. Mitchell. "Comments on Bode Compensator Design," *IEEE Trans. Autom. Control*, AC-22 (October 1977): 869.

3. C. L. Phillips and R. D. Harbor, *Feedback Control Systems*, 3rd ed. Englewood Cliffs, NJ: Prentice Hall, 1996.

PROBLEMS

9.1. Assume that a frequency-response design is to be employed for the systems of Figure P9.1. We require that the system characteristic equation be expressed as $1 + G_c(s)G_{eq}(s) = 0$. Find the transfer function $G_{eq}(s)$, for each system, that must be plotted as a Bode diagram for design purposes.

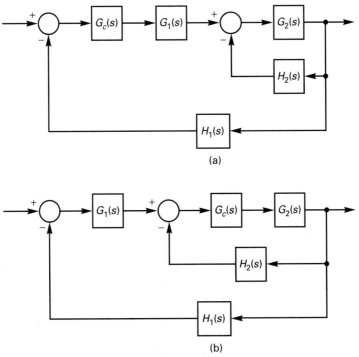

(a)

(b)

Figure P9.1

9.2. For the system of Figure P9.2, the plant transfer function is given by

$$G_p(s) = \frac{50}{(s+1)\,(s+2)\,(s+5)}$$

The frequency response $G_p(j\omega)$ is given in Table P9.2. Assume that the sensor gain is $H_k = 1.0$.

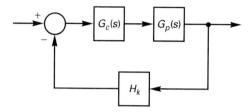

Figure P9.2

TABLE P 9.2   FREQUENCY RESPONSE FOR PROBLEM 9.2

ω	Magnitude	dB	Phase
.0100	4.99968	13.97884	−.97401
.0200	4.99871	13.97716	−1.94788
.0300	4.99710	13.97436	−2.92150
.0400	4.99485	13.97044	−3.89473
.0500	4.99195	13.96541	−4.86744
.0600	4.98842	13.95926	−5.83951
.0700	4.98425	13.95200	−6.81080
.0800	4.97946	13.94364	−7.78119
.0900	4.97403	13.94317	−8.75055
.1000	4.96799	13.92361	−9.71876
.2000	4.87467	13.75891	−19.31114
.3000	4.72764	13.49290	−28.66364
.4000	4.53773	13.13678	−37.68527
.5000	4.31708	12.70380	−46.31189
.6000	4.07739	12.34414	−54.50578
.7000	3.82885	11.66138	−62.25169
.8000	3.57956	11.07660	−69.55151
.9000	3.33553	10.46328	−76.41895
1.0000	3.10087	9.82967	−82.87499
2.0000	1.46805	3.33482	−130.23640
3.0000	.75207	−2.47482	−158.83880
4.0000	.42349	−7.46323	−178.05850
5.0000	.25751	−11.78402	−191.88870
6.0000	.16641	−15.57652	−202.29720
7.0000	.11291	−18.94538	−210.38680
8.0000	.07972	−21.96873	−216.83340
9.0000	.05817	−24.70599	−222.07640
10.0000	.04364	−27.20325	−226.41440

(a) Find the approximate gain and phase margins of the uncompensated system.
(b) Design a phase-lag compensator of unity dc gain that results in a phase margin of approximately 45°.
(c) Using MATLAB, for the designed system find
   (i) the gain and phase margins.
   (ii) the closed-loop poles.

(d) Find the system time constants.

(e) To see the effects of the long-time constant found in (d), run the system step response. Note the slow approach of the output to its final value.

9.3. Given the system of Figure P9.2, with

$$G_p(s) = \frac{10}{s(s+1)(s+2)}$$

and $H_k = 0.25$.

(a) We desire to design a compensator such that the information in Table 9.1 is useful. How should Table 9.1 be modified for use in this design?

(b) Design a unity-dc-gain phase-lag compensator by the analytical procedure that yields a phase margin of 50°.

(c) Repeat (b) for a phase-lead compensator, with the added specification of settling time $T_s < 4$ s.

(d) Use MATLAB to determine the gain and phase margins in both (b) and (c).

(e) For the system of (b), determine the input required to command the output to go to a constant value of 10.

(f) Simulate the systems of (b) and (c) with the input determined in (e), and plot the outputs. Give the rise time, the percent overshoot, and the settling time for each of the systems.

9.4. Consider the temperature-control system for a large test chamber shown in Figure P9.4. This system is described in Problem 6.12.

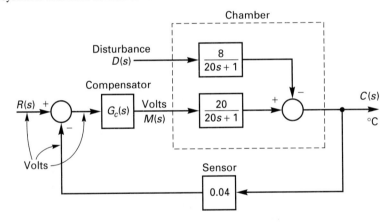

Figure P9.4

(a) Find the input voltage required to command the output to 50°C.

(b) Suppose that the compensator in Figure P9.4 is a gain K, that is, $G_c(s) = K$. Find the K such that the steady-state error for the input in (a) is not more than 1°C.

(c) Design a phase-lag compensator, with a dc gain of 61.25, such that the system phase margin is 130°.

(d) Use MATLAB to determine the gain and phase margins in both (b) and (c).

(e) The closed-loop system in (c) is second order. Use MATLAB to find the two system time constants. What physical characteristic make these two time constants large?

(f) Use SIMULINK to verify the time constants in (e).

9.5. The simplified block diagram of an automobile cruise-control system is given in Figure P9.5. This system is described in Problem 3.11. The plant transfer function is defined as

$G_p(s) = V(s)/M(s)$, and Table P9.5 gives the frequency response $G_p(j\omega)$. Assume that the sensor gain is $H_k = 1$, rather than the value of 0.03 as shown.

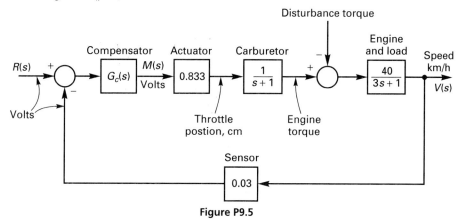

Figure P9.5

TABLE P 9.5   FREQUENCY RESPONSE FOR PROBLEM 9.5

ω	Magnitude	dB	Phase
0.1000	31.76883	30.04002	−22.41142
0.2000	28.02729	28.95162	−42.27622
0.3000	23.73049	27.50614	−58.68931
0.4000	19.81204	25.93858	−71.99866
0.5000	16.53682	24.36904	−82.87763
0.6000	13.88012	22.84786	−91.91159
0.7000	11.73955	21.39303	−99.53091
0.8000	10.01028	20.00893	−106.04200
0.9000	8.60446	18.69448	−111.66600
1.0000	7.45289	17.44649	−116.56680
2.0000	2.45048	7.78501	−143.97360
3.0000	1.16393	1.31858	−155.22550
4.0000	0.67132	−3.46146	−161.20060
5.0000	0.43481	−7.23411	−164.87640
6.0000	0.30394	−10.34414	−167.35820
7.0000	0.22420	−12.98722	−169.14390
8.0000	0.17210	−15.28419	−170.48930
9.0000	0.13623	−17.31466	−171.53890
10.0000	0.11049	−19.13376	−172.38040
20.0000	0.02774	−31.13897	−176.18280
30.0000	0.01234	−38.17593	−177.45430

(a) Verify two points on the frequency response to Table P9.5.
(b) Use Table P9.5 to find the approximate gain and phase margins of the uncompensated system.
(c) Design a constant-gain compensator $[G_c(s) = K]$ that results in a phase margin of approximately 45°.
(d) Find the compensated system gain margin using Table P9.5.
(e) Use SIMULINK to find the system rise time and percent overshoot for a step input.

9.6. Consider the automotive cruise-control system of Problem 9.5. Let $H_k = 1$ rather than the value of 0.03 as shown.

    (a) Design a phase-lag compensator of unity dc gain that results in a phase margin of approximately 45°.

    (b) Use MATLAB to find the compensated system phase margin.

    (c) Find the approximate percent overshoot using Figure 8.50.

    (d) Use SIMULINK to find the system rise time and percent overshoot for a step input.

9.7. **(a)** Repeat Problem 9.6 for a phase-lead compensator.

    (b) Compare rise times and the percent overshoots for the gain design of Problem 9.5, the phase-lead design of Problem 9.6, and the design of this problem.

9.8. Shown in Figure P9.8 is the block diagram of the servo-control system for one of the axes of a digital plotter. The input θ_r is the output of a digital computer, and the output θ_p is the position of the servomotor shaft. This system is described in Problem 6.13. For this problem, the compensation is to be phase lead rather than the rate feedback shown; thus $K_v = 0$ and K_d is replaced with $G_c(s)$. Design the lead compensator to have dc gain of 0.5 such that the system phase margin is 40°.

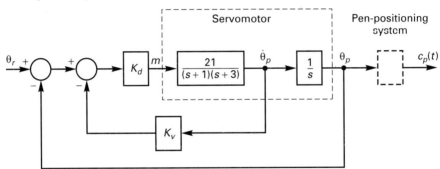

Figure P9.8

    (a) Use MATLAB to calculate the open-loop frequency response.

    (b) Verify by calculations at least one point on the frequency response in (a).

    (c) Design a phase-lead compensator to have a dc gain of 0.5 and to yield a phase margin of 40° and a settling time $T_s < 4$ s.

    (d) Verify the phase margin with MATLAB.

    (e) Find the percentage overshoot in the step response predicted by Figure 8.50.

    (f) Use SIMULINK to find the actual percent overshoot and settling time for the step input.

9.9. For the digital plotter servo of Problem 9.8, the rate feedback compensation shown is a form of PD compensation.

    (a) In Figure P9.7, let $K_v = 0$ and K_d be replaced by a PD compensator. Design the PD compensator for this plant that results in a phase margin of 40° and a settling time $T_s < 3.5$ s.

    (b) The PD compensator is to be realized by K_d and K_v, as shown in the figure. Using the results of (a), find values of K_d and K_v that result in the 40° phase margin.

    (c) Use MATLAB to verify the phase margin for the system of (b).

9.10. Consider the attitude-control system of a rigid satellite shown in Figure P9.10, which is discussed in Example 2.13 of Section 2.6.

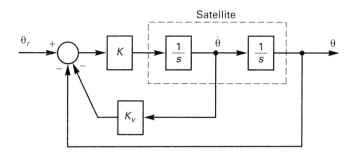

Satellite

Figure P9.10

(a) Can this system be stabilized with $K_v = 0$ (no rate feedback) and the gain K replaced with a phase-lag compensator? Justify your answer.

(b) With $K_v = 0$ and K replaced with a unity-gain phase-lead compensator, design the compensator such that the system phase margin is $50°$ and the settling time $T_s < 4$ s.

(c) Find the system gain margin for the compensated system.

(d) Repeat (b) for a PD compensator. Convert the PD compensator gain into the gains K and K_v of the rate feedback compensation of Figure P9.9.

(e) Find the percent overshoot in the step response predicted by Figure 8.50.

(f) Find the overshoot and the settling time for a step input, by simulation, for the two designs.

9.11. Shown in Figure P9.11 is the block diagram of the servo-control system for one of the joints of a robot. This system is described in Section 2.12.

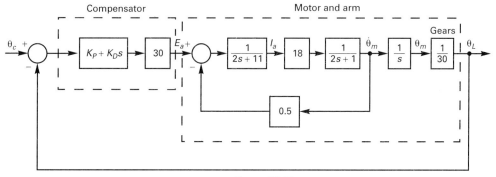

Figure P9.11

(a) Show that the plant transfer function is given by

$$\frac{\theta_L(s)}{E_a(s)} = \frac{0.15}{s(s+1)(s+5)}$$

(b) Design a unity-dc-gain phase-lead compensator (to replace the PD compensator shown) such that the system has a phase margin of $50°$ and a settling time $T_s < 4$ s. Note that in calculating the open-loop frequency response for the design, the gain of 30 must be included. Choose the phase-margin frequency to be $\omega_1 = 2$ rad/s.

(c) Repeat the design of (b) for the PD compensator.

(d) Add a high-frequency pole to the PD compensator of (a) to limit the high-frequency gain. Choose the pole such that the compensator low-frequency gain and zero do not change and the ratio of compensator high-frequency gain to low-frequency gain in not greater than 10. Give the compensator transfer function.

(e) Find the system phase margin for (d). Note the effect of the added pole.

(f) Use SIMULINK to find the rise times and the percents overshoot for the designed systems of (b) and (d).

9.12. Shown in Figure P9.12 is a closed-loop control rod positioning system for a nuclear reactor. The system positions control rods in a nuclear reactor to obtain a desired radiation level. The gain of 4.4 is the conversion factor from rod position to radiation level. The radiation sensor has a dc gain of unity and a time constant of 0.1s.

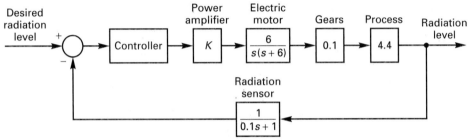

Figure P9.12

(a) Find the approximate gain and phase margin of the uncompensated system if the gain of the power amplifier, K, is unity.

(b) Suppose that the steady-state specifications require that the power amplifier gain be equal to 20. Design a unity dc gain phase-lag compensator that will yield a system phase margin of 50° for this case.

(c) Use Figure 8.50 to estimate the percent overshoot in the system step response.

(d) Repeat (b) for a PI compensator, using the same phase-margin frequency.

(e) Use MATLAB to verify the phase margins for the systems of (b) and (c).

(f) Use SIMULINK to find the actual percents overshoot in the step responses.

9.13. In Problem 9.3, a phase-lag compensator was designed for the phase-margin frequency $\omega_1 = 0.3$ rad/s, and a phase-lead compensator was designed for $\omega_1 = 2$ rad/s. The two compensator transfer functions were found to be

$$G_{ca} = \frac{27.5s + 10.9}{s + 10.9} \qquad G_{cb} = \frac{0.240s + 0.0258}{s + 0.0258}$$

(a) Which transfer function is for the phase-lag compensator? Why?

(b) Which designed system has the faster response? Why?

(c) Simulate two design systems with a step input. Determine the numerical increments needed in the simulations. Which system requires the shorter numerical increment? How is this related to the answer in (b)?

9.14. Repeat Problem 9.2 for the case that the compensator is PI.

9.15. (a) Design a PID compensator for the system of Problem 9.2 to yield a phase margin of 45°. Let $K_I = 0.04$.

(b) Verify the phase margin with MATLAB.

(c) Determine the approximate rise time and percent overshoot for the system step response by simulation.

9.16. (a) Design a PID compensator for the system of Problem 9.3 to yield a phase margin of 50°. Let $K_I = 0.04$.

(b) Verify the phase margin with MATLAB.

(c) Determine the approximate rise time and percent overshoot for the system step response by simulation.

(d) To see the effects of the integrator term, repeat (a) and (c) for
   (i) K_I equal to one-half the value used in (a)
   (ii) K_I equal to twice the value used in (a)
(e) It is seen that reducing K_I decreases the settling time. Why is the integrator term included?

9.17. Given the compensator transfer function $G_c(s) = K_{dc}(1+s/\omega_0)/(1+s/\omega_p)$, where K_{dc} is the compensator dc gain, ω_0 is the compensator zero, and ω_p is the pole, show that the step change on the compensator output, for a unit step input, is equal to $K_{dc}\omega_p/\omega_0$.

9.18. Given the compensator transfer function

$$G_c(s) = \frac{1.45s + 0.35}{s + 0.07}$$

(a) Find the dc gain of the compensator.
(b) Find the high-frequency gain of the compesator.
(c) Is the compensator phase-lead or phase-lag. Why?

9.19. Given the compensator transfer function

$$G_c(s) = \frac{0.5s + 0.005}{s + 0.02}$$

(a) Find the dc gain of the compensator.
(b) Find the high-frequency gain of the compensator.
(c) Is the compensator phase-lead or phase-lag. Why?

9.20. Given the phase-lag compensator

$$G_p(s) = \frac{(0.294s + 0.0706)}{(s + 0.0706)}$$

(a) Find a PI compensator such that both the high-frequency gain and the zero are the same as for the given phase-lag compensator.
(b) Compare the magnitude Bode diagrams of the two compensators.
(c) Give the dc gain of each of the two compensators.
(d) What is the advantage at high frequencies of the PI compensator, compared to the phase-lag compensator?
(e) What is the advantage at low frequencies of the PI compensator, compared to the phase-lag compensator?

9.21. In Example 9.7, a system with open-loop transfer function

$$G_p(s)H(s) = \frac{4}{s(s+1)(s+2)}$$

is considered. A PD compensator, $G_c(s) = 1.10 + 1.12s$, is designed to yield a phase margin of $50°$ at the frequency of 1.7 rad/s. Find a phase-lead compensator of the form of (9-28) to yield the same dc gain and the same zero as those of the PD compensator. In addition, if the PD compensator is replaced with the phase-lead compensator, the resulting system is to have a phase margin of at least $47°$.

9.22. For the phase-lead and phase-lag compensators, prove that $\omega_m = \sqrt{\omega_0\omega_p}$, where ω_m is the frequency at which the maximum phase shift occurs [see (9-24)].

9.23. A compensator has the transfer function $G(s) = K_d(1 + s/\omega_0)/(1 + s/\omega_p)$.
(a) Show that K_d is the compensator dc gain.
(b) Show that the step change on the compensator output, for a unit step input, is equal to $K_d\omega_0/\omega_p$.
(c) Consider the general transfer function

$$G(s) = \frac{C(s)}{E(s)} = \frac{b_n s^n + b_{n-1}s^{n-1} + \cdots + b_0}{s^n + a_{n-1}s^{n-1} + \cdots + a_0}$$

with the unit step input $[E(s) = 1/s]$. Show that the output will step with the value of the step equal to b_n. Hence a system will not have a step in the output for a step in the input if the order of the transfer-function numerator is less than that of the denominator, that is, if $b_n = 0$.

10

Modern Control Design

The design techniques presented in the preceding chapters are based on either frequency response or the root locus. These techniques are generally denoted as *classical*, or *traditional*, *methods*. The frequency-response techniques are very good methods of practical design, and most control systems are designed using variations of these methods. An important property of these methods, called *robustness* [1], is that the resultant closed-loop system characteristics tend to be insensitive to small inaccuracies in the system model. This characteristic is very important because of the difficulty in finding an accurate linear model of a physical system and also because many systems have significant nonlinear operation.

In the past several years much effort has been expended in developing new design methods, which are called *modern control methods* to differentiate them from the classical methods. These methods appear to be much more dependent on having an accurate system model for the design process. However, the modern methods offer a more complete control of a system; that is, the design can meet a larger number of specifications. The phase-lead, phase-lag, and PID controllers have been successfully utilized in physical closed-loop systems to such a degree that their usefulness is unquestioned. The controllers designed by modern methods have yet to gain the same degree of acceptance by controls engineers in industrial applications; however, modern methods are gaining in acceptance.

In this chapter we present a modern control design method known as pole placement, or pole assignment. This method is similar to the root-locus design, in that poles in the closed-loop transfer function may be placed in desired locations. However, pole-placement design allows all poles of the closed-loop transfer function to be placed in desirable locations, whereas the root-locus design procedure presented in Chapter 7 allowed only the two dominant poles to be placed. The cost of placing all poles is the required measurement of many system variables. In many applications, all needed system variables cannot be measured because of cost or because of the lack of suitable transducers. In these cases, those system signals that cannot be measured must be estimated from the ones that are measured. This estimation of system variables is the second topic of this chapter.

10.1 POLE-PLACEMENT DESIGN

In this section we present the design procedure known as pole placement. The classical design procedures developed in the preceding chapters are based on the transfer function of the system; pole-placement design is based on the state model of the system. Since we consider only linear time-invariant analog systems, the state model of the plant can be expressed as (see Chapter 3)

$$\dot{\mathbf{x}}(t) = \mathbf{A}\mathbf{x}(t) + \mathbf{B}u(t)$$
$$y(t) = \mathbf{C}\mathbf{x}(t)$$

(10-1)

We limit the development to single-input, single-output systems; thus $u(t)$ and $y(t)$ are scalars. Initially, we will assume that the *control system* input, which we denote as $r(t)$, is zero; the case that $r(t)$ is not zero is considered in Section 10.7.

In general, in modern control design, the plant input $u(t)$ is made a function of the states, of the form

$$u(t) = f[\mathbf{x}(t)]$$

(10-2)

This equation is called the control rule or the *control law*. In pole-placement design, the control law is specified as a linear function of the states, in the form

$$u(t) = \mathbf{K}\mathbf{x}(t)$$

(10-3)

where $\mathbf{K}$ is a $1 \times n$ vector of constant gains. We will show that this control law allows all poles of the closed-loop system to be placed in any desirable locations. This rule can be expressed as

$$u(t) = -K_1 x_1(t) - K_2 x_2(t) - \cdots - K_n x_n(t)$$

(10-4)

We see then that the signal fed back to the plant input is a weighted sum of all of the states of the system. The design problem is the specification of the desired root locations of the system characteristic equation, and the calculation of the gains K_i to yield these desired root locations.

The closed-loop system can be represented as shown in Figure 10.1. Since the closed-loop system has the input $r(t)$ equal to zero, we consider the purpose of this system to be to maintain the system output $y(t)$ at zero. A practical system will have disturbance inputs (not shown in Figure 10.1) that will tend to perturb the output of the system from zero; the feedback will tend to return the output (and all states) to zero in some desired manner. A system of this type (input equal to zero) is called a *regulator control system*. A linear system with a constant input can also be considered to be a regulator control system (see Problem 10.1).

We first introduce pole-placement design by an example; we then develop the general design procedure. Root-locus design was introduced in Section 7.6 with an example of a second-order plant (a satellite) with position and rate feedback. If the states of the plant are chosen as position and velocity, the feedback signal in this system is a weighted sum of the

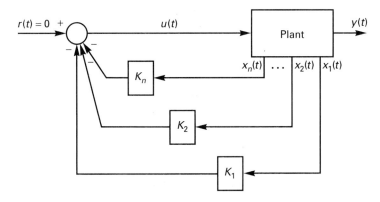

Figure 10.1   Pole-placement design.

states of the system in the form of (10-4); hence we were using pole-placement design for this system without identifying it as such. We now consider this same design from the pole-placement viewpoint.

Since the transfer function of the rigid satellite of Section 7.6 is $1/s^2$, the simulation diagram of the plant, with the states chosen as position and velocity, is as shown in Figure 10.2(a). The state equations are then

$$\dot{\mathbf{x}}(t) = \begin{bmatrix} 0 & 1 \\ 0 & 0 \end{bmatrix} \mathbf{x}(t) + \begin{bmatrix} 0 \\ 1 \end{bmatrix} u(t) \tag{10-5}$$

From (10-4) and as shown in Figure 10.2(b), the plant input signal is chosen to be

$$u(t) = -K_1 x_1(t) - K_2 x_2(t) = \begin{bmatrix} -K_1 & -K_2 \end{bmatrix} \mathbf{x}(t) = -\mathbf{K}\mathbf{x}(t) \tag{10-6}$$

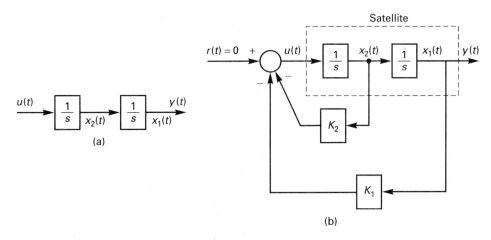

Figure 10.2   Simulation diagram for the satellite control system.

The state equations for the closed-loop system then become

$$\dot{\mathbf{x}}(t) = \mathbf{A}\mathbf{x}(t) + \mathbf{B}u(t)\big|_{u=-\mathbf{K}\mathbf{x}} = \mathbf{A}\mathbf{x}(t) - \mathbf{B}\mathbf{K}\mathbf{x}(t)$$

$$= \begin{bmatrix} 0 & 1 \\ 0 & 0 \end{bmatrix} \mathbf{x}(t) - \begin{bmatrix} 0 \\ 1 \end{bmatrix} \begin{bmatrix} K_1 & K_2 \end{bmatrix} \mathbf{x}(t) \tag{10-7}$$

$$= \begin{bmatrix} 0 & 1 \\ 0 & 0 \end{bmatrix} \mathbf{x}(t) - \begin{bmatrix} 0 & 0 \\ K_1 & K_2 \end{bmatrix} \mathbf{x}(t) = \begin{bmatrix} 0 & 1 \\ -K_1 & -K_2 \end{bmatrix} \mathbf{x}(t)$$

Hence the state equations of the closed-loop system can be expressed as

$$\dot{\mathbf{x}}(t) = \begin{bmatrix} 0 & 1 \\ -K_1 & -K_2 \end{bmatrix} \mathbf{x}(t) = \mathbf{A}_f \mathbf{x}(t) \tag{10-8}$$

where $\mathbf{A}_f$ is the system matrix of the closed-loop (feedback) system. The characteristic equation of the closed-loop system is, from (5-14),

$$|s\mathbf{I} - \mathbf{A}_f| = \begin{vmatrix} s & -1 \\ K_1 & s + K_2 \end{vmatrix} = s^2 + K_2 s + K_1 = 0 \tag{10-9}$$

Suppose that the design specifications for this system require that the two roots of the characteristic equation be placed at $-\lambda_1$ and $-\lambda_2$. Denoting the desired characteristic equation as $\alpha_c(s)$, we have

$$\alpha_c s = (s + \lambda_1)(s + \lambda_2) = s^2 + (\lambda_1 + \lambda_2)s + \lambda_1\lambda_2 = 0 \tag{10-10}$$

The design is completed by choosing the gains K_1 and K_2 such that the coefficients of (10-9) are equal to those of (10-10).

$$K_1 = \lambda_1\lambda_2$$
$$K_2 = \lambda_1 + \lambda_2 \tag{10-11}$$

For this example we see that the proper choice of the feedback gains allows the roots of the characteristic equation (poles of the closed-loop transfer function) for the system model to be placed anywhere in the s-plane. Note also that if λ_1 and λ_2 are complex, λ_2 must be the complex conjugate of λ_1. Thus the gains K_1 and K_2 are always real.

We now give a general development. For a linear time-invariant system, the state equations of the plant are given by

$$\dot{\mathbf{x}}(t) = \mathbf{A}\mathbf{x}(t) + \mathbf{B}u(t) \tag{10-12}$$

The control law is chosen to be

$$u(t) = -\mathbf{K}\mathbf{x}(t) \tag{10-13}$$

with

$$\mathbf{K} = \begin{bmatrix} K_1 & K_2 & \cdots & K_n \end{bmatrix}$$

and n is the order of the plant. Substitution of (10-13) into (10-12) yields

$$\dot{\mathbf{x}}(t) = \mathbf{A}\mathbf{x}(t) - \mathbf{B}\mathbf{K}\mathbf{x}(t) = (\mathbf{A} - \mathbf{B}\mathbf{K})\mathbf{x}(t) = \mathbf{A}_f\mathbf{x}(t) \tag{10-14}$$

where $\mathbf{A}_f = (\mathbf{A} - \mathbf{BK})$ is the system matrix for the closed-loop system. The characteristic equation for the closed-loop system is then

$$|s\mathbf{I} - \mathbf{A}_f| = |s\mathbf{I} - \mathbf{A} + \mathbf{BK}| = 0 \qquad (10\text{-}15)$$

Suppose that the design specifications require that the zeros of the characteristic equation be at $-\lambda_1, -\lambda_2, \ldots, -\lambda_n$. The desired characteristic equation for the system, which is denoted by $\alpha_c(s)$, is

$$\alpha_c(s) = s^n + \alpha_{n-1}s^{n-1} + \cdots + \alpha_1 s + \alpha_0$$
$$= (s + \lambda_1)(s + \lambda_2)\cdots(s + \lambda_n) = 0 \qquad (10\text{-}16)$$

The pole-placement design procedure results in a gain vector $\mathbf{K}$ such that (10-15) is equal to (10-16), that is,

$$|s\mathbf{I} - \mathbf{A} + \mathbf{BK}| = \alpha_c(s) = s^n + \alpha_{n-1}s^{n-1} + \cdots + \alpha_1 s + \alpha_0 \qquad (10\text{-}17)$$

In this equation there are n unknowns $(K_1, K_2, \ldots, K_n)$. Equating coefficients in this equation yields n equations in the n unknowns. Even though it is not obvious, these equations are linear; thus the equations may be solved for the unknown gain vector $\mathbf{K}$. In addition, the elements of $\mathbf{K}$ are always real for models of physical systems.

This procedure was used in the last example; however, it is not suited to high-order systems. A procedure that can be implemented on a digital computer for the calculation of the gain matrix $\mathbf{K}$ to satisfy (10-17) is given next.

10.2 ACKERMANN'S FORMULA

Ackermann [2] developed a formula for calculating the gain vector $\mathbf{K}$ to satisfy (10-17); this formula is appropriately known as Ackermann's formula. Ackermann's formula is not developed here, but an outline of its derivation is given.

Consider the simulation diagram, in *control-canonical* form (see Section 3.2), for an nth-order plant in Figure 10.3. Each block containing an integral symbol represents an integrator, with the transfer function s^{-1}. This plant has the transfer function

$$G_p(s) = \frac{b_{n-1}s^{n-1} + \cdots + b_1 s + b_0}{s^n + a_{n-1}s^{n-1} + \cdots + a_1 s + a_0} \qquad (10\text{-}18)$$

The state equations for this system are given by

$$\dot{\mathbf{x}}(t) = \begin{bmatrix} 0 & 1 & 0 & \cdots & 0 \\ 0 & 0 & 1 & \cdots & 0 \\ \vdots & \vdots & \vdots & \vdots & \vdots \\ 0 & 0 & 0 & \cdots & 1 \\ -a_0 & -a_1 & -a_2 & \cdots & -a_{n-1} \end{bmatrix} \mathbf{x}(t) + \begin{bmatrix} 0 \\ 0 \\ \vdots \\ 0 \\ 1 \end{bmatrix} u(t)$$

$$y(t) = \begin{bmatrix} b_0 & b_1 & b_2 & \cdots & b_{n-1} \end{bmatrix} \mathbf{x}(t) \qquad (10\text{-}19)$$

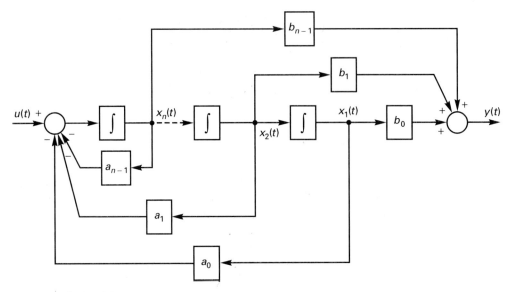

Figure 10.3 Control-canonical form.

Note the particular form of the system matrix **A**. All elements of this matrix are zero, except for the elements directly above the diagonal (which are unity) and the last row. The last row contains the negative of the coefficients of the system characteristic equation. If a system has a matrix **A** of this form, we can immediately write the system's characteristic equation.

The equation of the control law is, from (10-13),

$$u(t) = -\mathbf{K}\mathbf{x}(t) \tag{10-20}$$

Thus, in the closed-loop system matrix $(\mathbf{A} - \mathbf{BK})$, the term $\mathbf{BK}$ for the control-canonical form is seen to be

$$\mathbf{BK} = \begin{bmatrix} 0 \\ \vdots \\ 0 \\ 1 \end{bmatrix} \begin{bmatrix} K_1 & \cdots & K_{n-1} & K_n \end{bmatrix} = \begin{bmatrix} 0 & \cdots & 0 & 0 \\ \vdots & \vdots & \vdots & \vdots \\ 0 & \cdots & 0 & 0 \\ K_1 & \cdots & K_{n-1} & K_n \end{bmatrix}$$

The closed-loop system matrix $\mathbf{A}_f = \mathbf{A} - \mathbf{BK}$ is then

$$\mathbf{A}_f = \mathbf{A} - \mathbf{BK} = \begin{bmatrix} 0 & 1 & 0 & \cdots & 0 \\ 0 & 0 & 1 & \cdots & 0 \\ \vdots & \vdots & \vdots & \vdots & \vdots \\ 0 & 0 & 0 & \cdots & 1 \\ -a_0 - K_1 & -a_1 - K_2 & -a_2 - K_3 & \cdots & -a_{n-1} - K_n \end{bmatrix} \tag{10-21}$$

This matrix is of the standard form of the control-canonical form; thus the closed-loop system characteristic equation is

$$|s\mathbf{I} - \mathbf{A} + \mathbf{B}\mathbf{K}| = s^n + (a_{n-1} + K_n)s^{n-1} + \cdots + (a_1 + K_2)s + (a_0 + K_1) = 0$$

The desired closed-loop characteristic equation is given by, from (10-16),

$$\alpha_c(s) = s^n + \alpha_{n-1}s^{n-1} + \cdots + \alpha_1 s + \alpha_0 = 0$$

Equating coefficients for these two equations yields the general term

$$a_{i-1} + K_i = \alpha_{i-1}$$

Hence the gains are given by

$$K_i = \alpha_{i-1} - a_{i-1} \qquad i = 1, 2, \ldots, n \tag{10-22}$$

This equation is the general solution to the pole-placement design for single-input, single-output systems but requires that the system model be in the control canonical form. This requirement is difficult to satisfy, since the states of this form are generally not the natural states of a physical system and thus are not the states that would appear in a model that is derived from physical laws. Hence, the states of the control canonical form are not, in general, easily measured.

Ackermann's formula is based on a similarity transformation (see Section 3.5) that transforms a given state model into the control-canonical form, and the gains are then found using (10-22). Next the solutions for the feedback gains are transformed back so that they apply to the original system states. This derivation results in Ackermann's formula, which is given by

$$\mathbf{K} = \begin{bmatrix} 0 & 0 & \ldots & 0 & 1 \end{bmatrix} \begin{bmatrix} \mathbf{B} & \mathbf{A}\mathbf{B} & \ldots & \mathbf{A}^{n-2}\mathbf{B} & \mathbf{A}^{n-1}\mathbf{B} \end{bmatrix}^{-1} \alpha_c(\mathbf{A}) \tag{10-23}$$

where the notation $\alpha_c(\mathbf{A})$ represents the matrix polynomial formed with the coefficients of the desired characteristic equation $\alpha_c(s)$, that is,

$$\alpha_c(\mathbf{A}) = \mathbf{A}^n + \alpha_{n-1}\mathbf{A}^{n-1} + \ldots + \alpha_1\mathbf{A} + \alpha_0\mathbf{I} \tag{10-24}$$

The equation for $\mathbf{K}$, (10-23), may be evaluated by computer. We illustrate the use of this equation with an example.

Example 10.1

We consider the satellite design of Section 10.1. For this example, from (10-5), the plant model is given by

$$\dot{\mathbf{x}}(t) = \begin{bmatrix} 0 & 1 \\ 0 & 0 \end{bmatrix} \mathbf{x}(t) + \begin{bmatrix} 0 \\ 1 \end{bmatrix} u(t)$$

with the desired characteristic equation, from (10-10), given by

$$\alpha_c(s) = s^2 + (\lambda_1 + \lambda_2)s + \lambda_1\lambda_2 = 0$$

Ackermann's formula, (10-23), is now used to calculate the gain vector $\mathbf{K}$. First we evaluate $[\mathbf{B} \ \mathbf{A}\mathbf{B}]^{-1}$.

$$\mathbf{A}\mathbf{B} = \begin{bmatrix} 0 & 1 \\ 0 & 0 \end{bmatrix} \begin{bmatrix} 0 \\ 1 \end{bmatrix} = \begin{bmatrix} 1 \\ 0 \end{bmatrix}$$

Then

$$[\mathbf{B} \ \mathbf{AB}]^{-1} = \begin{bmatrix} 0 & 1 \\ 1 & 0 \end{bmatrix}^{-1} = \begin{bmatrix} 0 & 1 \\ 1 & 0 \end{bmatrix}$$

The evaluation of the matrix polynomial, (10-24), yields

$$\alpha_c(\mathbf{A}) = \mathbf{A}^2 + (\lambda_1 + \lambda_2)\mathbf{A} + \lambda_1\lambda_2\mathbf{I}$$

$$= \begin{bmatrix} 0 & 1 \\ 0 & 0 \end{bmatrix}\begin{bmatrix} 0 & 1 \\ 0 & 0 \end{bmatrix} + (\lambda_1 + \lambda_2)\begin{bmatrix} 0 & 1 \\ 0 & 0 \end{bmatrix} + \lambda_1\lambda_2\begin{bmatrix} 0 & 1 \\ 0 & 0 \end{bmatrix}$$

$$= \begin{bmatrix} 0 & 0 \\ 0 & 0 \end{bmatrix} + \begin{bmatrix} 0 & \lambda_1 + \lambda_2 \\ 0 & 0 \end{bmatrix} + \begin{bmatrix} \lambda_1\lambda_2 & 0 \\ 0 & \lambda_1\lambda_2 \end{bmatrix}$$

$$= \begin{bmatrix} \lambda_1\lambda_2 & \lambda_1 + \lambda_2 \\ 0 & \lambda_1\lambda_2 \end{bmatrix}$$

Thus, from (10-23), Ackermann's formula yields

$$\mathbf{K} = \begin{bmatrix} 0 & 1 \end{bmatrix}\begin{bmatrix} \mathbf{B} & \mathbf{AB} \end{bmatrix}^{-1}\alpha_c(\mathbf{A})$$

$$= \begin{bmatrix} 0 & 1 \end{bmatrix}\begin{bmatrix} 0 & 1 \\ 1 & 0 \end{bmatrix}\begin{bmatrix} \lambda_1\lambda_2 & \lambda_1 + \lambda_2 \\ 0 & \lambda_1\lambda_2 \end{bmatrix}$$

$$= \begin{bmatrix} 0 & 1 \end{bmatrix}\begin{bmatrix} 0 & \lambda_1\lambda_2 \\ \lambda_1\lambda_2 & \lambda_1 + \lambda_2 \end{bmatrix} = \begin{bmatrix} \lambda_1\lambda_2 & \lambda_1 + \lambda_2 \end{bmatrix}$$

Thus the required gain matrix is given by

$$\mathbf{K} = \begin{bmatrix} K_1 & K_2 \end{bmatrix} = \begin{bmatrix} \lambda_1\lambda_2 & \lambda_1 + \lambda_2 \end{bmatrix}$$

The solution is seen to be the same as that calculated in Section 10.1.

The next example gives results from using the design in Example 10.1.

Example 10.2

Figure 10.4 illustrates a practical implementation of the satellite control system designed above. As an example, suppose that the design specifications require a critically damped system (see Section 4.2) with a settling time of 1 s, that is, $4\tau = 1$ s. The required time constant is then $\tau = 0.25$ s, and the required two pole locations are at $s = -4$; thus $\lambda_1 = \lambda_2 = 4$. The desired characteristic equation is

$$\alpha_c(s) = (s + 4)^2 = s^2 + 8s + 16$$

From Example 10.1, the required gains are

$$K_1 = \lambda_1\lambda_2 = 16 \qquad K_2 = \lambda_1 + \lambda_2 = 8$$

The flow graph of the closed-loop system is shown in Figure 10.5, and the closed-loop transfer function, for the input shown, is seen to be

$$T(s) = \frac{\Theta(s)}{R(s)} = \frac{s^{-2}}{1 + 8s^{-1} + 16s^{-2}} = \frac{1}{s^2 + 8s + 16}$$

which has the required poles.

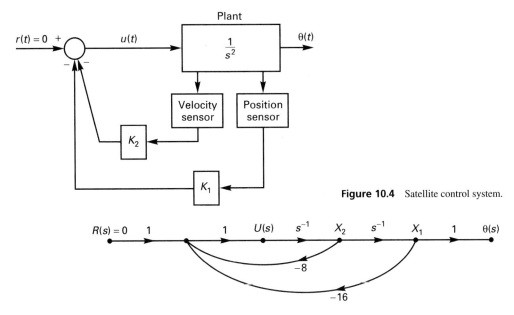

Figure 10.4   Satellite control system.

Figure 10.5   Satellite control system for Example 10.2.

The initial-condition response for the system is plotted in Figure 10.6, labeled with $\zeta = 1$, for the initial conditions $\mathbf{x}^T(0) = [1\ 0]$. The critically damped nature of the response is seen in the figure.

Suppose that, as a different design, the time constant used earlier, $\tau = 0.25$ s, is still required, but in addition we want $\zeta = 0.707$. Then the root locations needed are at $s = -4 \pm j4$, and

$$K_1 = \lambda_1 \lambda_2 = (4 + j4)(4 - j4) = 32$$
$$K_2 = \lambda_1 + \lambda_2 = 4 + j4 + 4 - j4 = 8$$

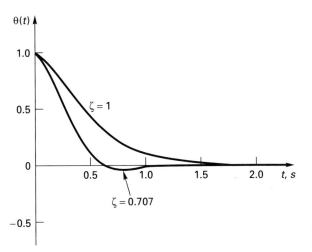

Figure 10.6   Initial-condition response for Example 10.2.

The closed-loop system characteristic equation is now

$$(s + 4 + j4)(s + 4 - j4) = s^2 + 8s + 32$$

The initial condition response of this system is also shown in Figure 10.6 for $\mathbf{x}^T(0) = [1 \ 0]$.

Note the differences in the two responses. Even though the time constants of the poles of the two systems are the same, the response of the system with $\zeta = 0.707$ decays much faster and with very little overshoot. From this example we can see one of the reasons for the popularity of choosing $\zeta = 0.707$ for the dominant poles of a system. A MATLAB program for performing the design and calculating the initial-condition response of Figure 10.6 for $\zeta = 0.707$ is given by

```
A = [0 1;0 0];  B = [0;1];  C = [1 0]; D = 0;
Pp=[-4+4*i -4-4*i];
K = acker(A, B, Pp), pause
e10p2=ss((A-B*K),B,C,D);
initial (e10p2, [1;0])
```

In this program, Pp is the vector of desired control-system pole locations and the statement *acker* calculates the gain vector K. The final two statements calculate the system initial-condition response.

It appears from the preceding example that we can choose the magnitude of the real part of the roots arbitrarily large, making the system response arbitrarily fast. For the system *model*, we can do this. However, as the time constant of the system becomes smaller, the gains increase. *This is true, in general, since to increase the rate at which a plant responds, the input signal to the plant must become larger.* As the magnitudes of the signals in a system increase, the likelihood of the system entering nonlinear regions of operation increases; for very large signals this nonlinear operation will occur for almost any physical system. Hence the linear model that is used in design no longer accurately models the physical system, and the linear model does not indicate the nature of the physical system response. Nevertheless, the mathematical model will respond as calculated, which is of little value to an engineer.

The discussion of the preceding paragraph can be applied directly to the satellite example. For the satellite to respond quickly, a large torque must be applied by the thrustors. However, physical thrustors can generate only a limited torque. If the control system calls for a torque larger than the maximum value possible, the thrustors will *saturate*, or *limit*, at the maximum value. This saturation is a nonlinear effect, and some of the effects of nonlinearities such as saturation are discussed in Chapter 14.

10.3 STATE ESTIMATION

The preceding sections introduced the modern design procedure of pole assignment. The control law for this design procedure is given by

$$u(t) = -\mathbf{K}\mathbf{x}(t) \tag{10-25}$$

which requires that all states of the plant be measured. The term *full state feedback* is used to denote systems in which all states are fed back. For a first- or second-order system, full state feedback is not unreasonable. Many second-order systems designed by classical methods use both position and rate feedback; hence these systems are using full state feedback. However, many other systems cannot be accurately modeled as a first- or second-order system. For most of these systems, full state measurements are not practical; to implement pole-assignment design in these systems, the states that cannot be measured must be estimated from the measurements that can be made on the system. The estimated states are then used in (10-25). The estimation of the states of a linear time-invariant system is the topic of this and the following two sections.

Suppose that we are given the single-input, single-output analog system described by the equations

$$\dot{\mathbf{x}}(t) = \mathbf{A}\mathbf{x}(t) + \mathbf{B}u(t)$$
$$y(t) = \mathbf{C}\mathbf{x}(t) \tag{10-26}$$

We wish to estimate the states of this system, $\mathbf{x}(t)$, with the estimate $\hat{\mathbf{x}}(t)$. We will use all available information in the estimation process; hence we will use the input function $u(t)$, the measurement $y(t)$, and the system matrices **A**, **B**, and **C**. We do not know $\mathbf{x}(t)$ at any time, and in particular we do not know the initial conditions $\mathbf{x}(0)$.

A block diagram of the estimation process is given in Figure 10.7. The *state estimator*, which is also called an *observer*, is given dynamics of the same type as those of the system. The estimator equations can then be expressed as

$$\dot{\hat{\mathbf{x}}}(t) = \mathbf{F}\hat{\mathbf{x}}(t) + \mathbf{H}u(t) + \mathbf{G}y(t) \tag{10-27}$$

We wish to choose the matrices **F, G,** and **H** in a manner such that $\hat{\mathbf{x}}(t)$ is an accurate estimate of $\mathbf{x}(t)$. Then, in the control system, the estimated states $\hat{\mathbf{x}}(t)$ are used to generate the feedback signal; that is, the feedback signal is $u(t) = -\mathbf{K}\hat{\mathbf{x}}(t)$.

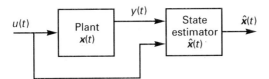

Figure 10.7   State estimation.

The equations for calculating the matrices **F, G,** and **H** may be developed in different ways; we use a transfer function approach here. The specification that will be used for the state estimator is the reasonable criterion that the transfer function from the input $u(t)$ to the state estimate $\hat{x}_i(t)$ must be equal to the transfer function from $u(t)$ to the state $x_i(t)$ for every state; that is,

$$\frac{\hat{X}_i(s)}{U(s)} = \frac{X_i(s)}{U(s)} \qquad i = 1, 2, ..., n \tag{10-28}$$

We now develop equations for **F, G,** and **H** to satisfy (10-28). The Laplace transform of the state equations (10-26) yields

$$s\mathbf{X}(s) - \mathbf{x}(0) = \mathbf{A}\mathbf{X}(s) + \mathbf{B}U(s)$$
$$Y(s) = \mathbf{C}\mathbf{x}(s)$$

as shown in the derivation of (3-18). Ignoring the initial conditions $\mathbf{x}(0)$, we solve for $\mathbf{X}(s)$.

$$\mathbf{X}(s) = (s\mathbf{I} - \mathbf{A})^{-1}\mathbf{B}U(s) \tag{10-29}$$

Note that $(s\mathbf{I} - \mathbf{A})^{-1}\mathbf{B}$ is a *transfer function matrix*, or *transfer matrix*, that is, a matrix of transfer functions. We want the transfer function matrix from $U(s)$ to $\hat{\mathbf{X}}(s)$ to be the same as that given in (10-29).

The Laplace transform of the estimator equations in (10-27) yields (ignoring the initial conditions),

$$s\hat{\mathbf{X}}(s) = \mathbf{F}\hat{\mathbf{X}}(s) + \mathbf{H}U(s) + \mathbf{G}Y(s) = \mathbf{F}\hat{\mathbf{X}}(s) + \mathbf{H}U(s) + \mathbf{G}\mathbf{C}\mathbf{X}(s) \qquad (10\text{-}30)$$

since $Y(s) = \mathbf{C}\mathbf{X}(s)$. Solving (10-30) for the estimated states $\hat{\mathbf{X}}(s)$ yields

$$\hat{\mathbf{X}}(s) = (s\mathbf{I} - \mathbf{F})^{-1}[\mathbf{H}U(s) + \mathbf{G}\mathbf{C}\mathbf{X}(s)]$$
$$= (s\mathbf{I} - \mathbf{F})^{-1}[\mathbf{H} + \mathbf{G}\mathbf{C}(s\mathbf{I} - \mathbf{A})^{-1}\mathbf{B}]\, U(s) \qquad (10\text{-}31)$$

using (10-29).

We want the transfer function matrix in (10-31) to be equal to that in (10-29).

$$(s\mathbf{I} - \mathbf{A})^{-1}\mathbf{B} = (s\mathbf{I} - \mathbf{F})^{-1}[\mathbf{H} + \mathbf{G}\mathbf{C}(s\mathbf{I} - \mathbf{A})^{-1}\mathbf{B}] \qquad (10\text{-}32)$$

Collecting coefficients of the term $(s\mathbf{I} - \mathbf{A})^{-1}\mathbf{B}$ yields

$$[\mathbf{I} - (s\mathbf{I} - \mathbf{F})^{-1}\mathbf{G}\mathbf{C}]\,(s\mathbf{I} - \mathbf{A})^{-1}\mathbf{B} = (s\mathbf{I} - \mathbf{F})^{-1}\mathbf{H}$$

Next $(s\mathbf{I} - \mathbf{F})^{-1}$ is factored from the left side of this equation.

$$(s\mathbf{I} - \mathbf{F})^{-1}[s\mathbf{I} - \mathbf{F} - \mathbf{G}\mathbf{C}]\,(s\mathbf{I} - \mathbf{A})^{-1}\mathbf{B} = (s\mathbf{I} - \mathbf{F})^{-1}\mathbf{H}$$

This equation is satisfied if the postmultiplying matrices of $(s\mathbf{I} - \mathbf{F})^{-1}$ are chosen to be equal.

$$[s\mathbf{I} - \mathbf{F} - \mathbf{G}\mathbf{C}]\,(s\mathbf{I} - \mathbf{A})^{-1}\mathbf{B} = \mathbf{H}$$

or

$$(s\mathbf{I} - \mathbf{A})^{-1}\mathbf{B} = (s\mathbf{I} - \mathbf{F} - \mathbf{G}\mathbf{C})^{-1}\mathbf{H}$$

This equation is satisfied if we choose $\mathbf{H} = \mathbf{B}$ and

$$\mathbf{F} + \mathbf{G}\mathbf{C} = \mathbf{A}$$

Hence, the matrices of the state estimator are given by

$$\mathbf{F} = \mathbf{A} - \mathbf{G}\mathbf{C}$$
$$\mathbf{H} = \mathbf{B} \qquad (10\text{-}33)$$

Note that the two transfer function matrices of (10-32) are equal if $\mathbf{F}$ and $\mathbf{H}$ are chosen as stated, with no requirement on $\mathbf{G}$. Hence we may choose $\mathbf{G}$ to satisfy other constraints; $\mathbf{G}$ is usually chosen to give an acceptable transient response or an acceptable frequency response for the state estimator.

We have derived the state estimator equations to be, from (10-27) and (10-33),

$$\dot{\hat{\mathbf{x}}} = (\mathbf{A} - \mathbf{G}\mathbf{C})\hat{\mathbf{x}}(t) + \mathbf{B}u(t) + \mathbf{G}y(t) \qquad (10\text{-}34)$$

where $\mathbf{G}$ is to be determined. It is of value to consider the errors in the estimation process. We define the error vector $\mathbf{e}(t)$ to be

$$\mathbf{e}(t) = \mathbf{x}(t) - \hat{\mathbf{x}}(t) \qquad (10\text{-}35)$$

or

$$\begin{bmatrix} e_1(t) \\ e_2(t) \\ \vdots \\ e_n(t) \end{bmatrix} = \begin{bmatrix} x_1(t) \\ x_2(t) \\ \vdots \\ x_n(t) \end{bmatrix} - \begin{bmatrix} \hat{x}_1(t) \\ \hat{x}_2(t) \\ \vdots \\ \hat{x}_n(t) \end{bmatrix}$$

The derivative of the error vector in (10-35) can be expressed as

$$\dot{\mathbf{e}}(t) = \dot{\mathbf{x}}(t) - \dot{\hat{\mathbf{x}}}(t)$$

Substituting (10-26) and (10-34) into this equation yields

$$\dot{\mathbf{e}}(t) = \mathbf{A}\mathbf{x}(t) + \mathbf{B}u(t) - (\mathbf{A} - \mathbf{GC})\hat{\mathbf{x}}(t) - \mathbf{B}u(t) - \mathbf{G}y(t) \tag{10-36}$$

Since $y(t) = \mathbf{C}\mathbf{x}(t)$, (10-37) can be expressed as

$$\dot{\mathbf{e}}(t) = \mathbf{A}\mathbf{x}(t) - (\mathbf{A} - \mathbf{GC})\hat{\mathbf{x}}(t) - \mathbf{GC}\mathbf{x}(t) = (\mathbf{A} - \mathbf{GC})\,[\mathbf{x}(t) - \hat{\mathbf{x}}(t)]$$

or

$$\dot{\mathbf{e}}(t) = (\mathbf{A} - \mathbf{GC})\mathbf{e}(t) \tag{10-37}$$

We see from this equation that the errors in the estimation of the states have the same dynamics as the state estimator, since the characteristic equation of the differential equation of (10-37) is the same as that of the state estimator of (10-34), which is

$$|s\mathbf{I} - \mathbf{A} + \mathbf{GC}| = 0 \tag{10-38}$$

The gain vector $\mathbf{G}$ is normally chosen to make the dynamics of the estimator faster than those of the system; a rule of thumb often stated is to make the estimator two to four times faster than the system.

The preceding derivation is mathematically correct but has many flaws with respect to the application to physical systems. According to the derivation, if we choose the estimator to be stable (which we will certainly do), the errors in estimation will decrease to zero with the dynamics as given in (10-38). Then we will have essentially perfect estimation. There must be flaws here, since we cannot do anything perfectly with respect to physical systems. Some of the flaws are as follows:

1. The system model was assumed to be exact. Hence, in the derivation of the error equation of (10-37), the matrices $\mathbf{A}$, $\mathbf{B}$, and $\mathbf{C}$ of the estimator were assumed to be equal to the matrices $\mathbf{A}$, $\mathbf{B}$, and $\mathbf{C}$ of the physical system, respectively. Since this cannot be true, the error equation of (10-37) is much more complex than shown.

2. Disturbances on the physical system have been ignored. A state model of a physical system that is more accurate than (10-26) is

$$\dot{\mathbf{x}}(t) = \mathbf{A}\mathbf{x}(t) + \mathbf{B}u(t) + \mathbf{B}_w\mathbf{w}(t)$$
$$y(t) = \mathbf{C}\mathbf{x}(t) + v(t) \tag{10-39}$$

In these equations $\mathbf{w}(t)$ is a vector of the disturbance inputs, and $v(t)$ represents the sensor errors. If this model is substituted into (10-36) instead of the model of (10-26), the error equation becomes

$$\dot{\mathbf{e}}(t) = (\mathbf{A} - \mathbf{GC})\mathbf{e}(t) + \mathbf{B}_w\mathbf{w}(t) - \mathbf{G}v(t) \tag{10-40}$$

We see then that the errors will not die out with increasing time, even if the modeling inaccuracies are ignored.

It should be noted that the use of estimation in physical systems can be dangerous. With estimators we are basing the control of the system not on measurements of system variables but on results of calculations that use the measurements. If the basis of those calculations is not well founded, the true values of the states can go in one direction and the results of the calculations can go in a different direction, that is, the errors of estimation can diverge. When we use a sensor output as the feedback signal, this signal is reasonably accurate except in the case of sensor failure. If we use estimation to generate the feedback signal, great care must be exercised to ensure that the effects of the estimator are well understood for all possible conditions of system operation.

10.3.1 Estimator Design

We now consider the design of state estimators. The equation of the estimator is given by

[eq. (10-34)] $\dot{\hat{\mathbf{x}}} = (\mathbf{A} - \mathbf{GC})\hat{\mathbf{x}}(t) + \mathbf{B}u(t) + \mathbf{G}y(t)$

The characteristic equation of the estimator is then

$$|s\mathbf{I} - \mathbf{A} + \mathbf{GC}| = 0 \tag{10-41}$$

As mentioned earlier, one method of designing state estimators is to make the estimator two to four times faster than the closed-loop system. Hence we choose a characteristic equation for the estimator, denoted by $\alpha_e(s)$, that reflects the desired speed of response:

$$\alpha_e(s) = s^n + \alpha_{n-1}s^{n-1} + \cdots + \alpha_1 s + \alpha_0 = 0 \tag{10-42}$$

Then the gain matrix $\mathbf{G}$ is calculated to satisfy

$$|s\mathbf{I} - \mathbf{A} + \mathbf{GC}| = \alpha_e(s) \tag{10-43}$$

Note the similarity of this design problem with that of pole placement, in (10-17). In fact, (10-43) can be transformed into an equation of the exact form of (10-17) (see Problem 10.5). Then Ackermann's formula for pole placement, (10-23), can be applied to the design of state estimators, with the resulting formula

$$\mathbf{G} = \alpha_e(\mathbf{A})\begin{bmatrix} \mathbf{C} \\ \mathbf{CA} \\ \vdots \\ \mathbf{CA}^{n-1} \end{bmatrix}^{-1}\begin{bmatrix} 0 \\ 0 \\ \vdots \\ 1 \end{bmatrix} \tag{10-44}$$

Given the desired estimator characteristic polynomial $\alpha_e(s)$ and the matrices $\mathbf{A}$ and $\mathbf{C}$, the gain matrix $\mathbf{G}$ can be calculated from (10-44). An example is used to illustrate this design.

Example 10.3

We will consider the satellite control system design of Example 10.2. For this case, the state equations of the plant are

$$\dot{\mathbf{x}}(t) = \begin{bmatrix} 0 & 1 \\ 0 & 0 \end{bmatrix} \mathbf{x}(t) + \begin{bmatrix} 0 \\ 1 \end{bmatrix} u(t)$$

$$y(t) = \begin{bmatrix} 1 & 0 \end{bmatrix} \mathbf{x}(t)$$

In Example 10.2, a controller was designed for the characteristic equation

$$\alpha_c(s) = s^2 + 8s + 32$$

which yielded a time constant of $\tau = 0.25$ s with $\zeta = 0.707$. We now choose the estimator to be critically damped with a time constant of $\tau = 0.1$ s. Hence

$$\alpha_e(s) = (s + 10)^2 = s^2 + 20s + 100$$

Then, in (10-44),

$$\alpha_e(\mathbf{A}) = \begin{bmatrix} 0 & 1 \\ 0 & 0 \end{bmatrix} \begin{bmatrix} 0 & 1 \\ 0 & 0 \end{bmatrix} + 20 \begin{bmatrix} 0 & 1 \\ 0 & 0 \end{bmatrix} + 100 \begin{bmatrix} 1 & 0 \\ 0 & 1 \end{bmatrix}$$

$$= \begin{bmatrix} 0 & 0 \\ 0 & 0 \end{bmatrix} + \begin{bmatrix} 0 & 20 \\ 0 & 0 \end{bmatrix} + \begin{bmatrix} 100 & 0 \\ 0 & 100 \end{bmatrix} = \begin{bmatrix} 100 & 20 \\ 0 & 100 \end{bmatrix}$$

Also, from (10-44),

$$\mathbf{CA} = \begin{bmatrix} 1 & 0 \end{bmatrix} \begin{bmatrix} 0 & 1 \\ 0 & 0 \end{bmatrix} = \begin{bmatrix} 0 & 1 \end{bmatrix}$$

and thus

$$\begin{bmatrix} \mathbf{C} \\ \mathbf{CA} \end{bmatrix}^{-1} = \begin{bmatrix} 1 & 0 \\ 0 & 1 \end{bmatrix}^{-1} = \begin{bmatrix} 1 & 0 \\ 0 & 1 \end{bmatrix}$$

Then

$$\mathbf{G} = \alpha_e(\mathbf{A}) \begin{bmatrix} \mathbf{C} \\ \mathbf{CA} \end{bmatrix}^{-1} \begin{bmatrix} 0 \\ 1 \end{bmatrix} = \begin{bmatrix} 100 & 20 \\ 0 & 100 \end{bmatrix} \begin{bmatrix} 1 & 0 \\ 0 & 1 \end{bmatrix} \begin{bmatrix} 0 \\ 1 \end{bmatrix}$$

$$= \begin{bmatrix} 100 & 20 \\ 0 & 100 \end{bmatrix} \begin{bmatrix} 0 \\ 1 \end{bmatrix} = \begin{bmatrix} 20 \\ 100 \end{bmatrix}$$

The design of the state estimator is now complete. This design is also performed by the MAT-LAB program

```
A=[0 1;0 0]; B=[0; 1]; C=[1 0]; D=0;
Pe = [-10 -10];
Gt = acker (A',C',Pe); G = Gt'
```

This program uses the *acker* statement and transposed matrices to calculate the estimator gains, since

$$(\mathbf{A} - \mathbf{BK})^T = \mathbf{A}^T - \mathbf{K}^T \mathbf{B}^T \Rightarrow \mathbf{A} - \mathbf{GC} \tag{10-45}$$

Hence, in *acker*, $\mathbf{A}^T$ replaces $\mathbf{A}$, $\mathbf{C}^T$ replaces $\mathbf{B}$, and the result is $\mathbf{G}^T$.

Example 10.4

In this example we consider the implementation of the control law and the state estimator designed in Examples 10.2 and 10.3, respectively. In Example 10.2 the gain matrix required to place the poles at $s = -4 \pm j4$ was calculated to be

$$\mathbf{K} = \begin{bmatrix} 32 & 8 \end{bmatrix}$$

and the state-estimator gain matrix was calculated in Example 10.3 to be

$$\mathbf{G} = \begin{bmatrix} 20 \\ 100 \end{bmatrix}$$

The estimator equations are, from (10-34),

$$\dot{\hat{\mathbf{x}}}(t) = (\mathbf{A} - \mathbf{GC})\hat{\mathbf{x}}(t) + \mathbf{B}u(t) + \mathbf{G}y(t) = (\mathbf{A} - \mathbf{GC} - \mathbf{BK})\hat{\mathbf{x}}(t) + \mathbf{G}y(t)$$

since $u(t) = -\mathbf{K}\hat{\mathbf{x}}(t)$. In the preceding equation,

$$\mathbf{GC} = \begin{bmatrix} 20 \\ 100 \end{bmatrix} \begin{bmatrix} 1 & 0 \end{bmatrix} = \begin{bmatrix} 20 & 0 \\ 100 & 0 \end{bmatrix}$$

and

$$\mathbf{BK} = \begin{bmatrix} 0 \\ 1 \end{bmatrix} \begin{bmatrix} 32 & 8 \end{bmatrix} = \begin{bmatrix} 0 & 0 \\ 32 & 8 \end{bmatrix}$$

Thus

$$\mathbf{A} - \mathbf{GC} - \mathbf{BK} = \begin{bmatrix} 0 & 1 \\ 0 & 0 \end{bmatrix} - \begin{bmatrix} 20 & 0 \\ 100 & 0 \end{bmatrix} - \begin{bmatrix} 0 & 0 \\ 32 & 8 \end{bmatrix}$$

$$= \begin{bmatrix} -20 & 1 \\ -132 & -8 \end{bmatrix}$$

The controller-estimator is then described by the equations

$$\dot{\hat{\mathbf{x}}}(t) = \begin{bmatrix} -20 & 1 \\ -132 & -8 \end{bmatrix} \hat{\mathbf{x}}(t) + \begin{bmatrix} 20 \\ 100 \end{bmatrix} y(t)$$

$$u(t) = \begin{bmatrix} -32 & -8 \end{bmatrix} \mathbf{x}(t)$$

In these equations, $y(t)$ is the *input* and $u(t)$ is the *output* of the estimator. A MATLAB program that calculates this state model is given by

```
A=[0 1;0 0]; B=[0;1]; C=[1 0]; D=0;
e10p4=ss(A,B,C,D);
Pp=[-4+4i -4-4*i];
K=acker(A,B,Pp)
Pe=[-10 -10];
Gt=acker(A',C',Pe);G=Gt'
rsys=reg(e10p5,K,G),pause
Gec=-tf(rsys), Gp=tf(e10p4), Gol=Gec*Gp, pause
[gm,pm,wg,wp]=margin(Gol)
```

This program combines the two earlier programs and uses the statement *reg* (for regulator) to calculate the controller-estimator system matrices. The last two statements are discussed below.

10.4 CLOSED-LOOP SYSTEM CHARACTERISTICS

In general, since we cannot measure all of the states of a system, it is necessary to employ a state estimator for pole-placement design. The design process has two steps—first we design the feedback gain matrix **K** to yield the *closed-loop* system characteristic equation $\alpha_c(s)$. Next we design the state estimator to yield the estimator characteristic equation $\alpha_e(s)$. In the implementation we place the estimator within the closed loop, as shown in Figure 10.8, and thus far we have not considered the effect on the closed-loop characteristic equation of adding the estimator. One effect is obvious—since the plant is nth order and the estimator is also nth order, the closed-loop characteristic equation is now of order $2n$. We now investigate the locations of these $2n$ roots.

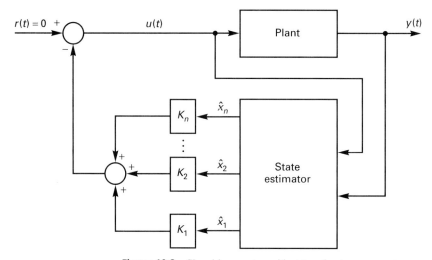

Figure 10.8   Closed-loop system with state estimator.

The characteristic equation of a system with full state feedback is, from (10-15),

$$\alpha_c(s) = |s\mathbf{I} - \mathbf{A} + \mathbf{BK}| = 0$$

and the characteristic equation of a state estimator is, from (10-41),

$$\alpha_e(s) = |s\mathbf{I} - \mathbf{A} + \mathbf{GC}| = 0 \qquad (10\text{-}46)$$

As shown in the example above, the controller-estimator state equations are

$$\dot{\hat{\mathbf{x}}}(t) = (\mathbf{A} - \mathbf{GC} - \mathbf{BK})\,\hat{\mathbf{x}}(t) + \mathbf{G}y(t)$$
$$u(t) = -\mathbf{K}\hat{\mathbf{x}}(t)$$

where $y(t)$ is the input and $u(t)$ is the output. Since these equations are in the standard state-equations form, we can calculate a transfer function, $-G_{ec}(s)$, for the controller-estimator using (3-42).

$$-G_{ec}(s) = -\mathbf{K}\,[s\mathbf{I} - \mathbf{A} + \mathbf{GC} + \mathbf{BK}]^{-1}\mathbf{G}$$

The reason for the negative sign on the left side is given below. For the last example, this equation yields the controller-estimator transfer function

$$G_{ec}(s) = \frac{1440s + 3200}{s^2 + 28s + 292}$$

This calculation is verified by the last next-to-last statement in the MATLAB program given in Example 10.4. The preceding development shows that the controller-estimator can be considered as a single transfer function of $G_{ec}(s)$, where

$$G_{ec}(s) = \mathbf{K}[s\mathbf{I} - \mathbf{A} + \mathbf{GC} + \mathbf{BK}]^{-1}\mathbf{G} \qquad (10\text{-}47)$$

This configuration is illustrated in Figure 10.9 in which a summing junction has been added to account for the negative sign. The system appears to be the standard single-loop unity feedback system covered in the preceding chapters. However, in the preceding example, the controller block is second order; in general it will be nth-order for an nth-order plant. The characteristic equation of the closed-loop system can be expressed as

$$1 + G_{ec}(s)G_p(s) = 0 \qquad (10\text{-}48)$$

Hence, a Bode diagram of the system can be plotted by plotting $G_{ec}(j\omega)G_p(j\omega)$. This plotting is done in the last statement in the MATLAB program in Example 10.4, and the phase margin was found to be 37.7°, which in some designs would be considered to be low. The next section considers the closed-loop system in greater detail.

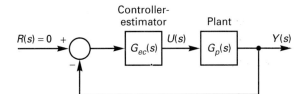

Figure 10.9 System with controller-estimator.

We now derive the characteristic equation of the closed-loop system of Figure 10.8. First consider the error variables of (10-35),

$$\mathbf{e}(t) = \mathbf{x}(t) - \hat{\mathbf{x}}(t) \qquad (10\text{-}49)$$

The plant equations of (10-4) are

$$\dot{\mathbf{x}}(t) = \mathbf{A}\mathbf{x}(t) + \mathbf{B}u(t)$$
$$y(t) = \mathbf{C}\mathbf{x}(t) \qquad (10\text{-}50)$$

with the control law $u(t) = -\mathbf{K}\hat{\mathbf{x}}(t)$. Then, from (10-49) and (10-50),

$$\dot{\mathbf{x}}(t) = \mathbf{A}\mathbf{x}(t) - \mathbf{B}\mathbf{K}\hat{\mathbf{x}}(t) = \mathbf{A}\mathbf{x}(t) - \mathbf{B}\mathbf{K}[\mathbf{x}(t) - \mathbf{e}(t)]$$
$$= (\mathbf{A} - \mathbf{B}\mathbf{K})\mathbf{x}(t) + \mathbf{B}\mathbf{K}\mathbf{e}(t) \qquad (10\text{-}51)$$

The state equations for the error variables are, from (10-37),

$$\dot{\mathbf{e}}(t) = (\mathbf{A} - \mathbf{GC})\mathbf{e}(t) \qquad (10\text{-}52)$$

If we adjoin the state variables of the plant with the error variables, from (10-51) and (10-52) we obtain

$$\begin{bmatrix} \dot{\mathbf{x}}(t) \\ \dot{\mathbf{e}}(t) \end{bmatrix} = \begin{bmatrix} \mathbf{A} - \mathbf{BK} & \mathbf{BK} \\ \mathbf{0} & \mathbf{A} - \mathbf{GC} \end{bmatrix} \begin{bmatrix} \mathbf{x}(t) \\ \mathbf{e}(t) \end{bmatrix} \tag{10-53}$$

We can consider the variables $[\mathbf{x}^T(t) \ \mathbf{e}^T(t)]^T$ to be the result of a similarity transformation of the variables $[\mathbf{x}^T(t) \ \hat{\mathbf{x}}^T(t)]^T$, since the error variables are a linear combination of the state and estimator variables (see Section 3.5). Hence the characteristic equation of the differential equations (10-53) is also the characteristic equation of the closed-loop system of Figure 10.8. From Appendix A, we see that the characteristic equation of (10-53) is given by

$$|s\mathbf{I} - \mathbf{A} + \mathbf{BK}||s\mathbf{I} - \mathbf{A} + \mathbf{GC}| = 0 \tag{10-54}$$

The $2n$ roots of the closed-loop characteristic equations are then the n roots of the pole-placement design plus the n roots of the estimator. This is fortunate; otherwise the roots from the pole-placement design would have been shifted by the addition of the estimator. Since the dynamics of the estimator are normally chosen to be faster than those of the control system with full state feedback, the pole-placement roots will tend to dominate.

The closed-loop state equations, with the states chosen as the system states plus the estimates of those states, will now be derived. From (10-1) and (10-45),

$$\dot{\mathbf{x}}(t) = \mathbf{A}\mathbf{x}(t) + \mathbf{B}\mathbf{u}(t) = \mathbf{A}\mathbf{x}(t) - \mathbf{BK}\hat{\mathbf{x}}(t)$$

$$\dot{\hat{\mathbf{x}}}(t) = (\mathbf{A} - \mathbf{GC} - \mathbf{BK})\hat{\mathbf{x}}(t) + \mathbf{G}y(t) = (\mathbf{A} - \mathbf{GC} - \mathbf{BK})\hat{\mathbf{x}}(t) + \mathbf{GC}\mathbf{x}(t)$$

The closed-loop state equations are then given by

$$\begin{bmatrix} \dot{\mathbf{x}}(t) \\ \dot{\hat{\mathbf{x}}}(t) \end{bmatrix} = \begin{bmatrix} \mathbf{A} & -\mathbf{BK} \\ \mathbf{GC} & \mathbf{A} - \mathbf{BK} - \mathbf{GC} \end{bmatrix} \begin{bmatrix} \mathbf{x}(t) \\ \hat{\mathbf{x}}(t) \end{bmatrix}$$

Hence the system matrix $\mathbf{A}_f$ for the closed-loop system is

$$\mathbf{A}_f = \begin{bmatrix} \mathbf{A} & -\mathbf{BK} \\ \mathbf{GC} & \mathbf{A} - \mathbf{BK} - \mathbf{GC} \end{bmatrix} \tag{10-55}$$

The next two examples illustrate closed-loop characteristics.

Example 10.5

In this example we investigate the time response of the closed-loop system of Example 10.4. In that example the plant model is given by

$$\dot{\mathbf{x}}(t) = \begin{bmatrix} 0 & 1 \\ 0 & 0 \end{bmatrix} \mathbf{x}(t) + \begin{bmatrix} 0 \\ 1 \end{bmatrix} u(t)$$

$$y(t) = \begin{bmatrix} 1 & 0 \end{bmatrix} \mathbf{x}(t)$$

and the controller-estimator model is given by

$$\dot{\hat{\mathbf{x}}}(t) = \begin{bmatrix} -20 & 1 \\ -132 & -8 \end{bmatrix} \hat{\mathbf{x}}(t) + \begin{bmatrix} 20 \\ 100 \end{bmatrix} y(t)$$

$$u(t) = \begin{bmatrix} -32 & -8 \end{bmatrix} \hat{\mathbf{x}}(t)$$

We can construct a flow graph of the system modeled after the block diagram in Figure 10.9; this flow graph is shown in Figure 10.10. We note that the system is fourth order, as is evident from the characteristic equation of (10-47). If we define $x_3(t) = \hat{x}_1(t)$ and $x_4(t) = \hat{x}_2(t)$, we can write a set of state equations for the closed-loop control system from Figure 10.10.

$$
\begin{bmatrix} \dot{x}_1(t) \\ \dot{x}_2(t) \\ \dot{x}_3(t) \\ \dot{x}_4(t) \end{bmatrix} = \begin{bmatrix} 0 & 1 & 0 & 0 \\ 0 & 0 & -32 & -8 \\ 20 & 0 & -20 & 1 \\ 100 & 0 & -132 & -8 \end{bmatrix} \begin{bmatrix} x_1(t) \\ x_2(t) \\ x_3(t) \\ x_4(t) \end{bmatrix} + \begin{bmatrix} 0 \\ 0 \\ 20 \\ 100 \end{bmatrix} r(t)
$$

$$
\theta(t) = \begin{bmatrix} 1 & 0 & 0 & 0 \end{bmatrix} \begin{bmatrix} x_1(t) & x_2(t) & x_3(t) & x_4(t) \end{bmatrix}^T
$$

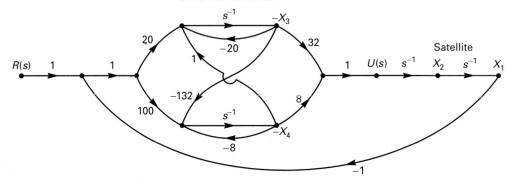

Figure 10.10   System with controller-estimator.

This calculation was verified and the system simulated with the MATLAB program

```
A=[0 1;0 0]; B=[0;1]; C=[1 0]; D=0;
K=[32 8]; G=[20;100];
Af=[A -B*K;G*C A-B*K-G*C]
e10p5=ss(Af,[B;0;0],[C 0 0],0); pause
initial(e10p5,[1 0 1 0],1.5), hold
initial(e10p5,[1 0 0 0],1.5), hold off
eig(Af)
```

The simulation results are given in Figure 10.11. In the first case, the initial conditions chosen were

$$
\mathbf{x}^T(0) = \begin{bmatrix} 1 & 0 & 1 & 0 \end{bmatrix}
$$

that is, the initial states of the estimator were chosen to be equal to the initial states of the plant. This response is labeled as $\mathbf{x}(0) = \hat{\mathbf{x}}(0)$ in the figure and is the same as the initial-condition response of the plant with full state feedback, as given in Figure 10.6. Since the plant model is known exactly and the initial state-estimation error is zero, the error equation (10-38) shows that this error will remain at zero. In the second case, which is somewhat more realistic, the initial conditions are chosen as

$$
\mathbf{x}^T(0) = \begin{bmatrix} 1 & 0 & 0 & 0 \end{bmatrix}
$$

that is, the plant initial conditions and the estimator initial conditions are different. The response for this case is denoted by $\mathbf{x}(0) \neq \hat{\mathbf{x}}(0)$ in Figure 10.11. The effects of the dynamics of the estimator are seen in this case; in addition, the 37.7° phase margin mentioned after Example 10.4 is evident from the large overshoot.

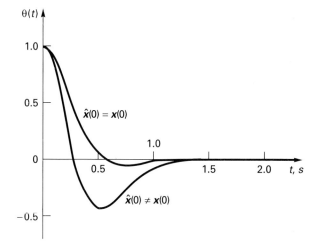

Figure 10.11 Initial-condition responses for Example 10.5.

Even though the design in the preceding examples seems reasonable, the phase margin is smaller than usual. In fact, it has been shown in controller-estimator designs such as this, the phase margin can become vanishingly small, even though the design specifications are reasonable [3]. Hence it is necessary to *always* check the design by calculating the phase margin of the closed-loop system, as just demonstrated.

Example 10.6

To illustrate the characteristic equation of (10-54), we consider the system of Example 10.5. For this system, the characteristic equation is

$$\alpha_c(s)\alpha_e(s) = (s^2 + 8s + 32)(s^2 + 20s + 100)$$

$$= s^4 + 28s^3 + 292s^2 + 1440s + 3200 = 0$$

We can also calculate the characteristic equation from (10-48), where the system is represented as shown in Figure 10.9:

$$1 + G_{ec}(s)G_p(s) = 1 + \left[\frac{1440s + 3200}{s^2 + 28s + 292}\right]\left[\frac{1}{s^2}\right] = 0$$

Clearing this equation yields

$$s^4 + 28s^3 + 292s^2 + 1440s + 3200 = 0$$

which is the same characteristic equation as just calculated. Another method of calculating the characteristic equation is to form the determinant

$$|s\mathbf{I} - \mathbf{A}_f| = 0$$

where $\mathbf{A}_f$ is the system matrix of the closed-loop system. This matrix was calculated in Example 10.5, and the determinant can be expressed as

$$|s\mathbf{I} - \mathbf{A}_f| = \begin{bmatrix} s & -1 & 0 & 0 \\ 0 & s & 32 & 8 \\ -20 & 0 & s+20 & -1 \\ -100 & 0 & 132 & s+8 \end{bmatrix}$$

The evaluation of this determinant yields the characteristic equation given earlier. The eigenvalues of $\mathbf{A}_f$ are calculated by the last statement in the MATLAB program given in Example 10.5.

10.5 REDUCED-ORDER ESTIMATORS

The estimator developed in the last section might be called a full-order estimator, since all states of the plant are estimated. However, usually the measurements available are states of the plant; for example, for the satellite considered in the sections above, the measurement is position, which is $x_1(t)$. We expect that the measurement of a state in general will be more accurate than any estimate of the state based on that measurement. Hence it is not logical in most cases to estimate states that we are measuring. One possible exception is the case in which a measurement is very noisy. The state estimator for this case may furnish some beneficial noise filtering; an examination of the frequency response of the estimator from the measurement input to that state estimate will indicate the degree of noise filtering present.

Since usually we will not want to estimate any state that we are measuring, we prefer to design an estimator that estimates only those states that are not measured. This type of estimator is called a *reduced-order estimator*. We develop design equations for such an estimator in this section. We consider only the case of one measurement; hence we are measuring only one state. It is assumed that the state variables are always chosen such that the state measured is $x_1(t)$; we can do this without loss of generality. The output equation then is always

$$y(t) = x_1(t) = \mathbf{C}\mathbf{x}(t) = \begin{bmatrix} 1 & 0 & \cdots & 0 \end{bmatrix} \mathbf{x}(t) \tag{10-56}$$

First we partition the state vector as

$$\mathbf{x}(t) = \begin{bmatrix} x_1(t) \\ \mathbf{x}_e(t) \end{bmatrix}$$

where $\mathbf{x}_e(t)$ are the states to be estimated. Next we write the partitioned state equations.

$$\begin{bmatrix} \dot{x}_1(t) \\ \dot{\mathbf{x}}_e(t) \end{bmatrix} = \begin{bmatrix} a_{11} & \mathbf{A}_{1e} \\ \mathbf{A}_{e1} & \mathbf{A}_{ee} \end{bmatrix} \begin{bmatrix} x_1(t) \\ \mathbf{x}_e(t) \end{bmatrix} + \begin{bmatrix} b_1 \\ \mathbf{B}_e \end{bmatrix} u(t) \tag{10-57}$$

From this equation we write the equation of the states to be estimated,

$$\dot{\mathbf{x}}_e(t) = \mathbf{A}_{e1}x_1(t) + \mathbf{A}_{ee}\mathbf{x}_e(t) + \mathbf{B}_e u(t) \tag{10-58}$$

and the equation of the state that is measured.

$$\dot{x}_1(t) = a_{11}x_1(t) + \mathbf{A}_{1e}\mathbf{x}_e(t) + b_1 u(t) \tag{10-59}$$

In these equations, $\mathbf{x}_e(t)$ is unknown and to be estimated, and $x_1(t)$ and $u(t)$ are known.

To derive the equations of the estimator, we manipulate (10-58) and (10-59) into the same form as the plant equations for full state estimation. Since we know the design equations for full state estimation, we then have the design equations for the reduced-order estimator. For full state estimation,

$$\dot{\mathbf{x}}(t) = \mathbf{A}\mathbf{x}(t) + \mathbf{B}u(t) \tag{10-60}$$

and for the reduced-order estimator, from (10-58),

$$\dot{\mathbf{x}}_e(t) = \mathbf{A}_{ee}\mathbf{x}_e(t) + [\mathbf{A}_{e1}x_1(t) + \mathbf{B}_e u(t)] \tag{10-61}$$

For full state estimation,

$$y(t) = \mathbf{C}\mathbf{x}(t) \tag{10-62}$$

and for the reduced-order estimator, from (10-59),

$$[\dot{x}_1(t) - a_{11}x_1(t) - b_1 u(t)] = \mathbf{A}_{1e}\mathbf{x}_e(t) \tag{10-63}$$

Comparing (10-60) with (10-61) and (10-62) with (10-63), we see that the equations are equivalent if we make the following substitutions.

$$\mathbf{x}(t) \leftarrow \mathbf{x}_e(t)$$
$$\mathbf{A} \leftarrow \mathbf{A}_{ee}$$
$$\mathbf{B}u(t) \leftarrow \mathbf{A}_{e1}x_1(t) + \mathbf{B}_e u(t) \tag{10-64}$$
$$y(t) \leftarrow \dot{x}_1(t) - a_{11}x_1(t) - b_1 u(t)$$
$$\mathbf{C} \leftarrow \mathbf{A}_{1e}$$

Making these substitutions into the equation for the full-order estimator,

[eq. (10-34)] $\dot{\hat{\mathbf{x}}}(t) = (\mathbf{A} - \mathbf{G}\mathbf{C})\hat{\mathbf{x}}(t) + \mathbf{B}u(t) + \mathbf{G}y(t)$

we obtain the equations of the reduced-order estimator.

$$\dot{\hat{\mathbf{x}}}_e(t) = (\mathbf{A}_{ee} - \mathbf{G}_e\mathbf{A}_{1e})\hat{\mathbf{x}}_e(t) + \mathbf{A}_{e1}y(t) + \mathbf{B}_e u(t)$$
$$+ \mathbf{G}_e[\dot{y}(t) - a_{11}y(t) - b_1 u(t)] \tag{10-65}$$

where $x_1(t)$ has been replaced with $y(t)$ and $\mathbf{G}_e$ is the gain matrix of the reduced-order estimator. We see then from (10-38) that the error dynamics are given by

$$\dot{\mathbf{e}}(t) = \dot{\hat{\mathbf{x}}}_e(t) - \dot{\hat{\mathbf{x}}}_e(t) = (\mathbf{A}_{ee} - \mathbf{G}_e\mathbf{A}_{1e})\mathbf{e}(t) \tag{10-66}$$

Hence the characteristic equation of the estimator and of the errors of estimation are given by

$$\alpha_e(s) = |s\mathbf{I} - \mathbf{A}_{ee} + \mathbf{G}_e\mathbf{A}_{1e}| = 0 \tag{10-67}$$

We then choose $\mathbf{G}_e$ to satisfy this equation, where we have chosen $\alpha_e(s)$ to give the estimator certain desired dynamics.

Ackermann's formula for full-order estimation is, from (10-45),

$$\mathbf{G} = \alpha_e(\mathbf{A}) \begin{bmatrix} \mathbf{C} \\ \mathbf{CA} \\ \vdots \\ \mathbf{CA}^{n-1} \end{bmatrix}^{-1} \begin{bmatrix} 0 \\ 0 \\ \vdots \\ 1 \end{bmatrix} \tag{10-68}$$

Making the substitutions indicated above yields the formula for the reduced-order estimator,

$$
\mathbf{G}_e = \alpha_e(\mathbf{A}_{ee})
\begin{bmatrix}
\mathbf{A}_{1e} \\
\mathbf{A}_{1e}\mathbf{A}_{ee} \\
\vdots \\
\mathbf{A}_{1e}\mathbf{A}_{ee}^{n-2}
\end{bmatrix}
\begin{bmatrix}
0 \\
0 \\
\vdots \\
1
\end{bmatrix}
\tag{10-69}
$$

Note that the order of this estimator is one less than that of the full-order estimator.

The preceding equations may be used to design a reduced-order estimator, but there is one possible problem. The reduced-order estimator requires the derivative of $y(t)$ as an input, as seen from (10-65). Although we can and do build circuits that differentiate signals (direct implementations of the PD controller, for example), the differentiator will amplify any high-frequency noise in $y(t)$. However, we can eliminate this need for differentiation by a change of variables, such that the calculation of $\dot{y}(t)$ is not required. First define a new variable that we denote as $\hat{\mathbf{x}}_{e1}$.

$$
\hat{\mathbf{x}}_{e1}(t) = \hat{\mathbf{x}}_e(t) - \mathbf{G}_e y(t)
\tag{10-70}
$$

or

$$
\hat{\mathbf{x}}_e(t) = \hat{\mathbf{x}}_{e1}(t) - \mathbf{G}_e y(t)
\tag{10-71}
$$

Substituting this expression into (10-65) yields

$$
\dot{\hat{\mathbf{x}}}_{e1}(t) + \mathbf{G}_e \dot{y}(t) = (\mathbf{A}_{ee} - \mathbf{G}_e\mathbf{A}_{1e})\,[\hat{\mathbf{x}}_{e1}(t) + \mathbf{G}_e y(t)] + \mathbf{A}_{e1}y(t)
$$
$$
+ \mathbf{B}_e u(t) + \mathbf{G}_e\,[\dot{y}(t) - a_{11}y(t) - b_1 u(t)]
$$

which can be written as

$$
\dot{\hat{\mathbf{x}}}_{e1}(t) = (\mathbf{A}_{ee} - \mathbf{G}_e\mathbf{A}_{1e})\hat{\mathbf{x}}_{e1}(t)
$$
$$
+ (\mathbf{A}_{e1} - \mathbf{G}_e a_{11} + \mathbf{A}_{ee}\mathbf{G}_e - \mathbf{G}_e\mathbf{A}_{1e}\mathbf{G}_e)y(t) + (\mathbf{B}_e - \mathbf{G}_e b_1)\,u(t)
\tag{10-72}
$$

This differential equation is solved for the variables $\hat{\mathbf{x}}_{e1}(t)$, and the estimated variables are then given by, from (10-71),

$$
\hat{\mathbf{x}}_e(t) = \hat{\mathbf{x}}_{e1}(t) - \mathbf{G}_e y(t)
\tag{10-73}
$$

The control input to the plant is

$$
u(t) = -K_1 y(t) - \mathbf{K}_e\hat{\mathbf{x}}_e(t) = -K_1 y(t) - \mathbf{K}_e\,[\hat{\mathbf{x}}_{e1}(t) + \mathbf{G}_e y(t)]
\tag{10-74}
$$

where the gain matrix $\mathbf{K}$ is partitioned as

$$
\mathbf{K} = \begin{bmatrix} K_1 & \mathbf{K}_e \end{bmatrix}
\tag{10-75}
$$

and $\mathbf{K}_e$ is $1 \times (n-1)$. An implementation of the closed-loop control system is shown in Figure 10.12. Since we have introduced an additional variable $\hat{\mathbf{x}}_{e1}(t)$, this implementation is not the only one that can be used.

In summary, the equations of the reduced-order estimator are (10-72) for the differential equation, (10-73) for the estimated states, and (10-74) for the control signal. An example is now given to illustrate the design of a reduced-order estimator.

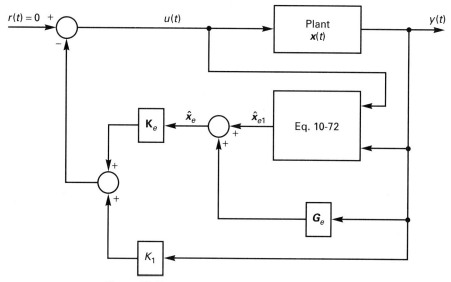

Figure 10.12   System with reduced-order state estimator.

Example 10.7

The satellite control system design of Example 10.6 is considered in this example. The equations of the plant are

$$\dot{\mathbf{x}}(t) = \begin{bmatrix} 0 & 1 \\ 0 & 0 \end{bmatrix} \mathbf{x}(t) + \begin{bmatrix} 0 \\ 1 \end{bmatrix} u(t)$$

$$y(t) = \begin{bmatrix} 1 & 0 \end{bmatrix} \mathbf{x}(t)$$

The controller designed for this system in Example 10.2 realized the characteristic equation

$$\alpha_c(s) = s^2 + 8s + 32$$

which resulted in a time constant of $\tau = 0.25$ s with $\zeta = 0.707$. As in Example 10.3, we design the estimator to have a time constant of $\tau = 0.1$ s.; hence the first-order estimator characteristic equation is

$$\alpha_e(s) = s + 10$$

Comparing the state equations of the satellite with those of the partitioned state matrices in (10-57) yields

$$\begin{array}{ccc} a_{11} = 0 & \mathbf{A}_{1e} = 1 & b_1 = 0 \\ \mathbf{A}_{e1} = 0 & \mathbf{A}_{ee} = 0 & \mathbf{B}_e = 1 \end{array}$$

From Ackermann's formula, (10-69),

$$\alpha_e(\mathbf{A}_{ee}) = 0 + 10 = 10$$

and

$$\mathbf{G}_e = \alpha_e(\mathbf{A}_{ee}) [\mathbf{A}_{1e}]^{-1} [1] = \frac{10}{1} = 10$$

Then, from (10-72),

$$\dot{\mathbf{x}}_{e1}(t) = (\mathbf{A}_{ee} - \mathbf{G}_e \mathbf{A}_{1e})\hat{\mathbf{x}}_{e1}(t)$$
$$+ (\mathbf{A}_{e1} - \mathbf{G}_e a_{11} + \mathbf{A}_{ee}\mathbf{G}_e - \mathbf{G}_e \mathbf{A}_{1e}\mathbf{G}_e)y(t) + (\mathbf{B}_e - \mathbf{G}_e b_1)u(t)$$

Thus, since this equation is first order,

$$\dot{\hat{x}}_{e1}(t) = [0 - (10)(1)]\hat{x}_{e1} + [0 - (10)(0) + (0)(10)$$
$$- (10)(1)(10)]y(t) + [1 - (10)(0)]u(t)$$

or

$$\dot{\hat{x}}_{e1}(t) = -10\hat{x}_{e1} - 100y(t) + u(t)$$

The estimated value of $x_2(t)$ is then, from (10-73),

$$\hat{x}_e(t) = \hat{x}_{e1}(t) + \mathbf{G}_e y(t) = \hat{x}_{e1} + 10y(t)$$

This completes the design of the reduced-order estimator in which the angular velocity of the satellite is estimated from the measurement of the angular position of the satellite. The design calculations for this example are performed by the MATLAB program

```
alphae = [1 10];
A = [0 1;0 0];   B = [0;1];
a11 = A(1,1); Ae1 = A(2,1);
A1e = A(1,2); Aee = A(2,2);
b1 = B(1); Be = B(2);
G = polyvalm(alphae,Aee)*inv(A1e)*1
```

Example 10.8

In this example we consider further the satellite-control system of Example 10.7. If we combine all plant equations with the estimator equations and the control equation

$$u(t) = -\mathbf{K}\big[y(t)\ \hat{x}_e(t)\big]^T = \big[-32\ -8\big]\big[y(t)\ \hat{x}_e(t)\big]^T$$
$$= -32y(t) - 8\hat{x}_e(t)$$

the flow graph of Figure 10.13 results. The characteristic equation of this flow graph, from Mason's gain formula, is given by

$$\Delta(s) = 1 + 10s^{-1} + 8s^{-1} + 32s^{-2} + 80s^{-2} - 800s^{-3}$$
$$+ (-10s^{-1})(-80s^{-2} - 32s^{-2}) = 1 + 18s^{-1} + 112s^{-2} + 320s^{-3} = 0$$

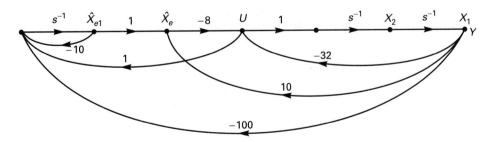

Figure 10.13 System for Example 10.8.

Multiplying this equation by s^3 and factoring yields

$$s^3 + 18s^2 + 112s + 320 = (s^2 + 8s + 32)(s + 10) = 0$$

which is the product of the characteristic equation for the control system, $\alpha_c(s)$, and the characteristic equation for the estimator, $\alpha_e(s)$.

We can also calculate the transfer function of the controller-estimator from the flow graph of Figure 10.13. We open the system in the unity gain branch following the node for $U(s)$ and then calculate the transfer function $-G_{ec}(s)$ from $Y(s)$ to $U(s)$.

$$\frac{U(s)}{Y(s)} = -G_{ec}(s) = \frac{-32(1 + 10s^{-1}) - 80(1 + 10s^{-1}) + 800s^{-1}}{1 + 10s^{-1} + 8s^{-1}}$$

$$= \frac{-112 - 320s^{-1}}{1 + 18s^{-1}} = \frac{-(112s + 320)}{s + 18}$$

Note that for this transfer function, $Y(s)$ is the input and $U(s)$ is the output. The system can then be represented as a unity feedback system, as shown in Figure 10.14. The characteristic equation for this system can then also be expressed as

$$1 + G_{ec}(s)G_p(s) = 1 + \frac{112s + 320}{s + 18}\frac{1}{s^2} = 0$$

or

$$s^3 + 18s^2 + 112s + 320 = 0$$

which checks the calculation using Mason's gain formula.

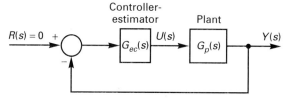

Figure 10.14 System for Example 10.8.

One final point is made in this example. The controller-estimator transfer function can be expressed in terms of its pole and zero:

$$G_{ec}(s) = \frac{112(s + 2.86)}{s + 18}$$

We see then that this design process results in a phase-lead compensator with a dc gain of 320/18, or 17.8. The phase-lead compensation is not surprising, since the Nyquist diagram of this plant shows that phase-lead compensation is *required* for stability. Hence any design procedure will result in a phase-lead compensator.

Example 10.9

As the final example in this section, the time response of the system designed in Example 10.7 is determined. In the flow graph of Figure 10.13, we denote $\hat{x}_{e1}(t)$ as $x_3(t)$. We can then write the state equations for the closed-loop system:

$$\dot{\mathbf{x}}(t) = \begin{bmatrix} 0 & 1 & 0 \\ -112 & 0 & -8 \\ -212 & 0 & -18 \end{bmatrix} \mathbf{x}(t)$$

These equations were integrated numerically, resulting in a simulation of the system. The initial condition $x_1(0)$ was chosen to be unity, and $x_2(0)$ was chosen to be zero. Since

$$x_3(t) = \hat{x}_{e1}(t) = \hat{x}_e(t) - 10y(t) = \hat{x}_2(t) - 10x_1(t)$$

the initial condition for $x_3(t)$ should be

$$x_3(0) = \hat{x}_2(0) - 10x_1(0) = -10$$

to give an initial estimation error of zero. Then, from (10-66), the error would remain at zero for all time. The simulation of this system with these initial conditions resulted in the same response as the full state feedback system of Example 10.2, shown in Figure 10.6, thus verifying the zero errors in estimation. Next, $x_3(0)$ was chosen to be zero rather than -10, and the initial condition response showed an overshoot of 30 percent in the negative direction. A calculation of the Nyquist diagram of the system as modeled in Figure 10.14 yielded a phase margin of 46°, indicating an improvement over the full-order estimation of Example 10.6 (38° phase margin).

10.6 CONTROLLABILITY AND OBSERVABILITY

In this section we consider certain problems that can occur in the design of control systems. In particular, we relate these problems to pole-placement design and to state-estimator design. The problems are evident in the two forms of Ackermann's formula in (10-23) and (10-45).

[eq. (10-23)] $\mathbf{K} = \begin{bmatrix} 0 & 0 & \cdots & 0 & 1 \end{bmatrix} \begin{bmatrix} \mathbf{B} & \mathbf{AB} & \cdots & \mathbf{A}^{n-1}\mathbf{B} \end{bmatrix}^{-1} \alpha_c(\mathbf{A})$

$$\mathbf{G} = \alpha_e(\mathbf{A}) \begin{bmatrix} \mathbf{C} \\ \mathbf{CA} \\ \vdots \\ \mathbf{CA}^{n-1} \end{bmatrix}^{-1} \begin{bmatrix} 0 \\ 0 \\ \vdots \\ 1 \end{bmatrix}$$

[eq. (10-45)]

We have not considered the possibility that either of the inverse matrices may not exist. If the inverse matrix of (10-23) does not exist, the plant is said to be *uncontrollable*, and all the poles of the closed-loop system cannot be placed. If the inverse matrix in (10-45) does not exist, the plant is said to be *unobservable*, and an estimator that estimates all of the states of the plant cannot be designed. We give better definitions of these two terms later.

First, however, we show that the same problems can occur in the classical design of control systems. Consider the closed-loop system of Figure 10.15. Note that the sensor has a pole coincident with a zero of the plant. The system characteristic equation can be expressed as

$$1 + G_c(s)G_p(s)H(s) = 0 \tag{10-76}$$

or

$$1 + \frac{N_c(s)}{D_c(s)} \frac{K(s+5)}{s(s+1)} \frac{5}{s+5} = 0 \tag{10-77}$$

To express the characteristic equation as a polynomial, we multiply by the denominator.

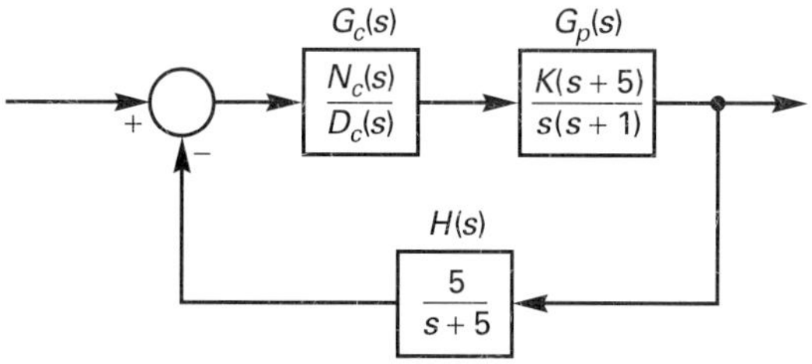

Figure 10.15 System with pole-zero cancellation.

$$D_c(s)\,[s(s+1)(s+5)] + N_c(s)\,[5K(s+5)] \;=\; 0$$

or

$$(s+5)\,[s(s+1)D_c(s) + 5KN_c(s)] \;=\; 0 \tag{10-78}$$

However, if the $(s+5)$ term is canceled before the multiplication by the denominator in (10-77), the resulting characteristic equation is the second factor in (10-78):

$$s(s+1)D_c(s) + 5KN_c(s) \;=\; 0 \tag{10-79}$$

Hence we have two different characteristic equations, (10-78) and (10-79), for this system. Which is correct? The answer is the one in (10-78). There will be a term in the transient response, e^{-5t} (we call this a *mode* of the system), that is excited by initial conditions and by some disturbances but *not* by the input shown in Figure 10.15. [This mode is shown in (10-78) but not in (10-79).] Thus this mode cannot be controlled by the input in Figure 10.15, and the compensator shown cannot be used to shift this characteristic-equation zero to a more desirable location; this property is independent of the method of designing the compensator. This particular system is considered further in an example.

We will now define the terms *controllable* and *observable*.

Controllable   The linear time-invariant system

$$\dot{\mathbf{x}}(t) \;=\; \mathbf{A}\mathbf{x}(t) + \mathbf{B}u(t)$$
$$y(t) \;=\; \mathbf{C}\mathbf{x}(t) \tag{10-80}$$

is said to be *controllable* if it is possible to find some input $u(t)$ that will transfer the initial state of the system $\mathbf{x}(0)$ to the origin of state space, $\mathbf{x}(t_0) = \mathbf{0}$, with t_0 finite.

The solution of the state equations (10-80) is, from (3-23),

$$\mathbf{x}(t) \;=\; \Phi(t)\mathbf{x}(0) + \int_0^t \Phi(t-\tau)\mathbf{B}u(\tau)\,d\tau \tag{10-81}$$

For the system of (10-80) to be controllable, a function $u(t)$ must exist that satisfies the equation

$$\mathbf{0} \;=\; \Phi(t_0)\mathbf{x}(0) + \int_0^{t_0} \Phi(t_0-\tau)\mathbf{B}u(\tau)\,d\tau \tag{10-82}$$

with t_0 finite. It can be shown this condition is satisfied if the matrix

$$\begin{bmatrix} \mathbf{B} & \mathbf{AB} & \cdots & \mathbf{A}^{n-2}\mathbf{B} & \mathbf{A}^{n-1}\mathbf{B} \end{bmatrix} \tag{10-83}$$

has an inverse [4].

Observable The linear time-invariant system (10-80) is said to be *observable* if the initial conditions $\mathbf{x}(0)$ can be determined from the output function $y(t), 0 \le t \le t_1$, where t_1 is finite.

From (10-80) and (10-81),

$$y(t) = \mathbf{C}\mathbf{x}(t) = \mathbf{C}\Phi(t)\mathbf{x}(0) + \mathbf{C}\int_0^t \Phi(t - \tau)\mathbf{B}u(\tau)d\tau \qquad (10\text{-}84)$$

Thus, given $u(t)$ and $y(t), 0 \le t \le t_1$ with t_1 some finite value, the system is observable if this equation can be solved for $\mathbf{x}(0)$. It can be shown that the system is observable if the matrix

$$\begin{bmatrix} \mathbf{C} \\ \mathbf{CA} \\ \vdots \\ \mathbf{CA}^{n-2} \\ \mathbf{CA}^{n-1} \end{bmatrix} \qquad (10\text{-}85)$$

has an inverse [4]. An example is given to illustrate these concepts.

Example 10.10

The system used as an example earlier in this section is considered again. From Figure 10.15, we can construct a flow graph of the plant and sensor, as shown in Figure 10.16. From this flow graph we write state equations for the plant and sensor:

$$\dot{\mathbf{x}}(t) = \begin{bmatrix} -1 & 1 & 0 \\ 0 & 0 & 0 \\ 5 & 0 & -5 \end{bmatrix}\mathbf{x}(t) + \begin{bmatrix} K \\ 5K \\ 0 \end{bmatrix}u(t)$$

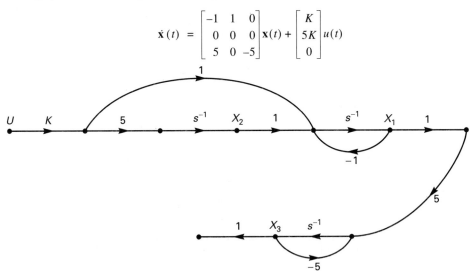

Figure 10.16 System for Example 10.10.

We first test the controllability:

$$\mathbf{AB} = \begin{bmatrix} -1 & 1 & 0 \\ 0 & 0 & 0 \\ 5 & 0 & -5 \end{bmatrix} \begin{bmatrix} K \\ 5K \\ 0 \end{bmatrix} = \begin{bmatrix} 4K \\ 0 \\ 5K \end{bmatrix}$$

and

$$\mathbf{A^2B} = \mathbf{AAB} = \begin{bmatrix} -1 & 1 & 0 \\ 0 & 0 & 0 \\ 5 & 0 & -5 \end{bmatrix} \begin{bmatrix} 4K \\ 0 \\ 5K \end{bmatrix} = \begin{bmatrix} -4K \\ 0 \\ -5K \end{bmatrix}$$

Thus

$$\begin{bmatrix} \mathbf{B} & \mathbf{AB} & \mathbf{A^2B} \end{bmatrix} = \begin{bmatrix} K & 4K & -4K \\ 5K & 0 & 0 \\ 0 & 5K & -5K \end{bmatrix}$$

Since the third column is equal to the second column multiplied by a constant (-1), the determinant of this matrix is zero, and the matrix does not have an inverse. Hence the system is uncontrollable, as was determined earlier.

We now investigate observability with respect to the sensor output, which is

$$x_3(t) = \begin{bmatrix} 0 & 0 & 1 \end{bmatrix} \mathbf{x}(t) = \mathbf{Cx}(t)$$

Then

$$\mathbf{CA} = \begin{bmatrix} 0 & 0 & 1 \end{bmatrix} \begin{bmatrix} -1 & 1 & 0 \\ 0 & 0 & 0 \\ 5 & 0 & -5 \end{bmatrix} = \begin{bmatrix} 5 & 0 & -5 \end{bmatrix}$$

and

$$\mathbf{CA^2} = \mathbf{CAA} = \begin{bmatrix} 5 & 0 & -5 \end{bmatrix} \begin{bmatrix} -1 & 1 & 0 \\ 0 & 0 & 0 \\ 5 & 0 & -5 \end{bmatrix} \begin{bmatrix} -30 & 5 & 25 \end{bmatrix}$$

Thus

$$\begin{bmatrix} \mathbf{C} \\ \mathbf{CA} \\ \mathbf{CA^2} \end{bmatrix} = \begin{bmatrix} 0 & 0 & 1 \\ 5 & 0 & -5 \\ -30 & 5 & 25 \end{bmatrix}$$

The determinant of this matrix is equal to 25, and the inverse does exist. This system is then observable. These calculations are verified by the MATLAB program

```
A=[-1 1 0; 0 0 0; 5 0 -5]; B=[1;5;0]; C=[0 0 1];
Co=ctrb(A,B)
det(Co), pause
Ob=obsv(A,C)
det(Ob)
```

The statement *ctrb* calculates the controllability matrix, and *obsv* the observability matrix.

In the preceding example, the system is uncontrollable but observable. Suppose that the transfer functions in Figure 10.15 are changed such that the pole-zero cancellation occurs between a zero of the sensor and a pole of the plant. Then the system will be controllable but unobservable (see Problem 10.15).

In most systems that are either uncontrollable or unobservable or both, pole-zero cancellation can occur. Hence the system can be modeled as a lower-order system when the cancellations are performed. A system in which the number of state variables is greater than the minimum order of the system has a nonminimal state model, and this state model will be uncontrollable, unobservable, or both. For example, consider the circuit of Figure 10.17. The loop equations for this circuit can be expressed as

$$Ri_1 + L\frac{di_1}{dt} + R_1(i_1 + i_2) = e$$

$$Ri_2 + L\frac{di_2}{dt} + R_1(i_1 + i_2) = e$$

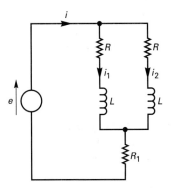

Figure 10.17   Uncontrollable circuit.

These equations can be written in the standard state format as

$$\frac{di_1}{dt} = -\left[\frac{R + R_1}{L}\right]i_1 - \frac{R_1}{L}i_2 + \frac{1}{L}e$$

$$\frac{di_2}{dt} = -\frac{R_1}{L}i_1 - \left[\frac{R + R_1}{L}\right]i_2 + \frac{1}{L}e$$

The matrices **A** and **B** of these state equations are given by

$$\mathbf{A} = \begin{bmatrix} -\dfrac{R + R_1}{L} & -\dfrac{R_1}{L} \\[3mm] -\dfrac{R_1}{L} & -\dfrac{R + R_1}{L} \end{bmatrix} \qquad \mathbf{B} = \begin{bmatrix} \dfrac{1}{L} \\[3mm] \dfrac{1}{L} \end{bmatrix}$$

Hence

$$\mathbf{AB} = \begin{bmatrix} -\dfrac{R + 2R_1}{L^2} \\[3mm] -\dfrac{R + 2R_1}{L^2} \end{bmatrix}$$

and the controllability matrix of (10-83) is given by

$$\begin{bmatrix} \mathbf{B} & \mathbf{AB} \end{bmatrix} = \begin{bmatrix} -\dfrac{1}{L} & -\dfrac{R+2R_1}{L^2} \\[2ex] -\dfrac{1}{L} & -\dfrac{R+2R_1}{L^2} \end{bmatrix}$$

The determinant of this matrix is obviously zero, and hence the system is uncontrollable.

Note that in the circuit just discussed, the two parallel RL circuits can be combined into one RL circuit with one resistance $R/2$ and one inductance $L/2$. Hence the transfer function from input voltage to the current $i(t)$ is

$$\frac{I(s)}{E(s)} = \frac{1}{(L/2)s + R/2 + R_1}$$

and the transfer function is first order, whereas the state model developed above is second order. This uncontrollability can also be seen from the circuit by noting that the control that $e(t)$ exerts of $i_1(t)$ is exactly the same as that over $i_2(t)$. If $e(t)$ forces a change in $i_1(t)$, exactly the same change occurs in $i_2(t)$. Hence the two states cannot be driven to a value of zero in the same finite time unless the two currents are initially equal, that is, unless $i_1(0) = i_2(0)$.

The preceding example illustrates that uncontrollability in physical systems can occur if certain types of symmetries occur. Some other reasons are if redundant states are defined, or if the model is inaccurate (the physical system is controllable but the model is not). An excellent discussion of controllability and observability in physical systems is given in [1].

It should be noted that the cancellation of poles and zeros can occur only in the model of a system, not in the system itself. A physical system has characteristics, and a model of that system has poles and zeros that describe those characteristics in some approximate sense. Hence, in the physical system, a characteristic of one part of a physical system may approximately negate a different characteristic of another part of the system. Generally exact cancellation of one characteristic by another will not occur. However, if the characteristics approximately cancel, it is very difficult to control and/or estimate these characteristics. Thus we can see the physical aspect of these problems, even though the definitions are strictly mathematical.

10.7 SYSTEMS WITH INPUTS

In the preceding sections only regulator control systems were considered. A regulator system has no input, and the purpose of the system is to return all state variables to values of zero when the states have been perturbed. However, many systems require that the system output track an input. For these cases, the design equations for the regulator systems must be modified; this modification is the topic of this section.

10.7.1 Full State Feedback

We first consider systems with full state feedback. For these systems, the plant is described by

$$\dot{\mathbf{x}}(t) = \mathbf{A}\mathbf{x}(t) + \mathbf{B}u(t)$$

with

$$u(t) = -\mathbf{Kx}(t) \tag{10-86}$$

The plant input $u(t)$ is the only function that we can modify; hence the input function $r(t)$ must be added to this function. Therefore, the equation for $u(t)$ becomes

$$u(t) = -\mathbf{Kx}(t) + K_r r(t) \tag{10-87}$$

Assuming that the gain matrix $\mathbf{K}$ has been chosen to yield a desired closed-loop system characteristic equation, the only remaining design parameter to be determined is the gain K_r.

The gain K_r can be determined to satisfy different design criteria. If we express (10-87) as

$$u(t) = -K_1 x_1(t) - K_2 x_2(t) - \cdots - K_n x_n(t) + K_r r(t) \tag{10-88}$$

we see that one method for choosing K_r is to combine the input $r(t)$ with a linear combination of the states. For example, suppose that state $x_1(t)$ is the system output that is to track the input $r(t)$. One logical choice for K_r is then the gain K_1, such that $u(t)$ becomes

$$u(t) = K_1 [r(t) - x_1(t)] - K_2 x_2(t) - \cdots - K_n x_n(t) \tag{10-89}$$

The block diagram for this system is shown in Figure 10.18, and the system is seen to be a unity feedback control system with the error signal $e(t)$ as shown. An example of this design is given next.

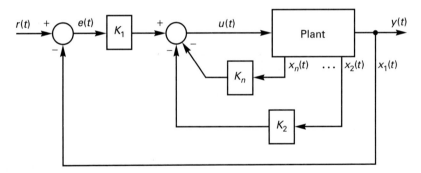

Figure 10.18   Pole-placement design with input.

Example 10.11

The system considered in this example is the satellite-control system designed in Example 10.2. The plant equations are

$$\dot{\mathbf{x}}(t) = \begin{bmatrix} 0 & 1 \\ 0 & 0 \end{bmatrix} \mathbf{x}(t) + \begin{bmatrix} 0 \\ 1 \end{bmatrix} u(t)$$

$$y(t) = \begin{bmatrix} 1 & 0 \end{bmatrix} \mathbf{x}(t)$$

with the gains

$$\mathbf{K} = \begin{bmatrix} 32 & 8 \end{bmatrix}$$

For the unity feedback system, the plant input equation becomes, from (10-89),

$$u(t) = 32\,[r(t) - x_1(t)] - 8x_2(t) = 32\,[r(t) - y(t)] - 8x_2(t)$$

and the system is as shown in Figure 10.19. The unit step response is given in Figure 10.20. This system is type 1 (see Section 5.5), and the steady-state error for a constant input is zero. The system step response is plotted by the MATLAB program

```
G=tf([0 0 32], [1 8 0]);
e10p11=minreal(G/(1+G));
step(e10p11)
```

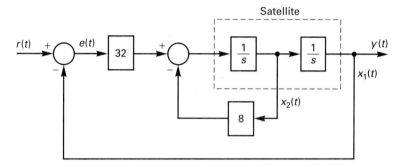

Figure 10.19   Satellite control system.

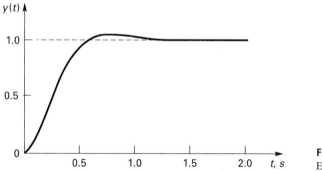

Figure 10.20   Time response for Example 10.11.

As an additional point concerning the unity feedback system, generally the gain of a sensor is not unity. If this gain of the sensor for $x_1(t)$ is H_{k1} (assuming no sensor dynamics), the system of Figure 10.18 can be modified to include this gain in the feedback path from output to input. Since the total gain of the feedback path for $x_1(t)$ must be K_1, the gain of the block labeled as K_1 in Figure 10.18 must be modified to have the gain K_1/H_{k1}. Of course, as a practical consideration, the gains of the other paths must be modified in the same manner to account for the sensor gains. For the case that the gain K_1 is changed to K_1/H_{k1}, the input function must then be scaled, since now the input and the output do not have the same units. Section 5.1 gives a discussion of the effects of the sensor gains.

This design procedure can be extended to the general case that the output that is to track the input is not the state $x_1(t)$ but instead is a linear combination of states, such that

$$y(t) = \mathbf{C}\mathbf{x}(t) = c_1 x_1(t) + c_2 x_2(t) + \cdots + c_n x_n(t) \tag{10-90}$$

We can then express $u(t)$ as

$$u(t) = -\mathbf{K}\mathbf{x}(t) + K_r r(t) = K_a [r(t) - y(t)] - \mathbf{K}_b \mathbf{x}(t) \tag{10-91}$$

since we want the system to be driven by the difference between the input and the output. Thus

$$u(t) = K_a r(t) - K_a \mathbf{C}\mathbf{x}(t) - \mathbf{K}_b \mathbf{x}(t) = K_a r(t) - [K_a \mathbf{C} + \mathbf{K}_b]\mathbf{x}(t) \tag{10-92}$$

Comparing (10-92) with (10-91) shows that $K_r = K_a$ and

$$K_a \mathbf{C} + \mathbf{K}_b = \mathbf{K}$$

This matrix equation can be expressed as

$$K_a c_1 + K_{1b} = K_1$$
$$K_a c_2 + K_{2b} = K_2$$
$$\vdots$$
$$K_a c_n + K_{nb} = K_n \tag{10-93}$$

where c_i are the elements of the known matrix $\mathbf{C}$, K_{ib} are elements of the unknown matrix $\mathbf{K}_b$, and K_i are elements of the known matrix $\mathbf{K}$. Equations (10-93) give n equations in the $(n+1)$ unknowns K_a, K_{1b}, K_{2b}, ..., K_{nb}. Hence one of the gains can be determined from other design criteria. Very often, the additional design criteria specified is that the steady-state difference between $r(t)$ and $y(t)$ should be small when $r(t)$ is constant.

10.7.2 Tracking Systems with Observers

For the case that a state estimator is required, there is more flexibility in the addition of an input to the system. The simplest case is illustrated in Figure 10.8, where the controller-estimator is considered to be a filter, and the input can be added as shown to produce a unity feedback system. As a first step, the controller-estimator equations are solved to give the filter transfer function, as shown in (10-46). Next the controller-estimator is realized as a filter, and finally the input is added as shown in Figure 10.8. An example of this case is given.

Example 10.12

We consider the satellite control system of Example 10.4. The plant transfer function is $1/s^2$, and the controller-estimator transfer function was found below Example 10.4 to be

$$G_{ec}(s) = \frac{1440s + 3200}{s^2 + 28s + 292}$$

The system, with input, is shown in Figure 10.21, and the design is complete. The unit-step response, with zero initial conditions, is given in Figure 10.22. The effects of the estimator are evident when Figures 10.20 and 10.22 are compared, in that the response of the system with the estimator has much more overshoot.

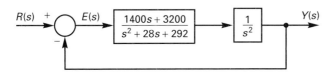

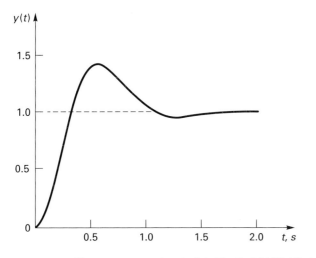

Figure 10.21   Satellite control system.

Figure 10.22   Time response for Example 10.12.

The step response is calculated by the MATLAB simulation

```
Gp=tf([1],[1 0 0]);
Gc=tf([1440 3200],[1 28 292]);
e10p11=minreal(Gc*Gp/(1+Gc*Gp));
step(e10p11,2)
```

We now consider the more general case of adding an input to regulator control systems with estimators. For the case that a full-state estimator is employed, the estimator equations are

[eq. (10-34)] $\dot{\hat{\mathbf{x}}} = (\mathbf{A} - \mathbf{GC})\hat{\mathbf{x}}(t) + \mathbf{B}u(t) + \mathbf{G}y(t)$

and the plant input is

$$u(t) = \mathbf{K}\hat{\mathbf{x}}(t) \tag{10-94}$$

We can add an input term to each of these equations, resulting in the general relationships.

$$\dot{\hat{\mathbf{x}}}(t) = (\mathbf{A} - \mathbf{GC})\hat{\mathbf{x}}(t) + \mathbf{B}u(t) + \mathbf{G}y(t) + \mathbf{M}r(t)$$
$$u(t) = -\mathbf{K}\hat{\mathbf{x}}(t) + Nr(t) \tag{10-95}$$

It is assumed that the gain matrices $\mathbf{K}$ and $\mathbf{G}$ have been calculated in the regulator design. Hence n gains in $\mathbf{M}$ and the gain N are to be determined by other considerations. Since there are $(n + 1)$ unknowns to be determined, there is much flexibility in this procedure. The last example is a design of this type, with M and N determined by the assumed form of unity feedback for the control system (see Problem 10.24).

10.7.3 Tracking with PI Controllers

An important specification for many control systems is that the system output be able to follow (track) a constant input with zero steady-state error. In classical design we accomplish this by adding a PI compensator for the case that the plant is type zero. We can use the same approach in pole placement design. Of course, the addition of the PI compensator will increase the order of the system by one; if the plant is of order n, the designed system is of order $n + 1$.

We will consider the designed system to be of the form of Figure 10.23. Note that the transfer function from $y(t)$ to $m(t)$ is

$$\frac{M(s)}{Y(s)} = -\left(K_1 + \frac{K_n + 1}{s} \right) = -G_c(s) \tag{10-96}$$

where $G_c(s)$ is the transfer function of a PI compensator. If desired, K_1 can be connected in the forward path in parallel with integrator and the gain K_{n+1}, without changing the system characteristic equation. This case will be considered further in the next example.

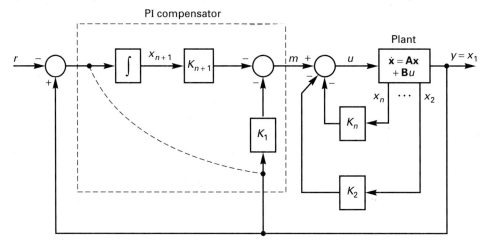

Figure 10.23   Pole placement with PI controller.

In Figure 10.23, the state equations for the plant are

$$\dot{\mathbf{x}}(t) = \mathbf{A}\mathbf{x}(t) + \mathbf{B}u(t)$$

$$y(t) = \mathbf{C}\mathbf{x}(t) = \begin{bmatrix} 1 & 0 & \cdots & 0 \end{bmatrix}\mathbf{x}(t) = x_1(t) \tag{10-97}$$

The equation for the input to the added integrator is

$$\dot{x}_{n+1}(t) = \mathbf{C}\mathbf{x}(t) - r(t) = x_1(t) - r(t) \tag{10-98}$$

The plant input is given by

$$u(t) = -\mathbf{K}\mathbf{x}(t) - K_{n+1}x_{n+1}(t) \tag{10-99}$$

where $\mathbf{K}$ and K_{n+1} are chosen to place the poles of the closed-loop transfer function in desirable locations.

The design procedure will now be developed. Substitution of (10-99) into (10-97) yields

$$\dot{\mathbf{x}}(t) = \mathbf{A}\mathbf{x}(t) - \mathbf{B}\mathbf{K}\mathbf{x}(t) - \mathbf{B}K_{n+1}x_{n+1}(t) \tag{10-100}$$

Next we adjoin $x_{n+1}(t)$ to $\mathbf{x}(t)$ to form the state vector of the feedback system $\mathbf{x}_a(t)$. Therefore, from (10-97), (10-98), and (10-100),

$$
\dot{\mathbf{x}}_a(t) = \begin{bmatrix} \dot{\mathbf{x}}(t) \\ \dot{x}_{n+1}(t) \end{bmatrix} = \begin{bmatrix} \mathbf{A} & \mathbf{0} \\ \mathbf{C} & 0 \end{bmatrix} \begin{bmatrix} \mathbf{x}(t) \\ x_{n+1}(t) \end{bmatrix}
$$
$$
- \begin{bmatrix} \mathbf{B}\mathbf{K} & \mathbf{B}K_{n+1} \\ 0 & 0 \end{bmatrix} \begin{bmatrix} \mathbf{x}(t) \\ x_{n+1}(t) \end{bmatrix} + \begin{bmatrix} 0 \\ -1 \end{bmatrix} r(t) \tag{10-101}
$$

We can express this equation as

$$\dot{\mathbf{x}}_a(t) = \mathbf{A}_a\mathbf{x}_a(t) - \mathbf{B}_a\mathbf{K}_a\mathbf{x}_a(t) + \begin{bmatrix} 0 \\ -1 \end{bmatrix} r(t) \tag{10-102}$$

where

$$\mathbf{A}_a = \begin{bmatrix} \mathbf{A} & \mathbf{0} \\ \mathbf{C} & 0 \end{bmatrix} \tag{10-103}$$

$$\mathbf{B}_a = \begin{bmatrix} \mathbf{B} \\ 0 \end{bmatrix} \qquad \mathbf{K}_a = \begin{bmatrix} \mathbf{K} & K_{n+1} \end{bmatrix}$$

Equation (10-102) has the characteristic equation $|s\mathbf{I} - \mathbf{A}_a + \mathbf{B}_a\mathbf{K}_a| = 0$, which is the standard form for pole-placement design. This characteristic equation is of order $(n + 1)$; hence we must specify the desired characteristic equation $\alpha_{ca}(s)$ as

$$\alpha_{ca}(s) = s^{n+1} + \alpha_n s^n + \cdots + \alpha_1 s + \alpha_0 = 0 \tag{10-104}$$

Therefore,

$$|s\mathbf{I} - \mathbf{A}_a + \mathbf{B}_a\mathbf{K}_a| = s^{n+1} + \alpha_n s^n + \cdots + \alpha_1 s + \alpha_0 \tag{10-105}$$

For this pole-placement design, the $(n + 1)$ values α_i are known, and the $(n + 1)$ gains K_i are unknown. The coefficients in (10-105) can be equated to give $(n + 1)$ linear equations, or Ackermann's formula, (10-23), can be utilized to solve for the unknown gains. Ackermann's formula for the solution of (10-105) is

$$\mathbf{K}_a = \begin{bmatrix} 0 & 0 & \cdots & 0 & 1 \end{bmatrix} \begin{bmatrix} \mathbf{B}_a & \mathbf{A}_a\mathbf{B}_a & \cdots & \mathbf{A}_a^n\mathbf{B}_a \end{bmatrix}^{-1} \alpha_{ca}(\mathbf{A}_a) \tag{10-106}$$

An example of this design will now be given.

Example 10.13

We now design a speed control system for a dc motor. The model of a dc motor is given in Figure 10.24, with a power amplifier with a gain of four added to drive the motor. This model was developed in Section 2.7, and the state model was developed in Section 3.2. The simulation diagram for the state model of the motor is shown in Figure 3.8 and is repeated in Figure 10.25. The states of the motor are $x_1(t) = \omega(t)$, the motor speed, and $x_2(t) = i_a(t)$, the armature current. The two states may be easily measured.

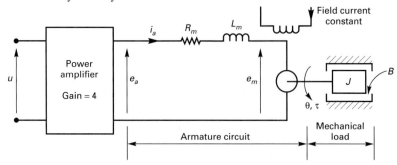

Figure 10.24   System for Example 10.13.

In Figure 10.25, the parameters of the motor have been assigned numerical values for this example. Suppose that the rated input voltage of the motor is 24 V and the speed is in rpm. From Figure 10.25, the motor transfer function is

$$G_p(s) = \frac{\Omega(s)}{E_a(s)} = \frac{50}{(s+1)(s+2)}$$

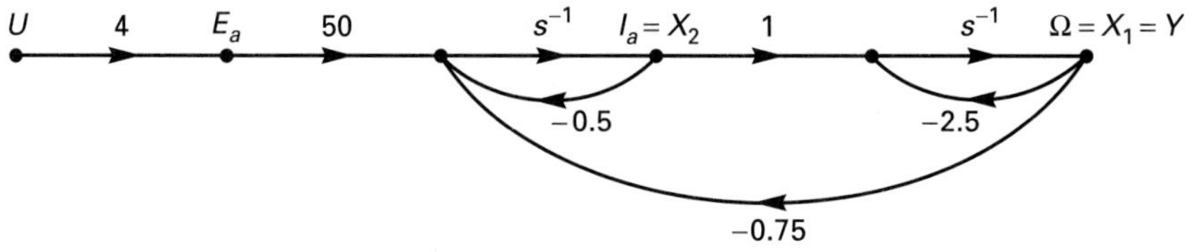

Figure 10.25   Simulation diagram for plant.

The dc gain of this transfer function is 25; hence the rated speed of the motor is $25 \times 24 = 600$ rpm. Since the gain of the power amplifier is 4, the rated voltage into the power amplifier is 6 V. The total plant transfer function is

$$\frac{\Omega(s)}{U(s)} = KG_p(s) = \frac{200}{(s+1)(s+2)}$$

and the state model is given by

$$\dot{\mathbf{x}}(t) = \begin{bmatrix} -2.5 & 1 \\ -0.75 & -0.5 \end{bmatrix} \mathbf{x}(t) + \begin{bmatrix} 0 \\ 200 \end{bmatrix} u(t)$$

$$y(t) = \begin{bmatrix} 1 & 0 \end{bmatrix}'\mathbf{x}(t)$$

Suppose that a settling time of 0.8 s is specified for the closed-loop system; we then choose the time constant of $\tau = 0.2$ s, and the desired system characteristic equation is chosen to be

$$\alpha_{ca}(s) = (s+5)^3 = s^3 + 15s^2 + 75s + 125$$

We next form the augmented matrices from (10-103) and the state model of the plant.

$$\mathbf{A}_a = \begin{bmatrix} \mathbf{A} & 0 \\ \mathbf{C} & 0 \end{bmatrix} = \begin{bmatrix} -2.5 & 1 & 0 \\ -0.75 & -0.5 & 0 \\ 1 & 0 & 0 \end{bmatrix} \qquad \mathbf{B}_a = \begin{bmatrix} \mathbf{B} \\ 0 \end{bmatrix} = \begin{bmatrix} 0 \\ 200 \\ 0 \end{bmatrix}$$

Then,

$$\mathbf{A}_a - \mathbf{B}_a\mathbf{K}_a = \begin{bmatrix} -2.5 & 1 & 0 \\ -0.75 & -0.5 & 0 \\ 1 & 0 & 0 \end{bmatrix} - \begin{bmatrix} 0 \\ 200 \\ 0 \end{bmatrix} \begin{bmatrix} K_1 & K_2 & K_3 \end{bmatrix}$$

$$= \begin{bmatrix} -2.5 & 1 & 0 \\ -0.75 - 200K_1 & -0.5 - 200K_2 & -200K_3 \\ 1 & 0 & 0 \end{bmatrix}$$

and the characteristic polynomial is, from (10-105),

$$\alpha_{ca}(s) = |s\mathbf{I} - \mathbf{A}_a + \mathbf{B}_a\mathbf{K}_a| = \begin{vmatrix} s + 2.5 & -1 & 0 \\ 0.75 + 200K_1 & s + 0.5 + 200K_2 & 200K_3 \\ -1 & 0 & s \end{vmatrix}$$

$$= s(s + 2.5)(s + 0.5 + 200K_2) + (0.75 + 200K_1)s + 200K_3$$

$$= s^3 + (3 + 200K_2)s^2 + (500K_2 + 2 + 200K_1)s + 200K_3$$

$$= s^3 + 15s^2 + 75s + 125$$

Hence, equating coefficients yields

$$K_2 = \frac{12}{200} = 0.06 \qquad K_3 = \frac{125}{200} = 0.625$$

$$200K_1 = 75 - 2 - 500(0.06) = 43 \Rightarrow K_1 = \frac{43}{200} = 0.215$$

This design is verified by the MATLAB program that employs Ackermann's formula:

```
Aa=[-2.5 1 0;-0.75 -0.5 0;1 0 0];   Ba=[0;200;0];
Pp=[-5 -5 -5];
Ka=acker(Aa,Ba,Pp), pause
B=[0;0;-1]; C=[1 0 0]; D=0
e10p13=ss((Aa-Ba*Ka),B,C,D);
step(e10p13)
```

The designed system is simulated by the last two statements of this program. The closed-loop system flow graph is given in Figure 10.26, and the state equations for the closed-loop system are

$$\dot{\mathbf{x}}_a(t) = \begin{bmatrix} -2.5 & 1 & 0 \\ -43.75 & -12.5 & -125 \\ 1 & 0 & 0 \end{bmatrix} \mathbf{x}_a(t) + \begin{bmatrix} 0 \\ 0 \\ -1 \end{bmatrix} r(t)$$

The system transfer function can be calculated directly from these state equations by computer, or by Mason's gain formula from Figure 10.26. In either case, the result is

$$T_1(s) = \frac{\Omega(s)}{R(s)} = \frac{125}{s^3 + 15s^2 + 75s + 125}$$

As mentioned above, the gain K_1 can be placed in parallel with the integrator, as shown by the dashed line in Figure 10.26. For this case, the closed-loop transfer function is

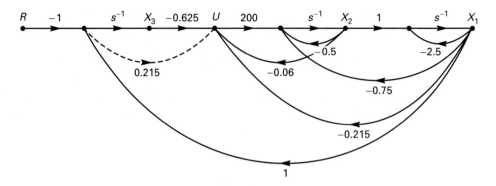

Figure 10.26   Simulation diagram for Example 10.13.

$$T_2(s) = \frac{\Omega(s)}{R(s)} = \frac{43s + 125}{s^3 + 15s^2 + 75s + 125}$$

The step responses of both realizations are given in Figure 10.27. In the second realization, $T_2(s)$, we expect the response to be faster, since there is a direct path from the step input to the motor voltage. In the first realization, $T_1(s)$, the step function must be integrated, so that the motor voltage initially approximates a ramp function. Both realizations are used; the first realization is used when we do not want to "bump" the plant with a step function.

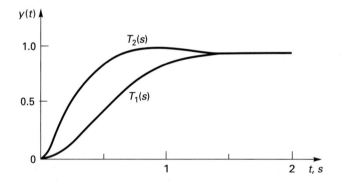

Figure 10.27   Unit step responses for Example 10.13.

10.8 SUMMARY

In this chapter a design procedure of modern control theory, pole-placement design, was developed. The direct implementation of this design requires that all the plant states be measured. In general, this is required by all modern design procedures. Because of the difficulty or impossibility of measuring all the states of complex systems, state estimators of some type are required in the implementation of the control systems. Hence the design of state estimators (observers) is also developed. Often the introduction of state estimators into a system degrades the stability margins of that system. In practical situations, the Nyquist diagram must be employed to ensure that this degradation of stability has not occurred.

Other modern design procedures are available that were not covered in this brief introduction to modern control theory [4,5]. One of the more popular procedures involves minimizing some mathematical function of the states and inputs of a system; these systems are called optimal control systems. One advantage of this design is that the procedure also applies to linear systems that are time-varying. All design procedures presented in this book apply only to linear time-invariant systems. The optimal-control design results in full state feedback in the same manner as pole placement, but the gain matrix **K** is time-varying for the optimal control system. An optimal state estimator can also be designed; the resulting estimator is called a *Kalman filter* [6].

REFERENCES

1. B. Friedland. *Control System Design.* New York: McGraw-Hill, 1986.

2. J. E. Ackermann. "Der Entwurf linearer regelungs Systems in Zustandstraum," *Regelungstech. Prozess-Datenverarb.* 7 (1972): 297–300.

3. P. S. Maybeck. *Stochastic Models, Estimation, and Control,* vol. 3. Orlando, FL: Academic Press, 1979.

4. W. L. Brogan. *Modern Control Theory,* 3rd ed. Englewood Cliffs, NJ: Prentice-Hall, 1991.

5. F. L. Lewis and Y. L. Syrmos. *Optimal Control,* 2nd ed. New York: Wiley, 1996.

6. P. S. Maybeck. *Stochastic Models, Estimation, and Control,* vol. 1. Orlando, FL: Academic Press, 1979.

PROBLEMS

10.1. Given a linear time-invariant stable system described by $\dot{\mathbf{x}}(t) = \mathbf{Ax}(t) + \mathbf{B}u(t)$.

    (a) Suppose that the system is excited by the input $u(t)$ and the initial condition vector $\mathbf{x}(0)$. Show that the solution of this equation, $\mathbf{x}(t)$, is composed of two independent components, one of which depends only on the initial conditions $\mathbf{x}(0)$ and the other depends only on the input $u(t)$.

    (b) Suppose that the input $u(t)$ is constant for a long time prior to $t = 0$ such that the system is in steady state. If the states at $t = 0$ are perturbed by an amount $\Delta \mathbf{x}(0)$, show that the change in states for $t > 0$ is a function only of $\Delta \mathbf{x}(0)$.

10.2. Consider the satellite system of Figure 10.2. Note that the only difference in this system and that of Example 10.1 is that the inertia of the satellite has been changed.

Figure P10.2   Satellite control system.

    (a) Design a control system by pole placement such that the closed-loop system has a time constant $\tau = 1.0$ s, with $\zeta = 0.707$. The two state variables are to be $\theta(t)$ and $d\theta(t)/dt$.

    (b) Write the state equations foor the closed-loop system.

    (c) Assume that the transfer function of the position sensor is a constant gain of 0.01, the gain of the velocity sensor is 0.4, and all gains required to realize the closed-loop control system are implemented by amplifiers. Draw a block diagram of the closed-loop system, showing the sensors, amplifiers, and the gains of the amplifiers.

    (d) Verify the results of this problem by computer.

10.3. Consider the plant of Figure P10.3. Assume that this is the model of a motor-driven positioning system and that the output $y(t)$ is position.

Figure P10.3

(a) Develop a state model of this plant such that the two states are position and velocity.
(b) Calculate the gains required to place the poles of the closed-loop system such that the system time constant is 0.5 s and the system is critically damped.
(c) Repeat (b) such that the system time constant is 0.5 s. and $\zeta = 0.707$.
(d) Write the state equations for the closed-loop systems of both (b) and (c).
(e) Verify all results by computer.

10.4. Consider the temperature-control system for a large test chamber shown in Figure P10.4. The input $u(t)$ is a voltage, the output $y(t)$ is the chamber temperature, and the sensor output is a voltage.

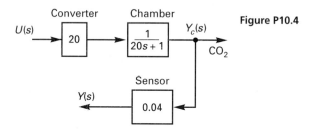

Figure P10.4

(a) Find a state model for the plant, with $u(t)$ as the input and $y(t)$ as the output.
(b) Design a pole-placement control that gives the closed-loop system a time constant of 10 s.
(c) Give a block diagram of the closed-loop system, if the sensor shown is used to measure the temperature.
(d) Verify all results by computer.

10.5. Show that Ackermann's formula for pole placement, (10-23), can be applied to state-estimator design, with the result as given in (10-44).

10.6. (a) Design a full state estimator for the system of Problem 10.2. Make the time constant of the estimator one-half that of the control system, with the estimator critically damped.
(b) Derive the transfer function of the controller-estimator, $G_{ec}(s)$, for the system configured as shown in Figure 10.8.
(c) Derive the characteristic equation of the closed-loop system from $1 + G_{ec}(s)G_p(s) = 0$. Show that this equation results in the product of the pole-placement characteristic equation and the estimator characteristic equation.
(d) Construct a flow graph of the closed-loop system of the form shown in Figure 10.9. From this flow graph, write the system state equations.
(e) Show that the system characteristic equation has the correct zeros.
(f) Verify the results of (a), (b), (c), and (e) with MATLAB.

10.7. (a) Design a full-state estimator for the system of Problem 10.3. Make the time constant of the estimator one-half that of the control system, with the estimator critically damped.
(b) Derive the transfer function of the controller-estimator, $G_{ec}(s)$, for the system configured as shown in Figure 10.8.

(c) Derive the characteristic equation of the closed-loop system from $1 + G_{ec}(s)G_p(s) = 0$. Show that this equation results in the product of the pole-placement characteristic equation and the estimator characteristic equation.

(d) Construct a flow graph of the closed-loop system in the form shown in Figure 10.9. From this flow graph, write the system state equations.

(e) Show that the system characteristic equation has the correct zeros.

(f) Verify the results of (a), (b), (c), and (e) with MATLAB.

10.8. (a) Design an estimator for the system of Problem 10.4. Make the time constant of the estimator one-third that of the control system, with the estimator critically damped.

(b) Derive the transfer function of the controller-estimator, $G_{ec}(s)$, for the system configured as shown in Figure 10.8.

(c) Derive the characteristic equation of the closed-loop system from $1 + G_{ec}(s)G_p(s) = 0$. Show that this equation results in the product of the pole-placement characteristic equation and the estimator characteristic equation.

(d) Construct a flow graph of the closed-loop system of the form shown in Figure 10.9. From this flow graph, write the system state equations.

(e) Show that the system characteristic equation has the correct zeros.

(f) Verify the results of (a), (b), (c), and (e) with MATLAB.

10.9. (a) Design a reduced-order estimator for the system of Problem 10.2, assuming that $x_1(t)$ is measured. Make the time constant of the estimator one-half that of the control system.

(b) Construct a flow graph of the closed-loop system of the form shown in Figure 10.13. From this flow graph, derive the transfer function of the controller-estimator, $G_{ec}(s)$, for the system configured as shown in Figure 10.14.

(c) Derive the closed-loop characteristic equation from $1 + G_{ec}(s)G_p(s) = 0$. Show that this equations results in the product of the pole-placement characteristic equation and the estimator characteristic equation.

(d) From the flow graph of (b), write the system state equations.

(e) Derive the characteristic equation of the closed-loop system from the system matrix found in (d). Show that this equation is the product of the pole-placement characteristic equation and the estimator characteristic equation.

10.10. (a) Design a reduced-order estimator for the system of Problem 10.3, assuming that $x_1(t)$ is measured. Make the time constant of the estimator one-half that of the control system.

(b) Construct a flow-graph of the closed-loop system of the form shown in Figure 10.13. From this flow graph, derive the transfer function of the controller-estimator, $G_{ec}(s)$, for the system configured as shown in Figure 10.14.

(c) Derive the closed-loop characteristic equation from $1 + G_{ec}(s)G_p(s) = 0$. Show that this equation results in the product of the pole-placement characteristic equation and the estimator characteristic equation.

(d) From the flow graph of (b), write the system state equations.

(e) Derive the characteristic equation of the closed-loop system from the system matrix found in (d). Show that this equation is the product of the pole-placement characteristic equation and the estimator characteristic equation.

(f) Verify the results of this problem by computer.

10.11. The observers considered in this chapter are based on the transfer function from the input to an estimated state being equal to the transfer function from the input to the actual state. Consider the satellite control system of Example 10.8 and Figure 10.13. Show that the transfer function $X_2(s)/U(s)$ is equal to $X_e(s)/U(s)$ in Figure 10.13, with the control gain **K** equal to zero, that is, with the system open loop.

10.12. Consider the satellite control system of Example 10.7, for which a reduced-order estimator was designed to estimate the velocity using the position measurement. Since velocity is the derivative of position, we expect the estimator to approximate a differentiator in some sense. Plot the Bode diagram of the transfer function $\hat{X}_e(s)/Y(s)$ for this estimator. On the same diagram, plot the Bode diagram of the transfer function of an ideal differentiator, and comment on both the similarities and the differences.

10.13. Consider the satellite control system of Example 10.7.
  (a) Design a reduced-order estimator with a time constant of 0.05 s, that is, let $a_e(s) = s + 20$.
  (b) Repeat Problem 10.12 for this estimator.
  (c) Compare the results of (b) with those of Problem 10.12, and comment on the effect on high-frequency noise filtering by the observer, if the observer is made faster (the time constant reduced).
  (d) Based on your answers in (c), how would you change the design of an observer to increase the high-frequency noise filtering?

10.14. In the system of Figure P10.14, write state equations for the system such that $x_1(t)$ is the state for the upper block and $x_2(t)$ is the state for the lower block.

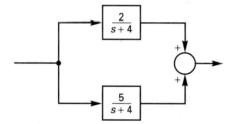

Figure P10.14

  (a) Determine if this system is controllable.
  (b) Determine if this system is observable.
  (c) Explain the results of (a) and (b) from the characteristics of the system (note the mathematics).
  (d) Verify the results of this problem with MATLAB.

10.15. Obtain the initial-condition response, using MATLAB, for the closed-loop systems in the following problems. For those systems with estimators, simulate the system both with the estimator initial conditions equal to the plant intial conditions and with the estimator initial conditions different from the plant initial conditions.
  (a) Problems 10.2 and 10.6
  (b) Problems 10.3 and 10.7
  (c) Problems 10.4 and 10.8

10.16. **(a)** For the system of Figure 10.15, replace the transfer functions given with the following:

$$G_p(s) = \frac{5}{s+3} \qquad H(s) = \frac{s+3}{(s+1)(s+2)}$$

Test this system for both controllability and observability, using the procedure of Example 10.10 to construct the state model.
  (b) Verify the results of this problem with MATLAB.

10.17. **(a)** For the system of Figure 10.15, replace the transfer functions given with the following:

$$G_p(s) = \frac{s+1}{s(s+3)} \qquad H(s) = \frac{s+3}{s+1}$$

Test this system for both controllability and observability, using the procedure of Example 10.10 to construct the state model.

(b) Verify the results of this problem with MATLAB.

10.18. (a) For the satellite control system of Example 10.5, suppose that the sensor that measures the attitude angle θ has a gain of 0.1 V/deg. Redraw the flow graph of Figure 10.9 with the sensor gain included, such that the control characteristic equation and the estimator characteristic equation are unchanged.

(b) Repeat (a) for Example 10.8 and Figure 10.13.

10.19. Consider the system described by the simulation diagram of Figure P10.19.

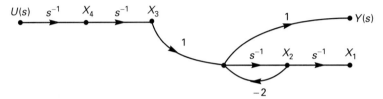

Figure P10.19

(a) Write the system state equations.

(b) Calculate the system transfer function using Mason's gain formula.

(c) Note that the state equations are fourth order, and the transfer function is second order. The realization of Figure P10.19 is called a nonminimal realization, since the state equations are of higher order than the transfer function. Determine if the system is controllable and observable.

(d) Verify the results of (c) with MATLAB.

10.20. Consider the circuit of Figure P10.20.

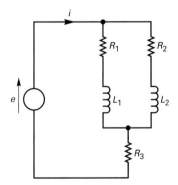

Figure P10.20

(a) Find the transfer function $I(s)/E(s)$ for this circuit.

(b) Write state equations for this circuit, with the currents in the inductors chosen to be the state variables.

(c) Determine the constraints on the resistances and the inductances such that the system is uncontrollable.

(d) Show that the conditions found in (c) are the same as those required for the transfer function of (a) to reduce to first order.

10.21. The satellite control system of Example 10.12 and Figure 10.21 is an implementation of the system equations (10-96), in which an input is shown. Find the values of **M** and N for this case. Figure 10.9 is useful in solving this problem.

11

Discrete-Time Systems

Three topics are covered in this chapter. First the concept of a discrete-time system is developed. Thus far in this book we have considered continuous-time, or analog, systems. These systems can be modeled by a set of differential equations. We have considered only the case that the differential equations of the model are linear with constant coefficients. In this chapter we consider *discrete-time systems,* which are systems that can be modeled by *difference equations.* We often refer to discrete-time systems as simply discrete systems. The transform used in the analysis of linear time-invariant analog systems is the Laplace transform; for linear time-invariant discrete systems, we use the z-transform. The second topic of this chapter is the introduction of the z-transform. As a final topic, the state equation models for discrete systems are developed.

As has been shown previously, we may model a linear time-invariant analog system by an nth-order transfer function in the Laplace transform variable s or by an nth-order differential equation, or by a set of n first-order coupled differential equations. In a like manner, we may model a linear time-invariant discrete system by an nth-order transfer function in the z-transform variable z, or by an nth-order difference equation, or by a set of n first-order coupled difference equations.

11.1 DISCRETE-TIME SYSTEM

To introduce the concept of a discrete-time system, suppose that a PI controller for a closed-loop system, such as has been discussed in preceding chapters, is to be implemented using a digital computer. We can add, multiply, and integrate numerically on a digital computer; hence we can implement the PI controller on a digital computer. In fact, we can implement any of the controllers considered thus far on a digital computer; these controllers are described by differential equations, and we can solve these equations numerically.

455

A digital control system of the type just described is shown in Figure 11.1. The digital computer is to perform the compensation function within the system. The interface at the input of the computer is an analog-to-digital (A/D) converter [1]; it is required to convert the error signal, which is a continuous-time signal, into a binary form that the computer can process. The interface at the computer output is required to convert the binary signal out of the computer to a voltage needed to drive the plant and is a digital-to-analog (D/A) converter.

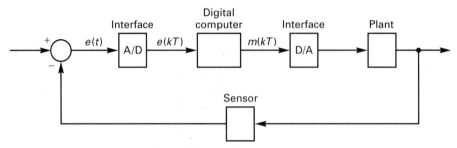

Figure 11.1   Digital control system.

For case of analysis and design, we allow the computer to perform only linear time-invariant operations; that is, we do not consider nonlinear or time-varying controllers. Recall that we also limited analog controllers to be linear and time invariant. Since the computer in Figure 11.1 is a digital device operating in real time, it can accept information only at discrete values of time. Assume that these values of time are regularly spaced every T seconds, beginning at $t = 0$. Then the signal into the computer that affects the computer output is described by the number sequence $e(0)$, $e(T)$, $e(2T)$, ..., which may be expressed by the notation $\{e(kT)\}$. Quite often the T is omitted, and the notation becomes $\{e(k)\}$.

It will be assumed here that negligible time is required for the computer to perform any calculations based on the input signal, which is often a valid assumption. Thus we assume that an input at $t = 0$ can produce an output at $t = 0$, an input at $t = T$ can produce an output at $t = T$, and so on.

Let the computer input at $t = 0$ be $e(0)$ and the output be $m(0)$. Since the computer operation must be linear and time invariant, we can express $m(0)$ as

$$m(0) = b_0 e(0) \tag{11-1}$$

where b_0 is a constant. If b_0 depends on either $e(0)$ or time, the equation will not be linear or time invariant, respectively. We can now store $e(0)$ and $m(0)$ in the computer. Thus $m(T)$ can be a function of $e(0)$, $m(0)$, and $e(T)$. For example, $m(T)$ might have the form

$$m(T) = b_0 e(T) + b_1 e(0) - a_1 m(0) \tag{11-2}$$

In a like manner, $m(2T)$ might appear as

$$m(2T) = b_0 e(2T) + b_1 e(T) + b_2 e(0) - a_1 m(T) - a_2 m(0) \tag{11-3}$$

These equations are called *difference equations*. The general form of an nth-order linear, time-invariant difference equation (with the T omitted) is

$$m(k) = b_0 e(k) + b_1 e(k-1) + \cdots + b_n e(k-n)$$
$$- a_1 m(k-1) - \cdots - a_n m(k-n) \tag{11-4}$$

We show later that if the analog plant in Figure 11.1 is also linear and time invariant, the entire system may be modeled by a difference equation of the form of (11-4) but, of course, of higher order than that of the controller. Compare (11-4) to a linear differential equation describing an nth-order analog system, with input $e(t)$ and output $y(t)$.

$$y(t) = \beta_0 e(t) + \beta_1 \frac{de(t)}{dt} + \cdots + \beta_n \frac{d^n e(t)}{dt^n} - \alpha_1 \frac{dy(t)}{dt} - \cdots - \alpha_n \frac{d^n y(t)}{dt^n} \qquad (11\text{-}5)$$

The describing equation of a linear, time-invariant analog compensator (filter) is also of the form of (11-5). The device that realizes this filter (often an RC network with operational amplifiers) can be considered to be an (analog) computer programmed to solve (11-5). In a like manner, (11-4) is the describing equation of a linear, time-invariant discrete filter, which is usually called a digital filter. The device that realizes this filter is a digital computer programmed to solve (11-4) or a special-purpose computer constructed to solve only (11-4).

Consider again the control system of Figure 11.1. The digital computer would be programmed to solve a difference equation of the form of (11-4). The control system designer must determine (1) T, the sampling period, (2) n, the order of the difference equation, and (3) a_i, b_i, the filter coefficients, such that the system has certain desired characteristics. There are additional problems in the realization of a digital filter, such as the computer word length required to keep system errors caused by round-off in the computer at an acceptable level. As an example, a digital filter has been designed and implemented to land aircraft automatically on U.S. Navy aircraft carriers [2]. In this design, $T = 0.04$ s, and the controller is eleventh order. The minimum word length required in the computer was found to be 32 bits, in order that system errors caused by round-off in the computer remain at an acceptable level. As an additional point, this controller is a PID filter with extensive noise filtering required principally because of the differentiation in the D section of the filter. The integration and the differentiation are performed numerically, as is discussed in Chapter 13.

11.2 TRANSFORM METHODS

In linear time-invariant continuous-time systems, the Laplace transform can be utilized in system analysis and design. For example, an alternate but equally valid description of the operation of a system described by the general differential equation (11-5) is obtained by taking the Laplace transform of this equation and solving for the transfer function.

$$\frac{Y(s)}{E(s)} = \frac{\beta_n s^n + \cdots + \beta_1 s + \beta_0}{\alpha_n s^n + \cdots + \alpha_1 s + 1} \qquad (11\text{-}6)$$

A transform is now defined that can be utilized in the analysis of discrete-time systems modeled by difference equations in the form of (11-4).

A transform for a number sequence is defined as follows.

z-transform   The z-transform of a number sequence $\{e(k)\}$ is defined as a power series in z^{-k} with coefficients equal to the values $e(k)$. This transform is then expressed as

$$E(z) = \mathfrak{z}[\{e(k)\}] = e(0) + e(1)z^{-1} + e(2)z^{-2} + \cdots \qquad (11\text{-}7)$$

where $\mathfrak{z}[\cdot]$ indicates the z-transform. Equation (11-7) can be written as

$$E(z) = \mathfrak{z}[\{e(k)\}] = \sum_{k=0}^{\infty} e(k)z^{-k} \qquad (11\text{-}8)$$

The z-transform is defined for any number sequence $\{e(k)\}$ and may be used in the analysis of any type of system described by difference equations. For example, the z-transform is used in discrete probability problems, in which the numbers in the sequence $\{e(k)\}$ are discrete probabilities [3].

If the sequence $\{e(k)\}$ is generated from a time function $e(t)$ by sampling every T seconds, $e(k)$ is understood to be $e(kT)$; that is, the T is dropped for convenience.

Three examples will now be given to illustrate the z-transform.

Example 11.1

Given $E(z)$ below. Find the sequence for $e(k)$.

$$E(z) = 1 - 2z^{-1} + 3z^{-3} + 0.5z^{-4} + \cdots$$

We see then, from (11-7), that

$$e(0) = 1 \qquad\qquad e(4) = 0.5$$
$$e(1) = -2 \qquad\qquad\quad \cdot$$
$$e(2) = 0 \qquad\qquad\quad\ \cdot$$
$$e(3) = 3 \qquad\qquad\quad\ \cdot$$

Next we find the z-transform of two useful number sequences. However, the following series expansion is needed to express certain z-transforms in closed form, which we will always do when possible. The series expansion for the reciprocal of $(1 - x)$ is [4]

Hardware for the simultaneous digital control of a number of processes (distributed control systems). (Courtesy of Rosemont, Inc., Eden Prairie, Minnesota.)

$$\frac{1}{1-x} = 1 + x + x^2 + x^3 + \cdots \qquad |x| < 1 \qquad (11\text{-}9)$$

Obviously the series is not convergent for $|x| \geq 1$. One method of deriving this expansion is through long division.

$$
\begin{array}{r}
1 + x + x^2 + \ldots \\
1 - x \overline{\smash{)}\,1} \\
\underline{1 - x} \\
x \\
\underline{x - x^2} \\
x^2 \\
\vdots
\end{array}
\qquad (11\text{-}10)
$$

This division is shown here, because the procedure will prove to be useful later in finding inverse z-transforms. The use of this power series in finding z-transforms is illustrated next.

Example 11.2

We find the z-transform of $e(k)$ for the case that $e(k)$ is equal to unity for all k. From the definition of the z-transform, (11-7),

$$E(z) = 1 + z^{-1} + z^{-2} + z^{-3} + \cdots$$

Compare this series with the power series of (11-9). If we let $z^{-1} = x$, we see that the z-transform can be expressed as

$$E(z) = \frac{1}{1 - z^{-1}} = \frac{z}{z - 1} \qquad |z^{-1}| < 1 \qquad (11\text{-}11)$$

Note that $\{e(k)\}$ may be generated by sampling a unit step function. Of course, there are many other time functions that have a value of unity every T seconds.

Example 11.3

Given that $e(k) = e^{-akT}$, we find its z-transform

$$
\begin{aligned}
E(z) &= 1 + e^{-aT}z^{-1} + e^{-2aT}z^{-2} + \cdots \\
&= 1 + (e^{-aT}z^{-1}) + (e^{-aT}z^{-1})^2 + \cdots \\
&= \frac{1}{1 - e^{-aT}z^{-1}} = \frac{z}{z - e^{-aT}} \qquad |e^{-aT}z^{-1}| < 1
\end{aligned}
$$

In this example note that the sequence $e(k)$ can be generated by sampling the function $e(t) = e^{-at}$ every T seconds, beginning at $t = 0$.

As in the case of the Laplace transform each z-transform has a region of existence in the complex plane. This region is of importance if an integral is used to find the inverse z-transform [5]. However, we will use tables for both the forward transform and the inverse transform, and generally will not state the region of existence when the transform is employed. Recall that we were not concerned with the region of existence of the Laplace transforms in Chapters 2 through 9.

As a final point, the z-transform defined in (11.8) is the *single-sided*, or *unilateral*, z-transform. If the lower limit of the summation in (11-8) is $k = -\infty$, the transform is then the *double-sided*, or *bilateral*, z-transform [5]. In this book, only the single-sided z-transform is used.

11.3 THEOREMS OF THE z-TRANSFORM

The usefulness of theorems of the Laplace transform has been amply demonstrated in the preceding chapters. We will see that z-transform theorems may be utilized in the same manner. In this section several theorems of the z-transform are given, and formal proofs are presented for these theorems.

11.3.1 Addition

Theorem.   The z-transform of the sum of number sequences is equal to the sum of the z-transforms of the number sequences; that is,

$$\mathfrak{z}[\{e_1(k)\} + \{e_2(k)\}] \ = \ E_1(z) + E_2(z) \tag{11-12}$$

Proof. From the definition of the z-transform,

$$\mathfrak{z}[\{e_1(k)\} + \{e_2(k)\}] \ = \ \sum_{k=0}^{\infty} [e_1(k) + e_2(k)] z^{-k}$$

$$= \ \sum_{k=0}^{\infty} e_1(k) z^{-k} + \sum_{k=0}^{\infty} e_2(k) z^{-k} \ = \ E_1(z) + E_2(z)$$

11.3.2 Multiplication by a Constant

Theorem.   The z-transform of a number sequence multiplied by a constant is equal to the constant multiplied by the z-transform of the number sequence; that is,

$$\mathfrak{z}[a\{e(k)\}] \ = \ a\mathfrak{z}[\{e(k)\}] \ = \ aE(z) \tag{11-13}$$

Proof. From the definition of the z-transform,

$$\mathfrak{z}[a\{e(k)\}] \ = \ \sum_{k=0}^{\infty} ae(k) z^{-k} = a \sum_{k=0}^{\infty} e(k) z^{-k} = aE(z)$$

11.3.3 Real Translation

Theorem.   Let n be a *positive* integer, and let $e(k)$ be zero for k less than zero. Further, let $E(z)$ be the z-transform of $\{e(k)\}$. Then

$$\mathfrak{z}[\{e(k-n)\}] \ = \ z^{-n}E(z) \tag{11-14}$$

and

$$\mathfrak{z}[\{e(k+n)\}] \ = \ z^{n}\left[E(z) - \sum_{k=0}^{n-1} e(k) z^{-k} \right] \tag{11-15}$$

Proof. From the definition of the z-transform,

$$\mathfrak{z}[\{e(k-n)\}] \;=\; e(-n) + e(-n+1)z^{-1} + \cdots + e(0)z^{-n} + e(1)z^{-(n+1)} + \cdots$$
$$= z^{-n}[e(0) + e(1)z^{-1} + e(2)z^{-2} + \cdots]$$
$$= z^{-n}E(z)$$

since we define $e(k) = 0$, $k < 0$. This definition is necessary for the same reason as for the definition that $e(t) = 0$, $t < 0$, for the shifting theorem of the Laplace transform, Appendix B. Otherwise we would not have the simple theorem that results from this assumption.

For (11-15),

$$\mathfrak{z}[\{e(k+n)\}] \;=\; e(n) + e(n+1)z^{-1} + e(n+2)z^{-2} + \cdots$$
$$= z^{n}\,[e(0) + e(1)z^{-1} + \cdots + e(n-1)z^{-n+1}$$
$$+ e(n)z^{-n} + e(n+1)z^{-n-1}$$
$$+ \cdots - e(0) - e(1)z^{-1} - \cdots - e(n-1)z^{-n+1}\,]$$

where we have added the first n terms of $\{e(k)\}$ and then subtracted the same n terms. Thus

$$\mathfrak{z}[\{e(k+n)\}] \;=\; z^{n}\!\left[E(z) - \sum_{k=0}^{n-1} e(k)z^{-k} \right]$$

This theorem cannot be expressed as simply as the first one. Note again that n must be a *positive integer* in this theorem.

To further illustrate this theorem, consider the sample number sequence given in Table 11.1. Note that sequence $e(k-2)$ is the sequence $e(k)$ delayed by two sample periods. No values of $e(k)$ have been lost in the delay; thus the z-transform of $e(k-2)$ can be expressed as a function of only $E(z)$. The sequence $e(k+2)$ is the sequence $e(k)$ advanced by two sample periods. Here samples of $e(k)$ have been lost (the first two); hence z-transform of $e(k+2)$ cannot be expressed as a simple function of $E(z)$.

TABLE 11.1   EXAMPLES OF SHIFTING

k	$e(k)$	$e(k-2)$	$e(k+2)$
0	1	0	0.3
1	0.5	0	0.2
2	0.3	1	0.15
3	0.2	0.5	—
4	0.15	0.3	—

Example 11.4

It was shown in Example 11.3 that

$$\mathfrak{z}[\{e^{-akT}\}] \;=\; \frac{z}{z - e^{-aT}}$$

Thus

$$\mathfrak{z}[\{e^{-a(k-3)T}u[(k-3)T]\}] \;=\; z^{-3}\!\left(\frac{z}{z - e^{-aT}} \right) = \frac{1}{z^{2}(z - e^{-aT})}$$

where $u(k)$ is the discrete unit step. Also,

$$\mathscr{z}[\{e^{-a(k+2)T}\}] = z^2\left(\frac{z}{z - e^{-aT}} - 1 - e^{-aT}z^{-1}\right)$$

The discrete unit step function, used in Example 11.4, is defined as

$$u(k) = \begin{cases} 1 & k \geq 0 \\ 0 & k < 0 \end{cases}$$

11.3.4 Initial Value

Theorem.   Given that the z-transform of $\{e(k)\}$ is $E(z)$. Then

$$e(0) = \lim_{z \to \infty} E(z) \tag{11-16}$$

Proof. Since

$$F(z) = e(0) + e(1)z^{-1} + e(2)z^{-2} + \cdots$$

equation (11-16) is seen by inspection.

11.3.5 Final Value

Theorem.   Given that the z-transform of $\{e(k)\}$ is $E(z)$. Then

$$\lim_{n \to \infty} e(n) = \lim_{z \to 1} (z - 1)E(z) \tag{11-17}$$

provided that the left-side limit exists.

Proof. Consider the transform

$$\mathscr{z}[\{e(k+1)\} - \{e(k)\}] = \lim_{n \to \infty}\left(\sum_{k=0}^{n} e(k+1)z^{-k} - \sum_{k=0}^{n} e(k)z^{-k}\right)$$

$$= \lim_{n \to \infty} [-e(0) + e(1)(1 - z^{-1}) + e(2)(z^{-1} - z^{-2}) + \cdots$$

$$+ e(n)(z^{-n+1} - z^{-n}) + e(n+1)z^{-n}]$$

Thus

$$\lim_{z \to 1} \mathscr{z}[\{e(k+1)\} - \{e(k)\}] = \lim_{n \to \infty} [e(n+1) - e(0)]$$

provided that the right-side limit exists. Also, from the real translation theorem,

$$\mathscr{z}[\{e(k+1)\} - \{e(k)\}] = z[E(z) - e(0)] - E(z)$$

$$= (z - 1)E(z) - ze(0)$$

Hence letting $z \to 1$ in this expression and equating it to the preceding expression gives

$$\lim_{n \to \infty} [e(n+1) - e(0)] = \lim_{z \to 1} [(z - 1)E(z) - ze(0)]$$

or

$$\lim_{n \to \infty} e(n) = \lim_{z \to 1} (z - 1)E(z)$$

provided that the left-side limit exists. As in the case of the final-value theorem of the Laplace transform, the right-side limit can give an incorrect value for the final value if the left-side limit does not exist.

Example 11.5

As shown in Example 11.2, if $e(k) = 1$ for all k, then $E(z) = z/(z-1)$. Thus

$$e(0) = \lim_{z \to \infty} E(z) = \lim_{z \to \infty} \frac{z}{z-1} = \lim_{z \to \infty} \frac{1}{1 - 1/z} = 1$$

and

$$\lim_{n \to \infty} e(n) = \lim_{z \to 1} (z-1)E(z) = \lim_{z \to 1} \frac{(z-1)z}{z-1} = \lim_{z \to 1} z = 1$$

We see then that both of these values are correct.

11.4 SOLUTION OF DIFFERENCE EQUATIONS

There are three basic techniques for solving linear time-invariant difference equations. The first method, the classical technique, consists of finding the complementary and the particular parts of the solution [6] in a manner similar to that used in the classical solution of linear differential equations. This technique is not discussed here. The second technique is a sequential procedure and is the technique used in the digital computer solution of difference equations. This technique is illustrated by the following example. The third technique is then presented.

Example 11.6

We solve for $m(k)$ for the equation

$$m(k) = e(k) - e(k-1) - m(k-1) \qquad k \ge 0$$

where

$$e(k) = \begin{cases} 1 & k \text{ even} \\ 0 & k \text{ odd} \end{cases}$$

and both $e(-1)$ and $m(-1)$ are zero. Then $m(k)$ can be determined by solving the difference equation first for $k = 0$, then for $k = 1$, then for $k = 2$, and so forth. Thus

$$m(0) = e(0) - e(-1) - m(-1) = 1 - 0 - 0 = 1$$
$$m(1) = e(1) - e(0) - m(0) = 0 - 1 - 1 = -2$$
$$m(2) = e(2) - e(1) - m(1) = 1 - 0 + 2 = 3$$
$$m(3) = e(3) - e(2) - m(2) = 0 - 1 - 3 = -4$$
$$m(4) = e(4) - e(3) - m(3) = 1 - 0 + 4 = 5$$

Using the technique of the last example, we can find $m(k)$ for any value of k. This technique is not practical for large values of k, except when implemented on a digital computer. For the last example, a MATLAB program that solves the difference equation $k = 0$, 1, ..., 5 is

```
mkminus1 = 0;  ekminus1 = 0;  ek = 1;
for k = 0:5
    mk = ek - ekminus1 - mkminus1;
    [k ek mk]
    mkminus1 = mk;
    ekminus1 = ek;
    ek = 1 - ek;
end
```

In this program, ekminus1 is $e(k-1)$, ek is $e(k)$, mkminus1 $= m(k-1)$, and mk is $m(k)$. The statements on the first row initialize the values in the difference equation. The programming of a digital filter is similar to this program.

A flow diagram of a practical implementation of a digital filter (difference equation) is given in Table 11.2. These steps can be compared with the preceding MATLAB implementation. The additional step of timing is required if the filter is to operate in real time and is usually implemented using a timing chip in the computer.

As a second example of the practical application of the sequential solution of difference equations, consider the numerical integration of a differential equation by Euler's method, as discussed in Section 3.6. Given the first-order differential equation

$$\dot{x}(t) = ax(t) + be(t)$$

For T small, $\dot{x}(t)$ may be approximated by

$$\dot{x}(t) \simeq \frac{x(t+T) - x(t)}{T}$$

Thus the differential equation can be expressed as (approximately)

$$\frac{x(t+T) - x(t)}{T} = ax(t) + be(t)$$

Solving this equation for $x(t+T)$, we obtain

$$x(t+T) = (1 + aT)x(t) + bTe(t)$$

Evaluation of this equation at $t = kT$ yields the difference equation

$$x[(k+1)T] = (1 + aT)x(kT) + bTe(kT)$$

We see then that Euler's integration for this differential equation leads to a difference equation. In fact, all numerical integration techniques may be expressed as difference equations [7] and may be programmed for solution on a digital computer as illustrated earlier. This topic is investigated further in problems at the end of this chapter.

The third technique for solving linear time-invariant difference equations, that of using the z-transform, is now presented. Consider the following nth-order difference equation, where it is assumed that the input sequence $\{e(k)\}$ is known:

$$m(k) + a_1 m(k-1) + \cdots + a_n m(k-n)$$
$$= b_0 e(k) + b_1 e(k-1) + \cdots + b_n e(k-n) \tag{11-18}$$

Using the real translation theorem, we can find the z-transform of this equation:

$$M(z) + a_1 z^{-1} M(z) + \cdots + a_n z^{-n} M(z)$$
$$= b_0 E(z) + b_1 z^{-1} E(z) + \cdots + b_n z^{-n} E(z) \tag{11-19}$$

TABLE 11.2 FLOW DIAGRAM OF A DIGITAL FILTER

1. Initialize the filter
2. Get the input
3. Solve the difference equation
4. Output the value
5. Update the memory
6. Perform timing
7. Go to step 2

or

$$M(z) = \frac{b_0 + b_1 z^{-1} + \cdots + b_n z^{-n}}{1 + a_1 z^{-1} + \cdots + a_n z^{-n}} E(z) \tag{11-20}$$

For a given $E(z)$, we find $m(k)$ by taking the inverse z-transform of (11-20). General techniques for finding the inverse z-transform are discussed in the next section. First an example is given to illustrate the z-transform method.

Example 11.7

Consider the difference equation of Example 11.6.

$$m(k) = e(k) - e(k-1) - m(k-1) \qquad e(k) = \begin{cases} 1 & k \text{ even} \\ 0 & k \text{ odd} \end{cases}$$

Using the real translation theorem, we find the z-transform of this equation:

$$M(z) = E(z) - z^{-1} E(z) - z^{-1} M(z)$$

or

$$M(z) = \frac{1 - z^{-1}}{1 + z^{-1}} E(z) = \frac{z-1}{z+1} E(z)$$

We see that $E(z)$ is given by

$$E(z) = 1 + z^{-2} + z^{-4} + \cdots = \left.\frac{1}{1-x}\right|_{x=z^{-2}} = \frac{1}{1-z^{-2}} = \frac{z^2}{z^2-1}$$

Thus

$$M(z) = \frac{z-1}{z+1} \frac{z^2}{z^2-1} = \frac{z^2}{z^2+2z+1}$$

We can expand $M(z)$ into a power series by dividing the numerator of $M(z)$ by its denominator, as demonstrated earlier.

$$
z^2 + 2z + 1 \overline{\smash{\big)}\,
\begin{array}{l}
1 - 2z^{-1} + 3z^{-2} - 4z^{-3} + \cdots \\
z^2 \\
\underline{z^2 + 2z + 1} \\
\quad -2z - 1 \\
\quad \underline{-2z - 4 - 2z^{-1}} \\
\qquad 3 + 2z^{-1} \\
\qquad \underline{3 + 6z^{-1} + 3z^{-2}} \\
\qquad\quad -4z^{-1} - 3z^{-2} \\
\qquad\qquad \vdots
\end{array}}
$$

Thus

$$M(z) = 1 - 2z^{-1} + 3z^{-2} - 4z^{-3} + \cdots$$

and the values of $m(k)$ are seen to be the same as those found using the sequential technique in Example 11.6.

Thus far we have considered only difference equations for which the initial conditions are zero. The solution presented in (11-20) represents only the forced part of the response. In order to include initial conditions in the solution of (11-18), first replace k with $(k + n)$. Then the equation becomes

$$m(k + n) + a_1 m(k + n - 1) + \cdots + a_n m(k)$$
$$= b_0 e(k + n) + b_1 e(k + n - 1) + \cdots + b_n e(k) \tag{11-21}$$

From the real translation theorem,

$$\mathscr{z}[\{m(k + i)\}] = z^i(M(z) - m(0) - m(1)z^{-1} - \cdots - m(i - 1)z^{-i+1}) \tag{11-22}$$

Thus the z-transform of (11-21) can be found using (11-22), and all initial conditions are included in the solution. Note that if all initial conditions are zero, the z-transform of (11-21) yields the same results as given in (11-20). Note also that the initial conditions on $m(k)$ for an nth-order difference equation are $m(0), m(1), \ldots, m(n - 1)$. The term *initial condition* has a somewhat different meaning here than for differential equations.

11.5 INVERSE z-TRANSFORM

The z-transform technique is useful in solving difference equations; however, techniques are needed for finding inverse z-transforms. There are many such techniques. Two are given here.

11.5.1 Power-Series Method

The power series method for finding the inverse z-transform involves dividing the denominator of $E(z)$ into the numerator such that a power series of the form

$$E(z) = e(0) + e(1)z^{-1} + e(2)z^{-2} + \cdots \tag{11-23}$$

is obtained. The values of the sequence $\{e(k)\}$ are seen to be the coefficients in the power series, from the definition of the z-transform. This technique was illustrated in Example 11.7. Another example is now given.

Example 11.8

We want to find the values of $\{e(k)\}$ for $E(z)$ given by

$$E(z) = \frac{z}{z^2 - 3z + 2}$$

By long division,

$$
\begin{array}{r l}
 & z^{-1} + 3z^{-2} + 7z^{-3} + 15z^{-4} + \ldots \\
z^2 + 3z + 2 \,\big) & z \\
 & \underline{z - 3 + 2z^{-1}} \\
 & 3 - 2z^{-1} \\
 & \underline{3 - 9z^{-1} + 6z^{-2}} \\
 & 7z^{-1} - 6z^{-2} \\
 & \underline{7z^{-1} - 21z^{-2} + 14z^{-3}} \\
 & 15z^{-2} - 14z^{-3} \\
 & \vdots
\end{array}
$$

We see then that

$$
\begin{aligned}
e(0) &= 0 \qquad & e(4) &= 15 \\
e(1) &= 1 & &\vdots \\
e(2) &= 2 & e(k) &= 2^k - 1 \\
e(3) &= 3 & &\vdots
\end{aligned}
$$

In this case, the value of $e(k)$ as a function of k can be recognized. In general, the inverse z-transform as a function of k cannot be recognized easily, when the power-series method is used.

11.5.2 Partial-Fraction Expansion Method

The function $E(z)$ may be expanded into partial fractions in the same manner as used with Laplace transforms. Then the z-transform tables may be used to find the inverse z-transform. A short table of z-transforms of number sequences is given in Table 11.3. A more complete table of the z-transforms of sampled time functions is given in Appendix C.

A description of the partial-fraction expansion technique was given in Appendix B and is not repeated here. However, one difference does occur in applying partial fractions to the z-transform. An examination of the functions in Table 11.3 shows that a factor z is required in the numerator of these terms. Hence we generally expand the function $E(z)/z$ into partial fractions, and then multiply by z to obtain the expansion in the proper form. This point is illustrated in the following example.

Example 11.9

Consider $E(z)$ as given in Example 11.8.

$$E(z) = \frac{z}{z^2 - 3z + 2} = \frac{z}{(z-1)(z-2)}$$

We expand $E(z)/z$ into partial fractions, with the results

$$\frac{E(z)}{z} = \frac{1}{(z-1)(z-2)} = \frac{-1}{z-1} + \frac{1}{z-2}$$

Thus

$$\mathfrak{z}^{-1}[E(z)] = \mathfrak{z}^{-1}\left[\frac{-z}{z-1}\right] + \mathfrak{z}^{-1}\left[\frac{z}{z-2}\right]$$

where $\mathfrak{z}^{-1}[\cdot]$ denotes the inverse z-transform. From Table 11.3,

TABLE 11.3 z-TRANSFORMS

$\{e(k)\}$	$E(z)$
$\{1\}$	$\dfrac{z}{z-1}$
$\{k\}$	$\dfrac{z}{(z-1)^2}$
$\{k^2\}$	$\dfrac{z(z+1)}{(z-1)^3}$
$\{a^k\}$	$\dfrac{z}{z-a}$
$\{ka^k\}$	$\dfrac{az}{(z-a)^2}$
$\{\sin ak\}$	$\dfrac{z\sin a}{z^2 - 2z\cos a + 1}$
$\{\cos ak\}$	$\dfrac{z(z-\cos a)}{z^2 - 2z\cos a + 1}$
$\{b^k \sin ak\}$	$\dfrac{bz\sin a}{z^2 - 2bz\cos a + b^2}$
$\{b^k \cos ak\}$	$\dfrac{z(z-b\cos a)}{z^2 - 2bz\cos a + b^2}$

$$e(k) = -1 + 2^k$$

which is the same result as found in Example 11.8. A MATLAB program that performs the partial-fraction expansion is

```
[r p k] = residue([0 0 1],[1 -3 2])
```

Note that the partial-fraction procedure is to expand $E(z)$ into the forms of the terms that appear in the z-transform tables. This procedure is then the same as for the Laplace transform. For the case that $E(z)$ has no factor z in the numerator and no poles at $z = 0$, it can be convenient to write $E(z)$ as $z^{-1}zE(z) = z^{-1}F(z)$. $F(z)$ will have the needed factor z in the numerator, and solution for $f(k)$ follows by partial fraction expansion. Application of the real translation theorem yields $e(k) = f(k-1)$. For the case that $E(z)$ has m poles at $z = 0$, $E(z)$ can be written as $(z^{-1}z)^{m+1}E(z) = z^{-(m+1)}F(z)$. As before, $F(z)$ will have the needed factor z in the numerator.

11.6 SIMULATION DIAGRAMS AND FLOW GRAPHS

In this section the basic elements used to simulate a system described by a linear difference equation are developed. Consider that the block shown in Figure 11.2(a) represents a shift register. Suppose that every T seconds a number is shifted into the register, and, at that instant, the number that was stored in the register is shifted out. Let $e(k)$ represent the number shifted into the register at $t = kT$. Then, at that instant, the number shifted out is $e(k-1)$.

Let the symbolic representation of this memory device be as shown in Figure 11.2(b). This symbol is used to represent any device that performs the above operation.

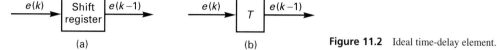

Figure 11.2 Ideal time-delay element.

An interconnection of these devices, along with multiplication by a constant and summation, can be used to represent a linear time-invariant difference equation. Consider the difference equation used in Example 11.6.

$$m(k) = e(k) - e(k - 1) - m(k - 1) \tag{11-24}$$

A simulation of this equation is shown in Figure 11.3. Electronic devices may be constructed to perform all of the operations shown in Figure 11.3. Suppose that such a construction exists. To solve Example 11.6 using this constructed machine, the numbers in both memory locations (shift registers) are set to zero. Then at $t = 0, 2T, 4T, \ldots$, the input $e(k)$ is set equal to 1, and at $t = T, 3T, 5T, \ldots$, $e(k)$ is set equal to 0. The solution $m(k)$ appears at the output terminal at $t = kT$. The reader may trace through the diagram for the first few iterations to determine that the output sequence is $1, -2, 3, -4, \ldots$ as previously derived.

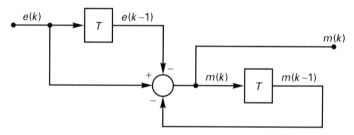

Figure 11.3 Simulation diagram for example.

The device just described is a special-purpose computer, capable of solving only the difference equation (11-24). Recall that the computer program given following Example 11.6 also solves this difference equation, but in this case a general-purpose computer is used. The general-purpose computer software commands the arithmetic registers, memory, and so forth, to perform the operations depicted in Figure 11.3.

Recall that in the simulation of continuous systems, the basic element is the integrator (see Sections 3.2 and 3.6). We see that in the simulation of discrete systems, the basic element is the time delay (or memory and shifting) of T seconds.

A somewhat different graphical representation of a difference equation is the signal flow graph representation. A block diagram, as illustrated in Figure 11.3, is simply a graphical representation of an equation or a set of equations. The signal flow graph may also be used to represent equations graphically. Thus a flow graph contains exactly the same information as does a block diagram.

Now consider the time-delay device shown in Figure 11.2(b). The z-transform of the input $e(k)$ is $E(z)$, and the z-transform of the output $e(k - 1)$ is $z^{-1}E(z)$. Thus the transfer function of the time delay is z^{-1}. (Recall that the transfer function of the integrator is s^{-1}.) Consider again the system shown in Figure 11.3. A flow graph representation of this system is as shown in Figure 11.4. The transfer function of this system may be obtained either from Figure 11.3, if each delay is replaced by its transfer function, or from Figure 11.4, by using Mason's gain formula (see Section 2.4).

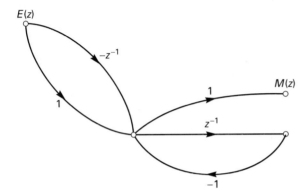

Figure 11.4   Signal flow graph for example.

The application of Mason's gain formula to Figure 11.4 yields the transfer function

$$\frac{M(z)}{E(z)} = \frac{1 - z^{-1}}{1 + z^{-1}} = \frac{z - 1}{z + 1}$$

which is the same as that obtained in Example 11.7.

Consider now a general nth-order difference equation of the form

$$m(k) + a_{n-1}m(k-1) + \cdots + a_0 m(k-n)$$
$$= b_n e(k) + b_{n-1}e(k-1) + \cdots + b_0 e(k-n) \tag{11-25}$$

Using the real translation theorem, we find the z-transform of this equation:

$$M(z) + a_{n-1}z^{-1}M(z) + \cdots + a_0 z^{-n}M(z)$$
$$= b_n E(z) + b_{n-1}z^{-1}E(z) + \cdots + b_0 z^{-n}E(z) \tag{11-26}$$

The transfer function $G(z)$ for this equation is

$$M(z) = \frac{b_n + b_{n-1}z^{-1} + \cdots + b_0 z^{-n}}{1 + a_{n-1}z^{-1} + \cdots + a_0 z^{-n}} E(z) = G(z)E(z) \tag{11-27}$$

The general system of (11-25) may be represented by the simulation diagram shown in Figure 11.5. This system may also be represented by other diagrams, since other diagrams can be constructed that have the transfer function in (11-27). This topic is discussed further in the next section of this chapter, when we consider state variables for discrete systems. The flow graph for Figure 11.5 is shown in Figure 11.6. Application of Mason's gain formula to this flow graph yields (11-27).

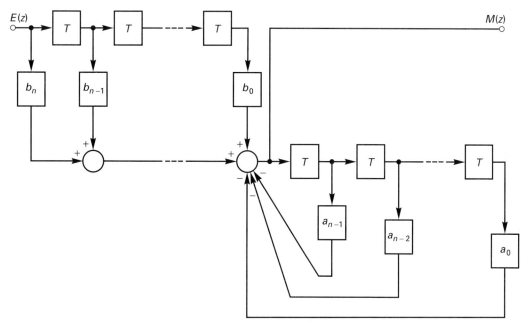

Figure 11.5 Simulation diagram for nth-order difference equation.

11.7 STATE VARIABLES

In the last section the topic of simulation diagrams was introduced. In Chapter 3 the simulation diagrams of analog systems led directly to state-variable models of these systems. Simulation diagrams of discrete systems also lead to state-variable models of discrete systems. In fact, much of the material needed to understand state-variable modeling of analog systems is directly applicable to the state modeling of discrete systems. Hence, the material of this section relies on the coverage of Chapter 3.

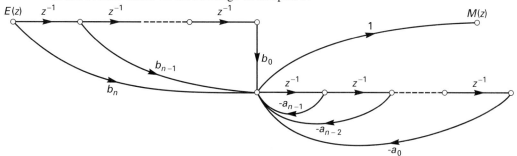

Figure 11.6 Flow graph for nth-order difference equation.

State models of discrete systems are introduced by an example. Suppose that a discrete system is described by the second-order difference equation

$$y(k) = 0.368u(k-1) + 0.264u(k-2) + 1.368y(k-1) - 0.368y(k-2) \qquad (11\text{-}28)$$

where $u(k)$ is the system input and $y(k)$ is the system output. We see that the z-transform of this equation yields the transfer function

$$\frac{Y(z)}{U(z)} = \frac{0.368z^{-1} + 0.264z^{-2}}{1 - 1.368z^{-1} + 0.368z^{-2}} = \frac{0.368z + 0.264}{z^2 - 1.368z + 0.368} \tag{11-29}$$

Many different simulation diagrams can be constructed for this system. One such diagram is of the form illustrated in Figure 11.5. A diagram of this form is called a *direct programming form* in the programming of digital filters [1]. A different simulation diagram for (11-29) is given in Figure 11.7(a). This form is called the *control canonical form* in the simulation of analog systems in Section 3.2. This form is also called a *canonical programming form* in the programming of digital filters. A flow graph of this simulation diagram is given in Figure 11.7(b). Application of Mason's gain formula to this flow graph shows that the simulation diagram's transfer function is (11-29).

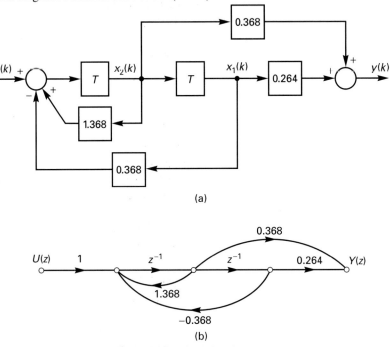

(a)

(b)

Figure 11.7 Simulation diagram.

A state-variable model of the system described by (11-28) can be developed by assigning a state variable to the output of each unit delay in the simulation diagram, Figure 11.7(a). We assign the variable $x_1(k)$ to the output of one of the delays and $x_2(k)$ to the other one, as shown in the figure. Then the input to the first delay must be $x_1(k + 1)$, and the input to the second one must be $x_2(k + 1)$. Thus the equations of the simulation diagram are

$$x_1(k + 1) = x_2(k)$$
$$x_2(k + 1) = -0.368x_1(k) + 1.368x_2(k) + u(k) \tag{11-30}$$
$$y(k) = 0.264x_1(k) + 0.368x_2(k)$$

These equations can be expressed in a vector-matrix form as

$$\begin{bmatrix} x_1(k+1) \\ x_2(k+1) \end{bmatrix} = \begin{bmatrix} 0 & 1 \\ -0.368 & 1.368 \end{bmatrix} \begin{bmatrix} x_1(k) \\ x_2(k) \end{bmatrix} + \begin{bmatrix} 0 \\ 1 \end{bmatrix} u(k)$$

$$y(k) = \begin{bmatrix} 0.264 & 0.368 \end{bmatrix} \begin{bmatrix} x_1(k) \\ x_2(k) \end{bmatrix}$$

(11-31)

We denote the various matrices and vectors of this equation in the standard format.

$$\mathbf{x}(k+1) = \mathbf{A}\mathbf{x}(k) + \mathbf{B}u(k)$$
$$y(k) = \mathbf{C}\mathbf{x}(k)$$

(11-32)

The procedure illustrated in the preceding paragraph can be used to find the state model of a discrete system given either the system's difference equation or the system's transfer function. First, a simulation diagram of the system is constructed. Next, the output of each time delay is assigned a state variable. As the final step, an equation is written for the input of each delay and for each system output, as functions of the outputs of the delays and the system inputs.

The general form of the state equations for a multivariable linear time-invariant discrete system is

$$\mathbf{x}(k+1) = \mathbf{A}\mathbf{x}(k) + \mathbf{B}\mathbf{u}(k)$$
$$\mathbf{y}(k) = \mathbf{C}\mathbf{x}(k) + \mathbf{D}\mathbf{u}(k)$$

(11-33)

where the state vector $\mathbf{x}(k)$ is of the order $n \times 1$, the input vector $\mathbf{u}(k)$ is $r \times 1$, and the output vector $\mathbf{y}(k)$ is $p \times 1$. Hence the system matrix $\mathbf{A}$ is $n \times n$, the input matrix $\mathbf{B}$ is $n \times r$, and the output matrix $\mathbf{C}$ is $p \times n$. The matrix $\mathbf{D}$, which represents direct coupling between the input and the output, is $p \times r$.

As just given, the model of (11-33) is linear and time invariant. If the various matrices in these equations are functions of k, the system is then linear and time varying. The general state-variable model of a *linear time-varying discrete system* is given by

$$\mathbf{x}(k+1) = \mathbf{A}(k)\mathbf{x}(k) + \mathbf{B}(k)\mathbf{u}(k)$$
$$\mathbf{y}(k) = \mathbf{C}(k)\mathbf{x}(k) + \mathbf{D}(k)\mathbf{u}(k)$$

(11-34)

11.8 SOLUTION OF STATE EQUATIONS

The first-order coupled difference equations that form the state model of a linear-time invariant discrete system are given by

$$\mathbf{x}(k+1) = \mathbf{A}\mathbf{x}(k) + \mathbf{B}\mathbf{u}(k)$$

(11-35)

The sequential method of Section 11.4 may be used to solve these equations. First we assume that $\mathbf{x}(0)$ and the inputs $\mathbf{u}(k)$ are known. Then the difference equations (11-35) are evaluated for $k = 0$, then for $k = 1$, then for $k = 2$, and so forth.

$$\mathbf{x}(1) = \mathbf{Ax}(0) + \mathbf{Bu}(0)$$

$$\mathbf{x}(2) = \mathbf{Ax}(1) + \mathbf{Bu}(1) = \mathbf{A}[\mathbf{Ax}(0) + \mathbf{Bu}(0)] + \mathbf{Bu}(1)$$

$$= \mathbf{A}^2\mathbf{x}(0) + \mathbf{ABu}(0) + \mathbf{Bu}(1)$$

$$\mathbf{x}(3) = \mathbf{Ax}(2) + \mathbf{Bu}(2) = \mathbf{A}[\mathbf{A}^2\mathbf{x}(0) + \mathbf{ABu}(0) + \mathbf{Bu}(1)] + \mathbf{Bu}(2)$$

$$= \mathbf{A}^3\mathbf{x}(0) + \mathbf{A}^2\mathbf{Bu}(0) + \mathbf{ABu}(1) + \mathbf{Bu}(2)$$

$$\vdots$$

$$\mathbf{x}(n) = \mathbf{A}^n\mathbf{x}(0) + \mathbf{A}^{n-1}\mathbf{Bu}(0) + \mathbf{A}^{n-2}\mathbf{Bu}(1) + \cdots + \mathbf{ABu}(n-2) + \mathbf{Bu}(n-1)$$

Hence the general solution of (11-35) can be expressed as

$$\mathbf{x}(n) = \mathbf{A}^n\mathbf{x}(0) + \sum_{k=0}^{n-1} \mathbf{A}^{n-1-k}\mathbf{Bu}(k) \tag{11-36}$$

The general solution of the state equations, (11-35), can also be found using the z-transform. The equations in (11-35) can be expressed as

$$x_1(k+1) = a_{11}x_1(k) + \cdots + a_{1n}x_n(k) + b_{11}u_1(k) + \cdots + b_{1r}u_r(k)$$

$$\vdots$$

$$x_n(k+1) = a_{n1}x_1(k) + \cdots + a_{nn}x_n(k) + b_{n1}u_1(k) + \cdots + b_{nr}u_r(k)$$

The z-transform of these equations yields

$$z[X_1(z) - x_1(0)] = a_{11}X_1(z) + \ldots + a_{1n}X_n(z) + b_{11}U_1(z) + \ldots + b_{1r}U_r(z)$$

$$\vdots$$

$$z[X_n(z) - x_n(0)] = a_{n1}X_1(z) + \ldots + a_{nn}X_n(z) + b_{n1}U_1(z) + \ldots + b_{nr}U_r(z)$$

These equations can be expressed in matrix form as

$$z[\mathbf{X}(z) - x(0)] = \mathbf{AX}(z) + \mathbf{BU}(z)$$

or,

$$[z\mathbf{I} - \mathbf{A}]\mathbf{X}(z) = z\mathbf{x}(0) + \mathbf{BU}(z)$$

Thus $\mathbf{X}(z)$ can be expressed as

$$\mathbf{X}(z) = z[z\mathbf{I} - \mathbf{A}]^{-1}\mathbf{x}(0) + [z\mathbf{I} - \mathbf{A}]^{-1}\mathbf{BU}(z) \tag{11-37}$$

The inverse z-transform of (11-37) yields the same results as (11-36), and thus the state transition matrix, $\Phi(k)$, for discrete state equations is given by

$$\Phi(k) = \mathscr{z}^{-1}(z[z\mathbf{I} - \mathbf{A}]^{-1}) = \mathbf{A}^k \tag{11-38}$$

The solution (11-36) can then be expressed as

$$\mathbf{x}(n) = \Phi(n)\mathbf{x}(0) + \sum_{k=0}^{n-1} \Phi(n-1-k)\mathbf{Bu}(k)$$

Two examples illustrate the solution of discrete state equations.

Example 11.10

As an example, consider the discrete system with the transfer function

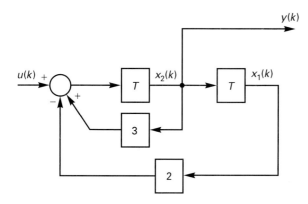

Figure 11.8 Simulation diagram for Example 11.10.

$$G(z) = \frac{Y(z)}{U(z)} = \frac{z}{z^2 - 3z + 2} = \frac{z}{(z-1)(z-2)}$$

To obtain a state model, the simulation diagram shown in Figure 11.8 is constructed and states assigned to each delay output. From this diagram we write the state equations

$$\mathbf{x}(k+1) = \begin{bmatrix} 0 & 1 \\ -2 & 3 \end{bmatrix} \mathbf{x}(k) + \begin{bmatrix} 0 \\ 1 \end{bmatrix} u(k)$$

$$y(k) = \begin{bmatrix} 0 & 1 \end{bmatrix} \mathbf{x}(k)$$

We now solve these equations by the sequential technique just given for the case that $\mathbf{x}(0) = \mathbf{0}$ and $u(k) = 1$, for all k. Since $\mathbf{x}(0) = \mathbf{0}$, then $y(0) = 0$. In the state equations above, for $k = 1$,

$$\mathbf{x}(1) = \mathbf{A}\mathbf{x}(0) + \mathbf{B}u(0)$$

$$= \begin{bmatrix} 0 & 1 \\ -2 & 3 \end{bmatrix} \begin{bmatrix} 0 \\ 0 \end{bmatrix} + \begin{bmatrix} 0 \\ 1 \end{bmatrix}(1) = \begin{bmatrix} 0 \\ 1 \end{bmatrix}, \qquad y(1) = 1$$

$$\mathbf{x}(2) = \mathbf{A}\mathbf{x}(1) + \mathbf{B}u(1)$$

$$= \begin{bmatrix} 0 & 1 \\ -2 & 3 \end{bmatrix} \begin{bmatrix} 0 \\ 1 \end{bmatrix} + \begin{bmatrix} 0 \\ 1 \end{bmatrix}(1) = \begin{bmatrix} 1 \\ 4 \end{bmatrix}, \qquad y(2) = 4$$

$$\mathbf{x}(3) = \begin{bmatrix} 0 & 1 \\ -2 & 3 \end{bmatrix} \begin{bmatrix} 1 \\ 4 \end{bmatrix} + \begin{bmatrix} 0 \\ 1 \end{bmatrix}(1) = \begin{bmatrix} 4 \\ 11 \end{bmatrix}, \qquad y(3) = 11$$

$$\mathbf{x}(4) = \begin{bmatrix} 0 & 1 \\ -2 & 3 \end{bmatrix} \begin{bmatrix} 4 \\ 11 \end{bmatrix} + \begin{bmatrix} 0 \\ 1 \end{bmatrix}(1) = \begin{bmatrix} 11 \\ 26 \end{bmatrix}, \qquad y(4) = 26$$

$$\vdots$$

The iterative procedure of this example is well suited to computer implementation. A general matrix MATLAB program, itsol.m, is given by

```
for k = 0:iter
    inputuk
    y(k+1) = C*x + D*u;
    [k y(k+1)]
    x1 = A*x + B*u;
    x = x1;
end
```

The m-file inputuk.m calculates $u(k)$. A calling program for itsol.m is given by

```
A = [0 1;-2 3];  B = [0;1];  C = [0 1];  D = 0;
x = [0;0];
iter = 4
   itsol
```

The number of iterations in the solution is iter.

Example 11.11

As a second example, the state equations of the system of the last example are solved by the z-transform method illustrated in (11-37). Thus

$$[z\mathbf{I} - \mathbf{A}] = \begin{bmatrix} z & -1 \\ 2 & z-3 \end{bmatrix} \qquad |z\mathbf{I} - \mathbf{A}| = z^2 - 3z + 2$$

and hence

$$[z\mathbf{I} - \mathbf{A}]^{-1} = \frac{1}{z^2 - 3z + 2}\begin{bmatrix} z-3 & 1 \\ -2 & z \end{bmatrix}$$

From (11-37), since $\mathbf{x}(0) = \mathbf{0}$,

$$\mathbf{X}(z) = [z\mathbf{I} - \mathbf{A}]^{-1}\mathbf{B}U(z)$$

$$= \frac{1}{z^2 - 3z + 2}\begin{bmatrix} z-3 & 1 \\ -2 & z \end{bmatrix}\begin{bmatrix} 0 \\ 1 \end{bmatrix}U(z)$$

$$= \frac{1}{z^2 - 3z + 2}\begin{bmatrix} 1 \\ z \end{bmatrix}U(z)$$

Since $U(z) = z/(z-1)$,

$$Y(z) = \mathbf{C}\mathbf{X}(z) = \begin{bmatrix} 0 & 1 \end{bmatrix}\begin{bmatrix} \dfrac{z}{(z-1)(z^2-3z+2)} \\ \dfrac{z^2}{(z-1)(z^2-3z+2)} \end{bmatrix}$$

or,

$$Y(z) = \frac{z^2}{(z-1)^2(z-2)} = \frac{-z}{(z-1)^2} + \frac{-2z}{z-1} + \frac{2z}{z-2}$$

Thus, from Table 11.3,

$$y(k) = -k - 2 + 2(2)^k$$

Hence the number sequence for $y(k)$ is 0, 1, 4, 11, 26, . . . , which verifies the values calculated in Example 11.10.

As a final topic, the state equations are related to the transfer function for the case of a single-input, single-output system. From (11-37), with the initial conditions $\mathbf{x}(0) = \mathbf{0}$, the states are related to the input by

$$\mathbf{X}(z) = [z\mathbf{I} - \mathbf{A}]^{-1}\mathbf{B}U(z) \qquad (11\text{-}39)$$

The *z*-transform of the output equation of (11-33) yields

$$Y(z) = \mathbf{C}\mathbf{X}(z) + DU(z) \tag{11-40}$$

Substitution of (11-39) into (11-40) yields the input–output equation

$$Y(z) = [\mathbf{C}(z\mathbf{I} - \mathbf{A})^{-1}\mathbf{B} + D]U(z)$$

and therefore the system transfer function is given by

$$G(z) = \frac{Y(z)}{U(z)} = \mathbf{C}(z\mathbf{I} - \mathbf{A})^{-1}\mathbf{B} + D \tag{11-41}$$

Hence the system transfer function can be found from the state equations by either the matrix procedure of (11-41) or the construction of the flow graph from the state equations, and the application of Mason's gain formula to the flow graph.

Example 11.12

This example illustrates the calculation of the transfer function from the state equations. We consider the system of Examples 11.10 and 11.11. For this system,

$$\mathbf{x}(k+1) = \begin{bmatrix} 0 & 1 \\ -2 & 3 \end{bmatrix} \mathbf{x}(k) + \begin{bmatrix} 0 \\ 1 \end{bmatrix} u(k)$$

$$y(k) = \begin{bmatrix} 0 & 1 \end{bmatrix} \mathbf{x}(k)$$

The matrix $[z\mathbf{I} - \mathbf{A}]^{-1}$ was found in Example 11.11 to be

$$[z\mathbf{I} - \mathbf{A}]^{-1} = \begin{bmatrix} \dfrac{z-3}{\Delta} & \dfrac{1}{\Delta} \\ \dfrac{-2}{\Delta} & \dfrac{z}{\Delta} \end{bmatrix} \qquad \Delta = z^2 - 3z + 2$$

Thus from (11-41), since $D = 0$, the system transfer function is given by

$$G(z) = \frac{Y(z)}{U(z)} = \mathbf{C}[z\mathbf{I} - \mathbf{A}]^{-1}\mathbf{B}$$

$$= \begin{bmatrix} 0 & 1 \end{bmatrix} \begin{bmatrix} \dfrac{z-3}{\Delta} & \dfrac{1}{\Delta} \\ \dfrac{-2}{\Delta} & \dfrac{z}{\Delta} \end{bmatrix} \begin{bmatrix} 0 \\ 1 \end{bmatrix}$$

$$= \begin{bmatrix} \dfrac{-2}{\Delta} & \dfrac{z}{\Delta} \end{bmatrix} \begin{bmatrix} 0 \\ 1 \end{bmatrix} = \frac{z}{z^2 - 3z + 2}$$

This result is seen to be the system transfer function given in Example 11.10. A MATLAB program for verifying the transfer function $G(z)$ is

```
A = [0 1;-2 3];   B = [0;1];   C = [0 1];
[Gnum,Gden] = ss2tf(A,B,C,0)
```

The last two sections give a brief introduction to state models of discrete systems. Many topics relating to state variables have not been covered; for example, the similarity transformations discussed in Section 3.5 for analog systems apply directly to discrete-system models. Those readers interested in a more complete coverage of state variables for discrete systems should see, for example, [1] and [8].

11.9 SUMMARY

Discrete-time systems were introduced in this chapter. A discrete-time system is one that is modeled by a difference equation. The transform that is used to solve difference equations, the z-transform, was defined; it is utilized in later chapters in the analysis and design of digital control systems. Useful theorems of the z-transform were developed, and two techniques for finding the inverse z-transform were presented. The representation of difference equations and thus of discrete-time systems by simulation diagrams and flow graphs was developed.

The final topic presented in this chapter was the state-variable modeling of discrete-time systems. The presentation of this topic is similar to that for state variables for analog systems in Chapter 3, and much of the material of Chapter 3 is directly applicable to discrete-time models. In the following chapters the topics developed in this chapter are applied to the analysis and design of digital control systems.

REFERENCES

1. C. L. Phillips and H. T. Nagle, Jr. *Digital Control System Analysis and Design,* 3rd ed. Upper Saddle River, NJ: Prentice Hall, 1995.

2. "Software Implementation ALS Computer Program," Contract N00421-75-C-0058, Bell Aerospace Corporation, Buffalo, NY, March 1975.

3. A. W. Drake. *Fundamentals of Applied Probability Theory.* New York: McGraw-Hill, 1967.

4. R. E. Johnson and F. L. Kiokemeister. *Calculus with Analytic Geometry.* Boston: Allyn and Bacon, 1969.

5. C. L. Phillips and J. M. Parr. *Signals, Systems, and Transform,* 2nd ed. Upper Saddle River, NJ: Prentice Hall, 1999.

6. F. Scheid. *Theory and Problems of Numerical Analysis.* New York: McGraw-Hill (Schaum's Outline Series), 1968.

7. C. F. Gerald. *Applied Numerical Analysis.* Reading, MA: Addison-Wesley, 1970.

8. G. F. Franklin, J. D. Powell, and M. Workman. *Digital Control of Dynamic Systems,* 3rd ed. Reading, MA: Addison-Wesley, 1998.

PROBLEMS

11.1. **(a)** Find the z-transform of the number sequence generated by sampling the time function $e(t) = u(t-2)$ at the sampling frequency of 5 Hz ($T = 0.2$ s). Express the z-transform both as a power series and in closed form.
   (b) Repeat (a) for $T = 1$ s.
   (c) Repeat (a) for $e(t) = t$, and for both $T = 0.1$ s and $T = 1$ s.
   (d) Repeat (c) for $e(t) = tu(t)$.

11.2. **(a)** The time function $e(t) = Ae^{-bt}$ is sampled at the rate of 20 Hz. The z-transform of the resulting number sequence is

$$E(z) = \frac{2z}{z - 0.9}$$

Find A and b.

(b) Using the results of (a), what function $e(t)$ sampled at 20 Hz would yield

$$E(z) = \frac{2}{z - 0.9}$$

(c) Verify the results in (a) and (b) by finding the z-transform of each $e(t)$.

11.3. Consider the function $e(t)$ which has the Laplace transform $\mathcal{L}[e(t)] = e^{-Ts}/[s(s + 1)]$.

(a) Find $e(t)$.

(b) The function $e(t)$ is sampled every T seconds. Express the z-transform of the resulting number sequence as both a power series and in closed form.

(c) Verify the result in (b) by finding the inverse z-transform of $E(z)$.

11.4. Given the z-transform

$$E(z) = \mathcal{z}[e(t)] = \frac{z}{z + 1} \qquad T = 0.01s$$

(a) Find $e(kT)$.

(b) Find a function $e_1(t)$ such that $e_1(kT) = e(kT)$.

(c) Find a second function $e_2(t)$ such that $e_2(kT) = e(kT)$ and $e_2(t) \neq e_1(t)$

(d) Sketch both $e_1(t)$ and $e_2(t)$ on the same time axis.

11.5. Given the z-transform

$$E(z) = \mathcal{z}[e(t)] = \frac{3z}{(z - 1)(z - 0.5)(z - 0.9)}$$

(a) By inspection we can see that $e(0) = e(1) = 0$ and $e(2) = 3$. Show that this statement is true.

(b) Solve for $e(k)$ as a function of k.

(c) Verify the partial fraction expansion in (b) using MATLAB.

(d) Show that the results in (b) satisfies the statement in (a).

(e) Will the final-value theorem give the correct final value?

(f) Find the correct final value of $e(k)$, using two different methods.

11.6. Find the inverse z-transform of each given $E(z)$ by the two methods given in Section 11.5. Compare the values of $e(k)$ for $k = 0, 1, 2,$ and 3 obtained by the two methods.

(a) $E(z) = \dfrac{z}{(z - 1)(z - 0.8)}$ **(b)** $E(z) = \dfrac{z(z + 1)}{(z - 1)(z - 0.8)}$

(c) $E(z) = \dfrac{1}{(z - 1)(z - 0.8)}$ **(d)** $E(z) = \dfrac{1}{z(z - 1)(z - 0.8)}$

(e) Verify the partial fractions using MATLAB.

11.7. (a) The sinusoidal function $e(t) = 5 \cos 20t$ is sampled every 0.05 s. Find the z-transform of the resultant number sequence, using the z-transform tables.

(b) Can the final-value theorem be applied to $E(z)$ in (a)? Why?

(c) The sinusoidal function $e(t) = A \cos \omega t$ is sampled every $T = 0.1$ s. The z-transform of the resultant number sequence is given by

$$E(z) = \frac{5z(z - 0.6967)}{z^2 - 1.3934z + 1}$$

Solve for A and ω.

(d) The function $e(t) = A_1 \cos \omega_1 t$ in (c) is sampled every $T = 0.2$ s, resulting in the same z-transform. How are A and A_1 related? How are ω and ω_1 related? Why?

11.8. Solve the given difference equation using the techniques listed. Verify that the two techniques yield the same result for $k = 2, 3, 4,$ and 5.

$$x(k+2) + x(k+1) + x(k) = 0$$

with $x(0) = 0$ and $x(1) = 2$.

(a) The sequential technique.

(b) The z-transform technique, with the inverse z-transform found using z-transform tables.

(c) Write a MATLAB program that will sequentially solve for $x(k)$, $k = 2, 3, 4, 5$.

11.9. Given the difference equation

$$x(k) - 3x(k-1) + 2x(k-2) = e(k)$$

where $x(-2) = x(-1) = 0$ and $e(k) = 1$ for $k \geq 0$.

(a) Solve for $x(k)$, $k = 0, 1, 2, 3,$ and 4, using the sequential method.

(b) Modify the MATLAB program below Example 11.6 to verify the results in (a).

(c) Find $X(z)$.

(d) Solve for $x(k)$, $k = 0, 1, 2, 3,$ and 4, using the power-series method.

(e) Solve for $x(k)$, as a function of k. Evaluate $x(k)$ for $k = 0, 1, 2, 3,$ and 4.

(f) Verify the partial-fraction expansion in (c) using MATLAB.

11.10. Given the difference equation

$$x(k) - 3x(k-1) + 2x(k-2) = e(k)$$

where $x(-2) = x(-1) = 0$ and

$$e(k) = \begin{cases} 1 & k = 0, 1 \\ 0 & k \geq 2 \end{cases}$$

(a) Solve for $x(k)$, $k = 0, 1, 2, 3,$ and 4, using the sequential method.

(b) Modify the MATLAB program below Example 11.6 to verify the results in (a).

(c) Find $X(z)$.

(d) Solve for $x(k)$, $k = 0, 1, 2, 3,$ and 4, using the power-series method.

(e) Solve for $x(k)$, as a function of k. Evaluate $x(k)$ for $k = 0, 1, 2, 3,$ and 4.

11.11. Determine the final value of the given functions. If applicable, use the final-value theorem of the z-transform.

(a) $\quad E(z) = \dfrac{0.1z}{z - 1.2}$ **(b)** $\quad E(z) = \dfrac{0.1z}{z - 0.8}$

(c) $\quad E(z) = \dfrac{0.1z}{z^2 - 1.7z + 0.7}$ **(d)** $\quad E(z) = \dfrac{0.1z}{z^2 - 1.7z + 1}$

11.12. The Euler rule for numerical integration is depicted in Figure P11.12. This rule approximates the integral of a function as the sum of the rectangular areas shown. Let $y(t)$ be the integral of $x(t)$.

(a) Write the difference equation that relates $y[(k+1)T]$, $y(kT)$, and $x(kT)$ for this integrator.

(b) Show that the transfer function of this integrator is given by

$$\frac{Y(z)}{X(z)} = \frac{T}{z - 1}$$

(c) Find $y(t)$ for

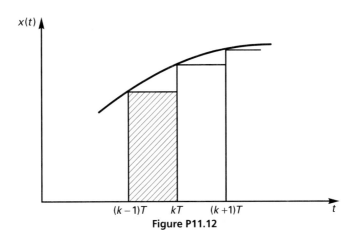

Figure P11.12

$$y(t) = \int_0^t e^{-\tau}d\tau$$

and evaluate $y(1)$.

(d) The integral in (c) is to be evaluated using the Euler rule. Find $X(z) = \mathfrak{z}[e^{-t}]$, for $T = 0.1$ s. Then find $y(kT)$ in (b) for $kT = 1$; that is, evaluate the integral of (c) at $t = 1$ s using the Euler rule.

(e) Compare the exact value in (c) with the value in (d).

(f) Repeat (d) and (e) for $T = 0.01$ s, and note the increased accuracy.

(g) Using MATLAB, program the difference equation in (a), and verify the results in both (d) and (f).

11.13. The trapezoidal rule for numerical integration is depicted in Figure P11.13. This rule approximates the integral of a function as the sum of the trapezoidal areas shown. Let $y(t)$ be the integral of $x(t)$.

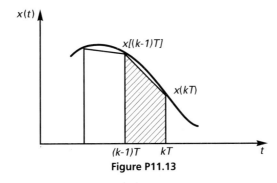

Figure P11.13

(a) Write the difference equation that relates $y[(k-1)T]$, $y(kT)$, $x[(k-1)T]$, and $x(kT)$ for this integrator.

(b) Show that the transfer function of this integrator is given by

$$\frac{Y(z)}{X(z)} = \frac{(T/2)(z+1)}{z-1}$$

(c) Repeat Problem 11.12(d), and (e) using the trapezoidal integrator.

11.14. The transfer function of an analog differentiator, s, is the reciprocal of the transfer function of an analog integrator, $1/s$. It seems reasonable that this should also be true for numerical differentiation and integration. The reciprocal of the Euler rule integrator in Problem 11.12 is given by

$$\frac{W(z)}{X(z)} = \frac{z-1}{T}$$

(a) If $W(z)$ is the output of this numerical differentiator with $X(z)$ its input, write the differentiator's difference equation.

(b) Draw a figure similar to Figure P11.12 illustrating this numerical differentiation.

11.15. Shown in Figure P11.15 are two different simulation diagrams for a second-order digital filter.

(a) Write the difference equations for realizations of the two filters; that is, write the difference equations for the simulation diagrams. The canonical programming form of (b) requires two difference equations.

(b) Determine the conditions on the coefficients such that the two filters have the same transfer function.

11.16. Write difference equations for each of the two given digital-filter transfer functions.

(a) $$\frac{M(z)}{E(z)} = \frac{0.1z}{z-0.9}$$ (b) $$\frac{M(z)}{E(z)} = \frac{0.333\,(z^2-1.7z+0.9)}{z^2-1.8z+0.9}$$

11.17. Given is a MATLAB program that solves the difference equation of a digital filter.

```
s = 0;
e = 1;
for k = 0:4
    f = e - 0.7*s;
    m = 0.8*f + 0.4*s;
    [k, m]
    s = f;
end
```

(a) Draw a simulation diagram for the filter, and identify all variables on the diagram.

(b) Find the transfer function of the filter.

(c) Find the z-transform of the filter input.

(d) Use the inverse z-transform to find the filter output.

(e) Run the MATLAB program to verify the results in (d).

11.18. A discrete system is described by the difference equation

$$m(k+1) - 0.8m(k) = 1.3e(k+1) - 0.9e(k)$$

where $e(k)$ is the input and $m(k)$ is the output.

(a) Draw a simulation diagram for this system.

(b) Use the diagram in (a) to find a state model for this system.

(c) Use (11-41) to find the system transfer function.

(d) Verify the results in (c) by calculating the transfer function directly from the difference equation.

(e) Verify the results in (c) from the simulation diagram and Mason's gain formula.

(f) Verify the results in (c) by MATLAB.

11.19. A discrete system is described by the difference equation

$$y(k+2) - 1.2y(k+1) + 0.6y(k) = 0.035u(k+1) + 0.03u(k)$$

where $u(k)$ is the input and $y(k)$ is the output.

(a) Draw a simulation diagram for this system.

(b) Use the diagram in (a) to find a state model for this system.

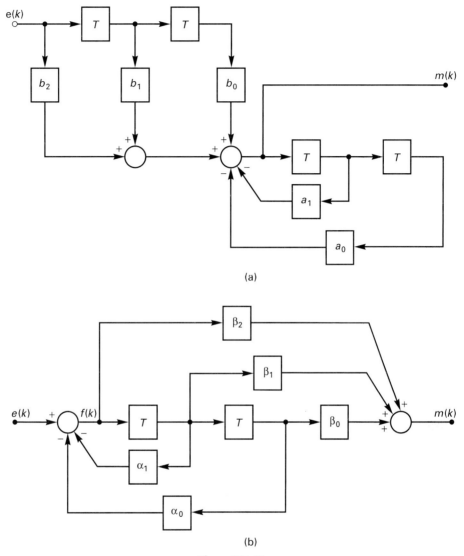

Figure P11.15

(c) Use (11-41) to find the system transfer function.

(d) Verify the results in (c) by calculating the transfer function directly from the difference equation.

(e) Verify the results in (c) from the simulation diagram and Mason's gain formula.

(f) Verify the results in (c) by MATLAB.

11.20. Find state models for the systems described by the following transfer functions.

(a) $\dfrac{z + 0.8}{(z - 1)^2}$

(b) $\dfrac{0.02z^2 + 0.5z + 0.025}{z^3 - 2.6z^2 + 1.9z - 0.4}$

(c) $\dfrac{z + 0.95}{z^2 - 1.9z + 0.93}$

(d) $\dfrac{0.5z^2}{(z - 1)^2}$

(e) Verify the state models by calculating the transfer functions using MATLAB.

11.21. A discrete system is described by the state equations

$$\mathbf{x}(k + 1) = \begin{bmatrix} 0 & 1 \\ -0.03 & 0.4 \end{bmatrix} \mathbf{x}(k) + \begin{bmatrix} 0 \\ 1 \end{bmatrix} u(k)$$

$$y(k) = \begin{bmatrix} 1 & 1 \end{bmatrix} \mathbf{x}(k)$$

(a) If $x_1(0) = x_2(0) = 0$ and $u(k) = 1, k \geq 0$, solve for $y(k), k = 0, 1, 2, 3$, directly from the state equations.

(b) Solve for $y(k)$ by taking the z-transform of the state equations, and verify the results of (a).

(c) Verify the time response of (b) using MATLAB.

(d) Find the transfer function $Y(z)/U(z)$.

(e) From the transfer function of (d), find the second-order system difference equation.

(f) Verify the transfer function of (d) using MATLAB.

11.22. For the discrete system described by the equations

$$\mathbf{x}(k + 1) = \mathbf{A}\mathbf{x}(k) = \begin{bmatrix} 0 & 1 \\ -3 & 4 \end{bmatrix} \mathbf{x}(k)$$

(a) Solve for the state transition matrix $\mathbf{A}^k$, using

$$\mathbf{A}^k = \mathscr{z}^{-1}[z(z\mathbf{I} - \mathbf{A})^{-1}]$$

(b) Verify the results in (a) by calculating $\mathbf{A}^k$ for $k = 0, 1, 2$, and 3 by (i) using $\mathbf{A}$ as given and (ii) using $\mathbf{A}^k$ from (a).

(c) Verify the results in (b) using MATLAB.

11.23. Given the unforced discrete system

$$\mathbf{x}(k + 1) = \begin{bmatrix} 1 & -1 \\ 1 & 3 \end{bmatrix} \mathbf{x}(k)$$

(a) Find the state transition matrix $\Phi(k)$ as a function of k.

(b) Let $\mathbf{x}^T(0) = [1 \ -2]$. Find $\mathbf{x}(3)$ directly from the system difference equation.

(c) Find $\mathbf{x}(3)$ using the results of (a); that is, use $\mathbf{x}(3) = \Phi(3)\mathbf{x}(0)$.

(d) Find $\mathbf{x}(3)$ using MATLAB.

12

Sampled-Data Systems

The mathematical definition of a discrete system was presented in Chapter 11. We define a discrete system as one whose operation is described (modeled) by a difference equation. A signal within a discrete system is described by a number sequence, for example, $\{e(k)\}$. For many systems, such as digital control systems, some of these number sequences are obtained by sampling a continuous-time signal. To understand the operation of these systems, it is necessary to understand the effects of sampling a continuous-time signal. In this chapter these effects are investigated. This investigation results in the mathematical models of digital control systems needed for analysis and design.

12.1 SAMPLED DATA

To introduce the effects of sampling, the system of Figure 12.1 is considered. In this system, the signal $e(t)$ is sampled. Sampling occurs in a system for many reasons. For example, the system of Figure 12.1 could be a radar tracking system, where the error is known only as each reflected electromagnetic signal returns from the target. Another example is the case that the error is measured by a sensor that gives binary outputs only at certain instants of time. For the present, it is assumed that there is no digital compensation in the system of Figure 12.1. Thus the system is continuous except for the single sampling operation. A system that is continuous except for one or more sampling operations, such as in Figure 12.1, is called a *sampled-data system*.

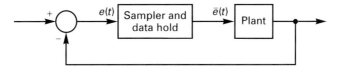

Figure 12.1 Sampled-data control system.

Obviously, the sampling operation causes a loss of information, since the sampled signal is not known at all instants of time. In an effort to reduce this loss of information, a data-reconstruction device, called a data-hold, is inserted into the system directly following the sampling operation. The purpose of the data-hold is to reconstruct the sampled signal into a form that closely resembles the signal before sampling. The simplest data-reconstruction device, and by far the most common one, is the zero-order hold. The operation of the sampler/zero-order hold combination is described by the signals shown in Figure 12.2, where T is the *sample period*. The zero-order hold clamps its output signal to a value equal to that of the sampled signal at the sampling instant.

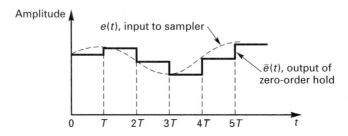

Figure 12.2   Input and output signals of sampler/data-hold.

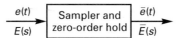

Figure 12.3   Sampler and data-hold.

The sampler and zero-order hold can be represented in block diagram form as shown in Figure 12.3. We can mathematically express the signal $\bar{e}(t)$ in Figure 12.2 as

$$\bar{e}(t) = e(0)\left[u(t) - u(t - T)\right] + e(T)\left[u(t - T) - u(t - 2T)\right]$$
$$+ e(2T)\left[u(t - 2T) - u(t - 3T)\right] + \cdots \tag{12-1}$$

The Laplace transform of $\bar{e}(t)$ is, then, from the shifting theorem in Appendix B,

$$\bar{E}(s) = e(0)\left[\frac{1}{s} - \frac{e^{-Ts}}{s}\right] + e(T)\left[\frac{e^{-Ts}}{s} - \frac{e^{-2Ts}}{s}\right] + e(2T)\left[\frac{e^{-2Ts}}{s} - \frac{e^{-3Ts}}{s}\right] + \cdots$$

$$= \frac{1 - e^{-Ts}}{s}\left[e(0) + e(T)e^{-Ts} + e(2T)e^{-2Ts} + \cdots\right] \tag{12-2}$$

$$= \frac{1 - e^{-Ts}}{s}\left[\sum_{n=0}^{\infty} e(nT)e^{-nTs}\right]$$

The second factor in (12-2) is seen to be a function of the input signal $e(t)$ and the sampling period T. The first factor is seen to be independent of $e(t)$. Thus the first factor can be considered to be a transfer function; therefore, the sample-hold operation can be represented as shown in Figure 12.4, where the function $E^*(s)$, called the *starred transform*, is *defined* as

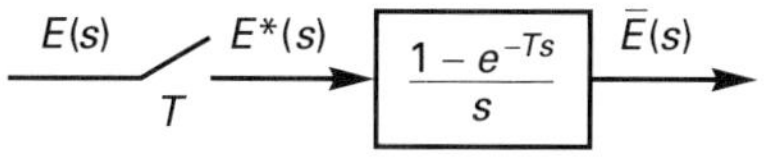

$$E(s) \quad E^*(s) \quad \boxed{\dfrac{1 - e^{-Ts}}{s}} \quad \bar{E}(s)$$

Figure 12.4 Representation of sampler/data-hold.

$$E^*(s) = \sum_{n=0}^{\infty} e(nT)e^{-nTs} \tag{12-3}$$

Hence (12-2) is satisfied by the representation in Figure 12.4. The operation denoted by the switch in Figure 12.4 is defined by (12-3) and is called an ideal sampler; the operation denoted by the transfer function is called the data-hold. It is to be emphasized that $E^*(s)$ *does not appear in the physical system* but appears as a result of factoring (12-2). The sampler (switch) in Figure 12.4 does not model a physical sampler and the block does not model a physical data-hold. However, the combination does accurately model the input–output characteristics of the sampler/data-hold device, as demonstrated earlier.

12.2 IDEAL SAMPLER

Even though $E^*(s)$ in (12-3) is not a physical signal and occurs as a result of factoring $\bar{E}(s)$, we now determine its characteristics. We will see later that it is convenient to use this signal in both analysis and design. The inverse Laplace transform of $E^*(s)$, from (12-3), is

$$\begin{aligned}
e^*(t) &= \mathcal{L}^{-1}[E^*(s)] \\
&= e(0)\delta(t) + e(T)\delta(t - T) + e(2T)\delta(t - 2T) + \cdots
\end{aligned} \tag{12-4}$$

where $\delta(t - t_0)$ is the unit impulse function (Dirac delta function) occurring at $t = t_0$. Then $e^*(t)$ is a train of impulse functions whose weights are equal to the values of the sampled signal at the instants of sampling. Thus $e^*(t)$ can be represented as shown in Figure 12.5, since the impulse function has infinite amplitude at the instant it occurs.

The sampler, which appears in the equivalent representation of a sampler/hold combination derived earlier, is usually referred to as an *ideal sampler*, since nonphysical signals (impulse functions) appear on its output. It is to be emphasized that the ideal sampler is not a physical sampler, but it appears in an equivalent representation because of mathematical manipulations.

To be mathematically consistent, we *define* the output signal of an ideal sampler as follows.

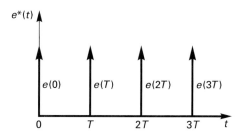

Figure 12.5 Representation of $e^*(t)$.

Ideal sampler output signal   The *output signal* of an ideal sampler is defined as the signal whose Laplace transform is

$$E^*(s) = \sum_{n=0}^{\infty} e(nT)e^{-nTs} \tag{12-5}$$

where $e(t)$ is the input signal to the sampler. If $e(t)$ is discontinuous at $t = kT$, k a positive integer, then $e(kT)$ is taken to be $e(kT^+)$. The notation $e(kT^+)$ indicates the value of $e(t)$ as t approaches kT from the right.

With the definition of the sampling operation as given in (12-5) and with the transfer function of the zero-order hold given by

$$G_{ho}(s) = \frac{1 - e^{-Ts}}{s} \tag{12-6}$$

the correct mathematical description of the sampler-hold operation is obtained. However, the ideal sampler does not model a physical sampler and the zero-order-hold transfer function, (12-6), does not model a physical data-hold. The *only* input function allowed for $G_{ho}(s)$ is the train of impulse functions $e^*(t)$, (12-4). An ideal sampler cannot appear in a system unless it is followed by $G_{ho}(s)$, and $G_{ho}(s)$ cannot appear in a system unless it is preceded by an ideal sampler. An example of finding $E^*(s)$ is now given.

Example 12.1

We wish to find $E^*(s)$ for $e(t) = u(t)$, the unit step. For the unit step, $e(nT) = 1$, $n = 0, 1, 2, \ldots$. Thus, from (12-5),

$$E^*(s) = \sum_{n=0}^{\infty} e(nT)e^{-nTs} = e(0) + e(T)e^{-Ts} + e(2T)e^{-2Ts} + \cdots$$

or

$$E^*(s) = 1 + e^{-Ts} + e^{-2Ts} + \cdots$$

As shown in Section 11.2, since

$$\frac{1}{1-x} = 1 + x + x^2 + \cdots, \qquad |x| < 1$$

then

$$E^*(s) = \frac{1}{1 - e^{-Ts}}, \qquad |e^{-Ts}| < 1$$

The reader may have noted the similarity of the starred transform, $E^*(s)$, and the z-transform, $E(z)$. The starred transform is defined as

$$E^*(s) = e(0) + e(T)e^{-Ts} + e(2T)e^{-2Ts} + \cdots$$

and the z-transform as defined as

$$E(z) = e(0) + e(1)z^{-1} + e(2)z^{-2} + \cdots$$

If the number sequence of the z-transform $\{e(k)\}$ is obtained by sampling the time function $e(t)$ every T seconds, we see that

$$E^*(s) = E(z)\big|_{z = e^{Ts}} \tag{12-7}$$

For this reason, we do not give a separate table of starred transforms; instead we use (12-7) and the table of z-transforms to find starred transforms.

Example 12.2

In Example 12.1, the starred transform of the unit step was derived. This same result can be found using the z-transform table of Appendix C. From this table, we see that the z-transform of the unit step is

$$E(z) = \mathcal{z}[u(t)] = \frac{z}{z - 1}$$

Thus from (12-7),

$$E^*(s) = E(z)\big|_{z = e^{Ts}} = \frac{z}{z - 1}\bigg|_{z = e^{Ts}} = \frac{e^{Ts}}{e^{Ts} - 1} = \frac{1}{1 - e^{-Ts}}$$

which checks the results derived in Example 12.1.

In the last example, the notation $\mathcal{z}[e(t)]$ is used to denote the ideal sampling of $e(t)$ every T seconds; that is, this notation implies that we are actually finding the z-transform of the number sequence $\{e(kT)\}$.

While (12-5) is the defining equation for the starred transform, this equation can be manipulated into other useful forms. We consider one of these other forms, and those interested in the derivation of this result should see [1]. The starred transform of (12-5) can also be expressed as

$$E^*(s) = \frac{1}{T} \sum_{n = -\infty}^{\infty} E(s + jn\omega_s) + \frac{e(0^+)}{2} \tag{12-8}$$

or, giving a few terms of the expansion,

$$E^*(s) = \frac{1}{T}[E(s) + E(s + j\omega_s) + E(s + j2\omega_s) + \cdots$$
$$+ E(s - j\omega_s) + E(s - j2\omega_s) + \cdots] + \frac{e(0^+)}{2}$$

In these equations, ω_s is the radian sampling frequency, that is, $\omega_s = 2\pi/T = 2\pi f_s$, where $f_s = 1/T$ is the sampling frequency in hertz. This form of the starred transform will prove to be very useful in deriving certain results. An example of this equation is now given.

Example 12.3

The starred transform of the unit step was found in Examples 12.1 and 12.2. From (12-8), this transform can also be expressed as

$$E^*(s) = \frac{1}{T} \sum_{n = -\infty}^{\infty} E(s + jn\omega_s) + \frac{e(0^+)}{2}$$

Since $E(s) = 1/s$,

$$E^*(s) = \frac{1}{T}\left[\frac{1}{s} + \frac{1}{s+j\omega_s} + \frac{1}{s+j2\omega_s} + \cdots + \frac{1}{s-j\omega_s} + \frac{1}{s-j2\omega_s} + \cdots\right] + \frac{1}{2}$$

It is not evident that this expression for $E^*(s)$ can be reduced to that derived in Example 12.1; however, this reduction is shown in [Ref. 2].

In the preceding example, note that certain terms in $E^*(s)$ can be combined in the form

$$\frac{1}{s+jk\omega_s} + \frac{1}{s-jk\omega_s} = \frac{2s}{s^2 + (k\omega_s)^2}$$

and thus $E^*(s)$ can be expressed as

$$E^*(s) = \frac{1}{T}\left[\frac{1}{s} + \frac{2s}{s^2+\omega_s^2} + \frac{2s}{s^2+(2\omega_s)^2} + \cdots\right] + \frac{1}{2}$$

Hence, from the Laplace transform tables, $e^*(t)$ contains the sinusoids $2\cos\omega_s t$, $2\cos 2\omega_s t$, $2\cos 3\omega_s t$, ... Since $e(t)$ is a unit step, we see that the sampling has generated sinusoidal signals of frequencies ω_s, $2\omega_s$, $3\omega_s$, ... We consider this topic further in the next section.

12.3 PROPERTIES OF THE STARRED TRANSFORM

Two important properties of the starred transform are now derived. These properties are needed to aid in understanding the effects of sampling in digital control systems.

1. $E^*(s)$ is periodic in s with period $j\omega_s$, that is, $E^*(s) = E^*(s+j\omega_s)$.

This property is proven using the definition of the starred transform, (12-5). From this equation,

$$E^*(s+j\omega_s) = \sum_{n=0}^{\infty} e(nT)e^{-nT(s+j\omega_s)} \tag{12-9}$$

Now, since $\omega_s = 2\pi/T$,

$$e^{-jnT\omega_s} = e^{-jnT2\pi/T} = e^{-jn2\pi} = 1\,\underline{/-n2\pi} = 1$$

since, from Euler's relation,

$$e^{-j2n\pi} = \cos 2n\pi - j\sin 2n\pi$$

From (12-9),

$$E^*(s+j\omega_s) = \sum_{n=0}^{\infty} e(nT)e^{-nTs}e^{-jnT\omega_s} = \sum_{n=0}^{\infty} e(nT)e^{-nTs} = E^*(s) \tag{12-10}$$

Note from this property that $E^*(s)$ is periodic in the complex variable s; that is, $E^*(s)$ is periodic in the complex plane.

2. Suppose that the function $E(s)$ has a pole at $s = s_1$. Then $E^*(s)$ has poles at $s = s_1 + jm\omega_s$, $m = 0, \pm 1, \pm 2, \ldots$.

This property can be proven using (12-8). Assume that $e(0^+) = 0$. Then

$$E^*(s) = \frac{1}{T} \sum_{n=-\infty}^{\infty} E(s + jn\omega_s) = \frac{1}{T}[E(s) + E(s + j\omega_s) + E(s + j2\omega_s)$$

$$+ \cdots + E(s - j\omega_s) + E(s - j2\omega_s) + \ldots]$$

(12-11)

If $E(s)$ has a pole at $s = s_1$, each term of (12-11) of the form $E(s - jm\omega_s)$ contributes a pole at $s = s_1 + jm\omega_s$, since $E(s - jm\omega_s)$ evaluated at $s = s_1 + jm\omega_s$ is equal to $E(s_1)$. This property is illustrated in Example 12.3, where (12-11) is evaluated for $e(t)$ equal to a unit step function.

No equivalent statement can be made concerning the zeros of $E^*(s)$; that is, the zero locations of $E(s)$ do not uniquely determine the zero locations of $E^*(s)$. These zero locations are determined by both the pole locations and the pole locations of $E(s)$ and by the sample period T. However, the zero locations are periodic with period $j\omega_s$, as indicated by the first property of $E^*(s)$. An example is given next.

Example 12.4

It was shown in Example 12.1 that the starred transform of a unit step is $E^*(s) = 1/(1 - e^{-Ts})$. Then, to determine the first property just given,

$$E^*(s + j\omega_s) = \frac{1}{1 - e^{-T(s + j\omega_s)}} = \frac{1}{1 - e^{-Ts}e^{-jT2\pi/T}}$$

$$= \frac{1}{1 - e^{-Ts}e^{-j2\pi}} = \frac{1}{1 - e^{-Ts}} = E^*(s)$$

Also, the poles of $E^*(s)$ occur at those values of s that make the denominator equal to zero. Thus the poles occur at the values of s that satisfy the relationship

$$e^{-Ts} = 1 = \cos 2m\pi - j\sin 2m\pi = e^{-j2m\pi} \qquad m = 0, \pm 1, \pm 2, \ldots$$

or

$$Ts = j2m\pi$$

Solving for s,

$$s = \frac{jm2\pi}{T} = jm\omega_s \qquad m = 0, \pm 1, \pm 2, \ldots$$

and the second property is demonstrated. The poles are periodic, with a pole located at $s = 0$, which is the location of the pole of $E(s)$. The pole locations are illustrated in Example 12.3 and plotted in Figure 12.6, which also shows the periodic nature of $E^*(s)$.

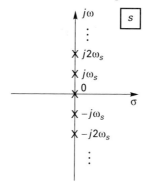

Figure 12.6 Pole plot for Example 12.4.

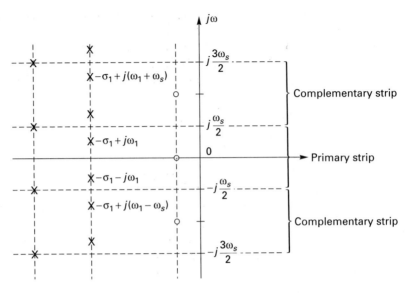

Figure 12.7 Pole–zero locations for $E^*(s)$.

A more complex example of pole–zero locations of $E^*(s)$ is given in Figure 12.7. The primary strip in the s-plane is defined as the strip for which $-\omega_s/2 \leq \omega \leq \omega_s/2$, as shown in Figure 12.7. Note that if the pole–zero locations are known for $E^*(s)$ in the primary strip, the pole–zero locations in the entire plane are also known.

In Figure 12.7, if $E(s)$ has a pole at $-\sigma_1 + j\omega_1$, the sampling operation will generate a pole in $E^*(s)$ at $-\sigma_1 + j(\omega_1 + \omega_s)$, by property 2. Conversely, if $E(s)$ has a pole at $-\sigma_1 + j(\omega_1 + \omega_s)$, the sampling will generate a pole at $-\sigma_1 + j\omega_1$. In fact, a pole location in $E(s)$ at $-\sigma_1 + j(\omega_1 + k\omega_s)$, k an integer, will result in identical pole locations in $E^*(s)$, regardless of the integer value of k. This conclusion is also demonstrated in the example of Figure 12.8. Note that both signals $e_1(t)$ and $e_2(t)$ have the same starred transform, since the two signals have the same value at each sampling instant. Now $\omega_s = 4\omega_1$, since $e_1(t)$ is sampled four

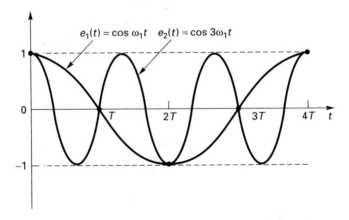

Figure 12.8 Two signals that have the same starred transform.

times for each period of the sinusoid. Hence one pole of $E_1(s)$ occurs at $s = j\omega_1 = j\omega_s/4$, and one of $E_2(s)$ occurs at $s = -j3\omega_1 = j(\omega_1 - \omega_s)$. The other pole of $E_1(s)$ occurs at $s = -j\omega_1 = -j\omega_s/4$, and the other pole of $E_2(s)$ occurs at $s = j3\omega_1 = -j(\omega_1 - \omega_s)$.

Suppose that $e(t)$, with the amplitude spectrum [3] shown in Figure 12.9(a), is sampled. Then $E^*(j\omega)$ has the amplitude spectrum shown in Figure 12.9(b). This can be seen by evaluating (12-8) for $s = j\omega$.

$$E^*(j\omega) = \frac{1}{T}[E(j\omega) + E(j\omega + j\omega_s) + E(j\omega + j2\omega_s)$$

$$+ \cdots + E(j\omega - j\omega_s) + E(j\omega - j2\omega_s) + \cdots] \tag{12-12}$$

Hence the effect of ideal sampling is to replicate the original spectrum at ω_s, at $2\omega_s$, at $-\omega_s$, at $-2\omega_s$, and so forth.

An ideal filter is a filter with a unity gain in the passband and zero gain outside the passband. Of course, such a filter is not physically realizable [3]. It is seen from Figure 12.9 that an ideal lowpass filter could completely recover $E(j\omega)$ [$e(t)$] if the bandwidth of the filter were $\omega_s/2$, for the case that the highest frequency component present in $E(j\omega)$ is less than $\omega_s/2$. This is, of course, essentially a statement of Shannon's sampling theorem [4].

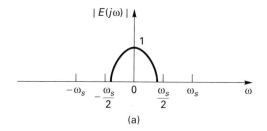

(a)

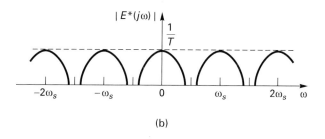

(b)

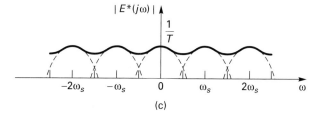

(c)

Figure 12.9   Frequency spectra for $E(j\omega)$ and $E^*(j\omega)$.

Suppose, in Figure 12.9(a), that ω_s is decreased until the higher-frequency components present in $E(j\omega)$ are greater than $\omega_s/2$. Then $E^*(j\omega)$ has the amplitude spectrum shown in Figure 12.9(c); for this case, no filtering scheme, ideal or realizable, will recover $e(t)$. Thus in choosing the sampling rate for a control system, the sampling frequency should be greater than twice the highest-frequency component of significant amplitude in the signal being sampled. As we will see later, other considerations require that the sampling frequency be significantly higher than this value.

It is to be recalled that the ideal sampler is not a physical device, and thus the frequency spectrum, as shown in Figure 12.9, is not the spectrum of a signal that appears in the physical system. The preceding developments will be extended to signals that do appear in physical systems after an investigation of data holds. First, however, consider the following example.

Example 12.5

Suppose that the signal $e(t) = \cos \omega_1 t$, with $\omega_1 = 3$ rad/s, is sampled at the frequency $\omega_s = 8$ rad/s. Since the pole locations of $E(s)$ are at $s = \pm j3$, the poles of $E^*(s)$ are, from property 2,

$$\pm j\omega_1 = \pm j3$$
$$\pm j\omega_1 \pm j\omega_s = \pm j3 \pm j8 = \pm j5, \pm j11$$
$$\pm j\omega_1 \pm j2\omega_s = \pm j3 \pm j16 = \pm j13, \pm j19$$
$$\pm j\omega_1 \pm j3\omega_s = \pm j3 \pm j24 = \pm j21, \pm j27$$
$$\vdots$$

Each set of conjugate imaginary poles represents a sinusoid. We see then that a sinusoid of frequency 3 rad/s, sampled at a rate of 8 rad/s, produced a sum of sinusoids of frequencies 3 rad/s, 5 rad/s, 11 rad/s, 13 rad/s, 19 rad/s, and so forth. In general, sampling a sinusoid of frequency ω_1 at a sampling frequency of ω_s produces a sum of sinusoids of frequencies ω_1 and $k\omega_s \pm \omega_1$, with $k = 1, 2, 3, \ldots$.

12.4 DATA RECONSTRUCTION

The *mathematical model* of the sampler/data-hold operation is given in Figure 12.4, and is repeated in Figure 12.10. Recall that neither of the devices in Figure 12.10 models physical components, but the input–output characteristics of the system in Figure 12.10 accurately model those of a physical sampler/data-hold device. The sampler of this figure is the ideal sampler, and the transfer function in the figure is that of the zero-order hold; that is,

$$G_{ho}(s) = \frac{1 - e^{-Ts}}{s} \tag{12-13}$$

is the transfer function of the zero-order hold. In Section 12-3 we investigated the characteristics of the ideal sampler; in this section we investigate the characteristics of the

Figure 12.10   Representation of sampler/data-hold.

zero-order hold. Then the input–output characteristics of the sampler/data-hold combination will be evident.

The frequency response of the zero-order hold is now derived in a convenient form. From (12-13), an expression for the frequency response can be developed by

$$G_{ho}(j\omega) = \frac{1 - e^{-j\omega T}}{j\omega} e^{j\omega T/2} e^{-j\omega T/2}$$

$$= \frac{2e^{-j\omega T/2}}{\omega} \left[\frac{e^{j\omega T/2} - e^{-j\omega T/2}}{2j} \right] = T \frac{\sin(\omega T/2)}{\omega T/2} e^{-j\omega T/2} \qquad (12\text{-}14)$$

Since

$$\frac{\omega T}{2} = \frac{\omega 2\pi}{2 \omega_s} = \frac{\pi\omega}{\omega_s} \qquad (12\text{-}15)$$

the frequency response of the zero-order hold, (12-14), can be expressed as

$$G_{ho}(j\omega) = T \frac{\sin(\pi\omega/\omega_s)}{\pi\omega/\omega_s} e^{-j\pi\omega/\omega_s}$$

Thus

$$|G_{ho}(j\omega)| = T \left| \frac{\sin(\pi\omega/\omega_s)}{\pi\omega/\omega_s} \right| \qquad (12\text{-}16)$$

and

$$\arg G_{ho}(j\omega) = \frac{-\pi\omega}{\omega_s} + \theta \qquad \theta = \begin{cases} 0 & \sin\dfrac{\pi\omega}{\omega_s} > 0 \\[2mm] \pi & \sin\dfrac{\pi\omega}{\omega_s} < 0 \end{cases} \qquad (12\text{-}17)$$

The magnitude and phase plots for $G_{ho}(j\omega)$ are shown in Figure 12.11.

A word is in order concerning the interpretation of the frequency response of the zero-order hold. First, it must be remembered that *the data-hold must be preceded by the ideal sampler.* Suppose that a sinusoid of frequency ω_1 is applied to the ideal sampler, where $\omega_1 < \omega_s/2$. The output of the sampler contains the frequencies as shown in Figure 12.12(b) (see Example 12.5). Thus the frequency response of the zero-order hold may be used to determine the spectrum of the data-hold output signal. The output signal components are shown in part (c) of the figure. Note that the output signal amplitude spectrum is the same as that shown in the figure if the input signal frequency is any of the frequencies given by $\omega - k\omega_s \pm \omega_1$, $k = 0, 1, 2, \ldots$. Note also that the signal spectrum given in Figure 12.12(c) is for a signal of the type shown in Figure 12.13, which is the output of a zero-order hold with a sinusoid applied to the sampler input.

The purpose of the data-hold is to reconstruct the output to an accurate approximation of the sampler input signal. Data-holds more complex than the zero-order hold can be constructed. However, these holds are seldom used in practice; in almost all digital control systems the zero-order hold is employed. Those interested in the description of more-complex data-holds can see Refs. 1 and 2.

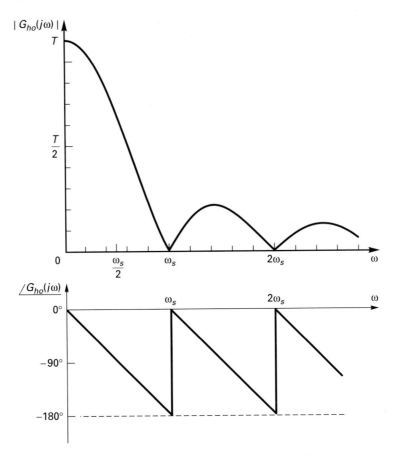

Figure 12.11   Frequency response of zero-order hold.

12.5 PULSE TRANSFER FUNCTION

In this section, a transfer function will be developed for an open-loop sampled-data system. To develop this transfer function, consider the system shown in Figure 12.14(a), where $G_p(s)$ is the plant transfer function. We denote the product of the plant transfer function and the zero-order hold transfer function as $G(s)$, as shown in part (b) of the figure; that is,

$$G(s) = \frac{1 - e^{-Ts}}{s} G_p(s)$$

Note that when a representation of a system as shown in Figure 12.14(b) is given, $G(s)$ *must* contain the transfer function of a data-hold. In general, we do not show the data-hold transfer function separately but combine it with the transfer function of that part of the system that follows the data-hold.

From Figure 12.14(b),

$$C(s) = G(s)E^*(s) \tag{12-18}$$

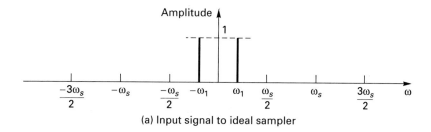

(a) Input signal to ideal sampler

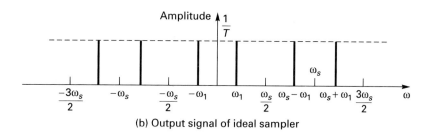

(b) Output signal of ideal sampler

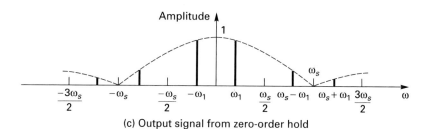

(c) Output signal from zero-order hold

Figure 12.12   Sinusoidal response of sampler/zero-order hold.

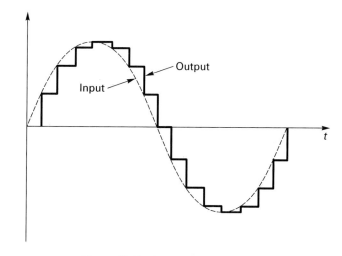

Figure 12.13   Sampler/data-hold response.

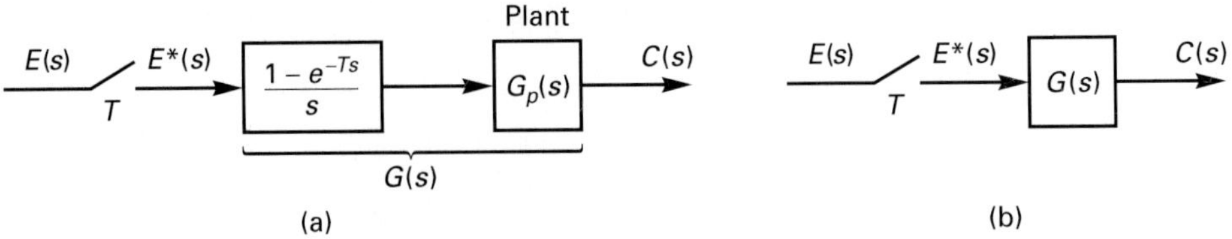

Figure 12.14 Open-loop sampled-data system.

Then, from (12-18) and (12-8), assuming that $c(0)$ is equal to zero,

$$C^*(s) = [G(s)E^*(s)]^* = \frac{1}{T}\sum_{n=-\infty}^{\infty} C(s + jn\omega_s) \tag{12-19}$$

where $[\cdot]^*$ denotes the starred transform of the function in the brackets. Thus, from (12-18) and (12-19),

$$C^*(s) = \frac{1}{T}\sum_{n=-\infty}^{\infty} G(s + jn\omega_s)E^*(s + jn\omega_s) \tag{12-20}$$

Since $E^*(s)$ is periodic, that is, since, from (12-10),

$$E^*(s + jn\omega_s) = E^*(s)$$

(12-20) becomes

$$C^*(s) = E^*(s)\frac{1}{T}\sum_{n=-\infty}^{\infty} G(s + jn\omega_s) = E^*(s)G^*(s) \tag{12-21}$$

In this equation, if we replace e^{Ts} with z, we have the functions as z-transforms, as seen from (12-7). Thus

$$C(z) = E(z)G(z) \tag{12-22}$$

Now $G(z)$ is called the *pulse transfer function* and is the transfer function between the sampled input and the output function *at the sampling instants*. Note that the pulse transfer function gives no information on the nature of the output, $c(t)$, between sampling instants. This information is not contained in either (12-21) or (12-22).

The derivation of (12-22) is general. Thus given *any* function that can be expressed as

$$A(s) = B(s)F^*(s) \tag{12-23}$$

where $F^*(s)$ must be expressible as

$$F^*(s) = f_0 + f_1 e^{-Ts} + f_2 e^{-2Ts} + \cdots \tag{12-24}$$

Then, from the preceding development,

$$A^*(s) = B^*(s)F^*(s) \tag{12-25}$$

and

$$A(z) = B(z)F(z) \tag{12-26}$$

In (12-26),

$$B(z) = \mathfrak{z}[B(s)] \qquad F(s) = F^*(s)\big|_{e^{Ts} = z} \tag{12-27}$$

These equations are general; given a function expressed as (12-23), the z-transform of this function is given by (12-26) and (12-27).

It was assumed in (12-19) that $c(0)$ is zero. This assumption is not necessary to prove (12-22), but it does result in a much simplified proof. Thus (12-22) applies to any system of the configuration shown in Figure 12.14, and (12-26) applies to any function of the form of (12-23). Two examples of the application of the preceding results are given next.

Example 12.6

We want to find the z-transform of the function

$$A(s) = \frac{1 - e^{-Ts}}{s(s + 1)} = \frac{1}{s(s + 1)}(1 - e^{-Ts})$$

Hence $A(s)$ is of the form of (12-23), where $B(s)$ and $F^*(s)$ are given by

$$B(s) = \frac{1}{s(s + 1)} \qquad F^*(s) = 1 - e^{-Ts}$$

Then, from the z-transform tables,

$$B(z) = \mathfrak{z}\left[\frac{1}{s(s + 1)}\right] = \frac{(1 - e^{-T})z}{(z - 1)(z - e^{-T})}$$

and

$$F(z) = 1 - e^{-Ts}\big|_{e^{Ts} = z} = 1 - z^{-1} = \frac{z - 1}{z}$$

Thus from (12-26),

$$A(z) = B(z)F(z) = \frac{(1 - e^{-T})z}{(z - 1)(z - e^{-T})}\frac{z - 1}{z} = \frac{1 - e^{-T}}{z - e^{-T}}$$

Example 12.7

As a second example, we want to find $C(z)$ in the system shown in Figure 12.15, with the input $e(t)$ equal to a unit step. Now, from (12-18),

$$C(s) = G(s)E^*(s) = \frac{1 - e^{-Ts}}{s(s + 1)}E^*(s)$$

and from (12-22),

$$C(z) = G(z)E(z)$$

The pulse transfer function for this system is, from Example 12.6,

$$G(z) = \mathfrak{z}\left[\frac{1 - e^{-Ts}}{s(s + 1)}\right] = \frac{1 - e^{-T}}{z - e^{-T}}$$

and from the z-transform tables

$$E(z) = \mathfrak{z}[u(t)] = \frac{z}{z - 1}$$

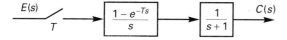

Figure 12.15   Sampled-data system.

Thus

$$C(z) = G(z)E(z) = \frac{(1 - e^{-T})z}{(z - e^{-T})(z - 1)}$$

We can solve for the time response of this system at the sampling instants by partial fractions:

$$\frac{C(z)}{z} = \frac{1 - e^{-T}}{(z - 1)(z - e^{-T})} = \frac{1}{z - 1} + \frac{-1}{z - e^{-T}}$$

and

$$C(z) = \frac{z}{z - 1} - \frac{z}{z - e^{-T}}$$

Hence

$$c(nT) = 1 - (e^{-T})^n = 1 - e^{-nT}$$

Note that the output rises exponentially to a final value of unity at the sampling instants, as shown in Figure 12.16. Note also that the z-transform analysis yields the response *only* at the sampling instants. Based on the preceding analysis, we know nothing about the response of the system between sampling instants. If this information is needed (it normally is in a practical situation), we usually find the complete response by simulation.

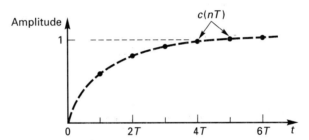

Figure 12.16   Response for Example 12.7.

A MATLAB program that calculates the pulse transfer function $G(z)$ from $G_p(s)$ is given by, for $T = 1$ s,

```
Gpnum = [0 1];   Gpden = [1 1];
Gp = tf(Gpnum,Gpden);
Gz = c2d(Gp,1)
```

Note that the c2d function presumes the presence of a zero–order hold.

A SIMULINK program that simulates this sytem is shown in Figure 12.17. Note that the ideal sampler does not appear explicitly but is integrated into the zero–order hold block.

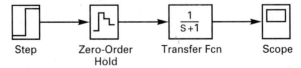

Figure 12.17   SIMULINK Model for Example 12.7.

We can make some additional points concerning the preceding example. First, note that the output of the zero-order hold *for this example* is the unit step; that is, the zero-order hold reconstructs a constant value exactly. *Thus, if the input to the sampler is constant, the sampler/data-hold combination has no effect on system operation and may be ignored.* Then, for this example the output is given by

$$C(s) = \frac{1}{s(s+1)} = \frac{1}{s} + \frac{-1}{s+1}$$

and

$$c(t) = 1 - e^{-t}$$

The z-transform analysis of the last example is then seen to be correct. Given $c(t)$, we can find $c(nT)$ by replacing t with nT. However, given $c(nT)$ from a z-transform analysis, we cannot replace nT with t and have the correct expression for $c(t)$, in general.

We now investigate open-loop systems of other configurations. Consider first the system of Figure 12.18(a). This system contains two analog systems, and both $G_1(s)$ and $G_2(s)$ contain the transfer functions of data-holds. For this system,

$$C(s) = G_2(s)A^*(s) \tag{12-28}$$

and thus

$$C(z) = G_2(z)A(z) \tag{12-29}$$

Also,

$$A(s) = G_1(s)E^*(s) \tag{12-30}$$

and thus

$$A(z) = G_1(z)E(z) \tag{12-31}$$

Then, from (12-29) and (12-31),

$$C(z) = G_1(z)G_2(z)E(z) \tag{12-32}$$

and the total transfer function is the product of the pulse transfer functions.

Consider now the system of Figure 12.18(b), which is the same as the one in Figure 12.18(a), except that the sampler between $G_1(s)$ and $G_2(s)$ has been removed. Of course, for this case, $G_2(s)$ does not contain a data-hold transfer function. Then

$$C(s) = G_1(s)G_2(s)E^*(s) \tag{12-33}$$

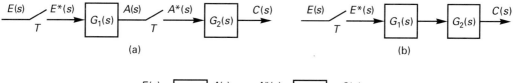

(a) (b) (c)

Figure 12.18 Some sampled-data systems.

and

$$C(z) = \overline{G_1 G_2}(z) E(z) \qquad (12\text{-}34)$$

where

$$\overline{G_1 G_2}(z) = \mathscr{Z}[G_1(s) G_2(s)] \qquad (12\text{-}35)$$

The bar above a product term indicates that the product occurs in the s-forms before the z-transform is taken. Note that

$$\overline{G_1 G_2}(z) \neq G_1(z) G_2(z) \qquad (12\text{-}36)$$

that is, the z-transform of a product of functions is not equal to the product of the z-transforms of the functions.

For the system of Figure 12.18(c),

$$C(s) = G_2(s) A^*(s) = G_2(s) \overline{G_1 E}^*(s) \qquad (12\text{-}37)$$

and

$$C(z) = G_2(z) \overline{G_1 E}(z) \qquad (12\text{-}38)$$

For this case, no transfer function can be found, since $E(z)$ cannot be factored from $\overline{G_1 E}(z)$. In general, no transfer function can be written for a system in which the input is applied to an analog component of the system before being sampled. This characteristic occurs because if the system input $e(t)$ is not sampled, the system output will be a function of $e(t)$ at all times and not just at the sampling instants. Since $E(z)$ does not contain a description of the signal $e(t)$ between sampling instants, the output cannot be expressed as a function of $E(z)$. For systems for which no transfer function can be written, the output can always be expressed as a function of the input. As we show later, this type of system presents no particular difficulties in either analysis or design for most systems.

12.6 OPEN-LOOP SYSTEMS CONTAINING DIGITAL FILTERS

In the preceding section we developed a transfer function technique for open-loop sampled-data systems. In this section, we extend this technique to cover the case that the open-loop sampled-data system contains a digital controller.

First consider the system in Figure 12.19. The A/D converter on the controller input converts the continuous-time signal $e(t)$ into a number sequence $\{e(kT)\}$. The digital controller processes this number sequence $\{e(kT)\}$ and calculates the output number sequence $\{m(kT)\}$. The number sequence $\{m(kT)\}$ is then converted into the continuous-time signal $m(t)$ by the D/A converter.

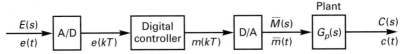

Figure 12.19   Open-loop system with digital controller.

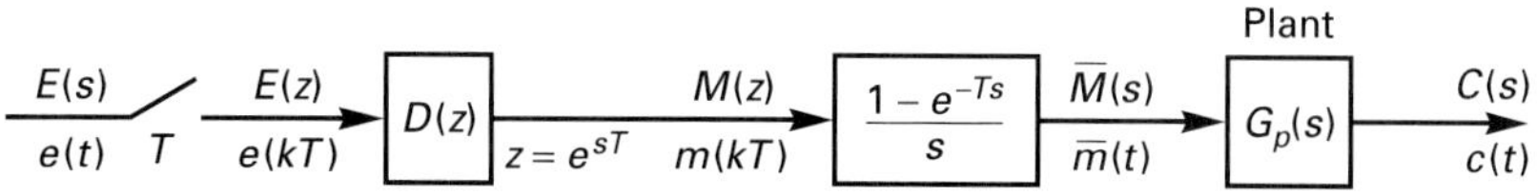

Figure 12.20   Model for open-loop system.

As was indicated in Chapter 11, the digital controller (filter) can be represented by a transfer function $D(z)$ such that

$$M(z) = D(z)E(z) \tag{12-39}$$

or, through the substitution of $z = e^{sT}$,

$$M^*(s) = D^*(s)E^*(s) \tag{12-40}$$

Thus the controller and associated A/D and D/A converters can be represented in block-diagram form as shown in Figure 12.20. In addition, the plant is shown in Figure 12.20. Hence

$$C(s) = G_p(s)\overline{M}(s) = G_p(s)\left[\frac{1 - e^{-Ts}}{s}\right]M^*(s) \tag{12-41}$$

Then

$$
\begin{aligned}
C(z) &= \mathscr{z}\left[\frac{1 - e^{-Ts}}{s}G_p(s)\right]M(z) \\
&= \mathscr{z}\left[\frac{1 - e^{-Ts}}{s}G_p(s)\right]D(z)E(z) = G(z)D(z)E(z)
\end{aligned}
\tag{12-42}
$$

An example illustrating this type of system is now presented.

Example 12.8

As an example, we find the unit step response of a system of the configuration of Figure 12.20. Suppose that the controller is described by the difference equation

$$m(kT) = 2e(kT) - e[(k-1)T]$$

As shown in Section 13.14, this is a discrete PD controller. The controller transfer function is given by

$$D(z) = \frac{M(z)}{E(z)} = 2 - z^{-1} = \frac{2z - 1}{z}$$

Suppose also that

$$G_p(s) = \frac{1}{s + 1}$$

and thus, from Example 12.7,

$$G(z) = \mathscr{z}\left[\frac{1 - e^{-Ts}}{s(s+1)}\right] = \frac{1 - e^{-T}}{z - e^{-T}}$$

Since $E(z) = z/(z - 1)$, we use (12-42) to obtain

$$C(z) = D(z)G(z)E(z)$$

$$= \frac{2z-1}{z}\frac{1-e^{-T}}{z-e^{-T}}\frac{z}{z-1} = \frac{(2z-1)(1-e^{-T})}{(z-1)(z-e^{-T})}$$

In accordance with Section 11.5,

$$C(z) = z^{-1}zC(z) = z^{-1}F(z) = z^{-1}\frac{z(2z-1)(1-e^{-T})}{(z-1)(z-e^{-T})}$$

By partial fractions,

$$\frac{F(z)}{z} = \frac{(2z-1)(1-e^{-T})}{(z-1)(z-e^{-T})} = \frac{1}{z-1} + \frac{1-2e^{-T}}{z-e^{-T}}$$

$$F(z) = \frac{z}{z-1} + \frac{(1-2e^{-T})z}{z-e^{-T}}$$

The inverse transform yields

$$f(k) = 1 + (1-2e^{-T})e^{-kT}$$

Then,

$$c(k) = f(k-1)u(k-1)$$

$$= [1 + (1-2e^{-T})e^{-(k-1)T}]u(k-1)$$

Note that the value $c(0) = 0$ is obvious from inspection of $C(z)$, since the order of the numerator of $C(z)$ is less than the order of the denominator (see Problem 11.5).

12.7 CLOSED-LOOP DISCRETE-TIME SYSTEMS

In the preceding sections, a technique was developed for finding the transfer functions of open-loop sampled-data systems. With sampled-data systems, special techniques are required for determining transfer functions, since a transfer function does not exist for the ideal sampler. The output of the ideal sampler, $E^*(s)$, cannot be expressed as a transfer function multiplied by the input $E(s)$, with the transfer function independent of $E(s)$.

In the following section a technique is developed for determining the transfer functions of closed-loop sampled-data systems. Other methods are available for finding these transfer functions [5]; however, it is felt that the method presented here is the simplest technique and the easiest to remember. This technique will be presented after the following derivation.

To illustrate the derivation of a closed-loop system transfer function, consider the system of Figure 12.21. For this system

$$C(s) = G(s)E^*(s) \tag{12-43}$$

where $G(s)$ includes the transfer function of a data-hold, and

$$E(s) = R(s) - H(s)C(s) \tag{12-44}$$

Substituting (12-43) into (12-44), we obtain

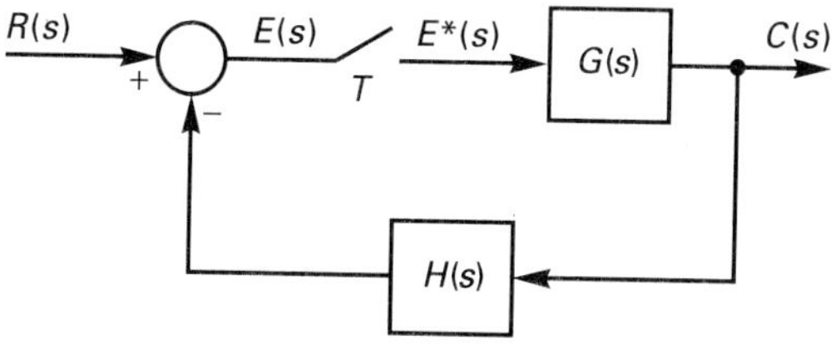

Figure 12.21 Closed-loop sampled-data system.

$$E(s) = R(s) - G(s)H(s)E^*(s) \tag{12-45}$$

and thus

$$E^*(s) = R^*(s) - \overline{GH}^*(s)E^*(s) \tag{12-46}$$

Solving for $E^*(s)$ yields

$$E^*(s) = \frac{R^*(s)}{1 + \overline{GH}^*(s)} \qquad E(z) = \frac{R(z)}{1 + \overline{GH}(z)} \tag{12-47}$$

From (12-43) and (12-47),

$$C(z) = G(z)E(z) = \frac{G(z)}{1 + \overline{GH}(z)}R(z) \tag{12-48}$$

which describes the system response at sampling instants. Also, from (12-43) and (12-47),

$$C(s) = \frac{G(s)R^*(s)}{1 + \overline{GH}^*(s)} \tag{12-49}$$

which describes the system response at all times. Generally we analyze systems via the z-transform because of the difficulties of analysis of equations of the form of (12-49), even though this expression for $C(s)$ yields the system characteristics at all instants of time.

Problems can be encountered in deriving the transfer function of a closed-loop system. This can be illustrated for the preceding case. Suppose that (12-44) had been starred and substituted into (12-43). Then

$$C(s) = G(s)R^*(s) - G(s)\overline{HC}^*(s) \tag{12-50}$$

and $C^*(s)$ is given by

$$C^*(s) = G^*(s)R^*(s) - G^*(s)\overline{HC}^*(s) \tag{12-51}$$

In general, $C^*(s)$ cannot be factored from $\overline{HC}^*(s)$. Thus (12-51) cannot be solved for $C^*(s)$.

To avoid the problem illustrated in (12-51), an equation should not be starred if an internal system variable is lost as a factor. However, for systems more complex than these, solving the system equations using the above procedure can be quite difficult. A better method of analysis is developed next.

12.8 TRANSFER FUNCTIONS FOR CLOSED-LOOP SYSTEMS

The determination of the transfer function of a closed-loop sampled-data system is difficult because the sampler does not have a transfer function. Consider again the system of

Taking the z-transform of the equation for $C(s)$ and substituting for $E(z)$ using the equation above yields

$$C(z) = \frac{D(z)G(z)}{1 + D(z)\overline{GH}(z)}R(z)$$

The transfer function of the single-loop digital control system of Figure 12.23 is then

$$T(z) = \frac{D(z)G(z)}{1 + D(z)\overline{GH}(z)}$$

In the design of digital control systems to be considered in the next chapter, we usually assume that our system has this transfer function. Comparing this equation to that of the analog control system considered in Chapters 7 and 9,

$$C(s) = \frac{G_c(s)G_p(s)}{1 + G_c(s)G_p(s)H(s)}R(s)$$

we see that the compensator transfer function enters the closed-loop system transfer function in the same manner in a digital control system as in an analog system. In the following chapter, we see many more similarities between analog control systems and digital control systems.

Example 12.10

Consider the system shown in Figure 12.24(a). The flow graph for this system is given in Figure 12.24(b). The system equations are, for step 4 of the procedure,

$$X_1 = R - G_2 X_2^*$$
$$X_2 = G_1 X_1^* - G_2 H X_2^*$$
$$C = G_2 X_2^*$$

Starring these equations yields

$$X_1^* = R^* - G_2^* X_2^*$$
$$X_2^* = G_1^* X_1^* - \overline{G_2 H}^* X_2^*$$
$$C^* = G_2^* X_2^*$$

We have three equations and three unknowns; these equations may be solved by any convenient method. In this example we construct a flow graph from the equations and apply Mason's gain formula to the flow graph. We use the first equation to construct the node for X_1^*, the second one for X_2^*, and the third one for C^*. The resulting flow graph is given in Figure 12.24(c). Also, the preceding equation for the continuous output is used to construct the $C(s)$ node. From the flow graph we write

$$C^* = \frac{G_1^* G_2^*}{1 + G_1^* G_2^* + \overline{G_2 H}^*}R^*$$

or

$$C(z) = \frac{G_1(z)G_2(z)}{1 + G_1(z)G_2(z) + \overline{G_2 H}(z)}R(z)$$

$R(s)$

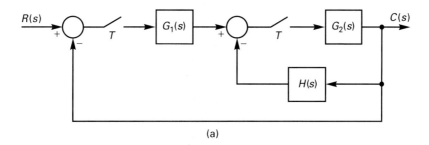

(a)

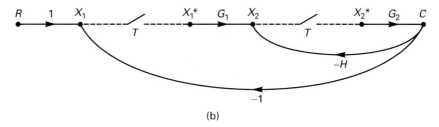

(b)

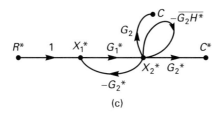

(c)

Figure 12.24   System for Example 12.10.

and the continuous-data output is

$$C(s) = \frac{G_2(s)G_1^*(s)}{1 + G_1^*(s)G_2^*(s) + \overline{G_2H}^*(s)} R^*(s)$$

We have computed the continuous-data output function, even though we usually do not use it.

Example 12.11

As a final example of closed-loop digital control systems, the system of Figure 12.25 is considered. In this system, the digital controller is in a feedback loop, and the system has one analog loop, that is, one loop that has no sampling. The system flow graph is given in part (b) of the figure, and from this flow graph we write

$$C(s) = \frac{G(s)R(s)}{1 + G(s)H_1(s)} - \frac{G(s)H_2(s)}{1 + G(s)H_1(s)} D^*(s)C^*(s)$$

The denominator in this equation appears because of the analog loop. Thus

$$C(z) = \left[\frac{GR}{1 + GH_1}\right](z) - \left[\frac{GH_2}{1 + GH_1}\right](z)D(z)C(z)$$

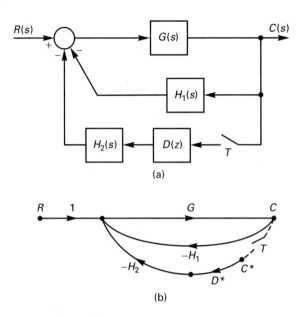

Figure 12.25 System for Example 12.11.

where the brackets are used to indicate that the products and quotients of the functions are performed in the s-variable before the z-transform is taken. Thus in this case, the brackets have the same meaning as the overbar in expressions such as that for $C(z)$ in the preceding example. We solve the preceding expression for $C(z)$, resulting in

$$C(z) = \frac{\left[\dfrac{GR}{1 + GH_1}\right](z)}{1 + \left[\dfrac{GH_2}{1 + GH_1}\right](z)D(z)}$$

Note that for the system in the preceding example, no transfer function can be written, since $R(z)$ cannot be factored from the equation for $C(z)$. We can see that this is the case from an examination of the system in Figure 12.25. In this system the input signal is applied to an analog part of the system $[G(s)]$ before being sampled, and thus the system response must be a function of $r(t)$ for all values of time and not just at the sampling instants.

12.9 STATE VARIABLES FOR SAMPLED-DATA SYSTEMS

In Section 11.7, state-variable models for discrete systems were introduced. As given in (11-32), the general form for the state model of a linear time-invariant discrete system is

$$\mathbf{x}(k + 1) = \mathbf{A}\mathbf{x}(k) + \mathbf{B}u(k)$$
$$\mathbf{y}(k) = \mathbf{C}\mathbf{x}(k)$$

(12-58)

For the single-input, single-output system, $u(k)$ and $y(k)$ are scalars, and the system transfer function is given by (11-40), which is

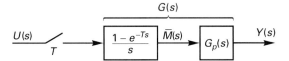

Figure 12.26   Sampled-data system.

$$G(z) = \frac{Y(z)}{U(z)} = \mathbf{C}(z\mathbf{I} - \mathbf{A})^{-1}\mathbf{B} \tag{12-59}$$

As shown in Section 11.7, given the transfer function $G(z)$, we can derive a state model of the system. The procedure involves first constructing a simulation diagram with the transfer function $G(z)$. Then a state variable is assigned to each unit delay output. As a final step, the equations of each unit delay input and of the system output are written, resulting in state equations of the form of (12-58).

Consider now the sampled-data system of Figure 12.26. As shown in the preceding sections, this system has the transfer function

$$G(z) = \frac{Y(z)}{U(z)} = \mathscr{z}\left[\frac{1 - e^{-Ts}}{s}G_p(s)\right] \tag{12-60}$$

Hence the procedure outlined in the last paragraph can be used to find a state model of this system. An example is now given.

Example 12.12

Suppose that the plant in Figure 12.26 has the transfer function $G_p(s) = 1/[s(s + 1)]$ and that the sample period T is 1 s. The pulse transfer function is

$$G(z) = \mathscr{z}\left[\frac{1 - e^{-Ts}}{s^2(s + 1)}\right] = \frac{z - 1}{z}\mathscr{z}\left[\frac{1}{s^2(s + 1)}\right]$$

From the z-transform table in Appendix C,

$$\mathscr{z}\left[\frac{1}{s^2(s + 1)}\right]_{T = 1} = \frac{z[(1 - 1 + e^{-1})z + (1 - e^{-1} - e^{-1})]}{(z - 1)^2(z - e^{-1})}$$

Thus

$$G(z) = \frac{0.368z + 0.264}{(z - 1)(z - 0.368)} = \frac{0.368z + 0.264}{z^2 - 1.368z + 0.368}$$

$$= \frac{0.368z^{-1} + 0.264z^{-2}}{1 - 1.368z^{-1} + 0.368z^{-2}}$$

A simulation diagram of this system is given in Figure 12.27, with the state variables denoted. We can then write the state equations as

$$\begin{bmatrix} x_1(k + 1) \\ x_2(k + 1) \end{bmatrix} = \begin{bmatrix} 0 & 1 \\ -0.368 & 1.368 \end{bmatrix}\begin{bmatrix} x_1(k) \\ x_2(k) \end{bmatrix} + \begin{bmatrix} 0 \\ 1 \end{bmatrix}u(k)$$

$$y(k) = \begin{bmatrix} 0.264 & 0.368 \end{bmatrix}\begin{bmatrix} x_1(k) \\ x_2(k) \end{bmatrix}$$

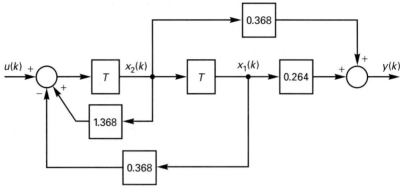

Figure 12.27 Simulation diagram for Example 12.12.

Of course, since many different simulation diagrams can be constructed that have the transfer function of this system, many different state models can be found for this system. The transfer function $G(z)$ can be calculated using the MATLAB program given in Example 12.7.

In the preceding example, suppose that the analog transfer function $G_p(s)$ is that of a dc motor (see Section 2.7). The plant output $y(t)$ is then the position of the motor shaft. In the discrete state model developed in the example, the output $y(k)$ is a linear combination of the two states, so neither state represents the physical variable of shaft position. Furthermore, neither state represents shaft velocity. In general, we prefer that the state variables represent physical variables, as much as is possible. However, if the state model is derived by the preceding simulation-diagram procedure, this will not be the case. A technique for finding the state model of a sampled-data system that will allow us to designate the states is developed next [1,6].

This technique is based on the analog state model of the plant. As shown in Chapter 3, we can develop a state model for the plant of the form [see (3-6)]

$$\dot{\mathbf{v}}(t) = \mathbf{A}_c\mathbf{v}(t) + \mathbf{B}_c\bar{m}(t)$$
$$y(t) = \mathbf{C}_c\mathbf{v}(t) \tag{12-61}$$

We denote the states of the analog plant by $\mathbf{v}(t)$ and the states of the discrete model of the plant by $\mathbf{x}(k)$. The matrices of the analog state equations have a subscript c, and those of the discrete model do not have subscripts. We now develop a procedure to derive the discrete state model of the plant (12-58) directly from the analog state model (12-61), while retaining the analog states as the discrete states.

The solution to the analog state equations (12-61) is given in (3-23), which is

$$\mathbf{v}(t) = \Phi_c(t)\mathbf{v}(0) + \int_0^t \Phi_c(t-\tau)\mathbf{B}_c\bar{m}(\tau)\,d\tau \tag{12-62}$$

where, from (3-21) and (3-32), the state-transition matrix $\Phi_c(t)$ is given by

$$\Phi_c(t) = \mathcal{L}^{-1}\left((s\mathbf{I}-\mathbf{A}_c)^{-1}\right) = \mathbf{I}+\mathbf{A}_ct+\mathbf{A}_c^2\frac{t^2}{2!}+\dots \tag{12-63}$$

Note that since $\bar{m}(t)$ is the output of a zero-order hold, then $\bar{m}(t)$ is constant at the value $m(0) = u(0)$ for $0 \le t < T$. Evaluating (12-62) at $t = T$,

$$\mathbf{v}(T) = \boldsymbol{\Phi}_c(T)\mathbf{v}(0) + \left[\int_0^T \boldsymbol{\Phi}_c(T - \tau) \, d\tau\right]\mathbf{B}_c u(0) \qquad (12\text{-}64)$$

Compare this equation to that of the discrete state variables in (12-58), evaluated for $k = 1$:

$$\mathbf{x}(1) = \mathbf{A}\mathbf{x}(0) + \mathbf{B}u(0) \qquad (12\text{-}65)$$

Now, in our notation, $\mathbf{x}(1) = \mathbf{x}(T)$. Thus if we make the substitutions in (12-65),

$$\mathbf{x} = \mathbf{v}$$
$$\mathbf{A} = \boldsymbol{\Phi}_c(T)$$
$$\mathbf{B} = \left[\int_0^T \boldsymbol{\Phi}_c(T - \tau) \, d\tau\right]\mathbf{B}_c \qquad (12\text{-}66)$$

then (12-65) becomes (12-64). Hence if we choose the discrete state variables to be the same as the analog state variables and if the matrices $\mathbf{A}$ and $\mathbf{B}$ are calculated as given in (12-66), we have a valid state model of the discrete system. From (12-61),

$$y(kT) = \mathbf{C}_c \mathbf{v}(kT)$$

and we see that from (12-58) the matrix $\mathbf{C}$ is equal to the matrix $\mathbf{C}_c$.

In summary, given the plant analog state equations (12-61) of the sampled-data system of Figure 12.26, a discrete state model of this system in the form of (12-58) is calculated by the equations

$$\mathbf{A} = \boldsymbol{\Phi}_c(T) = \mathcal{L}^{-1}[(s\mathbf{I} - \mathbf{A}_c)^{-1}]_{t=T} = \mathbf{I} + \mathbf{A}_c T + \mathbf{A}_c^2 \frac{T^2}{2!} + \cdots$$

$$\mathbf{B} = \left[\int_0^T \boldsymbol{\Phi}_c(T - \tau)d\tau\right]\mathbf{B}_c = \left[\mathbf{I}T + \mathbf{A}_c\frac{T^2}{2!} + \mathbf{A}_c^2\frac{T^3}{3!} + \cdots\right]\mathbf{B}_c \qquad (12\text{-}67)$$

$$\mathbf{C} = \mathbf{C}_c$$

With this evaluation of these matrices, the states of the discrete model are the same as those of the analog model. The infinite series expansion for the matrix $\mathbf{A}$ is obtained from (12-63), and the infinite series expansion for the matrix $\mathbf{B}$ may be derived by performing the integration indicated on the series expansion for $\boldsymbol{\Phi}_c(t)$. The series in (12-67) are especially useful in the digital computer evaluation of the discrete-model matrices, in which the series are truncated [6]. An example is given next.

Example 12.13

We develop a discrete model of the satellite discussed in Section 2.6. The system is given in Figure 12.28(a), and the analog state model is derived from the flow graph of Figure 12.28(b). From this flow graph,

$$\dot{\mathbf{v}}(t) = \begin{bmatrix} 0 & 1 \\ 0 & 0 \end{bmatrix}\mathbf{v}(t) + \begin{bmatrix} 0 \\ 1 \end{bmatrix}\bar{m}(t) = \mathbf{A}_c\mathbf{v}(t) + \mathbf{B}_c\bar{m}(t)$$

$$y(t) = \begin{bmatrix} 1 & 0 \end{bmatrix}\mathbf{v}(t) = \mathbf{C}_c\mathbf{v}(t)$$

Then

$$(s\mathbf{I} - \mathbf{A}_c) = \begin{bmatrix} s & -1 \\ 0 & s \end{bmatrix}, \qquad (s\mathbf{I} - \mathbf{A}_c)^{-1} = \begin{bmatrix} \dfrac{1}{s} & \dfrac{1}{s^2} \\ 0 & \dfrac{1}{s} \end{bmatrix}$$

and

$$\Phi_c(t) = \mathcal{L}^{-1}[(s\mathbf{I} - \mathbf{A}_c)^{-1}] = \begin{bmatrix} 1 & t \\ 0 & 1 \end{bmatrix}$$

The discrete model matrix $\mathbf{A}$ is then

$$\mathbf{A} = \Phi_c(T)|_{T=0.1} = \begin{bmatrix} 1 & 0.1 \\ 0 & 1 \end{bmatrix}$$

Also,

$$\int_0^T \Phi_c(T - \tau)\, d\tau = \begin{bmatrix} \displaystyle\int_0^T d\tau & \displaystyle\int_0^T (T - \tau)\, d\tau \\ 0 & \displaystyle\int_0^T d\tau \end{bmatrix} = \begin{bmatrix} T & \dfrac{T^2}{2} \\ 0 & T \end{bmatrix}$$

and since $T = 0.1$ s,

$$\mathbf{B} = \left[\int_0^T \Phi_c(T - \tau)\, d\tau \right] \mathbf{B}_c = \begin{bmatrix} 0.1 & 0.005 \\ 0 & 0.1 \end{bmatrix} \begin{bmatrix} 0 \\ 1 \end{bmatrix} = \begin{bmatrix} 0.005 \\ 0.1 \end{bmatrix}$$

$$\mathbf{C} = \mathbf{C}_c = \begin{bmatrix} 1 & 0 \end{bmatrix}$$

The discrete state model of the satellite is then

$$\mathbf{x}(k + 1) = \begin{bmatrix} 1 & 0.1 \\ 0 & 1 \end{bmatrix} \mathbf{x}(k) + \begin{bmatrix} 0.005 \\ 0.1 \end{bmatrix} u(k)$$

$$y(k) = \begin{bmatrix} 1 & 0 \end{bmatrix} \mathbf{x}(k)$$

The calculations in this example can be verified with the MATLAB program

```
Ac = [0 1; 0 0];   Bc = [0;1];   Cc = [1 0];
[A,B] = c2d(Ac,Bc,0.1)
```

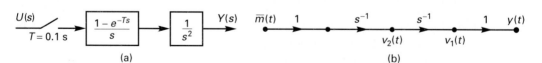

(a)   (b)

Figure 12.28 System for Example 12.13.

To illustrate a final point, the simulation diagram of the analog plant of the last example is given in Figure 12.29(a), and the simulation diagram of the discrete state model is given in Figure 12.29(b). Note that in the two simulation diagrams,

$$x_1(k) = v_1(t)|_{t = kT}$$

$$x_2(k) = v_2(t)|_{t = kT}$$

Even though the states, the input, and the output of the two diagrams are equal at the sampling instants, the two diagrams bear no resemblance to each other. In general, the two simulation diagrams for such a system are not similar.

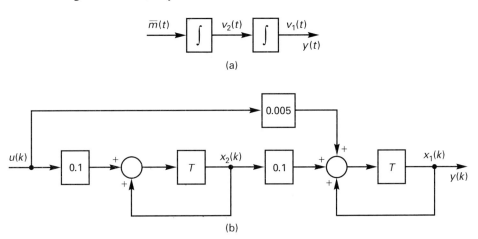

Figure 12.29   Simulation diagram for satellite.

12.10 SUMMARY

In this chapter the topics of sampling, of pulse transfer functions, and of state variables for sampled-data systems were considered. First the concept of ideal sampling, and the modeling of physical sampling by an ideal sampler and a data hold was demonstrated. The Laplace transform of the output of an ideal sampler, called the starred transform, was developed. The effects of sampling were presented in detail, and it was shown that for sampling to have negligible effects, the sampling rate should be much greater than the highest frequency present in the signal to be sampled. The most common data-reconstruction device, the zero-order hold, was investigated.

The principal problem in determining the transfer function of a sampled-data system is based on the characteristic that an ideal sampler cannot be represented by a transfer function. Thus the techniques for finding the transfer function of a sampled-data system are different from those used to find the transfer function of an analog system. A technique for finding transfer functions of sampled-data systems was presented in this chapter. For the case that an analog input signal is applied to an analog part of the system prior to being sampled, no system transfer function can be found. However, the technique presented for finding transfer functions also finds the system output expression as a function of the input for the case that no transfer function can be found.

Two methods were presented for finding a discrete state model of a sampled-data system. The first method developed the discrete state model from the pulse transfer function. The second method developed the discrete state model from the plant analog state model. The advantages of the second method are that the states may be specified as physical variables and also that the method may be implemented easily as a computer program. Hence the second method is used in almost all practical situations.

REFERENCES

1. C. L. Phillips and H. T. Nagle, Jr. *Digital Control System Analysis and Design*, 3rd ed. Upper Saddle River, NJ: Prentice Hall, 1995.

2. B. C. Kuo, *Analysis and Synthesis of Sampled-Data Control Systems.* Upper Saddle River, NJ: Prentice Hall, 1963.

3. C. L. Phillips and J. M. Parr. *Signals, Systems, and Transforms*, 2nd ed. Upper Saddle River, NJ: Prentice Hall, 1999.

4. R. M. Oliver, J. R. Pierce, and C. E. Shannon. "The Philosophy of Pulse Code Modulation," *Proc. IRE*, 36, no. 11 (November 1948): 1324–1331.

5. M. Sedlar and G. A. Bekey. "Signal Flow Graphs of Sampled Data Systems: A New Formulation," *IEEE Trans. Autom. Control*, AC-12, no. 2 (October 1967): 606–608.

6. G. F. Franklin, J. D. Powell, and M. Workman. *Digital Control of Dynamic Systems*, 3rd ed. Reading, MA: Addison-Wesley, 1998.

PROBLEMS

12.1. Find $E^*(s)$ for each of the following functions.

    (a) $e(t) = e^{-3t}$

    (b) $E(s) = e^{-Ts}/(s + 3)$

    (c) $e(t) = 1 - e^{-2t}$

    (d) $E(s) = (s + 3)/[s(s + 2)]$

12.2. Express the starred transform of $e(t - kT)u(t - kT)$, k a positive integer, in terms of the starred transform of $e(t)$.

12.3. Compare the pole-zero locations of $E^*(s)$ in the s-plane with those of $E(s)$ for those functions given in Problem 12.1. Let $T = 0.2$ s.

12.4. **(a)** Find $E^*(s)$ for $T = 0.2$ s, for each of the two functions given.

        (i) $e_1(t) = \cos(4\pi t)$

        (ii) $e_2(t) = \cos(14\pi t)$

    (b) Why are the two transforms the same?

    (c) Give a third time function that has the same starred transform as those in (a).

12.5. (a) A sinusoid with a frequency of 3 Hz is applied to a sampler-zero-order-hold combination. The sampling frequency is 12 Hz. List all the frequencies present in the data-hold output that are less than 40 Hz.

(b) Repeat (a) if the input sinusoid has a frequency of 9 Hz.

(c) Why are the results of (a) and (b) the same?

(d) List four other frequencies that will yield the same result as in (a) and (b).

12.6. A signal $e(t)$ is sampled by the ideal sampler as specified in (12-5).

(a) Give the conditions under which $e(t)$ can be *completely* recovered from $e*(t)$, that is, the conditions under which *no* loss of information by the sampling process occurs.

(b) State which of the conditions listed in (a) can occur in a physical system. Also, recall that the sampling operation itself is not physically realizable.

(c) Considering the answers in (b), state why we are able to successfully employ systems that use sampling.

12.7. A system is linear provided that the principle of superposition applies. Is the sampler-zero-order-hold device linear? Prove your answer.

12.8. Given $E*(s) = e^{-3Ts}$, $T = 0.5$ s.

(a) Find $e(kT)$ for all k.

(b) Can $e(t)$ be found from the available information? Justify your answer.

(c) Sketch two different time functions that satisfy (a).

12.9. The output of a sampler/zero-order hold is given by $\bar{E}(s) = (e^{-2Ts} - e^{-3Ts})/s$, with $T = 1$ s.

(a) Sketch $\bar{e}(t)$.

(b) Find $e(kT)$ for all of k.

(c) Can $e(t)$ be found from the available information? Justify your answer.

(d) Sketch two different time functions that satisfy (a).

12.10. The signal $e(t) = 5\sin(2t)$ is applied to a sampler-zero-order-hold combination, with $T = (\pi/6)$ s. Hence the zero-order-hold output is of the form shown in Figure 12.13.

(a) The output of the combination has a frequency component at $\omega = 2$ rad/s. Find the amplitude and phase of this component.

(b) Repeat (a) for the frequency component at $\omega = 14$ rad/s and at $\omega = 26$ rad/s.

12.11. Given the sampled-data system of Figure P12.11.

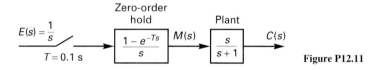

Figure P12.11

(a) Find the system response at the sampling instants for a unit step input.

(b) Verify the results in (a) by determining the input to the plant, $m(t)$, and then calculating $c(t)$ by continuous-data techniques.

(c) What is the effect of a sampler/zero-order hold in a system if the input signal to the sampler is constant? Why?

12.12. (a) Repeat Problem 12.11, but with the plant transfer function

$$G_p(s) = \frac{2s + 1}{(s + 1)(s + 2)}$$

(b) Verify the partial-fraction expansion of $C(s)$ using MATLAB.

12.13. (a) Find the system response at the sampling instants for a unit step input for the system of Figure P12.13. Plot $c(kT)$ versus time.

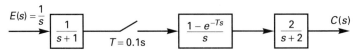

Figure P12.13

(b) Simulate the system with SIMULINK, and verify the plot of $c(kT)$.

12.14. (a) Find the transfer function $C(z)/E(z)$ for the system of Figure P12.14.

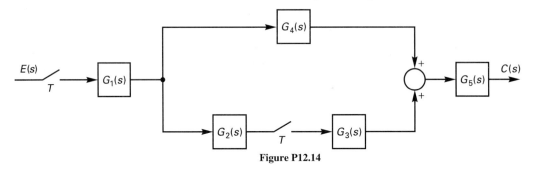

Figure P12.14

(b) Express $C(z)$ as a function of the input signal for the system of Figure P12.14 for the case that the sampler preceding $G_1(s)$ is removed.

(c) Find the transfer function $C(z)/E(z)$ for the system of Figure P12.14 for the case that the sampler preceding $G_3(s)$ is removed.

(d) For the systems of (a), (b), and (c), list the transfer functions that contain the transfer function of a data-hold as a factor.

12.15. (a) For the system of Figure P12.15, the digital controller solves the difference equation $m(k + 1) = m(k) + 0.5e(k)$, where $e(k)$ is the controller input and $m(k)$ is the controller output. Find the transfer function $C(z)/E(z)$ for the case that the sampling rate is 5 Hz.

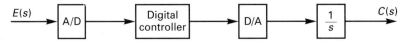

Figure P12.15

(b) Use the results of (a) to find a state model of the system. The model is second order.

(c) Verify that the state model in (b) has the correct transfer function.

12.16. For each of the systems of Figure P12.16, express the output $C(z)$ as a function of the input signal.

12.17. Consider the attitude control system of a rigid-body satellite shown in Figure P12.17. The model of the satellite is developed in Section 2.6. In this system, digital compensation has been added and the sample period is $T = 0.1$ s.

(a) Find the pulse transfer function for the satellite.

(b) With $D(z) = 1$, find the closed-loop transfer function.

(c) Suppose that the transfer function of the digital compensator is given by

$$D(z) = 1.2 + \frac{z-1}{z}$$

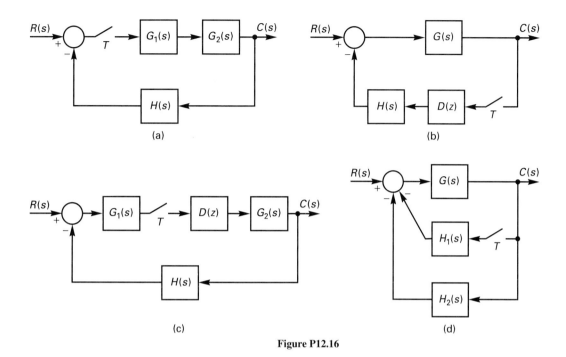

(a)

(b)

(c)

(d)

Figure P12.16

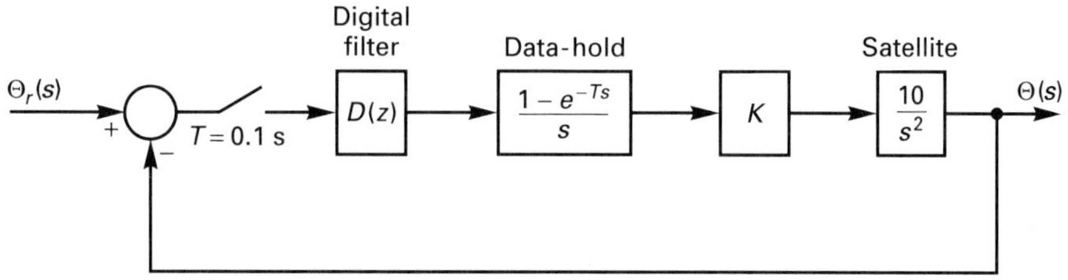

Figure P12.17

It is shown in Chapter 13 that this is the transfer function for a discrete PD compensator. Find the closed-loop system transfer function.

(**d**) Verify the transfer function in (a) by modifying the MATLAB program in Example 12.7.

12.18. Consider the attitude control system of a rigid-body satellite of Problem 12.17.

(**a**) Find a discrete state model for the plant such that one state is angular position and the second state is angular velocity.

(**b**) Verify the results in (a) with MATLAB.

(**c**) With $D(z) = 1$, use the results of (a) to find a closed-loop state model.

(**d**) Suppose that the transfer function of the digital compensator is given by

$$D(z) = 1.2 + \frac{z-1}{z}$$

It is shown in Chapter 13 that this is the transfer function for a discrete PD compensator. Use the results of (a) to find a closed-loop state model. The model will be third order.

(e) Use MATLAB to find the closed-loop transfer function using the results in (d) for $K = 1$. This will verify the results in Problem 12.17(c) for $K = 1$, if these results are available.

12.19. Consider the temperature-control system for the large test chamber shown in Figure P12.19. For this problem, ignore the disturbance input.

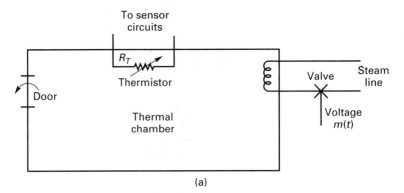

(a)

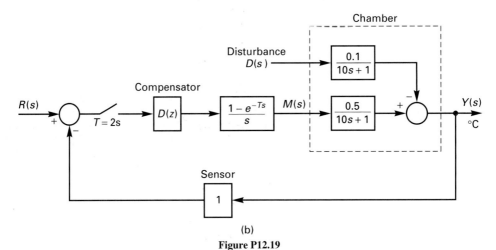

(b)

Figure P12.19

(a) Find the plant transfer function $G(z)$.
(b) Use MATLAB to verify $G(z)$.
(c) Using $G(z)$, draw a simulation diagram such that the state $x(k) = y(k)$.
(d) Use the simulation diagram of (c) to write state equations for the plant.
(e) Verify the results of (d) by calculating the plant transfer function from the state model.
(f) Verify the results of (e) using MATLAB.

12.20. Consider the joint control system for a robot arm shown in Figure P12.20. This system is described in Section 2.12. The plant transfer function is given by

$$\frac{\Theta_L(s)}{E_a(s)} = G_p = \frac{0.15}{s(s+1)(s+5)}$$

(a) Find only the poles of the plant transfer $G(z)$ for $T = 0.02$ s.
(b) From (a), find the third-order polynomial of the denominator of the plant transfer function G(z).

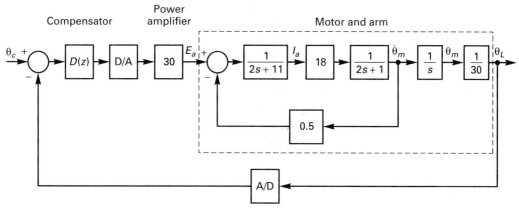

Figure P12.20

(c) Use MATLAB to find the discrete plant transfer function $G(z)$. The denominator must equal the polynomial found in (b).

(d) Use MATLAB to find a discrete state model of the plant.

12.21. Shown in Figure P12.21 is the block diagram of a carbon dioxide control system of an environmental plant chamber. The delay that appears in the sensor has been ignored.

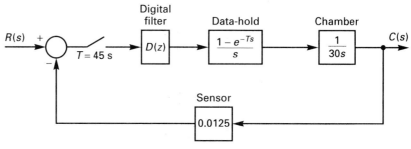

Figure P12.21

(a) Find an analog state model of the plant (chamber).

(b) Find the discrete state model of the plant, using the results in (a).

(c) Find the discrete state model of the closed-loop system, with $D(z) = 1$. Retain the same state as in the model in (b).

(d) Suppose that the transfer function of the digital filter is given by

$$D(z) + K_P + \frac{K_I T z}{z - 1}$$

which is a discrete PI compensator. The closed-loop system is now second order. Find a discrete state model of the closed-loop system, such that the state variable of (c) is one of the states of the closed-loop system.

12.22. Consider the satellite state model in Example 12.13.

(a) Use MATLAB to verify the state-model calculations.

(b) Use MATLAB to find the plant pulse transfer function $G(z)$ from the state model.

(c) Use MATLAB to find a different state model of the plant.

(d) Use MATLAB to verify the transfer function of the state model in (c).

12.23. Consider the digital-plotter servo depicted in Figure P12.23.

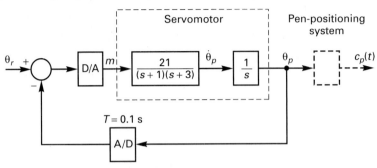

Figure P12.23

(a) Find only the poles of the plant transfer function $G(z)$.

(b) From (a), find the third-order polynomial of the denominator of $G(z)$.

(c) Use MATLAB to find the discrete plant transfer function $G(z)$. The denominator should check that in (b).

(d) Find an analog state model of the plant with the states

$$\mathbf{x}^T(t) = \left[\, \theta_p(t) \; \dot{\theta}_p(t) \; \ddot{\theta}_p(t) \,\right]$$

(e) Use MATLAB to find the discrete state model based on the analog model in (d).

(f) Use MATLAB to verify that the transfer function of (e).

12.24. This problem requires computer calculations for all parts. Consider the robot joint control system of Problem 12.20.

(a) From the plant transfer function $G_p(s)$ given in Problem 12.20, find an analog state model for the plant, using MATLAB.

(b) Use MATLAB to verify that the state model in (a) has the correct transfer function.

(c) Use MATLAB to find a discrete state model, using the results in (a).

(d) Use MATLAB to find the discrete plant transfer function, using the results of (c).

(e) Use MATLAB and the transfer function in (d) to find a different state model of the plant.

(f) Use MATLAB to verify that the state model in (e) has the correct transfer function.

13

Analysis and Design of Digital Control Systems

In Chapter 12 a technique was developed for finding the output expression for a general linear time-invariant discrete-time system. The technique was shown to apply to the special but important case of digital control systems. In this chapter the time response of discrete-time systems is investigated first. Stability criteria and stability tests are then developed. This naturally leads to the relationship of the zero locations of the system characteristic equation to the system transient response. Then it is shown that the root-locus technique as developed for continuous-time (analog) systems applies directly to discrete-time systems.

Next the classical frequency-response techniques are applied to discrete-time systems. The techniques to be considered are the Nyquist criterion, Bode diagrams, and gain-phase plots. In addition, it is shown that the Routh–Hurwitz criterion can be applied to discrete-time control systems by use of an important transformation. Finally, the design of digital control systems is considered. In the subsequent development, it is assumed that the reader is familiar with the previously mentioned techniques as applied to continuous-time systems (see Chapters 4 through 9).

13.1 TWO EXAMPLES

The time response of discrete-time systems is introduced through two examples.

Example 13.1

Consider the first-order system of Figure 13.1, for which the unit step response will be found. With the transfer function $G(s)$ defined in the figure, the output $C(s)$, from Section 12.7, is given by

$$C(s) = \frac{G(s)R^*(s)}{1 + G^*(s)} \tag{13-1}$$

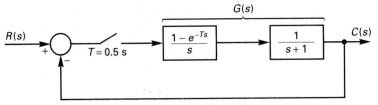

Figure 13.1   System for Example 13.1.

and $C(z)$ is given by

$$C(z) = \frac{G(z)}{1 + G(z)} R(z) \qquad (13\text{-}2)$$

We now find $C(z)$ for Figure 13.1. From the z-transform tables,

$$G(z) = \mathfrak{z}\left[\frac{1 - e^{-Ts}}{s(s+1)}\right] = (1 - e^{-Ts})\bigg|_{e^{Ts} = z} \ \mathfrak{z}\left[\frac{1}{s(s+1)}\right]$$

or

$$G(z) = \frac{z-1}{z}\left[\frac{(1 - e^{-T})z}{(z-1)(z - e^{-T})}\right]_{T = 0.5} = \frac{0.393}{z - 0.607}$$

and hence the closed-loop transfer function $T(z)$ is given by

$$T(z) = \frac{G(z)}{1 + G(z)} = \frac{0.393}{z - 0.607 + 0.393} = \frac{0.393}{z - 0.214} \qquad (13\text{-}3)$$

Since the input is the unit step function, $R(z) = z/(z-1)$, and from (13-2),

$$C(z) = \frac{0.393}{z - 0.214}\frac{z}{z - 1} = \frac{0.393z}{(z-1)(z-0.214)}$$

Hence

$$\frac{C(z)}{z} = \frac{0.393}{(z-1)(z-0.214)} = \frac{0.5}{z-1} + \frac{-0.5}{z-0.214}$$

Multiplying this expression by z, we see that the inverse z-transform of $C(z)$ is

$$c(nT) = 0.5[1 - (0.214)^n]$$

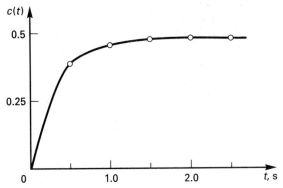

Figure 13.2   System time response.

This response is sketched in Figure 13.2 along with the continuous response $c(t)$, which was obtained by simulation. The interested reader may calculate $C(s)$ from (13-1) to find the mathematical expression for $c(t)$ and understand why we use simulations to solve for the continuous response. The step response of Figure 13.2 is easily verified with SIMULINK.

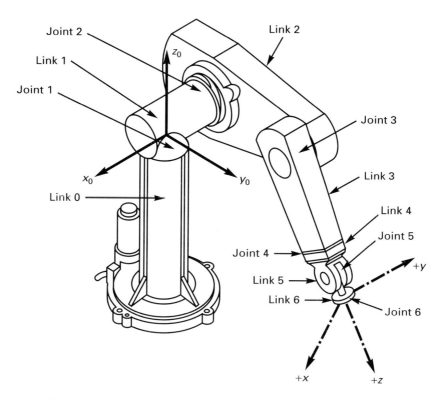

Figure 13.3   Industrial robot. (From K. S. Fu, R. C. Gonzalez, and C. S. G. Lee, *Robotics: Control, Sensing, Vision, and Intelligence,* McGraw-Hill Book Company, New York, 1987.)

Example 13.2

As a second example, consider a robot-arm joint control system for an industrial robot of the type depicted in Figure 13.3 (see Section 2.12). The inductance of the armature winding of the servomotor can be ignored (see Section 2.7), and the joint control system has the model given in Figure 13.4. The output $c(t)$ is the angle of the robot arm. Since this system is of the same configuration as the system in the last example, the system equations are the same. Hence, from (13-2),

$$C(z) = \frac{G(z)}{1 + G(z)} R(z)$$

where

$$G(z) = \frac{z-1}{z} \mathcal{z} \left[\frac{1}{s^2(s+1)} \right]$$

$$= \frac{z-1}{z} \left[\frac{z[(T-1+e^{-T})z + (1-e^{-T}-Te^{-T})]}{(z-1)^2(z-e^{-T})} \Bigg|_{T=1} \right] \tag{13-4}$$

$$= \frac{0.368z + 0.264}{z^2 - 1.368z + 0.368}$$

Then

$$\frac{G(z)}{1+G(z)} = \frac{0.368z + 0.264}{z^2 - 1.368z + 0.368 + 0.368z + 0.264}$$

$$= \frac{0.368z + 0.264}{z^2 - z + 0.632} \tag{13-5}$$

Since the input is the unit step, $R(z) = z/(z-1)$ and

$$C(z) = \frac{0.368z + 0.264}{z^2 - z + 0.632} \frac{z}{z-1}$$

$$= 0.368z^{-1} + 1.00z^{-2} + 1.40z^{-3}$$

$$+ 1.40z^{-4} + 1.15z^{-5} + 0.90z^{-6} + 0.80z^{-7} + 0.87z^{-8}$$

$$+ 0.99z^{-9} + 1.08z^{-10} + 1.08z^{-11} + 1.00z^{-12} + 0.98z^{-13} + \cdots$$

Since $C(z)$ has complex poles, no simple procedure exists for finding the inverse z-transform of $C(z)$. The values of $c(nT)$ just given were found by the power-series method. Note that the final value of $c(nT)$ is

$$\lim_{n \to \infty} c(nT) = \lim_{z \to 1} (z-1)C(z) = \frac{0.632}{0.632} = 1$$

The following MATLAB program allows us to verify both $G(z)$ and the closed-loop transfer function and to simulate the system.

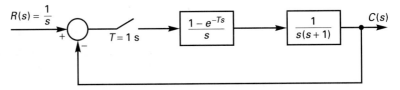

Figure 13.4 System for Example 13.2.

```
Gpnum = [0 0 1];   Gpden = [1 1 0];
[Ac,Bc,Cc,Dc] = tf2ss(Gpnum,Gpden);
[A,B] = c2d(Ac,Bc,1);
[Gznum,Gzden] = ss2tf(A,B,Cc,Dc)
pause
Tznum = Gznum, Tzden = Gznum + Gzden
pause
[A,B,C,D] = tf2ss(Tznum,Tzden);
sys = ss(A,B,C,D,1);
[c, t] = step(sys, 0:1:10)
```

The system step response is shown in Figure 13.5. The response between sampling instants was obtained by simulation. The response at the sampling instants can be verified from the power series for C(z) found previously. Also, the response of the same system *with the sampler and zero-order hold removed* was obtained by simulation and is also shown in Figure 13.5. The response of the sample-data system is probably unsatisfactory for a robot arm. Note that the sampling has a destabilizing effect on the system. This effect is discussed in detail later. The response of Figure 13.5 is readily verified with SIMULINK.

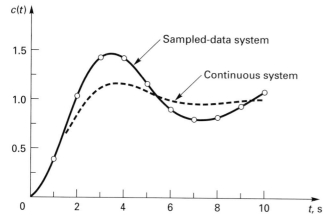

Figure 13.5 Step response for system analyzed in Example 13.2.

The system time response at sampling instants for the example above can also be calculated using a difference equation approach. From (13-5),

$$\frac{C(z)}{R(z)} = \frac{0.368z^{-1} + 0.264z^{-2}}{1 - z^{-1} + 0.632z^{-2}} \qquad (13\text{-}6)$$

or

$$C(z)[1 - z^{-1} + 0.632z^{-2}] = R(z)[0.368z^{-1} + 0.264z^{-2}]$$

Taking the inverse z-transform of this equation and rearranging terms yields the difference equation

$$c(kT) = 0.368r[(k-1)T] + 0.264r[(k-2)T] \\ + c[(k-1)T] - 0.632c[(k-2)T] \qquad (13\text{-}7)$$

This equation may be solved for $c(kT)$, $k \geq 0$, with $r(kT) = 1$, $k \geq 0$. The initial conditions for the equation are $r(-1) = r(-2) = c(-1) = c(-2) = 0$. Solving (13-7) for $c(kT)$ yields the same values as found in Example 13.2.

Programming (13-7) for a digital computer results in a digital simulation of the discrete system that is different from that given in Example 13.2, which is based on a state model of the system. For example, a MATLAB program for (13-7) is

```
rkminus1 = 0; rkminus2 = 0;
ckminus1 = 0; ckminus2 = 0;
rk = 1;
for k = 0:10
   ck = 0.368*rkminus1 + 0.264*rkminus2 + ckminus1 - 0.632*ckminus2;
   [k,ck]
   rkminus2 = rkminus1; rkminus1 = rk;
   ckminus2 = ckminus1; ckminus1 = ck;
end
```

where rkminus1 is $r(k-1)$ and so on. This system may also be simulated on a hybrid computer, where the sampler/data-hold is simulated by an A/D followed directly by a D/A.

The fourth method of simulating the system is a digital simulation in which the differential equations of the plant are integrated numerically. Logic must be included in this simulation, which updates the plant input only at the sampling instants, in order to model correctly the data-hold output.

As just illustrated, it is much easier to obtain the response of a sampled-data system by simulation than by calculations using the z-transform. As the order of the system increases, simulations become more attractive; for higher-order systems, simulation is the only practical technique for obtaining the time response.

13.2 DISCRETE SYSTEM STABILITY

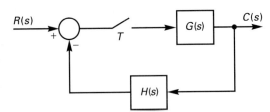

Figure 13.6   Sampled-data system.

To introduce stability, consider the sampled-data system shown in Figure 13.6. For this system,

$$C(z) = \frac{G(z)}{1 + \overline{GH}(z)} R(z) = \frac{K \prod\limits^{m} (z - z_i)}{\prod\limits_{n} (z - p_i)} R(z) \tag{13-8}$$

where p_i and z_i are the poles and the zeros of the closed-loop transfer function, respectively. Using the partial-fraction expansion, we can express $C(z)$ as

$$C(z) = \frac{k_1 z}{z - p_1} + \frac{k_2 z}{z - p_2} + \cdots + \frac{k_n z}{z - p_n} + C_R(z) \tag{13-9}$$

where $C_R(z)$ contains the terms of $C(z)$ that originate in the poles of $R(z)$. The first n terms of (13-9) are the *natural-response terms* of $C(z)$. If the inverse z-transform of these terms tend to zero as time increases, the system is bounded-input bounded-output (BIBO) stable, where BIBO stability is defined in Section 5.2. The inverse z-transform of the ith term of (13-9) at $t = nT$ yields

$$\mathscr{z}^{-1}\left[\frac{k_i z}{z - p_i}\right] = k_i (p_i)^k \tag{13-10}$$

Thus, if the magnitude of p_i is less than unity, this term approaches zero as k approaches infinity. Then the system of (13-8) is stable if the magnitude of each p_i is less than unity. The factors $(z - p_i)$ originate in the term $[1 + \overline{GH}(z)]$ in (13-8). Thus the characteristic equation of the system of Figure 13.6 is

$$1 + \overline{GH}(z) = 0 \tag{13-11}$$

The system is stable provided that all the roots of (13-11) lie inside the unit circle in the z-plane. Of course, (13-11) can also be expressed as

$$1 + \overline{GH}^*(s) = 0 \qquad (13\text{-}12)$$

and the roots of this equation must lie in the left half-plane of the s-plane for stability. Since the starred transform is a function of e^{Ts}, the roots of (13-12) in general are difficult to calculate.

13.3 JURY'S TEST

For analog systems, the Routh–Hurwitz criterion offers a simple and convenient technique for determining the stability of low-order systems. The Routh–Hurwitz criterion applies only to polynomials; thus it cannot be used for sampled-data systems with the characteristic equation expressed as a function of s, as in (13-12). Also, since the stability boundary in the z-plane is different from that in the s-plane, the Routh–Hurwitz criterion cannot be applied to sampled-data systems if the system characteristic equation is expressed as a function of z, as in (13-11). A stability criterion for sampled-data systems in the z-plane that is similar to the Routh–Hurwitz criterion is the Jury stability test [1].

Let the characteristic equation of a sampled-data system be expressed as the polynomial

$$Q(z) = a_n z^n + a_{n-1} z^{n-1} + \cdots + a_1 z + a_0 = 0 \qquad a_n > 0 \qquad (13\text{-}13)$$

To apply the Jury test, form the array shown in Table 13.1. Note that the elements of second row, fourth row, and so on, are the elements of the first row, third row, and so on, respectively, in reverse order. The elements of the first two rows are obtained from the characteristic equation of (13-13), and the elements of the remaining rows are calculated from the determinants

$$b_k = \begin{vmatrix} a_0 & a_{n-k} \\ a_n & a_k \end{vmatrix} \qquad c_k = \begin{vmatrix} b_0 & b_{n-1-k} \\ b_{n-1} & b_k \end{vmatrix}$$

$$d_k = \begin{vmatrix} c_0 & c_{n-2-k} \\ c_{n-2} & c_k \end{vmatrix} \qquad \vdots$$

TABLE 13.1   ARRAY FOR JURY'S STABILITY TEST

z^0	z^1	z^2	z^{n-k}	z^{n-1}	z^n
a_0	a_1	a_2	$\cdots$ a_{n-k} $\cdots$	a_{n-1}	a_n
a_n	a_{n-1}	a_{n-2}	$\cdots$ a_k $\cdots$	a_1	a_0
b_0	b_1	b_2	$\cdots$ b_{n-k} $\cdots$	b_{n-1}	
b_{n-1}	b_{n-2}	b_{n-3}	$\cdots$ b_{k-1} $\cdots$	b_0	
c_0	c_1	c_2	$\cdots$ c_{n-k} $\cdots$		
c_{n-2}	c_{n-3}	c_{n-4}	$\cdots$ c_{k-2} $\cdots$		
$\vdots$	$\vdots$	$\vdots$	$\vdots$ $\vdots$		
l_0	l_1	l_2	l_3		
l_3	l_2	l_1	l_0		
m_0	m_1	m_2			

The necessary and sufficient conditions for the polynomial $Q(z)$ to have no roots outside or on the unit circle are as follows:

$$Q(1) > 0 \qquad\qquad |d_0| > |d_{n-3}|$$
$$(-1)^n Q(-1) > 0 \qquad\qquad \vdots$$
$$|a_0| < a_n$$
$$|b_0| > |b_{n-1}| \qquad |m_0| > |m_2|$$
$$|c_0| > |c_{n-2}|$$

Note that for a second-order system, the array contains only one row, and thus no element calculations are required. For each additional order, two additional rows are added to the array. For a first-order system, the single root occurs at $z = -a_0/a_1$ and stability is obvious. Finally, note that for an nth-order system, with $n > 1$, there are a total of $n + 1$ constraints. An example of an application of Jury's test is given next.

Example 13.3

Consider again the robot-arm control system of Figure 13.4 and Example 13.2. Suppose that a gain factor K is added to the plant and that we wish to find the range of K for which the system is stable. Now, from (13-4) and (13-11), the system characteristic equation is

$$1 + KG(z) \;=\; 1 + \frac{(0.368z + 0.264)K}{z^2 - 1.368z + 0.368} \;=\; 0$$

or, in (13-13),

$$Q(z) \;=\; z^2 + (0.368K - 1.368)z + (0.368 + 0.264K) \;=\; 0$$

The Jury array is then

z^0	z^1	z^2
$0.368 + 0.264K$	$0.368K - 1.368$	1

The constraint $Q(1) > 0$ yields

$$Q(1) \;=\; 1 + (0.368K - 1.368) + (0.368 + 0.264K) > 0$$

which simplifies to $K > 0$. The constraint $(-1)^2 Q(-1) > 0$ yields

$$(-1)^2 Q(-1) \;=\; 1 - (0.368K - 1.368) + (0.368 + 0.264K) > 0$$

which simplifies to $K < 2.736/0.104 = 26.3$. The constraint $|a_0| < a_2$,

$$|0.368 + 0.264K| < 1$$

yields $K < 0.632/0.264 = 2.39$. Thus the system is stable for $0 < K < 2.39$. The following MATLAB program verifies the results of this example.

```
K = [1 2.39 3];
for k = 1:3
    q = [1 (0.368*K(k)-1.368) (0.264*K(k) + 0.368)];
    r = roots(q);
    rootmag = sqrt((real(r(1)))^2 + (imag(r(1)))^2)
end

results: rootmag =    0.7950    0.9995    1.0770
```

The roots of the system characteristic equation are found for $K = 1$ (stable), $K = 2.39$ (marginally stable), and $K = 3$ (unstable).

13.4 MAPPING THE *S*-PLANE INTO THE *Z*-PLANE

In investigating the transient response of analog systems in Section 5.6, we were able to assign time-response characteristics to closed-loop pole locations (characteristic-equation zero locations). We want to do the same for sampled-data systems.

To introduce this topic, consider a function $e(t)$, which is sampled with the resulting starred transform $E^*(s)$. At the sampling instants, the sampled signal is of the same nature as the continuous signal. For example, if $e(t)$ is exponential, then the sampled signal is exponential at the sampling instants, with the same amplitude and time constant. If $e(t) = e^{-at}$,

$$E(s) = \frac{1}{s+a} \qquad E^*(s) = \frac{e^{Ts}}{e^{Ts} - e^{-aT}} \qquad E(z) = \frac{z}{z - e^{-aT}}$$

Hence the pole at $s = -a$ results in the z-plane pole at $z = e^{-aT}$. In general, from the z-transform tables we see that a pole of $E(s)$ at $s = s_1$ results in a z-plane pole of $E(z)$ at $z = e^{s_1 T}$. This characteristic is also evident from the second property of starred transform given in Section 12.3. We use this characteristic in reverse. A z-plane pole at $z = z_1$ results in the transient response characteristics at the sampling instants of the equivalent s-plane pole s_1, where s_1 and z_1 are related by $z_1 = e^{s_1 T}$.

Consider first the mappings of the left-half-plane portion of the primary strip into the z-plane as shown in Figure 13.7. Along the $j\omega$-axis,

$$z = e^{sT} = e^{\sigma T} e^{j\omega T} = e^{j\omega T} = 1 \underline{/\omega T} \tag{13-14}$$

Hence pole locations on the unit circle in the z-plane are equivalent to pole locations on the imaginary axis in the s-plane. Thus pole locations on the unit circle in the z-plane signify a *marginally stable system*, that is, a system with a steady-state oscillation as its natural response. From (13-14), the frequency of oscillation is given by the angle of the pole (in radians) divided by T. Since for $\omega = \omega_s/2$, ωT is equal to π, and the $j\omega$-axis between $-j\omega_s/2$ and $j\omega_s/2$ maps into the unit circle in the z-plane. In fact, any portion of the $j\omega$-axis of length ω_s maps into the unit circle in the z-plane. The right-half-plane portion of the primary strip maps into the exterior of the unit circle, and the left-half-plane portion of the primary strip maps into the interior of the unit circle. Since the stable region of the s-plane is the left half-plane, the stable region of the z-plane is the interior of the unit circle, as was just shown.

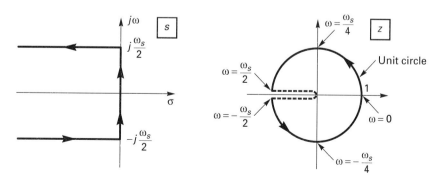

Figure 13.7   Mapping the primary strip into the z-plane.

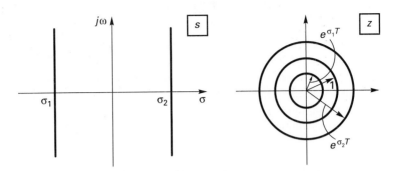

Figure 13.8   Mapping constant-damping loci into z-plane.

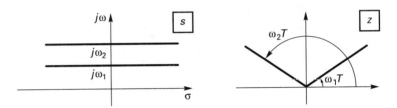

Figure 13.9   Mapping constant-frequency loci into z-plane.

Constant damping loci (σ constant) map into circles as shown in Figure 13.8. This can be seen using the relationship

$$z = e^{\sigma_1 T} e^{j\omega T} = e^{\sigma_1 T} = 1\,\underline{/\omega T}$$

Constant frequency loci (ω constant) map into rays, as shown in Figure 13.9. This can be seen from the preceding expression, with σ variable and ω constant. For constant damping-ratio loci, σ and ω are related by

$$\frac{\omega}{\sigma} = \tan\beta$$

Then

$$z = e^{sT} = e^{\sigma T}\,\underline{/\sigma T\tan\beta} \tag{13-15}$$

Since σ is negative in the second and third quadrants of the s-plane, (13-15) describes a logarithmic spiral whose amplitude decreases with σ increasing in magnitude. This is illustrated in Figure 13.10.

As just described, the characteristics of a sampled time function at the sampling instants are the same as those of the time function before sampling. Thus, using the mappings illustrated in Figures 13-7 through 13-10, we may assign time-response characteristics to characteristic-equation zero locations in the z-plane. This is illustrated by several examples in Figure 13.11. Since

$$z = e^{sT} = e^{\sigma T} e^{j\omega T} \tag{13-16}$$

the response characteristics are a function not only of s but also of T.

Consider the case in Figure 13.11 where poles occur at $s = \sigma \pm j\omega$. These poles result in a system transient-response term of the form $k_1 e^{\sigma t} \cos(\omega t + \phi)$. When sampling occurs, these s-plane poles result in z-plane poles at

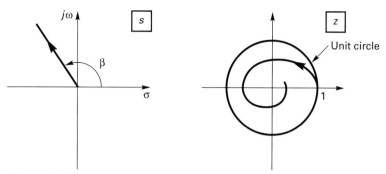

Figure 13.10   Mapping constant damping ratio loci into z-plane.

$$z = e^{sT}\big|_{s = \sigma \pm j\omega} = e^{\sigma T}e^{\pm j\omega T} = e^{\sigma T}\ \underline{/\pm\omega T} = r\ \underline{/\pm\theta}$$

Thus roots of the characteristic equation that appear at $z = r\ \underline{/\pm\theta}$ result in a transient response term of the form

$$k_1 e^{\sigma kT}\cos(\omega kT + \phi) = k_1(r)^k\cos(\theta k + \phi)$$

The above comments were made relative to a system for which a transfer function exists. Consider the system of Figure 13.12, for which no transfer function exists. It is easily shown using the techniques of Section 12.8 that the output of this system is given by

$$C(z) = \frac{\overline{GR}(z)}{1 + \overline{GH}(z)} \tag{13-17}$$

It is seen that the poles of $C(z)$ that contribute to the natural response originate in the roots of

$$1 + \overline{GH}(z) = 0 \tag{13-18}$$

Consequently, (13-18) is the system characteristic equation and the above analysis applies directly to systems of this type. A sampled-data system for which no transfer function exists presents no problems in terms of determining the system characteristic equation.

An example is given next to illustrate the relationship of z-plane pole locations of a transfer function to the transient-response characteristics.

Example 13.4

Consider Example 13.1. For $T = 0.5$ s, the closed-loop transfer function is given by

$$T(z) = \frac{0.393}{z - 0.214}$$

Thus the transfer-function pole at $z = 0.214$ has an equivalent s-plane location of

$$z = 0.214 = e^{sT} = e^{0.5s}$$

Hence $s = \ln(0.214)/0.5 = -3.08$. Since the time constant is equal to the reciprocal of the magnitude of the real part of an s-plane pole, for this system the time constant is $\tau = 1/3.08 = 0.325$ s. This time constant is seen in the system step response plotted in Figure 13.2. Note that the response between sampling instants is not of the same exponential nature.

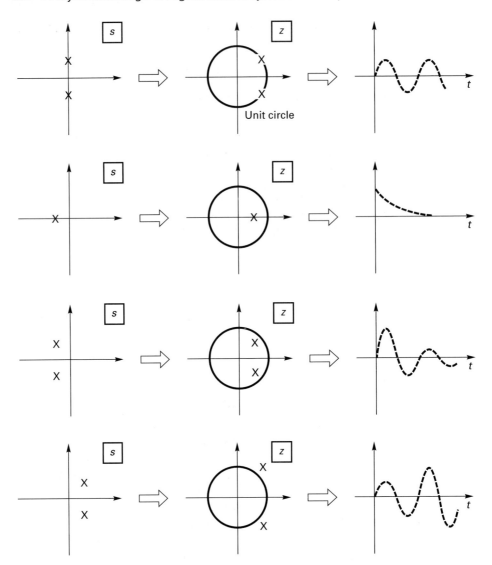

Figure 13.11   Examples relating pole locations to time response.

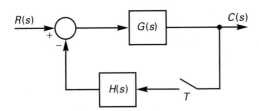

Figure 13.12   Sampled-data system.

Consider the robot-arm control system of Example 13.2. The closed-loop transfer function for this example is

$$T(z) = \frac{0.368z + 0.264}{z^2 - z + 0.632}$$

The poles of this transfer function are complex, and occur at $z = 0.5 \pm j0.618 = 0.795 \underline{/\pm 51.0°} = 0.795 \underline{/\pm 0.890}$ rad. Since

$$z = e^{\sigma T} \underline{/\pm \omega T} = 0.795 \underline{/\pm 0.890} \qquad T = 1 \ s$$

then the equivalent s-plane pole locations are given by $s = -0.229 \pm j0.890 = -\zeta\omega_n \pm j\omega_n\sqrt{1-\zeta^2}$, from (5-58). Solving for the damping ratio ζ and the natural frequency ω_n yields $\zeta = 0.249$ and $\omega_n = 0.919$. Since the system time constant is $\tau = 1/\zeta\omega_n$, then $\tau = 4.37$ s. From Figure 4.8, the step response has an overshoot of approximately 45 percent. All of these characteristics are illustrated in the step response in Figure 13.5. Note once again that these derived characteristics apply only at the sampling instants.

13.5 ROOT LOCUS

For the sampled-data system of Figure 13.13, the transfer function is given by

$$\frac{C(z)}{R(z)} = \frac{KG(z)}{1 + K\overline{GH}(z)} \qquad (13\text{-}19)$$

The system characteristic equation is then

$$1 + K\overline{GH}(z) = 0 \qquad (13\text{-}20)$$

By definition, the root locus for this system is a plot of the locus of the roots of (13-20) in the z-plane as a function of the gain K. Thus the rules of root-locus construction for discrete-time systems are identical to those for continuous-time systems. The roots of a polynomial $F(z)$ are the same as those of the polynomial $F(s)$, provided the coefficients are identical; the roots of a polynomial are determined by the polynomial coefficients and are independent of the labeling of the variable. Because the rules of root-locus construction were developed in Chapter 7, these rules are only listed in this chapter (Table 13.2).

However, although the rules for the construction of both s-plane and z-plane root loci are the same, there are important differences in the interpretation of the root loci. In the z-plane, the stable region is the interior of the unit circle. Also, root locations in the z-plane have different meanings from those in the s-plane in terms of the system's time response, as seen in Figure 13.11.

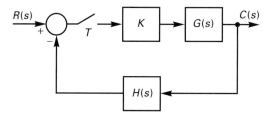

Figure 13.13 Sampled-data system.

Chapter 13

TABLE 13.2 RULES FOR ROOT-LOCUS CONSTRUCTION

---

1. The root locus is symmetrical with respect to the real axis.
2. Loci originate on poles of $\overline{GH}(z)$ and terminate on zeros of $\overline{GH}(z)$, including those zeros at infinity.
3. The number of asymptotes is equal to the number of poles of $\overline{GH}(z)$, n_p, minus the number of zeros of $\overline{GH}(z)$, n_z, with the angles given by

$$\text{angles of asymptotes} = \frac{(2k+1)\pi}{n_p - n_z} \qquad k = 0, 1, \ldots, (n_p - n_z - 1)$$

   The asymptotes intersect the real axis at σ_a, where

$$\sigma_a = \frac{\Sigma \text{ poles of } \overline{GH}(z) - \Sigma \text{ zeros of } \overline{GH}(z)}{n_p - n_z}$$

4. The root locus includes all points on the real axis to the left of an odd number of critical frequencies (poles and zeros).
5. The breakaway points appear among the roots of the polynomial obtained from either

$$\frac{d[\overline{GH}(z)]}{dz} = 0$$

   or, equivalently,

$$D(z)\frac{dN(z)}{dz} - N(z)\frac{dD(z)}{dz} = 0 \qquad \overline{GH}(z) = \frac{N(z)}{D(z)}$$

If the reader is not familiar with root-locus construction for analog systems, a study of the first four sections of Chapter 7 is recommended. Because of the similarities in the construction procedures, we do not review them here and immediately give an example.

Example 13.5

Consider again the robot-arm control system of Figure 13.4 and Example 13.2. For this system, the open-loop transfer function was calculated to be

$$KG(z) = K_3\left[\frac{1-e^{-Ts}}{s^2(s+1)}\right] = \frac{K(0.368z + 0.264)}{(z-1)(z-0.368)} = \frac{0.368K(z+0.717)}{(z-1)(z-0.368)}$$

Thus the root locus originates at $z = 1$ and $z = 0.368$ and terminates at $z = -0.717$ and $z = \infty$. There is one asymptote, at $180°$. The breakaway points occur at the roots of

$$D(z)N'(z) - N(z)D'(z)$$
$$= (z^2 - 1.368z + 0.368)(0.368) - (0.368z + 0.264)(2z - 1.368) = 0$$

or

$$-0.368z^2 - 0.528z + 0.497 = 0$$

The roots of this polynomial occur at $z = 0.648$ for $K = 0.196$ and $z = -2.08$ for $K = 15.0$. The root locus is sketched in Figure 13.14.

The points of the intersection of the root loci with the unit circle (points of marginal stability) may be found using the Jury test. From Example 13.3, the value of gain for marginal stability was found to be $K = 2.39$, and the system characteristic equation is

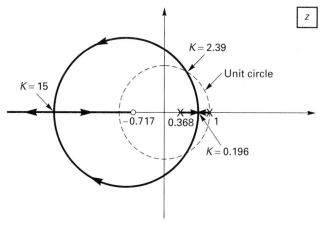

Figure 13.14   Root locus for Example 13.5.

$$z^2 + (0.368K - 1.368)z + (0.368 + 0.264K) = 0$$

For $K = 2.39$, the characteristic equation is given by

$$z^2 - 0.488z + 1 = 0$$

The roots of this equation are $z = 0.244 \pm j0.97 = 1 \angle \pm 75.9°$, and thus these are the points at which the root locus crosses the unit circle. Since the system is marginally stable for this value of gain, the frequency of oscillation for this condition can be found by the procedure

$$z = 0.244 \pm j0.97 = 1 \angle \pm 75.9° = 1 \angle \pm 1.33 \, \text{rad}$$
$$= 1 \angle \pm \omega T = e^{\pm j\omega T}$$

as seen from (13-14). For this system, T is 1 s, and the frequency of oscillation is $\omega = 1.33$ rad/s $= 2\pi/T_0$. Thus the period of oscillation is $T_0 = 2\pi/1.33 = 4.72$ s. The following MATLAB program plots the root locus for this example:

```
rlocus([0.368 0.264],[1 -1.368 0.368])
```

If we were testing the robot arm of Example 13.5, increasing the gain to $K = 2.39$ should result in the arm swinging back and forth with a period of approximately 5 s. However, we must always be *very* careful in pushing any physical system to its stability limits, because the models of these systems are imperfect and we cannot be certain how the physical system will respond.

13.6 NYQUIST CRITERION

To develop the Nyquist criterion for sampled-data systems, consider the two systems shown in Figure 13.15. In Figure 13.15(b), $G(s)$ contains the transfer function of a data-hold. We first review the Nyquist criterion for analog systems, because of the complexity of the criterion. A complete development for this topic is contained in Chapter 8, and those readers unfamiliar with the Nyquist criterion should study Chapter 8 before proceeding with this section.

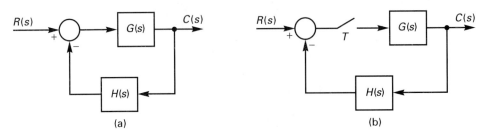

Figure 13.15   Analog and sampled-data systems.

The transfer function for the analog system of Figure 13.15(a) is

$$\frac{C(s)}{R(s)} = \frac{G(s)}{1 + G(s)H(s)} \tag{13-21}$$

and, for the sampled-data system of Figure 13.15(b), is

$$\frac{C^*(s)}{R^*(s)} = \frac{G^*(s)}{1 + \overline{GH}^*(s)} \tag{13-22}$$

Thus the characteristic equation for the analog system is

$$1 + G(s)H(s) = 0 \tag{13-23}$$

and for the sampled-data system it is

$$1 + \overline{GH}^*(s) = 0 \tag{13-24}$$

Making the substitution $e^{Ts} = z$ in (13-24) yields the characteristic equation for the sampled-data system as a function of z.

$$1 + \overline{GH}(z) = 0 \tag{13-25}$$

The analog system is stable if the roots of (13-23) are all in the left half-plane. The sampled-data system is stable if the roots of (13-24) also are all in the left half-plane or if the roots of (13-25) are all within the unit circle.

The Nyquist criterion for analog systems is reviewed next. The Nyquist path in the s-plane for analog systems encloses the right half-plane, as shown in Figure 13.16. The Nyquist criterion states that if the complex-plane plot of $G(s)H(s)$ is made for values of s on the Nyquist path, then

$$N = Z - P \tag{13-26}$$

or

$$Z = N + P \tag{13-27}$$

In this expression, Z is the number of zeros of the characteristic equation, (13-23), enclosed by the Nyquist path, N is the number of clockwise encirclements of the -1 point made by the plot of $G(s)H(s)$ in the complex plane, and P is the number of poles of the characteristic equation enclosed by the Nyquist path. The plot of $G(s)H(s)$ is called the Nyquist diagram.

Note then that the Nyquist diagram is a mapping of the Nyquist path into the complex plane, with the mapping function equal to $G(s)H(s)$. In (13-27), P is the number of poles of

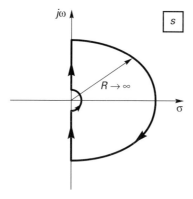

jω

s

$R \rightarrow \infty$

σ

Figure 13.16   Nyquist path in *s*-plane.

$G(s)H(s)$ in the right half-plane, and the closed-loop system is stable if and only if $Z = 0$. An example of the application of the Nyquist criterion will now be given.

Example 13.6

We apply the Nyquist criterion to the system of Figure 13.17. The Nyquist path and the resultant Nyquist diagram are given in Figure 13.18. The small detour (denoted by I) taken by the Nyquist path around the origin is necessary since $G(s)$ has a pole at the origin (see Section 8.5). Along this detour, let $s = \rho e^{j\theta}$, where $\rho \ll 1$. Then

$$G(s)\big|_{s = \rho e^{j\theta}} = \frac{K}{\rho e^{j\theta}(1 + \rho e^{j\theta})} \simeq \frac{K}{\rho e^{j\theta}} = \frac{K}{\rho} \angle{-\theta}$$

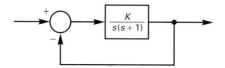

+

$\dfrac{K}{s(s+1)}$

–

Figure 13.17   System for Example 13.6.

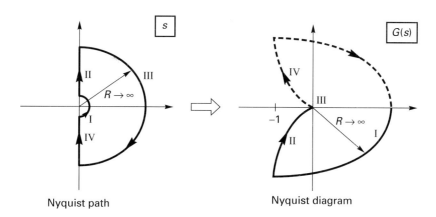

s

II   III

$R \rightarrow \infty$

I

IV

Nyquist path

$G(s)$

IV

III

−1

$R \rightarrow \infty$

II

I

Nyquist diagram

Figure 13.18   Nyquist diagram for Example 13.6.

Thus the small detour around the origin on the Nyquist path generates the large arc on the Nyquist diagram. For the Nyquist path along the $j\omega$-axis, denoted by II in Figure 13.18,

$$G(s)|_{s = j\omega} = \frac{K}{j\omega(1 + j\omega)} = \frac{K}{\omega(1 + \omega^2)^{1/2}} \; \underline{/-90° - \tan^{-1}\omega}$$

This portion of the Nyquist diagram is simply a polar plot of the frequency response of $G(s)$.

Along the large arc of the Nyquist path, denoted by III in Figure 13.18, $G(s) \approx 0$. Along part IV of the Nyquist path, $s = -j\omega$, and the Nyquist diagram is the complex conjugate of that of part II. The complete Nyquist diagram is then as shown in Figure 13.18, and the -1 point is not encircled. Thus N in (13-27) is zero. The poles of $G(s)$ occur at 0 and at -1; thus P in (13-27) is zero. Therefore,

$$Z = N + P = 0$$

and the closed-loop system characteristic equation has no zeros in the right half-plane. Hence the system is stable.

Consider now the sampled-data system of Figure 13.15(b). The characteristic equation is given in (13-24), and the Nyquist diagram for the system may be generated by using the s-plane Nyquist path of Figure 13.16, which is the same path as that for analog systems. However, recall from Chapter 12 that $\overline{GH}^*(s)$ is periodic in s with period $j\omega_s$. Thus it is necessary that $\overline{GH}^*(j\omega)$ be plotted only for $-\omega_s/2 \leq \omega \leq \omega_s/2$, in order to plot the complete frequency response. With reference to (12-8), consider the relationship

$$\overline{GH}^*(j\omega) = \frac{1}{T} \sum_{n = -\infty}^{\infty} GH(j\omega + jn\omega_s) = \frac{1}{T}[GH(j\omega) + GH(j\omega + j\omega_s)$$

$$+ GH(j\omega - j\omega_s) + GH(j\omega + j2\omega_s) + GH(j\omega - j2\omega_s) + \cdots \,] \tag{13-28}$$

Since physical systems are generally low pass, $\overline{GH}^*(j\omega)$ may be approximated by only a few terms of (13-28). A digital computer program may be written for (13-28), and thus the Nyquist diagram of a sampled-data system may be obtained without calculating the z-form of the transfer function. This is important for high-order systems, since calculating the z-transform transfer function can be difficult.

The Nyquist diagram may also be calculated directly in the z-plane. The Nyquist path for the z-plane is the unit circle, and the path direction is counterclockwise, as shown in Figure 13.19. To demonstrate this, let Z_i and P_i be the zeros and poles of the characteristic equation, respectively, inside the Nyquist path (the unit circle). Also, let Z_0 and P_0 be the zeros and poles of the characteristic equation, respectively, outside the Nyquist path. Since the Nyquist path encloses the interior of the unit circle,

$$N = -(Z_i - P_i) \tag{13-29}$$

where N is the number of clockwise encirclements of the -1 point made by the Nyquist diagram of $\overline{GH}(z)$. The minus sign appears in (13-29) because the Nyquist path is counterclockwise, and the number of encirclements of the -1 point are counted in a clockwise direction. In general, the order of the numerator and the order of the denominator of $1 + \overline{GH}(z)$ are the same. Let this order be n. Then

$$Z_0 + Z_i = n$$
$$P_0 + P_i = n \tag{13-30}$$

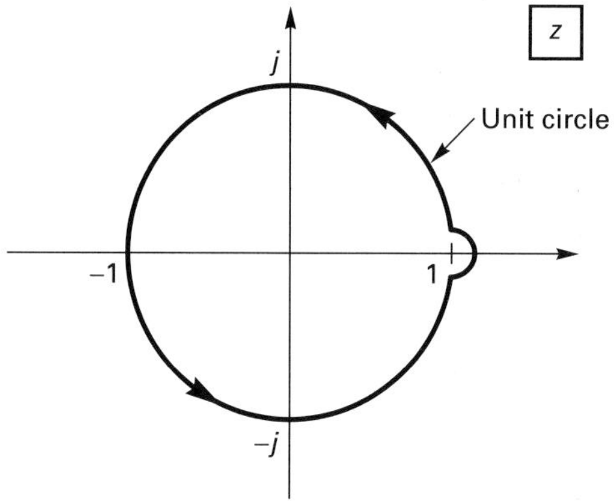

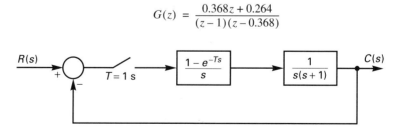

Figure 13.19   Nyquist path in z-plane.

Solving these equations for Z_i and P_i and substituting the results in (13-29) yields

$$N = Z_0 - P_0 \tag{13-31}$$

Thus the Nyquist criterion is given by (13-31), with the Nyquist path given in Figure 13.19. In (13-31), Z_0 must be zero for stability.

The Nyquist diagram for the sampled-data system of Figure 13.15(b) can then be obtained by plotting $\overline{GH}(z)$ for values of z on the unit circle. Then, omitting the subscripts in (13-31),

$$Z = N + P \tag{13-32}$$

where Z and P are the number of zeros and poles, respectively, of the characteristic equation outside the unit circle and N is the number of clockwise encirclements of the -1 point on the Nyquist diagram. An example of the Nyquist diagram of a sampled-data system is now given.

Example 13.7

We illustrate the Nyquist criterion using the system of Figure 13.20. From Example 13.2,

$$G(z) = \frac{0.368z + 0.264}{(z-1)(z-0.368)}$$

Figure 13.20   System for Example 13.7.

The Nyquist path and the Nyquist diagram are both given in Figure 13.21. The detour around the point $z = 1$ on the Nyquist path is necessary since $G(z)$ has a pole at this point. On this detour,

$$z = 1 + \rho e^{j\theta}, \qquad \rho \ll 1 \qquad \text{and} \qquad -90° \leq \theta \leq 90°$$

and

$$G(z)\big|_{z=1+\rho e^{j\theta}} = \frac{0.368(1+\rho e^{j\theta})+0.264}{(1+\rho e^{j\theta}-1)(1+\rho e^{j\theta}-0.368)}$$

$$\simeq \frac{0.632}{\rho e^{j\theta}(0.632)} = \frac{1}{\rho}\underline{/-\theta}$$

since ρ is very small. Thus the detour generates the large arc, denoted by **I** in the Nyquist diagram. For z on the unit circle,

$$G(z)\big|_{z=e^{j\omega T}} = \frac{0.368e^{j\omega T}+0.264}{(e^{j\omega T}-1)(e^{j\omega T}-0.368)}$$

In this function, ω varies from $-\omega_s/2$ to $\omega_s/2$. Since $G(e^{j\omega T})$ for $0 > \omega > -\omega_s/2$ is the complex conjugate of $G(e^{j\omega T})$ for $0 < \omega < \omega_s/2$, it is necessary to calculate $G(e^{j\omega T})$ only for $0 < \omega < \omega_s/2$. This calculation, which for this example was performed by computer, results in the frequency response for $G(z)$ and part II of the Nyquist diagram in Figure 13.21. Note that $G(-1) = -0.0381$.

The Nyquist diagram of Figure 13.21 has no encirclements of the -1 point; therefore N in (13-32) is zero. In addition, $G(z)$ has no poles outside the unit circle, and thus P is zero. Hence Z in (13-32) is zero, and the system is stable. Note, however, that if a gain factor of $K = 1/0.418 = 2.39$ is added to the plant, the Nyquist diagram passes through the -1 point and the system becomes marginally stable. The same conclusion resulted from the Jury test in Example 13.3 and the root locus in Example 13.5. The following MATLAB program lists the frequency ω, and the magnitude and the phase of $G(z)$, for 20 equally spaced values of $z = e^{j\omega T}$ between zero and π on the unit circle.

```
format short e
Gznum = [0 0.368 0.264];   Gzden = [1 -1.368 0.368];
[Gz,w] = freqz(Gznum,Gzden,20);
Gzmag = abs(Gz);   Gzphase = (180/pi)*angle(Gz);
[w,Gzmag,Gzphase]
```

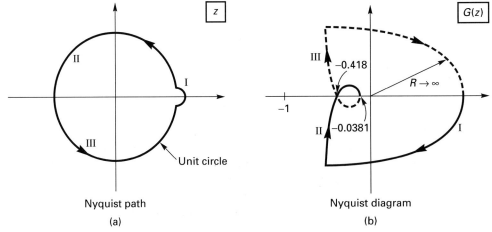

Nyquist path

(a)

Nyquist diagram

(b)

Figure 13.21   Nyquist path and Nyquist diagram for Example 13.7.

The system of Example 13.7 can be made unstable by increasing the system gain. However, the same system without sampling, as shown in Example 13.6, is stable for all positive values of gain. Thus, for this system, the sampling effect is destabilizing. This destabilizing effect can be seen from the frequency response of the zero-order hold. It is to be recalled that the zero-order hold introduces phase lag into a system (see Figure 12.11). Phase lag is generally destabilizing since it rotates the Nyquist diagram toward the −1 point.

In this section we see that, for a sampled-data system, the Nyquist criterion may be applied in either the s-plane or the z-plane. The resulting Nyquist diagram is the same in either case. In the next section a third plane, the w-plane, is introduced, and the Nyquist criterion may also be applied in this plane, resulting in the same Nyquist diagram as in the s-plane and the z-plane applications.

13.7 BILINEAR TRANSFORMATION

Many of the analysis and design techniques for continuous-data systems, such as the Routh–Hurwitz criterion and Bode techniques, are based on the property that in the s-plane, the stability boundary is the imaginary axis. Thus these techniques cannot be applied to sampled-data systems in the z-plane, since there the stability boundary is the unit circle. However, through the use of the bilinear transformation defined as

$$z = \frac{1 + (T/2)w}{1 - (T/2)w} \tag{13-33}$$

or solving for w,

$$w = \frac{2}{T}\frac{z-1}{z+1} \tag{13-34}$$

the unit circle of the z-plane maps into the imaginary axis of the w-plane.

This mapping can be seen through the following development. On the unit circle in the z-plane, $z = e^{j\omega T}$ and

$$w = \frac{2}{T}\frac{z-1}{z+1}\bigg|_{z = e^{j\omega T}} = \frac{2}{T}\frac{e^{j\omega T}-1}{e^{j\omega T}+1} = \frac{2}{T}\frac{e^{j\omega T/2}-e^{-j\omega T/2}}{e^{j\omega T/2}+e^{-j\omega T/2}}$$

Thus, by Euler's identity,

$$w = j\frac{2}{T}\tan\left(\frac{\omega T}{2}\right) \tag{13-35}$$

For $\omega = 0$, $w = j0$ and, as ω approaches $\omega_s/2$, $\omega T/2$ approaches $\pi/2$ and w approaches $j\infty$. Hence the line $0 \le j\omega < j\omega_s/2$ in the s-plane maps into the upper half of the unit circle in the z-plane and into the upper half of the imaginary axis in the w-plane. The mappings of the primary strip of the s-plane into both the z-plane and the w-plane are shown in Figure 13.22, and we note that the stable region of the w-plane is the left half-plane.

Let $j\omega_w$ be the imaginary part of w. We refer to ω_w as the w-plane frequency. Then (13-35) can be expressed as

$$\omega_w = \frac{2}{T}\tan\left(\frac{\omega T}{2}\right) \tag{13-36}$$

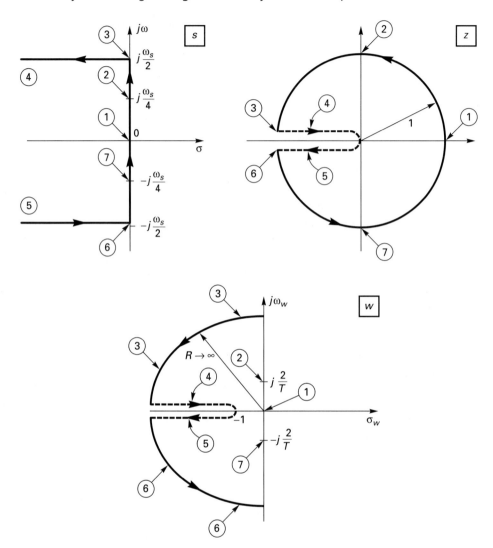

Figure 13.22 Mapping from s-plane to z-plane to w-plane.

This expression gives the relationship between frequencies in the s-plane and frequencies in the w-plane.

The bilinear transformation may be used in generating the Nyquist diagram. Since the stability regions are the same in the s-plane and the w-plane, the Nyquist path in the w-plane is then the same as that in the s-plane, which is shown in Figure 13.16. Hence the techniques given in Chapter 8 for constructing Nyquist diagrams for analog systems also apply here. An example of the construction of the Nyquist diagram in the w-plane is given next.

Example 13.8

Consider again the robot-arm control system of Example 13.7, which has been investigated in several examples in this chapter. From Example 13.7, the plant transfer function is given by

$$G(z) = \frac{0.368z + 0.264}{z^2 - 1.368z + 0.368} \qquad T = 1\,\text{s}$$

Since $T/2 = 0.5$, $G(w)$ is given by, from (13-33),

$$G(w) = \frac{0.368\left(\dfrac{1 + 0.5w}{1 - 0.5w}\right) + 0.264}{\left(\dfrac{1 + 0.5w}{1 - 0.5w}\right)^2 - 1.368\left(\dfrac{1 + 0.5w}{1 - 0.5w}\right) + 0.368}$$

$$= \frac{-0.0381(w - 2)(w + 12.14)}{w(w + 0.924)}$$

The Nyquist diagram can be obtained from this function by allowing w to assume values between $j0$ and $j\infty$. Since $G(w)$ has a pole at the origin, the Nyquist path must detour around this point. The Nyquist diagram generated from $G(w)$ is identical with that generated using $G(z)$, as given in Figure 13.21(b). The interested reader can show this.

Three different transfer functions may be used to calculate the Nyquist diagram: $G^*(s)$, $G(z)$, and $G(w)$. Of course, the Nyquist diagrams are identical in each case. Table 13.3 gives the functions and the range of variables required for the calculation of the Nyquist diagram. The range of variables gives only the upper half of the Nyquist path. If a pole of the transfer function occurs in the range of the variable, a detour must be made around the point that the pole occurs.

Table 13.3 makes one important point regarding the Nyquist diagram for sampled-data systems. For an analog system, the Nyquist diagram generally approaches 0 (the origin) as frequency approaches infinity. For sampled-data systems, the endpoint of the Nyquist diagram occurs for a frequency of $\omega = \omega_s/2$, and the gain of the open-loop function is generally not zero at this frequency. This point occurs for the z-variable equal to -1, and thus the open-loop function $G(z)$ has a real value. Hence the high-frequency endpoint of the Nyquist diagram for a sampled-data system occurs on the real axis of the complex plane, but usually not at the origin (for example, see Figure 13.21).

Table 13.3 FUNCTIONS FOR
CALCULATING THE NYQUIST DIAGRAM

Function	Range of variable
$G^*(s)$	$s = j\omega$, $j0 \le j\omega \le j\omega_s/2$
$G(z)$	$z = e^{j\omega T}$, $0 \le \omega T \le \pi$
$G(w)$	$w = j\omega_w$, $j0 \le j\omega_w < j\infty$

For small values of real frequency (s-plane frequency) such that ωT is small, (13-36) becomes

$$\omega_w = \frac{2}{T}\tan\left(\frac{\omega T}{2}\right) \approx \frac{2}{T}\frac{\omega T}{2} = \omega \qquad (13\text{-}37)$$

Thus w-plane frequency is approximately equal to s-plane frequency for this case. The approximation is valid for those values of frequency for which $\tan(\omega T/2) \approx \omega T/2$. For

$$\frac{\omega T}{2} \le \frac{\pi}{10} \qquad \omega \le \frac{2\pi}{10T} = \frac{\omega_s}{10} \tag{13-38}$$

the error in this approximation is less than 4 percent. Because of the phase lag introduced by the zero-order hold, we usually choose the sample period T such that (13-38) is satisfied over most if not all of the system bandwidth. At $\omega = \omega_s/10$, the zero-order hold introduces a phase lag of 18° (see Figure 12.11), which is an appreciable amount.

For an analog system, the gain margin is defined as the factor by which the gain must be changed to force the system to marginal stability. The phase margin is defined as the angle through which the Nyquist diagram must be rotated such that the diagram intersects the −1 point (see Section 8.6). The gain and phase margins are defined in exactly the same manner for sampled-data systems. Consider the Nyquist diagram for a sampled-data system shown in Figure 13.23. The gain margin is $1/a$, and the phase margin is ϕ_m.

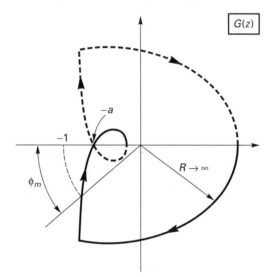

Figure 13.23 Nyquist diagram illustrating stability margins.

13.8 ROUTH–HURWITZ CRITERION

The Routh–Hurwitz criterion may be used to determine if any roots of a given polynomial are in the right half-plane (see Chapter 6). If this criterion is applied to the characteristic equation of a discrete system, expressed as a function of z, no useful information on stability is obtained. However, if the characteristic equation is expressed as a function of w, the stability of the system may be determined using the Routh–Hurwitz criterion. Since the Routh–Hurwitz criterion is covered in detail in Chapter 6, we immediately illustrate the use of the Routh–Hurwitz criterion with an example.

Example 13.9

Consider again the robot-arm control system of Example 13.7, with a gain factor K added to the plant. The system characteristic equation is then given by

$$1 + KG(w) = 0$$

where, from Example 13.8,

$$1 + KG(w) = 1 + \frac{-0.0381K(w-2)(w+12.14)}{w(w+0.924)}$$

$$= \frac{(1 - 0.0381K)w^2 + (0.924 - 0.386K)w + 0.924K}{w(w+0.924)}$$

Thus the characteristic equation may be expressed as

$$(1 - 0.0381K)w^2 + (0.924 - 0.386K)w + 0.924K = 0$$

The Routh array is then

$$
\begin{array}{c|cc}
w^2 & 1 - 0.0381K & 0.924K \\
w^1 & 0.924 - 0.386K & \\
w^0 & 0.924K &
\end{array}
$$

For stability, the signs of all elements in the first column must be the same; thus

$$K < \frac{1}{0.0381} = 26.2 \qquad K < \frac{0.924}{0.386} = 2.39 \qquad K > 0$$

Hence the system is stable for $0 < K < 2.39$. The result checks those of the Jury test of Example 13.3, the root locus of Example 13.5 and the Nyquist criterion of Example 13.7.

We see from the preceding example and Example 13.5 that the Routh–Hurwitz criterion may be utilized to determine the value of K for which the root locus crosses the unit circle in the z-plane. This value is, of course, the value of K for which the system becomes unstable. The Jury stability test may also be used to determine this value of K.

13.9 BODE DIAGRAM

The convenience of frequency-response plots for analog systems in the form of the Bode diagram stems from the straight-line approximations that are made; these approximations are based on the independent variable, $j\omega$, being imaginary. Thus Bode diagrams for discrete systems may be plotted, using straight-line approximations, provided that the w-plane form of the transfer function is used. The Bode-diagram development in Chapter 8 applies directly to transfer functions expressed in the w-plane variable, and that development is not repeated here. An example of the application of Bode diagrams to sampled-data systems is now given.

Example 13.10

Consider again the robot-arm control system of Example 13.9. For this system,

$$G(w) = \frac{-0.0381(w-2)(w+12.14)}{w(w+0.924)}$$

and

$$G(j\omega_w) = \frac{-0.0381(j\omega_w - 2)(j\omega_w + 12.14)}{j\omega_w(j\omega_w + 0.924)}$$

$$= \frac{(1 - j\omega_w/2)(1 + j\omega_w/12.14)}{j\omega_w(1 + j\omega_w/0.924)}$$

We see that the numerator break frequencies are $\omega_w = 2.0$ and $\omega_w = 12.14$, and the denominator frequencies are $\omega_w = 0$ and $\omega_w = 0.924$. The Bode diagram for this system, using straight-line approximations, is shown in Figure 13.24. Both the gain and phase margins of the system are shown on the diagram. Note that the magnitude and phase of the Nyquist diagram, shown in Figure 13.21(b), are evident from the Bode diagram. A MATLAB program that plots the Bode diagram is given by

```
Gnum1 = [1 -2];  Gnum2 = [1 12.14];
Gnum = -0.0381*conv(Gnum1,Gnum2);  Gden = [1 0.924 0];
  bode(Gnum,Gden)
```

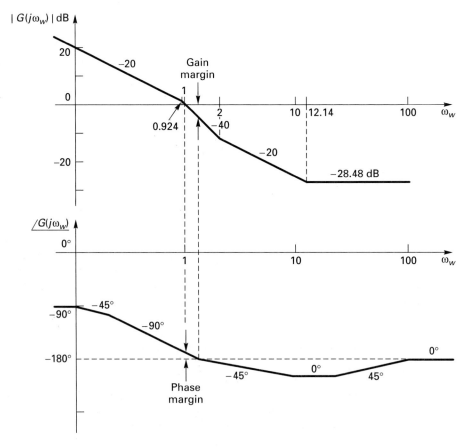

Figure 13.24   Bode diagram for Example 13.10.

As noted in the preceding example, the frequency-response information of the open-loop function is shown on both the Nyquist diagram and the Bode diagram. The two diagrams exhibit the same information but are plotted on different axes. A third method of showing this information is the gain–phase plot, as discussed in Section 8.6. For the gain–phase plot, the frequency response is plotted as gain in decibels versus phase on rectangular axes, with frequency as a parameter. Thus the frequency response of a discrete system may be plotted on gain-phase axes as well as on a Bode diagram. As an example, the gain-phase plot of the system of Example 13.10 is shown in Figure 13.25. Many controls engineers prefer to work with gain-phase plots rather than Bode diagrams when considering frequency responses.

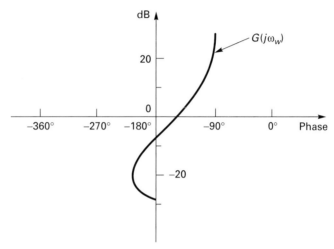

Figure 13.25   Gain–phase plot for Example 13.10.

Throughout the preceding sections we have used the term frequency response with respect to discrete systems. The physical meaning of a system frequency response in relation to an analog system is well known (see Section 4.4). It can be shown that the frequency response of a discrete system has the same meaning [2]. Consider a stable discrete system described by the relationship

$$C(z) = G(z)E(z)$$

If $e(t) = \sin \omega t$, it can be shown that in the steady state [2],

$$c_{ss}(kT) = |G(e^{j\omega T})| \sin (\omega kT + \theta), \qquad \theta = \arg G(e^{j\omega T}) \quad . \tag{13-39}$$

Thus the frequency response of a discrete system describes the steady-state response at the sampling instants for a sinusoidal input; this response is sinusoidal. Note that the response between sampling instants is not given; in fact it is not sinusoidal and is very difficult to calculate except by simulation. If the sampling rate is chosen large compared to the system bandwidth, the steady-state response between sampling instants is approximately sinusoidal for a sinusoidal input.

In the example in this section, the frequency response for the Bode diagram was calculated by first finding $G(w)$. This approach is generally not a practical approach, except for low-order systems, because of the difficulty of calculating $G(w)$. Instead, the frequency response as a function of real frequency ω is calculated by computer, using one of the procedures given in Section 13.6. For each real frequency ω for which the response is calculated, the equivalent w-plane frequency ω_w is also calculated by (13-36):

$$\omega_w = \frac{2}{T} \tan\left(\frac{\omega T}{2}\right)$$

The Bode diagram is obtained then by plotting the calculated frequency response versus ω_w. Note then that it is not necessary to find $G(w)$ in order to evaluate $G(j\omega_w)$.

13.10 STEADY-STATE ACCURACY

The steady-state accuracy of analog systems was considered in Section 5.5. The results derived there are based on the final-value theorem of the Laplace transform. In this section we derive equivalent results for sampled-data systems, based on the final-value theorem of the z-transform, which, from Section 11.3, is given by

$$\lim_{k \to \infty} e(kT) = \lim_{z \to 1} (z-1)E(z) \tag{13-40}$$

provided that the left-side limit exists. The left-side limit exists if all poles of $E(z)$ are inside the unit circle, with the possible exception of one pole at $z = 1$.

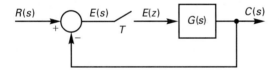

Figure 13.26 Sampled-data system.

We now investigate steady-state accuracy for sampled-data systems. We assume that the system being investigated is stable. Only unity-gain feedback systems are considered. The techniques developed in Chapter 5 for nonunity-gain feedback analog systems apply directly to sampled-data systems. We define the error in the system of Figure 13.26 as the difference between the system input and the system output *at the sampling instants*. Hence

$$e(kT) = r(kT) - c(kT)$$

and thus the z-transform of this error is

$$E(z) = R(z) - C(z) = R(z) - \frac{G(z)}{1+G(z)}R(z) = \frac{R(z)}{1+G(z)} \tag{13-41}$$

From the final-value theorem, the steady-state error is given by

$$e_{ss}(kT) = \lim_{k \to \infty} e(kT) = \lim_{z \to 1} \frac{(z-1)R(z)}{1+G(z)} \tag{13-42}$$

Consider first the steady-state error for a step input. For this case, $R(z) = Az/(z-1)$, where A is the amplitude of the step and

$$e_{ss}(k) = \lim_{z \to 1} \frac{A(z)}{1 + G(z)} = \frac{A(z)}{1 + \lim_{z \to 1} G(z)} = \frac{A}{1 + K_p} \tag{13-43}$$

where K_p is called the *position error constant* and is given by

$$K_p = \lim_{z \to 1} G(z) \tag{13-44}$$

Note that if $G(z)$ has at least one pole at $z = 1$, K_p is unbounded and the steady-state error for a step input is zero. We call the number of poles of the open-loop function at $z = 1$ the *system type*. Hence, if a system is type 1 or higher, the steady-state error for a step input is zero. Otherwise, the steady-state error is nonzero and is given by (13-43).

For the case that the input is the ramp function $r(t) = At$, then $R(z) = ATz/(z - 1)^2$. From (13-42) the steady state error is given by

$$e_{ss}(k) = \lim_{z \to 1} \frac{ATz}{(z - 1) + (z - 1)G(z)} = \frac{AT}{\lim_{z \to 1}(z - 1)G(z)} \tag{13-45}$$

We define the *velocity error constant* K_v to be

$$K_v = \lim_{z \to 1} \frac{1}{T}(z - 1)G(z) \tag{13-46}$$

and the steady-state error for a ramp input is given by

$$e_{ss}(k) = \frac{A}{K_v} \tag{13-47}$$

If the open-loop function has two or more poles at $z = 1$, the error constant K_v is unbounded, and the steady-state error for a ramp input is zero. Thus for a type 2 or higher system, the steady-state error for a ramp input is zero.

These steady-state error equations can also be applied to the case that a digital compensator, $D(z)$, follows the sampler in Figure 13.26. For this case, $G(z)$ is replaced with $D(z)G(z)$ in (13-44) and (13-46), and (13-43) and (13-47) apply directly.

As a final point, the steady-state error in this section is denoted by $e_{ss}(k)$, to emphasize that this value applies only at the sampling instants. The steady-state response between sampling instants has not been determined and cannot be determined via the z-transform, since the z-transform description gives system characteristics only at the sampling instants. An example is given to illustrate steady-state errors in sampled-data control systems.

Example 13.11

The first-order system given in Figure 13.1 with the open-loop function

$$G(z) = \mathscr{z}\left[\frac{1 - e^{-Ts}}{s(s + 1)}\right] = \frac{0.393}{z - 0.607} \qquad T = 0.5s$$

is considered. This system is type 0, since the plant has no poles at $z = 1$, and the position error constant, from (13-44), is

$$K_p = \lim_{z \to 1} G(z) = \frac{0.393}{1 - 0.607} = 1.0$$

The steady-state error for a unit-step input, from (13-43), is

$$e_{ss}(k) = \frac{1}{1 + K_p} = \frac{1}{2}$$

This value of steady-state error is seen in the unit-step response of this system plotted in Figure 13.2.

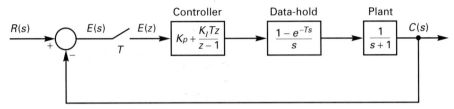

Figure 13.27   System for Example 13.11.

Since this steady-state error is large, we reduce it by introducing a digital PI controller to the system. It is shown later in this chapter that the transfer function of a commonly used PI controller is

$$D(z) = K_p + \frac{K_I T z}{z - 1}$$

where K_P and K_I are two gains to be determined in the design process. The system is now assumed to be of the form given in Figure 13.27. The open-loop function is

$$D(z)G(z) = \frac{0.393\,[(K_p + K_I T)z - K_p]}{(z - 1)(z - 0.607)}$$

This system is type 1, and the steady-state error for a step input is zero, *provided that the system is stable*. The design process must choose the gains K_P and K_I such that (1) the system is stable, (2) the system has an acceptable transient response, and (3) other design specifications as may be given are met in some acceptable manner. Design techniques are presented in the remainder of this chapter.

13.11   DESIGN OF DIGITAL CONTROL SYSTEMS

Thus far we have been concerned primarily with analysis of discrete control systems. We have assumed that the control system is given, and we analyze the system to determine stability margins, time response, frequency response, and so forth.

Some simple designs were considered—for example, the determination of gains required to satisfy steady-state error specifications.

In this and the four following sections we consider the design problem: how we design a digital controller transfer function (or difference equation) that will satisfy design specifications for a given control system. In these sections the emphasis is on classical design techniques (frequency-response and root-locus techniques).

The design of control systems involves the changing of system parameters and/or the addition of digital subsystems (called compensators, controllers, or filters) to achieve certain desired system characteristics. The desired characteristics, or performance specifications, generally relate to steady-state accuracy, transient response, relative stability, sensitivity to changes in system parameters, and disturbance rejection. These performance specifications were investigated in detail in Section 9.1 for analog systems and should be reviewed by readers not familiar with these topics.

The conclusions drawn with respect to the design of analog systems in Section 9.1 apply directly to digital control systems. For a digital control system, to improve steady-state accuracy we add poles at $z = 1$ to increase the type number of the system, and/or we increase the loop gain. Of course, adequate stability margins must be maintained. To increase the speed of the system response, the system bandwidth must be increased, since the product of rise time multiplied by bandwidth is approximately constant. In terms of the closed-loop pole locations, the poles must be moved towards the origin in the z-plane to increase the speed of response. To reduce the overshoot in the transient response, the stability margins generally must be increased, which will increase the equivalent damping ratio ζ of the dominant complex poles. The effect of increasing stability margins is to reduce any peaks in the closed-loop frequency response.

Adequate stability margins (gain and phase margins) are required to give a reasonable transient response and to ensure stability in the event of small parameter changes in the plant and of certain nonlinear operations. A high loop gain decreases the sensitivity of the system characteristics to parameter variations in the plant. This high loop gain also reduces the response of the system to disturbances, provided that the high gain does not appear in the direct path from the disturbance input to the system output.

Generally, we prefer a system with a high loop gain, at least at low frequencies. If the high loop gain can be maintained at the higher frequencies, the system will respond faster, and the effects of the higher frequencies in the disturbance inputs will be reduced. Furthermore, there will be fewer effects from parameter variations. However, efforts to increase the loop gain can lead to reduced stability margins, which will degrade the transient response and the system response if nonlinear operation occurs. In many control systems, high-frequency noise is present, and the response to this noise presents problems. In these systems, it may be necessary to reduce the loop gain to reduce the noise response, leading to conflicting specifications. In these cases, some trade-offs are necessary (in practical situations trade-offs are almost always necessary).

We assume that the system to be designed is of the configuration of Figure 13.28 (a). The digital compensator is generally realized by an A/D converter, a digital computer, and a D/A converter, as illustrated in Figure 12.19. For the system of Figure 13.28(a) the closed-loop transfer function is given by

$$\frac{C(z)}{R(z)} = \frac{D(z)G(z)}{1 + D(z)\overline{GH}(z)} \tag{13-48}$$

(see Section 12.8), and hence the characteristic equation is

$$1 + D(z)\overline{GH}(z) = 0 \tag{13-49}$$

We call the compensation of the type shown in Figure 13.28(a) cascade, or series, compensation. The effects of this compensation on system characteristics are given by the characteristic equation (13-49).

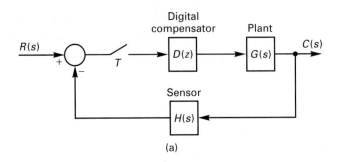

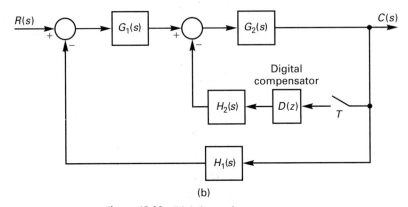

Figure 13.28 Digital control systems.

Sometimes it is more feasible to place the compensator within an inner loop in the system, such as illustrated in Figure 13.28(b). For this system,

$$C(z) = \frac{\mathscr{Z}\left[\dfrac{G_1 G_2 R}{1 + G_1 G_2 H_1}\right]}{1 + D(z)\mathscr{Z}\left[\dfrac{G_2 H_2}{1 + G_1 G_2 H_1}\right]} \tag{13-50}$$

The characteristic equation for this system is then

$$1 + D(z)\mathscr{Z}\left[\frac{G_2 H_2}{1 + G_1 G_2 H_1}\right] = 0 \tag{13-51}$$

This type of compensation is termed feedback, parallel, or minor-loop compensation. For this system, the effects of compensation on system characteristics are given by (13-51).

In the following two sections, we basically consider compensation by a first-order device. Thus the compensator transfer function is of the form

$$D(z) = \frac{K_d(z - z_0)}{z - z_p} \tag{13-52}$$

For root-locus design, we directly determine the pole, the zero, and the gain factor K_d for this compensator. However, for frequency-response design, the design plane is not the

z-plane, but the w-plane. Hence we must express the compensator transfer function as a function of w. The transformation of $D(z)$ to the w-plane yields $D(w)$, that is,

$$D(w) = D(z)\big|_{z = [1 + (T/2)w]/[1 - (T/2)w]} \tag{13-53}$$

Thus $D(w)$ is first order, and we initially assume it to be of the form

$$D(w) = \frac{1 + w/\omega_{w0}}{1 + w/\omega_{wp}} \tag{13-54}$$

where $-\omega_{w0}$ is the zero location and $-\omega_{wp}$ is the pole location, in the w-plane.

The dc gain of the compensator is found in (13-52) by letting $z = 1$ or in (13-54) by letting $w = 0$. Thus, in (13-54), we are assuming a unity dc gain for the compensator. A non-unity dc gain is obtained by multiplying the right side of (13-54) by a constant equal to the value of the desired dc gain.

To realize a compensator, the transfer function must be expressed in z, as in (13-52). From (13-54),

$$
\begin{aligned}
D(z) &= \frac{1 + \dfrac{w}{\omega_{w0}}}{1 + \dfrac{w}{\omega_{wp}}} \Bigg|_{w = (2/T)[(z-1)/(z+1)]} \\[2mm]
&= \frac{\omega_{wp}(2/T + \omega_{w0})}{\omega_{w0}(2/T + \omega_{wp})} \left[\frac{z - \dfrac{2/T - \omega_{w0}}{2/T + \omega_{w0}}}{z - \dfrac{2/T - \omega_{wp}}{2/T + \omega_{wp}}} \right]
\end{aligned}
\tag{13-55}
$$

Hence, in (13-52), the controller parameters in the z-plane as functions of the w-plane parameters are given by

$$K_d = \frac{\omega_{wp}(2/T + \omega_{w0})}{\omega_{w0}(2/T + \omega_{wp})} \qquad z_0 = \frac{2/T - \omega_{w0}}{2/T + \omega_{w0}} \qquad z_p = \frac{2/T - \omega_{wp}}{2/T + \omega_{wp}} \tag{13-56}$$

As in the case of analog compensators, the digital compensator of (13-54) is classified by the location of the zero, ω_{w0}, relative to that of the pole, ω_{wp}. If $\omega_{w0} < \omega_{wp}$, the compensation is called *phase-lead*. If $\omega_{w0} > \omega_{wp}$, the compensator is called *phase-lag*. The phase-lag compensator is discussed first. From (13-56), note that if the zero z_0 is closer to $z = 1$ than is the pole z_p, $D(z)$ is phase lead; otherwise it is phase lag.

The frequency-response design procedures presented in the chapter are identical to those presented in Chapter 9 of analog controllers. The stability region in the w-plane for discrete control systems is the same as that in the s-plane for analog control systems, as shown in Section 13.7. However, pole locations in the w-plane do not yield the same transient-response characteristics as do equal pole locations in the s-plane. An s-plane pole location transforms into the equivalent w-plane pole location by the transformation (13-34), which is repeated here:

$$w = \frac{2}{T}\left[\frac{z - 1}{z + 1}\right]_{z = e^{sT}} = \frac{2}{T}\left[\frac{e^{sT} - 1}{e^{sT} + 1}\right] \tag{13-57}$$

However, the mechanics of frequency-response design are the same for both analog and digital control systems. For this reason, the reader should review the design procedures for analog controllers presented in Chapter 9. Only the basic outlines are presented here, along with examples of the applications of the procedures to digital control systems.

Before design procedures for digital controllers are presented, we will list some of the advantages of digital controllers over analog controllers [3].

1. A digital controller is more flexible, since changing the parameters of a digital controller usually involves changing only a number in a memory location. Changing a parameter in an analog controller usually involves replacing at least one circuit element.

2. Digital signals are less susceptible to noise and to parameter variations in instrumentation, since the data can be represented, generated, transmitted, and processed as binary words.

3. High accuracy and speed are possible through digital signal processing. Digital signal processing can be implemented in hardware rather than software for even greater speed.

4. Complex signal-processing algorithms can be accurately implemented digitally, since the accuracy of the parameters is limited by the wordlength in the digital processor. The parameters of analog signal processors depend, in general, on values of resistors, capacitors, and so on.

13.12 PHASE-LAG DESIGN

For the phase-lag compensator, $\omega_{w0} > \omega_{wp}$ in the compensator transfer function

$$D(w) = \frac{1 + w/\omega_{w0}}{1 + w/\omega_{wp}} \tag{13-58}$$

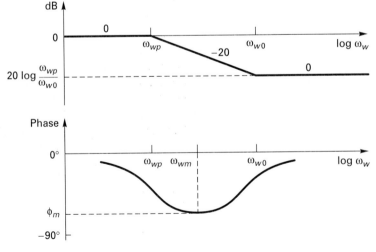

Figure 13.29   Phase-lag digital filter frequency-response characteristics.

The Bode diagram of this transfer function is given in Figure 13.29 and is seen to be identical to the Bode diagram of an analog phase-lag compensator, in Figure 9.12, which is plotted versus the real frequency variable ω. Note that the phase of the compensator of Figure 13.29 is negative, which accounts for the compensator's name. The dc gain is unity, and the high-frequency gain is given by

$$\text{(high-frequency gain)}_{\text{dB}} = 20 \log \frac{\omega_{wp}}{\omega_{w0}} \tag{13-59}$$

or the magnitude of the high-frequency gain is the ratio ω_{wp}/ω_{w0}, which is less than unity. Hence the phase-lag compensator is employed to reduce high-frequency gain relative to low-frequency gain and increases stability margins without degrading the low-frequency response. Of course, the reduction in high-frequency gain reduces the closed-loop system bandwidth, resulting in a slower system response.

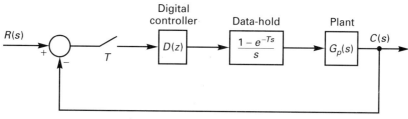

Figure 13.30 Digital control system.

Design using phase-lag compensation is discussed relative to the system of Figure 13.30. The system characteristic equation is given by

$$1 + D(z)G(z) = 0 \tag{13-60}$$

where

$$G(z) = \mathscr{Z}\left[\frac{1 - e^{-sT}}{s}G_p(s)\right] \tag{13-61}$$

For a system configuration differing from that of Figure 13.30, the characteristic equation is formed as in (13-60). Next the transfer function that multiplies $D(z)$ is calculated. From that point, the design procedure follows that given next.

As discussed earlier and as is seen in Figure 13.29, phase-lag compensators introduce both reduced gain and negative phase into the frequency response of the open-loop function, which as a polar plot is the Nyquist diagram. Since, in general, negative phase tends to destabilize a system by rotating the Nyquist diagram toward the -1 point, the break frequencies ω_{w0} and ω_{wp} must be chosen carefully such that the added negative phase does not occur in the vicinity of the $-180°$ crossover of the plant frequency response $G(j\omega_w)$, where

$$G(w) = G(z)\big|_{z = (1 + Tw/2)/(1 - Tw/2)} \tag{13-62}$$

and

$$G(j\omega_w) = G(w)\big|_{w = j\omega_w} \tag{13-63}$$

for the system of Figure 13.30. However, to increase the system stability margins, it is necessary for the compensator to introduce the reduced gain in the vicinity of the $-180°$ crossover. Thus both ω_{w0} and ω_{wp} must be much smaller than the frequency at which the $-180°$ phase occurs.

Figure 13.31 illustrates design by phase-lag compensation. In this figure both the system gain margin and the system phase margin have been increased, increasing relative stability. In addition, the low-frequency gain has not been reduced, and thus the steady-state response has not been degraded to achieve the improved relative stability. The closed-loop bandwidth has been decreased, which generally results in a slower system time response.

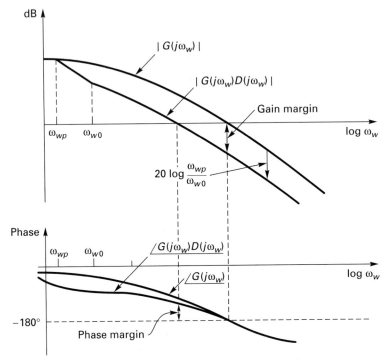

Figure 13.31   Design using phase-lag compensation.

A technique for determining ω_{wp} and ω_{w0} to yield a desired phase margin is now given. This technique is identical to that given in Section 9.4 for analog controllers. The design procedure is as follows:

1. Adjust the dc gain of $G(z)$ by the factor K_c to satisfy low-frequency specifications.
2. Find the frequency ω_{w1} at which the angle of $G(j\omega_w)$ is equal to $(-180° + \phi_m + 5°)$, where ϕ_m is the desired phase margin.
3. Calculate the zero from

$$\omega_{w0} = 0.1\omega_{w1} \tag{13-64}$$

4. Calculate the pole from

$$\frac{\omega_{wp}}{\omega_{w0}} = \frac{1}{|G(j\omega_{w1})|} \tag{13-65}$$

5. Calculate the compensator transfer function from

$$D(w) = \frac{K_c(1 + w/\omega_{w0})}{1 + w/\omega_{wp}} \tag{13-66}$$

where K_c is the factor by which the dc gain of $G(w)$ was adjusted in step 1.

The 5° factor in step 2 is added to account for the phase lag of approximately 5° added by the compensator at the frequency ω_{w1}. Once $D(w)$ is calculated, $D(z)$ is calculated from (13-56), where the gain K_d must be multiplied by K_c. An example is now given.

Example 13.12

We consider the radar tracking system as was considered in the examples of the design of analog controllers in Chapter 9. The plant transfer function is given by

$$G_p(s) = \frac{4}{s(s+1)(s+2)}$$

The system configuration is as given in Figure 13.30. Note that the faster time constant of the plant is 0.5 s, and the sample period is normally chosen smaller than this value to minimize the effects of sampling. However, for this example, T is chosen to be 0.5 s, so that the effects of the sampling will be more evident when compared to the analog designs of Chapter 9. The plant pulse transfer function is then

$$\begin{aligned}
G(z) &= \frac{z-1}{z} \mathscr{Z}\left[\frac{4}{s^2(s+1)(s+2)} \right] \\
&= \frac{z-1}{z} \mathscr{Z}\left[\frac{2}{s^2} + \frac{-3}{s} + \frac{4}{s+1} + \frac{-1}{s+2} \right] \\
&= \frac{z-1}{z} \left[\frac{z}{(z-1)^2} - \frac{3z}{z-1} + \frac{4z}{z-0.6065} - \frac{z}{z-0.3679} \right] \\
&= \frac{0.05824z^2 + 0.1629z + 0.02753}{z^3 - 1.9744z^2 + 1.1975z - 0.2231}
\end{aligned}$$

Although the partial-fraction expansion of $G(z)$ is given, the final result was obtained using MATLAB. Generally, numerical-accuracy problems occur when combining terms in $G(z)$ as shown earlier, because the numerator coefficients result from the difference of numbers that are much larger than the coefficients. The frequency response of the plant is given in Table 13.4, and this response was also calculated by computer.

As in the examples in Chapter 9, the desired phase margin is chosen to be 50°. It is assumed that the loop dc gain is satisfactory; hence step 1 of the design process requires no adjustment of the frequency response of Table 13.4. In step 2, the angle

$$-180° + \phi_m + 5° = -180° + 50° + 5° = -125°$$

By additional computer calculations, it was found that $G(j0.36) \approx 5.1 \; \underline{/-125°}$. Thus $\omega_{w1} = 0.36$, and from (13-64), $\omega_{w0} = 0.1\omega_{w1} = 0.036$. From (13-65),

$$\omega_{wp} = \frac{\omega_{w0}}{|G(j\omega_{w1})|} = \frac{0.036}{5.1} = 0.00706$$

The compensator transfer function is then

$$D(w) = \frac{1 + w/0.036}{1 + w/0.00706}$$

From (13-56),

TABLE 13.4 FREQUENCY RESPONSE FOR EXAMPLE 13.11

ω_w	ω	Magnitude	dB	Phase
0.0100	0.0100	199.98770	46.02007	−91.00266
0.0200	0.0200	99.97540	39.99786	−92.00517
0.0300	0.0300	66.62981	36.47337	−93.00739
0.0400	0.0400	49.95088	33.97086	−94.00919
0.0500	0.0500	39.93864	32.02786	−95.01043
0.0600	0.0600	33.25977	30.43838	−96.01097
0.0700	0.0700	28.48569	29.09253	−97.01067
0.0800	0.0800	24.90213	27.92473	−98.00938
0.0900	0.0900	22.11226	26.89266	−99.00697
0.1000	0.1000	19.87800	25.96745	−100.00330
0.2000	0.1998	9.76153	19.79036	−109.86910
0.3000	0.2994	6.32192	16.01698	−119.47430
0.4000	0.3987	4.56235	13.18378	−128.72070
0.5000	0.4974	3.48441	10.84260	−137.54000
0.6000	0.5956	2.75491	8.80214	−145.89240
0.7000	0.6930	2.23008	6.96639	−153.76290
0.8000	0.7896	1.83697	5.28203	−161.15450
0.9000	0.8853	1.53411	3.71714	−168.03290
1.0000	0.9799	1.29589	2.25133	−174.57150
2.0000	1.8546	0.36206	−8.82430	−221.18550
3.0000	2.5740	0.16135	−15.84445	−248.18390
4.0000	3.1416	0.09408	−20.52963	−265.76170
5.0000	3.5842	0.06474	−23.77598	−278.22040
6.0000	3.9312	0.04955	−26.09867	−287.60740
7.0000	4.2066	0.04071	−27.80672	−295.00050
8.0000	4.4286	0.03510	−29.09268	−301.01390
9.0000	4.6103	0.03133	−30.08075	−306.02200
10.0000	4.7612	0.02866	−30.85354	−310.26750

$$K_d = \frac{\omega_{wp}(2/T + \omega_{w0})}{\omega_{w0}(2/T + \omega_{wp})} = \frac{0.00706(2/0.5 + 0.036)}{0.036(2/0.5 + 0.00706)} = 0.1975$$

$$z_0 = \frac{2/T - \omega_{w0}}{2/T + \omega_{w0}} = \frac{2/0.5 - 0.036}{2/0.5 + 0.036} = 0.9822$$

$$z_p = \frac{2/T - \omega_{wp}}{2/T + \omega_{wp}} = \frac{2/0.5 - 0.00706}{2/0.5 + 0.00706} = 0.99648$$

Thus the compensator transfer function as a function of z is

$$D(z) = \frac{0.1975(z - 0.9822)}{z - 0.99648} = \frac{0.1975z - 0.1940}{z - 0.99648}$$

A digital simulation of the system yields the unit step response as given in Figure 13.32. Also given in this figure is the step response from Example 9.2, which is an analog phase-lag design also for a 50° phase margin. Note that the sampling introduces some destabilization. If the sample period were chosen larger than 0.5 s, the destabilizing effects would be greater, whereas choosing T less than 0.5 s would reduce the destabilizing effects.

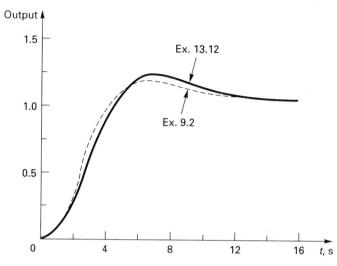

Figure 13.32   Response for Example 13.12.

13.13 PHASE-LEAD DESIGN

Design by phase-lead compensation is now discussed. For a phase-lead compensator, in (13-54) $\omega_{w0} < \omega_{wp}$, and the compensator frequency response is as shown in Figure 13.33. This characteristic was discussed in detail in Section 9.5 for analog compensation, and that discussion applies directly to the digital compensator. However, the frequency axis in Figure 13.33 is w-plane frequency, which at lower frequencies is approximately the same as real frequency. At high frequencies, the w-plane frequency axis is a great distortion of the real-frequency axis, with $\omega_w \to \infty$ equivalent to $\omega \to \omega_s/2$. Thus when we speak of high frequencies in the w-plane, the equivalent s-plane frequencies are usually high relative to the bandwidth of the plant. However, the equivalent s-plane frequencies are always less than $\omega_s/2$.

We see from Figure 13.33 that the phase-lead compensator will increase the high-frequency gain and hence the bandwidth of the closed-loop system, which leads to a faster system response. As discussed in Section 9.5, the zero and pole of the compensator $D(w)$ of (13-54) must be placed in the vicinity of the 180° crossover frequency of the plant frequency response (see Figure 9.19). One approach to phase-lead design is the trial-and-error placement of the compensator zero and pole, with many iterations on these placements. This procedure is discussed in Section 9.5.

In this section we present only the analytical design procedure of Section 9.6. Since this procedure forces the compensated Nyquist diagram through the point $1\underline{/-180° + \phi_m}$ with ϕ_m the phase margin, the equations given in Section 9.6 apply directly to the design of digital control systems. This procedure is now given.

Let ω_{w1} be the w-plane frequency at which the phase margin occurs. At this frequency, for the system of Figure 13.30,

$$D(j\omega_{w1})G(j\omega_{w1}) = 1\underline{/-180° + \phi_m} \tag{13-67}$$

where

$$D(w) = \frac{a_1 w + a_0}{b_1 w + 1} \tag{13-68}$$

Note that the assumed form of the compensator is now of the general form of the phase-lag compensator of (13-66), even though the notation is different. This compensator of (13-68) has a dc gain of a_0. The design equations are [4]

$$a_1 = \frac{1 - a_0|G(j\omega_{w1})|\cos\ \theta}{\omega_{w1}|G(j\omega_{w1})|\sin\ \theta}$$

$$b_1 = \frac{\cos\ \theta - a_0|Gj\omega_{w1}|}{\omega_{w1}\sin\ \theta} \tag{13-69}$$

where, from (13-67),

$$\theta = \arg D(j\omega_{w1}) = -180° + \phi_m - \arg G(j\omega_{w1}) \tag{13-70}$$

and

$$|D(j\omega_{w1})| = \frac{1}{|G(j\omega_{w1})|} \tag{13-71}$$

If the compensator coefficients satisfy the preceding equations, the Nyquist diagram will pass through the point $1\underline{/-180° + \phi_m}$. If the designed system is stable, this system has the required phase margin. However, nothing in the procedure guarantees stability. Thus, once the coefficients are calculated, the Bode diagram (or Nyquist diagram) must be calculated to determine if the closed-loop system is stable.

This design procedure requires that the dc gain a_0 and the phase-margin frequency ω_{w1} be chosen. Then (13-69) determines the compensator coefficients a_1 and b_1. The dc gain of the compensator is determined by the steady-state specifications of the control system.

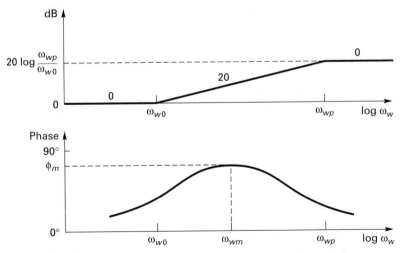

Figure 13.33   Phase-lead digital filter frequency-response characteristics.

The frequency ω_{w1} can be approximately determined in the following manner. Since the compensator is to be phase lead, the angle of the compensator, θ, must be positive (see Figure 13.33). Thus, from (13-70),

$$\theta = -180° + \phi_m - \arg G(j\omega_{w1}) > 0°$$

or

$$\arg G(j\omega_{w1}) < -180° + \phi_m \qquad (13\text{-}72)$$

Thus ω_{w1} is not arbitrary and must be chosen to satisfy (13-72). In general, higher values of ω_{w1} result in larger closed-loop system bandwidths, provided that the designed system is stable. A more complete discussion of the choice of the phase-margin frequency is given in Section 9.6. An example is now given.

Example 13.13

The radar-tracking-system plant of Example 13.12 is used in this example. Recall that

$$G(s)H(s) = \frac{4}{s(s+1)(s+2)}$$

with $T = 0.5$ s. The system configuration is given in Figure 13.30. The same plant is used in the analog phase-lead design of Example 9.3, and in that example, a phase-lead design was performed with $\phi_m = 50°$. The phase-margin frequency was chosen to be $\omega = 1.7$ rad/s. From (13-36), the equivalent w-plane frequency is

$$\omega_{w1} = \frac{2}{T}\tan\left(\frac{\omega_1 T}{2}\right) = \left(\frac{2}{0.5}\right)\tan\left[\frac{(1.7)(0.5)}{2}\right] = 1.810$$

With a unity dc gain ($a_0 = 1$) and choosing $\omega_{w1} = 1.810$, the application of (13-69) results in an unstable controller. Different choices of ω_{w1} also resulted in unstable controllers. This instability results from the phase-lag introduced by the zero-order hold. Hence a phase margin of $50°$ cannot be achieved with a unity dc gain controller.

To reduce the effects of the sampling process, the sample period was reduced to 0.1 s, that is, the sampling frequency was increased by a factor of 5, from 2 Hz to 10 Hz. This increase results in the plant pulse transfer of

$$G(z) = \frac{(0.6189z^2 + 2.2984z + 0.5327)10^{-3}}{z^3 - 2.7234z^2 + 2.4644z - 0.7408}$$

which was calculated by computer. For this sample period, the real frequency $\omega = 1.7$ rad/s is equivalent to the w-plane frequency of $\omega_w = 1.704$. However, the design equations again resulted in an unstable controller. Hence a phase margin of $45°$ was deemed to be acceptable, since we could not obtain the $50°$ phase margin. As above, the phase-margin frequency was chosen to be $\omega_{w1} = 1.704$. Evaluation by computer, which is necessary for this case, yields

$$G(j\omega_{w1})\big|_{\omega_{w1} = 1.704} = 0.4540 \;\underline{/-194.8°}$$

From (13-70),

$$\theta = \arg D(j\omega_{w1}) = -180° + \phi_m - \arg G(j\omega_{w1})$$
$$= -180° + 45° + 194.8° = 59.8°$$

From (13-69), with $a_0 = 1$,

$$a_1 = \frac{1 - a_0|G(j\omega_{w1})|\cos\theta}{\omega_{w1}|G(j\omega_{w1})|\sin\theta} = \frac{1 - (1)(0.454)\cos(59.8°)}{(1.704)(0.454)\sin(59.8°)} = 1.154$$

$$b_1 = \frac{\cos\theta - a_0|G(j\omega_{w1})|}{\omega_{w1}\sin\theta} = \frac{\cos(59.8°) - (1)(0.454)}{1.704\sin(59.8°)} = 0.03329$$

The compensator transfer function is then

$$D(w) = \frac{1.154w + 1}{0.03329w + 1} = \frac{1 + w/0.8666}{1 + w/30.04}$$

Conversion of this transfer function to the z-plane by (13-56) yields

$$D(z) = \frac{14.45(z - 0.9169)}{z + 0.2006} = \frac{14.45z - 13.25}{z + 0.2006}$$

The unit step response of this system is given in Figure 13.34, along with that of the analog design in Example 9.3. Since the phase margin in Example 9.3 is 50° and the phase margin in this design is 45°, the overshoot with the digital controller is greater. Also given in the figure is the step response of the phase-lag digital control system of Example 13.12, to illustrate the difference in employing a phase-lead controller and a phase-lag controller.

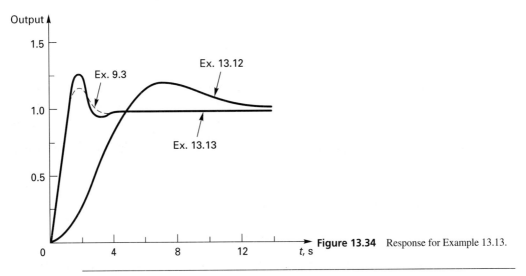

Figure 13.34 Response for Example 13.13.

13.14 DIGITAL PID CONTROLLERS

The analog PID controller is described in Sections 7.10 and 9.8. Those readers unfamiliar with this controller should review those sections before continuing with this section. If $e(t)$ is the input to the controller and $m(t)$ is the output, the PID controller is described by the equation

$$m(t) = K_P e(t) + K_I \int e(t)\,dt + \frac{K_D\,de(t)}{dt} \qquad (13\text{-}73)$$

The design of the controller results in the determination of the three gains K_P, K_I, and K_D. Procedures for designing PID analog controllers are given in Chapters 7 and 9.

Digital PID controllers also satisfy equation (13-73), except the multiplication, integration, and differentiation are performed numerically in a digital computer. If the numerical integration and differentiation are performed accurately, there is little difference in the system response for the analog implementation and the system response for the digital implementation of (13-73). Some techniques for numerical integration and differentiation will now be presented.

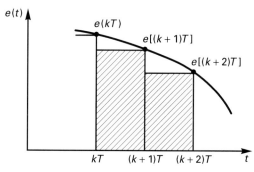

Figure 13.35   Euler method for numerical integration.

While many different methods are available for the numerical integration of a signal, only the simplest procedure is covered here. This procedure is known as the Euler method, or the right-side rectangular rule. This rule is illustrated in Figure 13.35. The area under each segment of the curve is approximated by the area of the rectangle shown. If $m(t)$ is to be the integral of $e(t)$, the value of the integral at $t = (k + 1)T$ is equal to the value at $t = kT$ plus the area under the curve for $e(t)$ between kT and $(k + 1)T$. Thus, by Euler's rule,

$$m[(k + 1)T] = m(kT) + Te[(k + 1)T] \qquad (13\text{-}74)$$

This is the difference equation for the Euler rule for numerical integration. The z-transform of this equation yields

$$z[M(z) - m(0)] = M(z) + Tz[E(z) - e(0)] \qquad (13\text{-}75)$$

Ignoring the initial conditions, we can find the transfer function of this integrator:

$$M(z) = \frac{Tz}{z - 1} E(z) \qquad (13\text{-}76)$$

Other numerical integration methods yield different transfer functions for the integrators [2].

Numerical differentiation can be performed by the method illustrated in Figure 13.36. With this method it is assumed that the slope of $e(t)$ at $t = (k + 1)T$ is equal to the slope of the straight line that connects $e(kT)$ with $e[(k + 1)T]$. If $m(t)$ is to be the derivative of $e(t)$, this numerical differentiation procedure is described by

$$m[(k + 1)T] = \frac{e[(k + 1)T] - e(kT)}{T} \qquad (13\text{-}77)$$

The z-transform of this equation yields the transfer function

$$M(z) = \frac{z - 1}{Tz} E(z) \qquad (13\text{-}78)$$

Note that this transfer function is the reciprocal of that of the Euler-rule integrator. It can be shown that the reciprocal of the transfer functions of certain other numerical integrators give the transfer functions of numerical differentiators, as in the analog case.

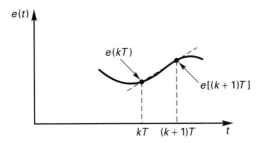

Figure 13.36 Illustration of numerical differentiation.

Two approaches to the design of digital PID controllers can be taken. In one approach, the sample period T, which is also the numerical integration and differentiation increment, can be chosen sufficiently small to ensure accurate integration and differentiation. Then the analog procedures given in Chapters 7 and 9 can be employed to design the three gains K_P, K_I, and K_D.

In the other approach, the transfer functions of the numerical integrator and the numerical differentiator are used with (13-73) to form a controller transfer function $D(z)$:

$$M(z) = D(z)E(z) = \left[K_P + \frac{K_I Tz}{z-1} + \frac{K_D(z-1)}{Tz} \right] E(z) \qquad (13\text{-}79)$$

This digital PID controller is shown in Figure 13.37. The controller-transfer function may then be transformed to the w-plane, and frequency-response techniques very similar to those of Chapter 9 may then be employed to design the gains K_P, K_I, and K_D. This procedure is exact in that the approximations of the numerical integration and the numerical differentiation are accounted for. This procedure must be used if the sample period T is so large that the result is inaccurate numerical procedures. See Ref. 2 for the details of this method.

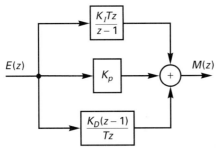

Figure 13.37 Digital PID filter.

Recall that many algorithms for numerical integration and numerical differentiation exist. Hence (13-79) is not unique for a digital PID controller. However, (13-79) is the transfer function for many of the commercial PID controllers available.

In this book we assume that the T can be chosen small enough so that the numerical procedures are sufficiently accurate. Hence the analog design techniques presented in Sections 7.11 and 9.9 may be used.

Regardless of the procedure used to determine the gains of the PID controller, we realize the controller using the difference equations of (13-74) and (13-77), resulting in the controller difference equation

$$m(k+1) = K_P e(k+1) + K_I m_i(k+1) + \frac{K_D[e(k+1) - e(k)]}{T} \tag{13-80}$$

where $m_i(k+1)$ is given by

$$m_i(k+1) = m_i(k) + Te(k+1)$$

Two examples are now presented.

Example 13.14

The radar tracking system used in the phase-lag design of Example 13.12 is used here in a PI controller design. The sample period is $T = 0.5$ s. This plant is the same as used in the analog PI controller design in Example 9.6, where the desired phase margin was 50°. In Example 9.6, the controller gains were calculated to be $K_P = 0.220$ and $K_I = 0.0088$. The analog controller transfer function is

$$G_c(s) = 0.220 + \frac{0.0088}{s}$$

and thus the discrete controller transfer function is, from (13-79),

$$D(z) = 0.220 + \frac{0.0044z}{z-1}$$

The unit step response of the digitally compensated closed-loop system is plotted in Figure 13.38, along with the step response of the analog system designed in Example 9.6. Note that both closed-loop systems are compensated with PI compensators with the same compensator gains. However, in this example, the integration is performed numerically with a numerical integration increment of 0.5 s. In the system of Example 9.6, the integration was exact. The effects of the sampling and the inaccurate numerical integration are seen to be destabilizing.

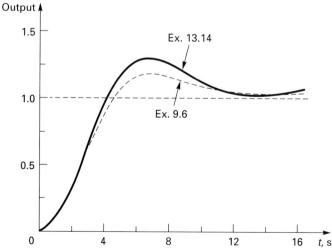

Figure 13.38   Response for Example 13.14.

To investigate this effect further, the open-loop frequency responses of both of the compensated systems were calculated and are plotted in Figure 13.39. These responses are plotted against *real frequency* ω. The magnitude characteristics are so nearly the same that they cannot be plotted separately. However, the phase characteristics are noticeably different. The phase margin of the analog system is 52°, and that of the digital control system is only 45°. If the sample period were reduced, the phase margin and the step response of the digital control system would approach that of the analog system. The step responses and the frequency responses may be calculated with MATLAB.

Example 13.15

The radar tracking system used in the last example is also used in this example, which is the design of a PID controller. From Example 13.13, we saw that a sample period of 0.5 s was not sufficient for phase-lead compensation. Since the derivative term of the PID compensator is phase-lead, the sample period for this example will be reduced to 0.1 s, as in Example 13.13. Example 9.8 presents a PID design for this plant. The required phase margin in this example is 50°. The resulting gains were found to be $K_P = 1.100$, $K_I = 0.005$, and $K_D = 1.123$. These gains will be used with the digital compensator of (13-79), resulting in the transfer function

$$D(z) = 1.100 + \frac{0.0005z}{z-1} + \frac{11.23(z-1)}{z}$$

The unit-step response is given in Figure 13.40. Also given is the unit-step response of the analog system of Example 9.8. Once again we see the destabilizing effects of the sampling operation and of the inaccurate numerical integration and differentiation.

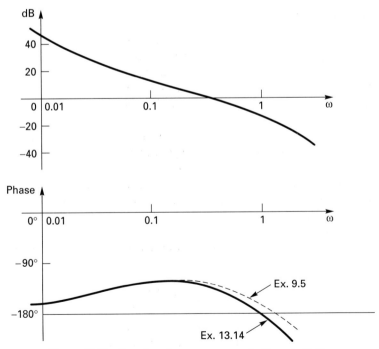

Figure 13.39 Open-loop frequency response for Example 13.14.

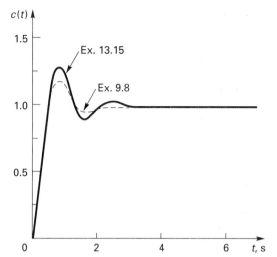

Figure 13.40 Response of Example 13.15

In the preceding examples, a digital implementation of an analog PID compensation was realized, and the digital control systems show more overshoot than the analog systems. This effect results from the addition of phase-lag from the sampling operation, which reduces the system phase margin, and from the inaccurate numerical differentiation and integration. If the PID controllers had been designed by the exact methods of design [2], this effect would not have been as great and in many cases is not significant.

13.15 ROOT-LOCUS DESIGN

Root-locus analysis of discrete systems was considered in Section 13.5. That section stated that the rules of root-locus construction for a discrete system are identical with those for an analog system; the difference is in the meaning of the root locations in terms of the transient response. The relationship of characteristic-equation root locations in the z-plane to the transient response of a discrete system is developed in Section 13.4.

As shown in Chapters 7 and 9 for analog systems, the techniques for root-locus design result in the same types of controllers as do the frequency-response design techniques. Hence the root-locus design of analog systems results in phase-lead, phase-lag, and PID compensators. This is also the case for discrete systems. In fact, since the purpose of root-locus design is to place poles of a closed-loop transfer function in certain locations in the complex plane, the techniques of root-locus design are *identical* for analog systems and for discrete systems. Once again, the only difference for the two types of systems is in the meaning of the root locations.

The analytical design procedure of Section 7.8 can be applied directly in the w-plane, with $D(w)$ given by (13-68).

$$D(w) = \frac{a_1 w + a_0}{b_1 w + 1} \tag{13-81}$$

For this case, the dominant pole locations must be transformed from the z-plane to the w-plane, and design equations (7-53) may be used as given. Note that in (13-81) a_0 is the compensator dc gain.

For design directly in the z-plane, we assume the compensator to be of the form

$$D(z) = \frac{\alpha_1 z + \alpha_0}{\beta_1 z + 1} \tag{13-82}$$

Design equations (7-53) also apply directly for these parameters; however, for this case, the compensator dc gain is not α_0 but is given by

$$\text{dc gain} = D(1) = \frac{\alpha_1 + \alpha_0}{\beta_1 + 1} \tag{13-83}$$

Hence, in (7-53), if the substitutions $a_0 = \alpha_0$, $a_1 = \alpha_1$, and $b_1 = \beta_1$ are made, these two equations and (13-83) may be solved simultaneously for the three coefficients of the digital compensator (13-82) as a function of the dominant pole locations and the compensator dc gain. These equations are not solved here.

Since the methods of root-locus design given in Chapter 7 can be applied to digital control systems in the w-plane, only a simple example of z-plane design is given here.

Example 13.16

In this example we consider the first-order system of Section 13.1 and Figure 13.1. For this system

$$G(z) = \frac{0.393}{z - 0.607} \qquad T = 0.5\text{s}$$

As seen in Figure 13.2, this system has a steady-state error of 50 percent for a step input. We employ a PI compensator to eliminate this error. From (13-79), the transfer function of a digital PI compensator can be expressed as

$$D(z) = K_P + \frac{K_I T z}{z - 1} = \frac{(K_P + K_I T)z - K_P}{z - 1} = \frac{K_d(z - z_0)}{z - 1}$$

where

$$K_d = K_P + K_I T \qquad z_0 = \frac{K_P}{K_P + K_I T} = \frac{K_P}{K_d}$$

The open-loop function of the compensated system is then

$$D(z)G(z) = \frac{0.393 K_d(z - z_0)}{(z - 1)(z - 0.607)}$$

Since, for a phase-lag compensation, the zero must be chosen close to the pole, we choose z_0 to be 0.90. The root-locus of the system is then as shown in Figure 13.41. We assume that the transient-response characteristics of the uncompensated closed-loop system in System 13.1 are satisfactory; hence we place the dominant pole at $z_1 = 0.214$. For the point z_1 to be on the root locus, $D(z_1)G(z_1) = -1$. Thus

$$\frac{0.393 K_d(0.214 - 0.90)}{(0.214 - 1)(0.214 - 0.607)} = -1$$

or $K_d = 1.146$. From the preceding equations,

$$K_P = K_d z_0 = (1.146)(0.90) = 1.031$$

$$K_I = \frac{K_d - K_P}{T} = \frac{1.146 - 1.031}{0.5} = 0.230$$

and the compensator transfer function is given by

$$D(z) = 1.031 + \frac{0.115z}{z - 1}$$

The step response of the compensated system is given in Figure 13.42. For comparison purposes, the step response of the uncompensated system of Figure 13.2 is also given.

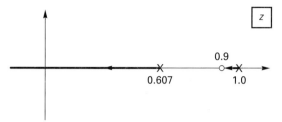

Figure 13.41   Root locus for Example 13.16.

Note the unusual transient response of the compensated system. We designed the dominant pole of the closed-loop compensated system to be the same as that of the uncompensated system; however, the response is much slower. The reason for this slow response can be seen from the root locus in Figure 13.41. Since the compensated system is second order, we must consider both closed-loop transfer function poles. One pole was designed to occur at $z = 0.214$; the second can be calculated to be at $z = 0.924$. The time constant of this pole, given by

$$z = e^{-T/\tau} = e^{-0.5/\tau} = 0.924$$

is found to be 6.33 s. We see in this case that the pole added by the compensator is very significant. A phase-lag compensator always adds a pole with a large time constant; often this pole does not dominate the time response as it does in this example.

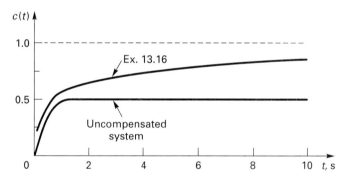

Figure 13.42   Step responses for examples.

13.16 SUMMARY

This chapter presented the basic principles of digital control system analysis and design. First stability of discrete systems was defined, and several methods of stability analysis of linear time-invariant discrete systems were given. Almost all these methods are also applicable to analog systems and have been covered previously in the chapters on analog control. For this reason, the methods were not developed in detail but relied on the reader's knowledge obtained from earlier chapters. One new procedure, the Jury test, was presented for stability analysis in the z-plane. Both time-response and frequency-response methods of analysis were given.

The second major topic developed in this chapter was the design of digital control systems. The controllers considered were the phase-lead, the phase-lag, and the PID controllers. Both frequency-response and time-response (root-locus) design procedures were given. As with stability analysis, the design methods are the same procedures as developed for analog systems, and the developments in this chapter relied on the material of the chapters on analog design.

As was stated previously, we are considering the *models* of physical systems in all the analysis and design procedures presented in this book. We are not working with the physical systems themselves. For a given model, the model may or may not accurately describe the physical system's characteristics. If the model is accurate, it will be accurate only over a limited range of operation of the physical system. Hence we should be very careful in attaching undue significance to the analysis and design techniques presented here or elsewhere. We can design a model to have 10 percent overshoot for a step input, to have very low steady-state errors, and so forth. However, all this is meaningless until it is tested using the physical system; generally the response of the physical system will not be as good as that of the model. In fact, many iterations on the design and on improving the system model are often needed before there is even a resemblance between the model response and the physical system response. The point to be made here is that all analytical work is with the *model* of a system; the major part of a controls engineer's work is in applying the results of analysis and design to the physical system (see Figure 1.4).

REFERENCES

1. E. I. Jury. *Theory and Application of the z-Transform Method.* New York: Wiley, 1964.

2. C. L. Phillips and H. T. Nagle, Jr. *Digital Control System Analysis and Design,* 3rd ed. Upper Saddle River, NJ: Prentice Hall, 1995.

3. C. W. deSilva. *Control Sensors and Actuators.* Upper Saddle River, NJ: Prentice Hall, 1989.

4. C. L. Phillips and R. D. Harbor. *Feedback Control Systems,* 3rd ed. Upper Saddle River, NJ: Prentice Hall, 1996.

PROBLEMS

13.1. Figure P13.1 gives the block diagram of a control system in which the temperature of a vat of liquid is controlled (see Section 2.9). Note that the plant has a time constant $\tau = 0.5$ s. Determine the unit step responses of the system, with $K = 1$, for the following cases:

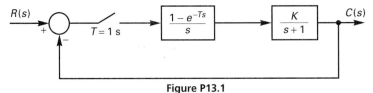

Figure P13.1

(a) The sample period $T = 1$ s.

(b) The sample period $T = 0.1$ s.

(c) The sampler/data-hold is removed, resulting in an analog system.

(d) Plot the time responses of (a), (b), and (c) on the same graph.

(e) In (a), (b), and (c), the steady-state outputs are equal. Why is this true?

(f) Use the MATLAB program of Example 13.2 to verify the response in (a).

13.2. For the three systems of Problem 13.1:

(a) Find the characteristic equations, and the ranges of the gain K for stability.

(b) For $K = 8$, determine the range of the sample period for stability.

(c) Verify the results of (a) and (b) by simulating the system with SIMULINK and observing the effects on the system response and stability of varying T and K.

13.3. For the system of Problem 13.1, let $G_p(s)$ be replaced with $K/(s + 1)$, and let $T = 0.1$ s and $K = 1$.

(a) Find the foward path transfer function $G(z)$.

(b) Verify the results in (a) with MATLAB.

(c) Find the closed-loop system transfer function.

(d) Calculate the unit step response of the system.

(e) From (c), find the system difference equation.

(f) Write a MATLAB program that solves this difference equation.

(g) Run the MATLAB program in (f) for a unit step input, to verify the results in (d). This program is a simulation of the system.

13.4. The system of Figure P13.1 models the pen-position control system of a certain digital plotter, if the transfer function $K/(s + 2)$ is replaced with $G_p(s) = K/[s(s + 0.5)]$. For this problem, let $K = 1$.

(a) Calculate $G(z)$ for $T = 1$ s.

(b) Calculate $G(z)$ for $T = 0.1$ s, and note the numerical problems with subtracting numbers that are almost equal.

(c) Verify the calculations in (a) and (b) with MATLAB.

(d) Calculate the closed-loop transfer functions in (a) and (b).

(e) Calculate the time constants of the closed-loop systems.

(f) The system is type 1, since the plant transfer function $G_p(s)$ has a pole at $s = 0$. Hence the dc gain of the closed-loop systems are equal to unity. Show that this is true.

13.5. Consider the pen-position control system for a digital plotter given in Problem 13.4, with $T = 1$ s. The results of Problem 13.4 are useful.

(a) Determine the range of K for stability.

(b) Find the value of K that results in the system being in steady-state oscillation (marginally stable).

(c) Find the frequency of oscillation for (b).

(d) Use MATLAB to simulate the marginally stable system to verify the results in (b) and (c) (see Example 13.2). In addition, plot the time response.

(e) Let $K = 0.6$. Find the effective values of ζ, ω_n, and τ for this system.

13.6. Use the Jury test to determine the range of K for which the system of Figure P13.6 is stable for the case that

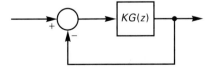

Figure P13.6

(a) $G(z) = \dfrac{0.1z + 0.06}{(z-1)(z-0.7)}$

(b) $G(z) = \dfrac{0.1z + 0.06}{z(z-1)(z-0.7)}$

(c) Verify the results in (a) and (b) using MATLAB to find the roots of the characteristic equation, using a value of K that results in marginal stability. List the magnitudes of the roots.

13.7. For the temperature-control system of Problem 13.1, plot the root locus for each case listed. For each case, also determine the range of K for stability *using the information on the root locus.*
(a) $T = 1$ s.
(b) $T = 0.1$ s.
(c) The sampler/data-hold removed from the system; that is, the system is analog.
(d) Use MATLAB to plot the root loci to verify your results.

13.8. For the system of Problem 13.4, let $T = 1$ s.
(a) Plot the root locus.
(b) Find the roots of the characteristic equation for the value of K for which the system is marginally stable.
(c) Find the largest value of K for which there is no overshoot in the system step response.
(d) Plot a root locus using MATLAB to verify your results in (a).
(e) Verify the results in (b) using MATLAB.
(f) Verify the results in (c) by running a step response using MATLAB.

13.9. Plot the root locus for the system of Figure P13.6 for the case that
(a) $G(z) = \dfrac{0.1z + 0.06}{(z-1)(z-0.7)}$

(b) $G(z) = \dfrac{0.1z + 0.06}{z(z-1)(z-0.7)}$

(c) Use MATLAB to verify the root loci in (a) and (b).

13.10. Plot the Nyquist diagrams on the same graph for the system of Figure 13.1 for the following cases.
(a) $T = 1$ s.
(b) $T = 0.1$ s.
(c) The sampler/data-hold removed from the system; that is, the system is analog.
(d) Use the Nyquist diagrams to determine the ranges of the gain K for stability.
(e) Use MATLAB to plot the Nyquist diagrams to verify your results.

13.11. For the pen-position control system of Problem 13.4, with $T = 1$ and $K = 1$,
(a) Plot the Nyquist diagram.
(b) Determine the range of K for which the system is stable, using the Nyquist diagram.
(c) Determine the frequency at which the system will oscillate for the system marginally stable.
(d) Use MATLAB to plot the Nyquist diagram to verify the results in (a).
(e) Use MATLAB to verify the results in (b).
(f) Run the step response with MATLAB for the case that the system is marginally stable, to verify the results in (b) and (c).

13.12. Use the Routh–Hurwitz criterion to determine the range of K for stability for the temperature-control system of Problem 13.1, for the following cases. Compare these results with those of Problem 13.2, if available.
(a) The sample period $T = 1$ s.
(b) The sample period $T = 0.1$ s.
(c) The sampler/data-hold is removed, resulting in an analog system.

13.13. (a) Use the Routh–Hurwitz criterion to determine the range of K for which the system of Problem 13.4 is stable, for $T = 1$ s.

(b) Verify the results in (a) using MATLAB, by finding the roots of the system characteristic equation of the marginally stable system.

13.14. Consider the temperature-control system of Figure P13.1, with $K = 1$. Plot the Bode diagram and indicate the gain and phase margins for the following cases.

(a) The sample period $T = 1$ s.

(b) The sample period $T = 0.1$ s.

(c) The sampler/data-hold is removed, resulting in an analog system.

(d) Use MATLAB to verify the Bode plots.

13.15. Consider the system of Problem 13.4, with $T = 1$ s and $K = 1$.

(a) Plot the Bode diagram and indicate the gain and phase margins.

(b) Use MATLAB to verify the Bode diagram in (a).

13.16. Consider the system of Figure P13.1, with $K = 1$. Determine the steady-state errors for both a unit step input and a unit ramp input for the following cases.

(a) The sample period $T = 1$ s.

(b) The sample period $T = 0.1$ s.

(c) The sampler/data-hold is removed, resulting in an analog system.

(d) Why are the steady-state errors for a unit step input in (a), (b), and (c) equal?

13.17. (a) Consider the system of Problem 13.4, with $T = 1$ s and $K = 1$. Calculate the steady-state errors for both a unit step input and a unit ramp input.

(b) Repeat (a) for the sampler/zero-order hold removed, that is, for the analog system.

(c) Verify the results in (a) for the ramp response of the system using a MATLAB simulation. It will be found that the simulation must run for approximately 250 s of system time for steady state to occur. Why?

13.18. Shown in Figure P13.18 is the block diagram for an attitude for a rigid-body satellite. This system was described in Section 2.6. In the system of Figure P13.18 digital compensation has been added. However, for this problem assume that $D(z) = 1$.

(a) Use the Jury test to determine the range of K for which the system is stable.

(b) Use the Nyquist criterion to verify the results of (a).

(c) Basing your answer on the Nyquist diagram in (b), can a phase-lag compensator be used to stabilize the system?

(d) Design a unity dc gain phase-lead compensator that will result in a phase margin of $50°$. *Hint:* A value of $\omega_{w1} = 10$ will stabilize the system.

(e) Determine the gain and phase margins of the compensated system by computer.

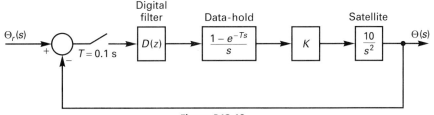

Figure P13.18

13.19. Shown in Figure P13.19 is the block diagram of a carbon dioxide control system for an environmental plant chamber. This chamber is used by plant scientists in the study of plant growth. The sensor has a time delay of 45 s, which is the time required for the instrumentation to analyze an air sample to determine its carbon dioxide content. In this problem we ignore the time delay, resulting in $H(s) = 0.0125$. Also, assume that $K = 50$.

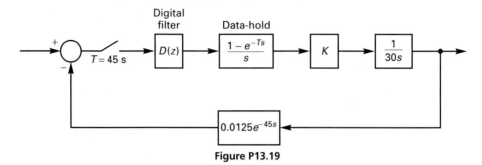

Figure P13.19

(a) Calculate the open-loop function $\overline{GH}(z)$, and verify the result with MATLAB.

(b) Plot the Bode diagram for the uncompensated system, indicating the gain and phase margins. Verify this Bode diagram with MATLAB.

(c) Plot the Nyquist diagram for the uncompensated system, and from this diagram determine the system stability. Verify the Nyquist diagram with MATLAB.

(d) Plot the root locus for the uncompensated system, and from this root locus determine system stability. Verify the root locus with MATLAB.

(e) Design a unity dc gain phase-lag compensator that will yield a phase margin of 70°.

(f) Can a satisfactory unity dc gain phase-lead compensator be designed? *Hint:* Consider the gain margin.

13.20. Consider the carbon dioxide control system of Problem 13.19. With the sensor transfer function containing the time delay of 45 s as given and $K = 50$, repeat Problem 13.19. Note that the open-loop transfer function is changed in a simple manner.

13.21. Consider an analog control system with the plant transfer function

$$G_p(s) = \frac{50}{(s+1)(s+2)(s+5)}$$

The system characteristic equation is $1 + G_c(s)G_p(s) = 0$ and a phase margin of 45° is desired. After the design of an analog compensator, the compensator realization is to be by the discrete transfer function $D(z)$, with $T = 0.1$ s. Computer calculations are to be used for this problem.

(a) An analog PI compensator is designed using the procedures given in Chapter 9, resulting in the transfer function

$$G_c(s) = 0.679 + \frac{0.114}{s}$$

This design yields a phase margin of 45°. Give the transfer function of an equivalent discrete PI compensator [see (13-79)], and find the loss in phase margin because of the digital realization, using MATLAB. Recall that

$$G(z) = \frac{z-1}{z} \mathscr{Z}\left[\frac{G_p(s)}{s}\right]$$

must be used for the plant transfer function.

(b) An analog PD compensator is designed, yielding the transfer function

$$G_c(s) = 1.73 + 0.403s$$

This design results in a 45° phase margin. Give the transfer function of an equivalent discrete PD compensator [see (13-79)], and find the loss in phase margin because of the digital realization, using MATLAB.

(c) Why is the phase-margin loss in (b) greater than in (a)?

13.22. Use SIMULINK to simulate the systems and verify the step responses of the systems in Examples 13.12, 13.13, 13.14, and 13.15.

13.23. For the pen-position control system of Problem 13.4, let $T = 1$ and $K = 1$. A digital compensator is to be added to the loop, such that the system characteristic equation is given by

$$1 + D(z)G(z) = 0$$

(a) Design a unity dc gain phase-lag compensator that yields a system phase margin of 40°.

(b) Repeat (a) for a unity dc gain phase-lead compensator.

(c) Run frequency responses for the closed-loop systems of (a) and (b) to verify the results, using MATLAB.

(d) Run step responses for the compensated systems and compare rise times, using MATLAB.

(e) Repeat (d) using SIMULINK.

13.24. Shown in Figure P13.24 is the block diagram of the servo control system for one of the joints of a robot. This system is described in Section 2.12.

(a) Show that the plant transfer function is given by

$$G_p(s) = \frac{\Theta_L(s)}{E_a(s)} = \frac{0.475}{s(s+1)(s+10)}$$

(b) Find the pulse transfer function of $G(z)$ by computer for $T = 0.25$ s.

(c) Use MATLAB to plot a Bode diagram for $G(z)$.

(d) Determine the gain and phase margins from (c), with $D(z) = 1$. Note that the gain of the power amplifier must be included.

(e) Design a phase-lead compensator that has a dc gain of 5 and yields a phase margin of 40°. *Hint:* Use $\omega_{w1} = 3$.

(f) Simulate the step response of the system using SIMULINK, and determine the rise time and percent overshoot.

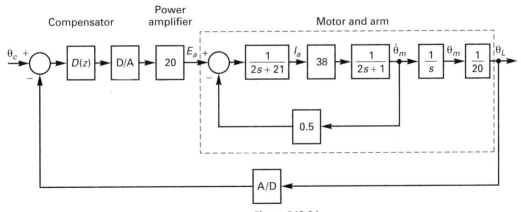

Figure P13.24

14

Nonlinear System Analysis

In this chapter we consider nonlinear analysis systems for the first time. As has been stated previously, we cannot model a physical system exactly. Usually, by increasing the order of the linear model, the accuracy of the model is increased. However, the point is reached at which adding orders to the model will not significantly improve the model. For those cases in which the model accuracy is still not sufficient, it will be necessary to add nonlinearities. The purpose of this chapter is to consider some analysis techniques for models that contain nonlinearities.

In all the preceding chapters we define stability of a linear control system in a BIBO sense. The output of a stable linear system will remain bounded for any bounded input and for any bounded initial conditions. No such statement can be made concerning nonlinear systems. Some nonlinear systems may be stable for certain inputs and then may become unstable if different inputs are applied. Some nonlinear systems may be stable for certain sets of initial conditions but be unstable for different sets of initial conditions. This is the nature of the physical world, since all physical systems are nonlinear.

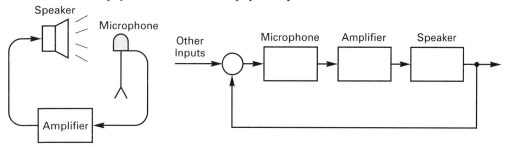

Figure 14.1 Example of a nonlinear system.

input $r_1(t)$ but unstable for the input $r_2(t)$. At this point we do not attempt to define stability for nonlinear systems (there are many such definitions), but we instead rely on an intuitive feeling for stability. Later, as the need arises, we will give the required definitions.

Some general characteristics of nonlinear systems are now presented.

1. **Limit cycle.** A periodic oscillation in a nonlinear system is called a limit cycle. In general, limit cycles are nonsinusoidal. A periodic oscillation in a linear time-invariant system is sinusoidal with the amplitude of oscillation a function of both the amplitude of the system excitation and the initial conditions. In certain nonlinear systems, the amplitude of oscillation is independent of the system excitation or initial conditions.

2. **Subharmonic and harmonic response under a periodic input.** A nonlinear system with a periodic input may exhibit a periodic output whose frequency is either a subharmonic or a harmonic of the input frequency. For example, an input of frequency 10 Hz may result in an output of 5 Hz for the subharmonic case or 30 Hz for the harmonic case.

3. **Jump phenomenon.** A jump phenomenon is illustrated in Figure 14.3. Here the frequency response of a resonant linear system is shown; also shown is a nonlinear-system frequency response that exhibits the jump phenomenon, which is also called *jump resonance*. Suppose that the nonlinear system input is a sinusoid of constant amplitude. Then, as the frequency of the input sinusoid is increased, a discontinuity (jump) occurs in the amplitude of the response. As the frequency is decreased, a jump again occurs but at a different frequency.

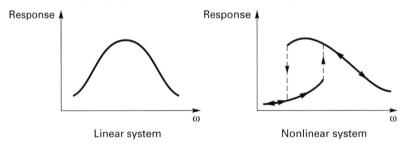

Linear system   Nonlinear system

Figure 14.3   Jump resonance in frequency response.

4. **Multiple equilibrium states.** In a linear stable system, all states approach zero (the origin of state space) as time increases, with no system input. For a nonlinear stable system, there may be a number of different states, other than the origin, that the system can approach as time increases, for no system input. These different states are called equilibrium states, and the one that the system approaches is determined by the system initial conditions. This condition is illustrated in a physical system that, when perturbed (disturbed), can settle to a number of different states, depending on the disturbance.

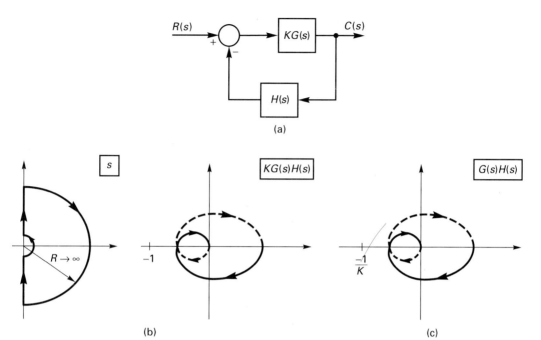

Figure 14.4 Application of Nyquist criterion.

14.2 REVIEW OF THE NYQUIST CRITERION

Before introducing nonlinear system analysis by the describing function, a short review of the Nyquist criterion is presented. The Nyquist criterion is covered in detail in Chapter 8 and is applicable to systems operating in linear regions only. Later we apply the Nyquist criterion in describing-function analysis.

Consider the linear system of Figure 14.4(a). The characteristic equation of this system is

$$1 + KG(s)H(s) = 0 \qquad (14\text{-}1)$$

To apply the Nyquist criterion, the Nyquist path in the s-plane is mapped into the complex plane through the mapping $KG(s)H(s)$, as shown in Figure 14.4(b). Then

$$Z = N + P \qquad (14\text{-}2)$$

where Z is the number of zeros of the characteristic equation, (14-1), inside the Nyquist path (in the right half of the s-plane), P is the number of poles of the characteristic equation inside the Nyquist path, and N is the number of clockwise encirclements of the plot of $KG(s)H(s)$ of the -1 point in the complex plane.

Equation (14-2) was originally derived for a plot of the characteristic equation (14-1), with N the encirclements of the origin, that is, $[1 + KG(s)H(s)]$ was plotted (see Section 8.4). Subtracting 1 from both sides of (14-1) yields

$$KG(s)H(s) = -1 \qquad (14\text{-}3)$$

If $KG(s)H(s)$ is plotted, N is the number of encirclements of the -1 point. We now take this process one step further. Suppose that (14-3) is divided by K, that is,

$$G(s)H(s) = -\frac{1}{K} \qquad (14\text{-}4)$$

and $G(s)H(s)$ is plotted. Then N in (14-2) is the number of encirclements of the point $-1/K$, as illustrated in Figure 14.4(c). We later apply this variation of the Nyquist criterion.

14.3 DESCRIBING FUNCTION

The describing function as considered here is applicable to systems that contain only one nonlinearity; we call this function the *ordinary describing function*. A system that contains only one nonlinearity is depicted in Figure 14.5. Note that the system input is zero. A system with a zero input is called a *free* system. If, in addition, the system is time-invariant, the system is called an *autonomous* system. Even though $G(s)$ in Figure 14.5 is denoted as the plant, in the general case $G(s)$ can also include the compensator transfer function and the sensor transfer function.

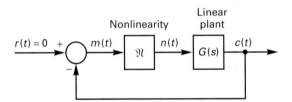

Figure 14.5   Nonlinear system.

We now develop the describing function by applying a sinusoid to the nonlinearity input; that is, we let

$$m(t) = M \sin \omega t \qquad (14\text{-}5)$$

Then, in the steady-state, $n(t)$ is periodic and, in general, nonsinusoidal. Thus $n(t)$ can be represented by a Fourier series [4]:

$$n(t) = \frac{A_0}{2} + \sum_{k+1}^{\infty} A_k \cos k\omega t + \sum_{k+1}^{\infty} B_k \sin k\omega t \qquad (14\text{-}6)$$

The Fourier coefficients are given by

$$A_k = \frac{2}{T} \int_{t_0}^{t_0 + T} n(t) \ \cos k\omega t \, dt \qquad (14\text{-}7)$$

$$B_k = \frac{2}{T} \int_{t_0}^{t_0 + T} n(t) \ \sin k\omega t \, dt \qquad (14\text{-}8)$$

where T is the period of the input sinusoid $m(t)$ $[T = 2\pi/\omega]$ and t_0 is any time. We will generally consider only the case that A_0 is zero. This value of A_0 results if the nonlinearity is symmetric with respect to the amplitude of the input.

Suppose that, under the conditions just described, $G(s)$ is low-pass with respect to the harmonics in $n(t)$ that is, $|G(j\omega)|$ is small for all other components of $n(t)$ in (14-6) compared to its value for the fundamental component. Then the output $c(t)$ can be expressed as

$$c(t) \simeq C \sin(\omega t + \theta) \qquad (14\text{-}9)$$

This assumption is the foundation of describing-function analysis. The harmonics in $n(t)$ are then not important, since these harmonics have very little effect on $c(t)$. The harmonics can be ignored, and $n(t)$ can be approximated as

$$n(t) \simeq A_1 \cos \omega t + B_1 \sin \omega t$$
$$= A_1 \sin(\omega t + 90°) + B_1 \sin \omega t = N_1 \sin(\omega t + \phi) \qquad (14\text{-}10)$$

where, by using trigonometric identities or, more simply, phasors [4],

$$N_1 \underline{/\phi} = B_1 + jA_1 \qquad (14\text{-}11)$$

From (14-10), we see that $n(t)$ can be approximated as a sinusoid of the same frequency as $m(t)$ but not of the same amplitude and phase; thus the nonlinearity can be replaced with a *complex gain* of

$$\text{Gain of nonlinearity} = N(M, \omega) = \frac{B_1 + jA_1}{M} = \frac{N_1 \underline{/\phi}}{M} \qquad (14\text{-}12)$$

This equivalent gain can be represented as shown in Figure 14.6 and is called the describing function. *The describing function $N(M, \omega)$ in general is a function of both the amplitude and the frequency of the input sinusoid.*

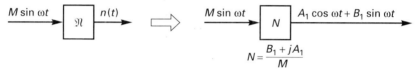

Figure 14.6   Describing-function representation.

Note the assumptions in the definition of the describing function. The describing function has meaning only if

1. The input to the nonlinearity is a sinusoid.
2. The linear system following the nonlinearity is sufficiently low-pass so as to attenuate the second and higher harmonics to such values as to be negligible.

Note also that the describing function is an equivalent gain, which applies only under the very restrictive conditions just given. However, the equivalent gain is *not* linear, since its value is a function of the amplitude of its input.

14.4 DERIVATIONS OF DESCRIBING FUNCTIONS

The describing functions of some simple nonlinearities are now derived. The techniques presented here may be used to derive the describing functions for other nonlinearities.

14.4.1 Cubic Nonlinearity

Consider first a nonlinearity that can be approximated by a cubic function, such that, for Figure 14.5,

$$n(t) = m^3(t) \qquad (14\text{-}13)$$

The input must be assumed to be the sinusoid (under assumption 1)

$$m(t) = M \sin \omega t \tag{14-14}$$

Thus

$$n(t) = M^3 \sin^3 \omega t = M^3 \sin \omega t \frac{1 - \cos 2\omega t}{2}$$

$$= \frac{M^3}{2}\left(\sin \omega t + \frac{1}{2} \sin \omega t - \frac{1}{2} \sin 3\omega t \right) = \frac{M^3}{4}(3 \sin \omega t - \sin 3\omega t) \tag{14-15}$$

since

$$\sin^2 \omega t = \frac{1}{2}(1 - \cos 2\omega t) \tag{14-16}$$

and

$$\sin \omega t \cos 2\omega t = \frac{1}{2}[\sin(\omega t - 2\omega t) + \sin(\omega t + 2\omega t)] \tag{14-17}$$

Hence in (14-15), the third harmonic is ignored (assumption 2), and

$$n(t) = \frac{3M^3}{4} \sin \omega t = A_1 \cos \omega t + B_1 \sin \omega t$$

Thus the describing function for this nonlinearity, from (14-12), is given by

$$N(M, \omega) = \frac{B_1 + jA_1}{M} = \frac{3M^3/4 + j0}{M} = \frac{3M^2}{4} \tag{14-18}$$

Note that, even though this is an equivalent gain for the nonlinearity, it is proportional to the input amplitude squared and is nonlinear itself. For this nonlinearity, (14-15) is the Fourier series of the output $n(t)$ and was easily derived. In general, the Fourier series required is quite difficult to derive.

The describing function for the cubic function, (14-18), is proportional to the square of the amplitude of the input signal. Since the describing function is a type of gain, this gain then increases with increasing signal amplitude. Suppose that this nonlinearity is placed in a linear system that has stability problems for high loop gains. We might then anticipate stability problems for the nonlinear system if the amplitude of the signal into the nonlinearity becomes large, even though this input signal may not be sinusoidal.

In general, the describing functions for nonlinearities are functions of both the amplitude and the frequency of the input sinusoid. We indicate this by the notation $N(M, \omega)$, as in (14-18), even though in this case the describing function is not a function of frequency. Two additional examples of deriving describing functions are given next.

14.4.2 Ideal Relay

Consider first the *ideal relay* characteristic shown in Figure 14.7. It is seen from this figure that for a sinusoidal input, the output is a square wave. Since the output is an odd function, A_1 of (4-10) is zero [4], and

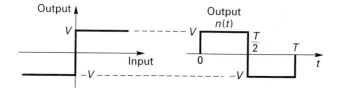

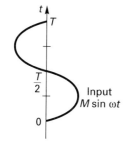

Figure 14.7   Ideal relay characteristic.

$$B_1 \; = \; \frac{2}{T}\int_0^T n(t) \, \sin \omega t \, dt \; = \; \frac{4V}{T}\int_0^{T/2} \sin \omega t \, dt$$

$$= \; \frac{4V}{T}\left[\frac{-\cos \omega t}{\omega}\right]_0^{T/2}$$

$$(14\text{-}19)$$

Since $\omega = 2\pi/T$, and

$$\omega t\big|_{t \,=\, T/2} \; = \; \frac{2\pi}{T}\left(\frac{T}{2}\right) \; = \; \pi$$

then

$$B_1 \; = \; \frac{2V}{\pi}(-\cos \pi + 1) \; = \; \frac{4V}{\pi}$$

$$(14\text{-}20)$$

The describing function for the ideal relay is then

$$N(M, \omega) \; = \; \frac{B_1 + jA_1}{M} \; = \; \frac{4V}{\pi M}$$

$$(14\text{-}21)$$

Thus, under the assumption made, the nonlinearity can be replaced by the equivalent gain of (14-21), as shown in Figure 14.8. Note that the equivalent gain is inversely proportional to the amplitude of the input sinusoid and becomes smaller as the input amplitude M becomes larger. This decreasing gain is also seen from the nonlinearity itself, since the amplitude of the output remains constant with increasing input amplitude.

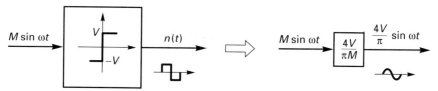

Figure 14.8   Ideal relay describing function.

14.4.3 Limiter

Consider next the *saturation*, or *limiter*, characteristic shown in Figure 14.9. This nonlinearity appears often in physical systems. Examples are electronic amplifiers that saturate, mechanical stops in systems with translational and rotational motion, such as rudders in aircraft and ships, and so forth. For small signals, the operation is linear. However, if the input amplitude is sufficiently large to cause saturation, the output is a clipped sine wave for a sinusoidal input, as shown in Figure 14.9.

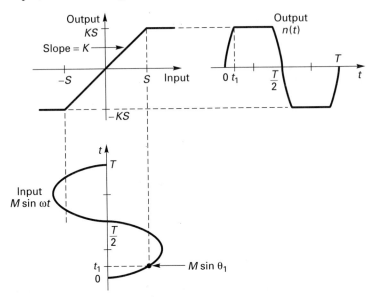

Figure 14.9   Saturation characteristic.

For the describing function, in Figure 14.9,

$$S = M \sin \theta_1 = M \sin \omega t_1 \tag{14-22}$$

where S is the amplitude of the input signal at the time t_1 that saturation occurs, and θ_1 is the angle of the sinusoid at t_1. Or

$$\theta_1 = \omega t_1 = \sin^{-1} \frac{S}{M} \tag{14-23}$$

The saturation output is given by

$$n(t) = \begin{cases} KM \sin \omega t & 0 < t \le t_1 \\ KS & t_1 < t \le \dfrac{T}{4} \end{cases} \tag{14-24}$$

Since $n(t)$ is odd, A_1 in (14-10) is zero. Also, $n(t)$ has quarter-wave symmetry; thus

$$B_1 = \frac{4}{T} \int_0^{T/2} n(t) \sin \omega t \, dt = \frac{8}{T} \int_0^{T4} n(t) \sin \omega t \, dt \tag{14-25}$$

From (14-24) and (14-25),

$$B_1 = \frac{8}{T}\left[\int_0^{t_1} KM \sin^2 \omega t\, dt + \int_{t_1}^{T/4} KS \sin \omega t\, dt\right]$$

$$= \frac{8}{T}\left[\int_0^{t_1} KM\left(\frac{1}{2} - \frac{1}{2}\cos 2\omega t\right) dt + \int_{t_1}^{T/4} KS \sin \omega t\, dt\right] \qquad (14\text{-}26)$$

$$= \frac{8}{T}\left[KM\left(\frac{t}{2} - \frac{1}{4\omega}\sin 2\omega t\right)\Big|_0^{t_1} - \frac{KS}{\omega}\cos \omega t\,\Big|_{t_1}^{T/4}\right]$$

Now, since

$$\omega t\big|_{t_1} = \theta_1 \qquad \omega t\big|_{T/4} = \frac{\pi}{2} \qquad (14\text{-}27)$$

(14-26) becomes

$$B_1 = \frac{8}{T\omega}\left[KM\left(\frac{\theta_1}{2} - \frac{1}{4}\sin 2\theta_1\right) + KS \cos \theta_1\right]$$

$$= \frac{8}{T\omega}\left[\frac{KM\theta_1}{2} - \frac{KM}{4}(2 \sin \theta_1 \cos \theta_1) + K(M \sin \theta_1) \cos \theta_1\right]$$

$$= \frac{4}{\pi}\left[\frac{KM\theta_1}{2} + \frac{1}{2}KM \sin \theta_1 \cos \theta_1\right] \qquad (14\text{-}28)$$

$$= \frac{2}{\pi}[KM\theta_1 + KS \cos \theta_1]$$

Equation (14-28) applies only for $M > S$, since for $M \leq S$ saturation does not occur, and the operation is linear. For linear operation, the gain for a sinusoidal input (as well as for any other input) is K.

For saturation, from (14-12) and (14-28), the describing function is given by

$$N(M, \omega) = \frac{B_1 + jA_1}{M} = \frac{2K}{\pi}\left[\theta_1 + \frac{S}{M}\cos \theta_1\right] \qquad (14\text{-}29)$$

The complete describing function for saturation, which is also called limiting, is given by

$$N(M, \omega) = \begin{cases} \dfrac{2K}{\pi}\left[\theta_1 + \dfrac{S}{M}\cos \theta_1\right] & M > S \\ K & M \leq S \end{cases} \qquad (14\text{-}30)$$

Note again that the describing function for a linear gain is simply that gain.

14.4.4 Describing Function Tables

The describing functions just derived, plus some other commonly used ones, are given in Table 14.1. The describing functions just derived are all real, that is, they introduce no phase shift. Note that some of the describing functions listed in Table 14.1 are complex, and these do introduce phase shift.

To simplify the use of some of the describing functions of Table 14.1, a term that is common to many of these functions, $N_s(x)$, is listed in Table 14.2. This function is defined as

$$N_s(x) = \frac{2}{\pi}\left[\sin^{-1}\frac{1}{x} + \frac{1}{x}\cos\left(\sin^{-1}\frac{1}{x}\right) \right] \tag{14-31}$$

The function is the describing function of the limiter of Figure 14.9, with $K = 1$ (unity-gain limiter), $S = 1$ (limits at an input amplitude of unity), and $x = M$.

To further simplify the analysis of systems for some of the complex describing functions, the describing functions for the ideal relay with dead zone and for backlash are given in Figures 14.10 and 14.11, respectively. Note that these two describing functions can be normalized and plotted. Note also that the describing function for backlash is complex and introduces phase lag (which is destabilizing). Backlash occurs in gears, because gears cannot mesh perfectly. When the drive gear changes direction, the play in the gears must be taken up before the load gear can begin moving again. This characteristic results in time delay and the resultant phase-lag.

TABLE 14.1 DESCRIBING FUNCTIONS: $N_s(x) = \frac{2}{\pi}\left[\sin^{-1}\frac{1}{x} + \frac{1}{x}\cos\left(\sin^{-1}\frac{1}{x}\right)\right]$

Nonlinearity	$N(M,\omega)$	
	$K,$	$M \le A$
	$K_1 + (K - K_1)\, N_s\!\left(\dfrac{M}{A}\right),$	$(M > S)$
	$K,$	$M \le S$
Limiter	$KN_s\!\left(\dfrac{M}{S}\right),$	$M > S$
Ideal relay	$\dfrac{4A}{\pi M}$	
	$0,$	$M \le A$
Dead zone	$K\left[1 - N_s\!\left(\dfrac{M}{A}\right)\right]$	$(M > S)$

TABLE 14.1   DESCRIBING FUNCTIONS: $N_s(x) = \frac{2}{\pi}\left[\sin^{-1}\frac{1}{x} + \frac{1}{x}\cos\left(\sin^{-1}\frac{1}{x}\right)\right]$ (CONTINUED)

Nonlinearity	$N(M,\omega)$
	$K + \dfrac{4A}{\pi M}$
Ideal relay with dead zone	$0, \qquad M \leq A$ $\dfrac{4B}{\pi M}\sqrt{1 - \left(\dfrac{A}{M}\right)^2}, \qquad M > A$
Backlash	$0, \qquad M \leq A$ $\dfrac{K}{2}\left[1 - N_s\left(\dfrac{M/A}{2 - M/A}\right)\right] - j\dfrac{4KA(M-A)}{\pi M^2}, \qquad M > A$
Hystersis	$0, \qquad M \leq A$ $\dfrac{4B}{\pi M}\sqrt{1 - \left(\dfrac{A}{M}\right)^2} - j\dfrac{4AB}{\pi M^2}, \qquad M > A$
Relay	$0, \qquad M \leq A + B$ $\dfrac{2C}{\pi M}\left[\sqrt{1 - \left(\dfrac{B-A}{M}\right)^2} + \sqrt{1 - \left(\dfrac{B+A}{M}\right)^2}\right] - j\dfrac{4AC}{\pi M^2}, \qquad M > A + B$
	$0, \qquad M \leq A$ $K\left[1 - N_s\left(\dfrac{M}{A}\right)\right] + \dfrac{4B}{\pi M}\sqrt{1 - \left(\dfrac{A}{M}\right)^2}, \qquad M > A$

TABLE 14.1   DESCRIBING FUNCTIONS: $N_s(x) = \frac{2}{\pi}\left[\sin^{-1}\frac{1}{x} + \frac{1}{x}\cos\left(\sin^{-1}\frac{1}{x}\right)\right]$ (CONTINUED)

Nonlinearity	$N(M,\omega)$
	$0 - j\dfrac{4A}{\pi M}$

TABLE 14.2   VALUES OF $N_s(x)$

x	$N_s(x)$	x	$N_s(x)$
1.0	1.000	9.0	0.141
1.5	0.781	9.5	0.134
2.0	0.609	10.0	0.127
2.5	0.495	10.5	0.121
3.0	0.416	11.0	0.116
3.5	0.359	11.5	0.111
4.0	0.315	12.0	0.106
4.5	0.281	12.5	0.102
5.0	0.253	13.0	0.0978
5.5	0.230	14.0	0.0909
6.0	0.211	15.0	0.0848
6.5	0.195	19.0	0.0670
7.0	0.181	25.0	0.0509
7.5	0.169	30.0	0.0424
8.0	0.159	50.0	0.0255
8.5	0.149	100.0	0.0127

To summarize these developments, the describing function of a nonlinearity is defined for a sinusoidal input of $M \sin \omega t$ and is given by

$$N(M, \omega) = \frac{B_1 + jA_1}{M} \tag{14-32}$$

where

$$B_1 = \frac{2}{T}\int_{t_0}^{t_0 + T} n(t) \sin \omega t \, dt$$

$$\tag{14-33}$$

$$A_1 = \frac{2}{T}\int_{t_0}^{t_0 + T} n(t) \cos \omega t \, dt$$

and $n(t)$ is the nonlinearity output.

14.5 USE OF THE DESCRIBING FUNCTION

The use of the describing function in the stability analysis of nonlinear systems is discussed in this section. Recall the assumptions made in deriving the describing function.

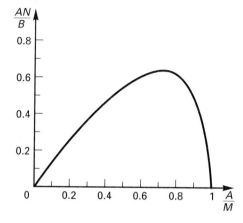

Figure 14.10 Describing function for ideal relay with dead zone.

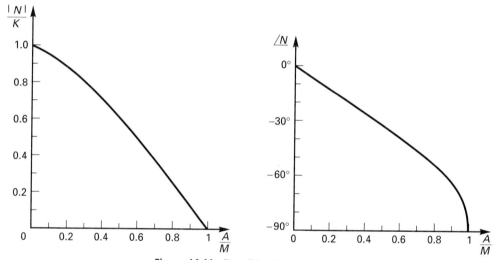

Figure 14.11 Describing function for backlash.

1. The input to the nonlinearity is a sinusoid.
2. The transfer function of the linear system that closes the loop between the output and the input of the nonlinearity is lowpass with respect to the second and higher harmonics.

In describing-function analysis, we have further assumed that the system has no inputs; that is, we are considering only the initial-condition response.

Consider the nonlinear system of Figure 14.12(a). It is assumed that for this system, the describing-function analysis is applicable, resulting in the equivalent system of Figure 14.12(b). In this figure, $N(M, \omega)$ is the equivalent gain, or the describing function, of the nonlinearity of Figure 14.12(a). Now, let

$$N(M, \omega) = |N(M, \omega)|\underline{/\phi(M, \omega)} \tag{14-34}$$

and

$$G(j\omega) = |G(j\omega)| \underline{/\theta(\omega)} \qquad (14\text{-}35)$$

A steady-state sinusoidal analysis (called an ac analysis in circuits [4]) is performed next. The output of the nonlinearity $n(t)$ in Figure 14.12(b) can be expressed as (see Section 4.4)

$$n(t) = |N(M, \omega)| M \sin(\omega t + \phi) \qquad (14\text{-}36)$$

and hence the system output is

$$c(t) = |G(j\omega)||N(M, \omega)| M \sin(\omega t + \phi + \theta) \qquad (14\text{-}37)$$

We now state the basic assumption of describing function analysis.

The nonlinear system will have a limit cycle, with the input to the nonlinearity approximately sinusoidal, if the sinusoid at the nonlinearity input regenerates itself in the loop (loop gain = 1).

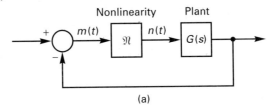

(a)

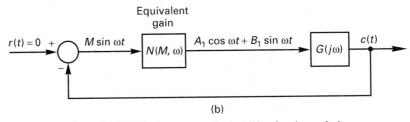

(b)

Figure 14.12   Nonlinear system for describing-function analysis.

Hence we wish to determine if there is an amplitude M and a frequency ω such that the open-loop gain from the nonlinearity input back to the nonlinearity input is equal to unity, with the nonlinearity replaced with the describing function. Thus, from the preceding assumption, since $m(t)$ is equal to $-c(t)$, the system of Figure 14.12 will have a limit cycle if

$$M \sin \omega t = -|G(j\omega)||N(M, \omega)| M \sin(\omega t + \phi + \theta) \qquad (14\text{-}38)$$

Equating phasors for the sinusoids in this expression, we obtain

$$M = -G(j\omega)N(M, \omega)M$$

or

$$1 = -G(j\omega)N(M, \omega)$$

This equation can be expressed as

$$1 + N(M, \omega)G(j\omega) = 0 \qquad (14\text{-}39)$$

If this equation can be satisfied for some value of M and some value of ω, a limit cycle is *predicted* for the nonlinear system. Because of the many approximations in the development of this method of analysis, we can only predict limit cycles. It is necessary to simulate the system to determine if a limit cycle actually occurs in the model. The occurrence of a limit cycle in the physical system can be determined only by tests on the physical system.

Note the similarity of this equation with the characteristic equation of a linear feed-back system. However, (14-39) applies only if the nonlinear system is in a steady-state limit cycle. Thus a describing-function analysis predicts only the presence or the absence of a limit cycle and cannot be applied to analysis for other types of time responses.

Since $N(M, \omega)$ is generally a mathematically complex function, (14-39) cannot be solved directly to yield values for M and ω. Instead, we usually take a semigraphical approach to the solution. We express (14-39) as

$$G(j\omega) = -\frac{1}{N(M, \omega)} \tag{14-40}$$

and plot both sides of this equation in the complex plane. If an intersection occurs for the two functions for the same value of ω, (14-40) is satisfied at that intersection and a limit cycle is predicted for the nonlinear system. The values of M and ω at the intersection give the amplitude and frequency of the sinusoid at the input to the nonlinearity. Note that we are plotting the frequency response of the linear part of the system in plotting the left side of (14-40). We now demonstrate describing-function analysis through three examples.

Example 14.1

Consider the system of Figure 14.13, which contains an ideal-relay, or comparator, nonlinearity. From Table 14.1, the right side of (14-40) is given by

$$-\frac{1}{N(M, \omega)} = -\frac{\pi M}{4V}$$

Shown in Figure 14.14(a) is a plot of this function, as M varies from zero to infinity, along with a plot of $G(j\omega)$. Note that

$$G(j1) = \frac{4}{j(1 + j)^2} = \frac{4}{1\underline{/90°}(\sqrt{2}\underline{/45°})^2} = -2$$

Thus, from (14-40), at the point of intersection in Figure 14.14,

$$-2 = -\frac{\pi M}{4V} \qquad \omega = 1$$

Hence,

$$M = \frac{8V}{\pi}$$

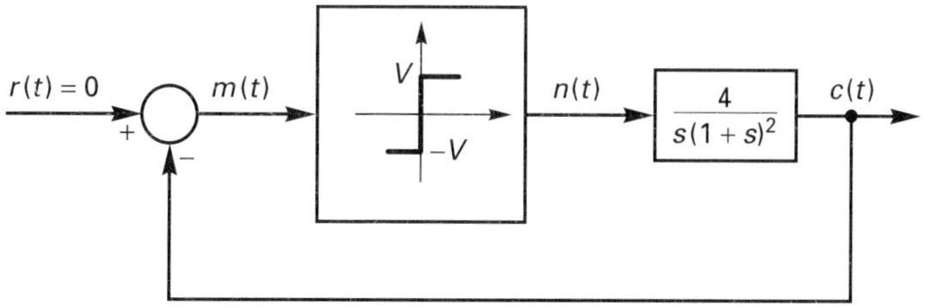

Figure 14.13   System for Example 14.1.

For this system, the describing-function analysis predicts a limit cycle, with the system output given by

$$c(t) = -m(t) = -M \sin \omega t = -\frac{8V}{\pi} \sin t$$

The output of the ideal relay is a square wave of amplitude V, as shown in Figure 14.7. The analysis in this example can be verified by the SIMULINK program of Figure 14.14(b). The state model was calculated with the MATLAB program

```
G = tf([0 0 0 4], [1 2 1 0]);
[A,B,C,D] = ss(Gnum,Gden)
```

In the SIMULINK program, $V = 1$, and thus the predicted amplitude of the limit cycle $c(t)$ is $8/\pi = 2.55$ and the period is $T = 2\pi/\omega = 6.28$. The simulation produced a limit cycle with amplitude of approximately 2.62 and period of approximately 8 s. Thus, for this system, the describing-function analysis gives relatively accurate results.

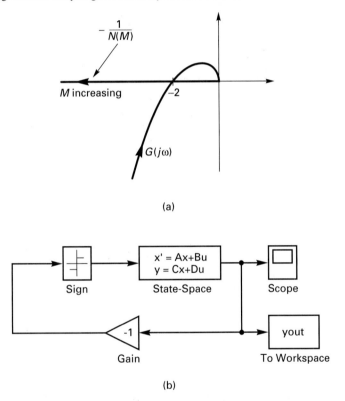

(a)

(b)

Figure 14.14 (a) Describing-function analysis and (b) Simulation for Example 14.1.

Example 14.2

This example is a continuation of Example 14.1. In this example we determine if the assumptions for the describing-function analysis are reasonable for this system.

The output of the nonlinearity, assuming that a limit cycle does occur, is a square wave, as just stated. The square wave has half-wave symmetry; hence the second-harmonic ($\omega = 2$ rad/s) amplitude is zero. We now solve for the approximate amplitudes of the first and third harmonics.

For a square wave of amplitude V, the amplitude of the first harmonic is $4V/\pi$ and of the third harmonic is $4V/3\pi$ [4]. The plant gain at the first harmonic frequency is 2, as just determined. The gain of the plant at the third harmonic frequency is given by

$$|G(j3)| = \left| \frac{4}{j3(1+j3)^2} \right| = \frac{4}{30}$$

Thus the first-harmonic amplitude at the plant output, which is the same as at the nonlinearity input (see Figure 14.13), is equal to the gain times the input, $2(4V/\pi)$, or $8V/\pi$. The third-harmonic amplitude at the nonlinearity input, by the same procedure, is $(4/30)(4V/3\pi)$, or $8V/45\pi$. Hence the amplitude of the third harmonic at the nonlinearity input is $1/45$ of the amplitude of the first harmonic. It seems reasonable then to ignore the third harmonic, and we expect the describing-function analysis to yield good results. Execution of the MATLAB program given in Example 14.1 shows a limit cycle of approximately the predicted amplitude and frequency.

Example 14.3

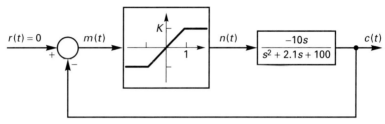

Figure 14.15   System for Example 14.3.

As a second example of describing-function analysis, consider the system of Figure 14.15, which is a common form for an electronic oscillator as used in laboratories. For this system, from Table 14.1,

$$N(M, \omega) = KN_s\left(\frac{M}{S}\right) = \frac{2K}{\pi}\left[\sin^{-1}\frac{1}{M} + \frac{1}{M}\cos\left(\sin^{-1}\frac{1}{M}\right)\right]$$

for $M > 1$. For $M = 1$, which is the limiting value for linear operation,

$$N(M, \omega) = \frac{2K}{\pi}[\sin^{-1}1 + \cos(\sin^{-1}1)] = K$$

and for M very large,

$$\lim_{M \to \infty} N(M, \omega) = \frac{2K}{\pi}\left(0 + \frac{1}{M}\right) = 0$$

These values are also evident from Table 14.2. The plot of $-1/N(M,\omega)$ is then as shown in Figure 14.16(a). Also shown in this figure is a plot of $G(j\omega)$. Note that

$$G(j10) = \frac{-j100}{-100 + j21 + 100} = -4.76$$

Thus we see that a limit cycle is predicted if $1/K < 4.76$, or if $K > 0.21$.

Suppose, for example, that K is given a value of 0.25. The frequency of the predicted sinusoid is $\omega = 10$ rad/s, from the point of intersection. The amplitude, M, is determined by

$$G(j10) = -4.76 = \frac{1}{N(M, \omega)} = -\frac{1}{KN_s(M/S)}$$

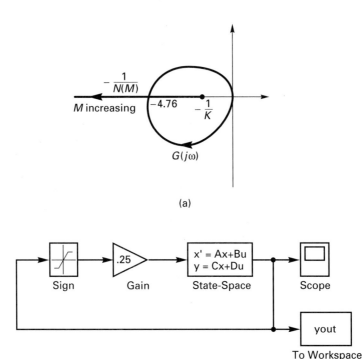

(a)

(b)

Figure 14.16 (a) Describing function analysis and (b) Simulation for Example 14.3.

or

$$N_s(M/S) = \frac{1}{4.76K} = \frac{1}{4.76(0.25)} = 0.84$$

Linear interpolation in Table 14.2 yields a value of $x = 1.36$ for $N_s(x) = 0.84$. Since $S = 1$ and since $x = M/S$, then $M = 1.36$. Thus

$$m(t) = 1.36 \sin 10t$$

Note that, by the Nyquist criterion, the system is marginally stable for $K = 0.21$ and is stable for $K < 0.21$ if the system is operating in the linear region. For $K > 0.21$, the system is unstable when operating in the linear region. Thus for this system, operation cannot occur completely in the linear region in the steady state, and if a limit cycle occurs, its amplitude must be greater than unity. The analysis in this example can be verified by the SIMULINK program of Figure 14.16(b). From the simulation, the amplitude of the limit cycle was approximately 1.39 (predicted value of 1.36), and the period T of approximately 0.63 s (predicted value of 0.628 s). Thus the describing function analysis is quite accurate for this example.

We make one further point concerning the last example. A practical oscillator of any type must be a nonlinear system with a limit cycle, as just shown. Since we can do nothing exactly in physical systems, we cannot place poles of a linear analog system exactly on the $j\omega$-axis. Hence we cannot construct a marginally stable linear system. If this construction is attempted, the system output will eventually begin to die out or to grow, because the poles are either slightly in the left half-plane or slightly in the right half-plane. Thus, to construct a system that has truly steady-state oscillation, it is necessary that the system operate in nonlinear regions.

14.6 STABILITY OF LIMIT CYCLES

Two types of limit cycles can appear in a nonlinear system. A limit cycle that returns to its original form after being perturbed in amplitude is called a *stable limit cycle*. Otherwise the limit cycle is *unstable*. For example, certain nonlinear systems respond with unstable limit cycles such that if the amplitude is caused to decrease, the limit cycle will die out. Or, if the amplitude is caused to increase, the amplitude will then continue to increase without limit or possibly will approach another limit cycle of different amplitude and/or different frequency. Unstable limit cycles are difficult to observe in simulations, since inherent noise in the simulation prevents the unstable limit cycle from maintaining a steady-state oscillation.

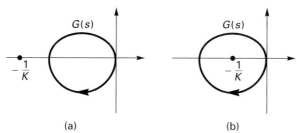

Figure 14.17   Nyquist diagrams for example.

The describing function can be used to investigate the stability of certain limit cycles. The technique to be presented here uses the Nyquist criterion. Recall from Section 14.2 that for a linear system with an open-loop function of $KG(s)H(s)$, we can count the number of encirclements of the point—$1/K$ if the Nyquist diagram of $G(s)H(s)$ is plotted. To illustrate this point, consider that, in the system of Figure 14.15, the saturation nonlinearity is replaced with a linear gain K. Then if the Nyquist diagram does not encircle the $-1/K$ point, as shown in Figure 14.17(a), the linear system is stable. However, if the Nyquist diagram does encircle the point $-1/K$, as shown in Figure 14.17(b), the linear system is unstable.

These concepts are now extended to the nonlinear system of Example 14.3 and Figure 14.15. The describing-function analysis of this system is repeated in Figure 14.18. First, assume that the nonlinear system has a limit cycle and thus is operating at the point shown. Next, we assume that we can linearize the system for some small region (possibly very small) about this operating point and that we can apply linear system theory in this small region. Now, assume that a perturbation in system operation occurs such that the amplitude of the limit cycle, M, decreases slightly. As seen from Figure 14.18, if M decreases, the operating point moves inside the Nyquist diagram and the system is now unstable, as shown in Figure 14.17(b). Thus M will increase, moving the operating point back to its original position in Figure 14.18.

Next assume that a perturbation causes M to increase slightly. Then the operating point moves outside the Nyquist diagram, as in Figure 14.17(a), and the system is now stable. Thus M decreases and the operating point moves back to its original position in Figure 14.18. Then for this system, we say that the limit cycle is stable. Simulation of this system supports this analysis.

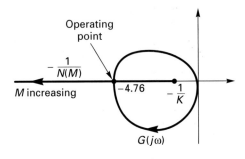

Figure 14.18   Describing-function analysis for the system of Figure 14.15.

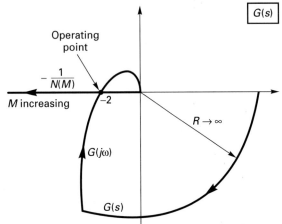

Figure 14.19   Describing-function analysis for the system of Figure 14.13.

Consider next the system of Example 14.1 and Figure 14.13. The describing-function analysis for this system is repeated in Figure 14.19. Assume that the system has a limit cycle at the operating point shown. For this case, a decrease in M causes the linearized system to become unstable, thereby causing M to increase back to its original value. An increase in M causes the system to become stable, thereby causing M to decrease back to its original value. Thus the limit cycle is stable. An example of an unstable limit cycle is now given

Example 14.4

Consider the system of Figure 14.20. The nonlinearity in this system is a dead zone, which, from Table 14.1, has the describing function of

$$N(M) = \begin{cases} 0 & M \le 1 \\ K[1 - N_s(M)] & M > 1 \end{cases}$$

Note that, for $M \gg 1$, the effect of the dead zone becomes negligible and $N(M)$ approaches the value of K. Thus $0 \le N(M) \le K$. For K greater than 0.5, the describing-function analysis is as shown in Figure 14.21. Assume that the system has a limit cycle at the operating point shown. An increase in M causes the system to become unstable and thus M continues to increase; a decrease in M causes the system to be stable and thus M continues to decrease. Hence the limit cycle is unstable.

This unstable limit cycle can be argued from physical considerations. For small signals with $|m(t)| < 1$, no signal is transmitted through the dead zone, and the system is certainly stable. For somewhat larger signals, the effective gain of the nonlinearity is small, and the system is

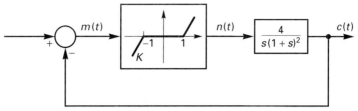

Figure 14.20 System for Example 14.4.

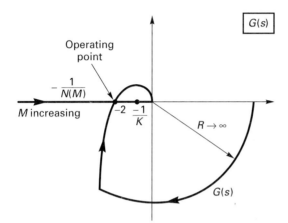

Figure 14.21 Describing-function analysis for system of Figure 14.20.

still stable. For very large signals, the dead-zone effect is negligible, and the nonlinearity appears as a linear gain K. For this linear gain, the system is unstable, and the system response will grow without limit.

Example 14.5

As a final example of the stability of limit cycle, we consider the system of Figure 14.22(a). In this system, the nonlinearity is backlash, which introduces phase lag in the system. The backlash describing function is given in Table 14.1 and Figure 14.11; we see that for the backlash characteristic in Figure 14.22(a), $A = 1$ and $K = 1$. A table of values for this describing function is given in Table 14.3, along with values of $G(j\omega)$. Sketches of $-1/N$ and $G(j\omega)$ are given in Figure 14.22(b); these sketches are not to scale. It is seen from Table 14.3 and Figure 14.22(b) that two intersections occur between the plots $-1/N$ and $G(j\omega)$. The first intersection, labeled as (a) in Figure 14.22(b), occurs for $M \approx 1.25$ and $\omega \approx 0.41$, and thus a limit cycle of

$$m(t) = 1.25 \sin 0.41t$$

TABLE 14.3 VALUES FOR EXAMPLE 4.5

M	$-1/N$	$G(j\omega)$	ω
1.00	$\infty\,\underline{/-90°}$	—	—
1.11	$8.33\,\underline{/-115°}$	$6.31\,\underline{/-115°}$	0.30
1.25	$4.17\,\underline{/-124°}$	$4.39\,\underline{/-124°}$	0.41
1.67	$2.08\,\underline{/-141°}$	$2.41\,\underline{/-141°}$	0.65

TABLE 14.3 VALUES FOR EXAMPLE 4.5

M	$-1/N$	$G(j\omega)$	ω
2.50	$1.43\angle{-153°}$	$1.68\angle{-153°}$	0.86
3.75	$1.25\angle{-162°}$	$1.26\angle{-162°}$	1.00
5.00	$1.14\angle{-167°}$	$1.04\angle{-167°}$	1.12
10.00	$1.04\angle{-174°}$	$0.83\angle{-174°}$	1.27

is predicted. However, note that for operation at this point, increasing M leads to an encircle-ment of the operating point by the Nyquist diagram. Hence M will continue to increase, and this limit cycle is unstable limit. The second intersection, labeled as (b) in Figure 14.22(b), occurs for $M \approx 3.75$ and $\omega \approx 1.0$, and thus a limit cycle of

$$m(t) = 3.75 \sin t$$

is predicted. For this limit cycle, increasing M leads to no encirclements of the operating point on the Nyquist diagram, and this limit cycle is stable.

A SIMULINK simulation of this system is depicted in Figure 14.22(c). It is worthwhile for the reader to experiment with this simulation, since it illustrates stable and unstable limit cycles. The following results can be observed:

1. Large initial conditions result in a stable limit cycle with an amplitude of approximately 4 and a period of approximately 6.2 s, which is reasonably accurate.
2. Small initial conditions result in the output decaying without overshoot.
3. Proper choices of initial conditions result in an oscillation that will either approach result 1 or result 2 above.

14.7 DESIGN

Using the describing-function analysis techniques just developed, we can see how compen-sators may be designed for a nonlinear system in order to eliminate limit cycles. The pur-pose of the compensation is assumed to be the elimination of limit cycles; hence the compensator must be designed to prevent intersections between the plots of $-1/N$ and the frequency response of the remainder of the system, which is labeled as $G(j\omega)$ in the pre-ceding systems. However, since models of physical systems may be quite inaccurate, we usually want the plots of $-1/N$ and $G(j\omega)$ to be separated by an adequate margin of safety. For linear systems, we label these margins of safety as the phase margin and gain margin.

Design procedures are introduced for the system of Figure 14.23. Comparing this sys-tem to that of Figure 14.12(a), we see that the describing-function analysis applies if we replace $G(j\omega)$ in (14-40) with $G_c(j\omega)G_p(j\omega)$. Hence (14-40) becomes, for the system of Figure 14-23,

$$G_c(j\omega)G_p(j\omega) = -\frac{1}{N(M, \omega)} \qquad (14\text{-}41)$$

We introduce design using the system of one of the examples just given. Consider the system of Example 14.4, which is repeated in Figure 14.24(a), with the describing-function analysis repeated in Figure 14-24(b). We see that the intersection between $-1/N$ and $G(j\omega)$ can be eliminated by adding a P-type compensator (a pure gain), with the gain determined by

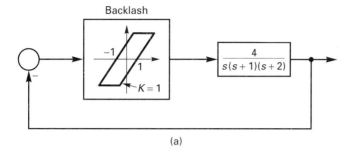

(a)

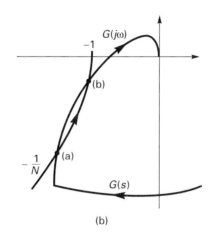

(b)

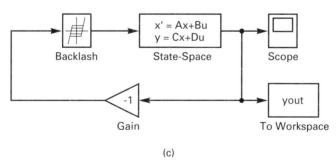

(c)

Figure 14.22 (a) System with backlash, (b) describing function analysis, and (c) simulation for Example 14.5.

$$\frac{1}{KK_p} > 2$$

or $K_P < 1/(2K)$. For example, for a 6-dB gain margin, we could choose $K_P = 1/(4K)$.

This procedure can be extended to include dynamic compensators. The frequency-response design procedures of Chapter 9 can be applied directly, if the −1 point in linear design is replaced by the point −1/K for the system of Figure 14.24. Perhaps a better procedure is to absorb the gain K of the nonlinearity into the plant transfer function, such that the gain on the nonlinearity is now unity. The design procedure now is with respect to the −1 point, and the plant transfer function is now given by

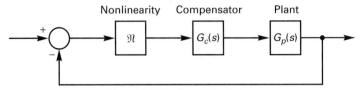

Figure 14.23 System for design.

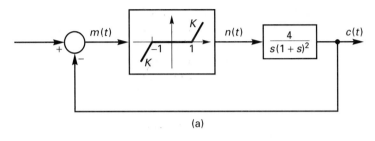

(a)

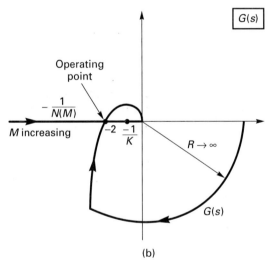

(b)

Figure 14.24

$$G_P(s) \ = \ \frac{4K}{s(s+1)^2}$$

The dynamic compensator can now be designed using the exact procedures of Chapter 9, using either a phase-lead or a phase-lag compensator to give a particular phase margin with respect to the −1 point. However, the testing of any compensators in an accurate simulation and in the physical system becomes critical in this case, since the design procedures were developed for a linear system and the operation of this system is always nonlinear.

The design procedure just described may be applied to some other types of nonlinearities, but as in the case of the analysis of nonlinear systems, no general design procedures apply to all nonlinear systems. For the case that the basic assumptions of the describing function are not valid, the design procedure described above is without justification.

14.8 APPLICATION TO OTHER SYSTEMS

The describing-function analysis was derived for single-loop systems of the configuration shown in Figure 14.12. For this system, the input $r(t)$ is zero, and thus the system can be represented as shown in Figure 14.25. Any system with only one nonlinearity can be represented as shown in this figure, where $-G(s)$ is the transfer function of the linear portion of the system from the output of the nonlinearity back to its input. Thus the describing-function technique applies to any time-invariant system with a single nonlinearity that satisfies the constraint of being sufficiently low-pass to filter out the harmonics of any limit cycle. An example is given to illustrate this point.

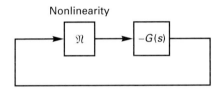

Nonlinearity

Figure 14.25   General system with one nonlinearity.

Example 14.6

To illustrate the preceding discussion, consider the system of Figure 14.26(a), which had a nonlinearity in the rate sensor. Thus the nonlinearity appears in the rate feedback path. To obtain the transfer function from the nonlinearity output to the nonlinearity input, consider the system redrawn in Figure 14.26(b), with the nonlinearity omitted. The desired transfer function is

$$-G(s) = \frac{E_o(s)}{E_i(s)} = \frac{-Ks^{-1}}{1 + as^{-1} + Ks^{-2}} = \frac{-Ks}{s^2 + as + K}$$

Thus the system may be represented as shown in Figure 14.27, and the describing-function analysis is directly applicable.

14.9 LINEARIZATION

The second technique of nonlinear analysis to be presented is *linearization*. The reader has used linearization many times in the past, perhaps without realizing it. The small-signal analysis of electronic circuits is a linearization analysis. *All* physical systems are inherently nonlinear; thus, when we use a linear model of a physical system, we are employing some form of linearization. In many cases, fortunately, the linearized models yield accurate results; that is, the linear models accurately model the physical system. Linear circuit models, such as resistance, inductance, and capacitance, present one such example.

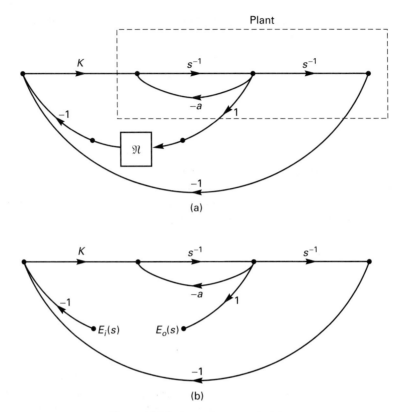

Plant

(a)

(b)

Figure 14.26   System in nonstandard form.

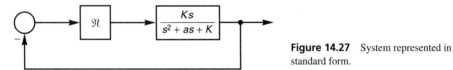

Figure 14.27   System represented in standard form.

In other cases the linearized model is a very poor approximation to the physical system and cannot be used with any confidence. In these cases other techniques of analysis, such as the describing function, must be employed using a nonlinear model. For some systems, no valid analysis techniques have been found, and the system characteristics must be determined through simulation. An additional point should be made here concerning simulation. If a valid analysis technique is available, the results of the application of this technique to the nonlinear system should always be verified through simulation if at all possible.

In order to introduce linearization, we consider the nonlinear gain characteristic of Figure 14.28, which has an input of x and an output of $f(x)$. We assume that the operating point on the gain curve is at the input value x_0, as shown. By this statement we mean that the steady-state input to the nonlinearity is x_0. Suppose that a small perturbation, Δx, occurs in

x. Now the input to the nonlinearity is $x = x_0 + \Delta x$. The ratio of the change in $f(x)$ to the change in x is approximately

$$\frac{\Delta f(x)}{\Delta x} \simeq \frac{df(x)}{dx}\bigg|_{x = x_0} \tag{14-42}$$

or

$$\Delta f(x) \simeq \frac{df(x)}{dx}\bigg|_{x = x_0} \Delta x \tag{14-43}$$

This equation is linear, since the derivative of a function evaluated at a point is a constant. We have linearized the nonlinear gain of Figure 14.28 in the vicinity of the operating point x_0. The accuracy of (14-43) is a function of the magnitude of Δx and of the curvature (or smoothness) of $f(x)$ in the vicinity of $f(x_0)$.

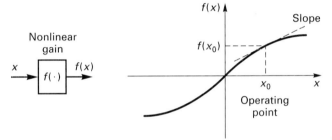

Figure 14.28   Nonlinear gain characteristic.

A more rigorous derivation of this linearization procedure is given next. This derivation is based on the Taylor series expansion of $f(x)$ about the point $f(x_0)$ [5].

$$f(x) = f(x_0) + \frac{df(x)}{dx}\bigg|_{x = x_0} (x - x_0) + \frac{d^2 f(x)}{dx^2}\bigg|_{x = x_0} \frac{(x - x_0)^2}{2!} + \cdots \tag{14-44}$$

We can rewrite this expansion, with $\Delta x = (x - x_0)$, as

$$f(x) - f(x_0) = \Delta f(x) = \frac{df(x)}{dx}\bigg|_{x = x_0} \Delta x + \frac{d^2 f(x)}{dx^2}\bigg|_{x = x_0} \frac{\Delta x^2}{2!} + \cdots \tag{14-45}$$

Comparing (14-43) and (14-45), we see that the linearization in (14-43) is simply the first term of the Taylor series expansion of $f(x)$. For (14-43) to be a valid approximation for (14-45), obviously the higher-order derivative terms in the expansion of $f(x)$ must be negligible; that is, $f(x)$ must be a smooth function and/or Δx must be small.

We now consider the derivation of linear models for nonlinear differential equations, and introduce this topic by an example. Consider the second-order nonlinear equation

$$\ddot{x} + \dot{x} + \dot{x}x = u \tag{14-46}$$

where all variables in this equation are understood to be functions of time. First a state model of this system is found. The state variables x_1 and x_2 are defined by the equations

$$x_1 = x$$
$$x_2 = \dot{x} = \dot{x}_1$$

and, from (14-46),

$$\dot{x}_2 = -x_1 x_2 - x_2 + u$$

Then the (nonlinear) state equations are

$$\dot{x}_1 = x_2 = f_1(x_1, x_2, u)$$
$$\dot{x}_2 = -x_1 x_2 - x_2 + u = f_2(x_1, x_2, u) \tag{14-47}$$

where the functions f_1 and f_2 are defined by these equations. We may express these equations in the general vector form

$$\dot{\mathbf{x}} = \mathbf{f}(\mathbf{x}, \mathbf{u}) \qquad \mathbf{f} = \left[f_1(\mathbf{x}, \mathbf{u}) \; f_2(\mathbf{x}, \mathbf{u}) \right]^T \tag{14-48}$$

Because of the presence of the nonlinear terms in (14-48), the state equations cannot be written in the standard vector-matrix format,

$$\dot{\mathbf{x}} = \mathbf{A}\mathbf{x} + \mathbf{B}\mathbf{u} \tag{14-49}$$

as was the case for linear systems in Chapter 3.

Equation (14-48) is the general model for a nonlinear, time-invariant system. To describe an nth-order system, both $\mathbf{x}$ and $\mathbf{f}$ are $n \times 1$ matrices (nth-order vectors), and $\mathbf{u}$ is an $r \times 1$ matrix. Now, let $\mathbf{x}^0$ be the operating point of the nth-order nonlinear system and $\mathbf{u}^0$ be the (constant) input that produces system operation at $\mathbf{x}^0$. Suppose that a perturbation occurs such that

$$\mathbf{u} = \mathbf{u} + \delta\mathbf{u}$$
$$\mathbf{x} = \mathbf{x}^0 + \delta\mathbf{x} \tag{14-50}$$

For example, for a second-order system with two inputs,

$$\mathbf{u} = \begin{bmatrix} u_1^0 + \delta u_1 \\ u_2^0 + \delta u_2 \end{bmatrix} \qquad \mathbf{x} = \begin{bmatrix} x_1^0 + \delta x_1 \\ x_2^0 + \delta x_2 \end{bmatrix}$$

Then, from (14-48),

$$\frac{d}{dt}(\mathbf{x}^0 + \delta\mathbf{x}) = \dot{\mathbf{x}}^0 + \delta\dot{\mathbf{x}} = \mathbf{f}(\dot{\mathbf{x}}^0 + \delta\mathbf{x}, \mathbf{u}^0 + \delta\mathbf{u}) \tag{14-51}$$

Consider a Taylor series expansion of the jth equation in (14-51), retaining only the linear terms.

$$\dot{x}_j^0 + \delta\dot{x}_j = f_j(\mathbf{x}^0, \mathbf{u}^0) + \frac{\partial f_j}{\partial x_1}\bigg|_{\mathbf{x}^0, \mathbf{u}^0} \delta x_1 + \cdots + \frac{\partial f_j}{\partial x_n}\bigg|_{\mathbf{x}^0, \mathbf{u}^0} \delta x_n$$
$$+ \frac{\partial f_j}{\partial u_1}\bigg|_{\mathbf{x}^0, \mathbf{u}^0} \delta u_1 + \cdots + \frac{\partial f_j}{\partial u_r}\bigg|_{\mathbf{x}^0, \mathbf{u}^0} \delta u_r \tag{14-52}$$

Since, from (14-48),

$$\dot{x}_j^0 = f_j(\mathbf{x}^0, \mathbf{u}^0) \tag{14-53}$$

(14-52) becomes

$$\delta \dot{x}_j = \left.\frac{\partial f_j}{\partial x_1}\right|_{\mathbf{x}^0, \mathbf{u}^0} \delta x_1 + \cdots + \left.\frac{\partial f_j}{\partial x_n}\right|_{\mathbf{x}^0, \mathbf{u}^0} \delta x_n$$

$$+ \left.\frac{\partial f_j}{\partial u_1}\right|_{\mathbf{x}^0, \mathbf{u}^0} \delta u_1 + \cdots + \left.\frac{\partial f_j}{\partial u_r}\right|_{\mathbf{x}^0, \mathbf{u}^0} \delta u_r$$

(14-54)

Thus we can express each equation in (14-51) in this form, and the linearized equations for (14-48) become

$$\delta \dot{\mathbf{x}} = \mathbf{A}\,\delta \mathbf{x} + \mathbf{B}\,\delta \mathbf{u}$$

(14-55)

where

$$\mathbf{A} = \begin{bmatrix} \dfrac{\partial f_1}{\partial x_1} & \dfrac{\partial f_1}{\partial x_2} & \cdots & \dfrac{\partial f_1}{\partial x_n} \\[2mm] \dfrac{\partial f_2}{\partial x_1} & \dfrac{\partial f_2}{\partial x_2} & \cdots & \dfrac{\partial f_2}{\partial x_n} \\[2mm] \vdots & \vdots & & \vdots \\[2mm] \dfrac{\partial f_n}{\partial x_1} & \dfrac{\partial f_n}{\partial x_2} & \cdots & \dfrac{\partial f_n}{\partial x_n} \end{bmatrix}_{\mathbf{x}^0, \mathbf{u}^0} = \left.\frac{\partial \mathbf{f}}{\partial \mathbf{x}}\right|_{\mathbf{x}^0, \mathbf{u}^0}$$

(14-56)

and

$$\mathbf{B} = \begin{bmatrix} \dfrac{\partial f_1}{\partial u_1} & \dfrac{\partial f_1}{\partial u_2} & \cdots & \dfrac{\partial f_1}{\partial u_r} \\[2mm] \dfrac{\partial f_2}{\partial u_1} & \dfrac{\partial f_2}{\partial u_2} & \cdots & \dfrac{\partial f_2}{\partial u_r} \\[2mm] \vdots & \vdots & & \vdots \\[2mm] \dfrac{\partial f_n}{\partial u_1} & \dfrac{\partial f_n}{\partial u_2} & \cdots & \dfrac{\partial f_n}{\partial u_r} \end{bmatrix}_{\mathbf{x}^0, \mathbf{u}^0} = \left.\frac{\partial \mathbf{f}}{\partial \mathbf{u}}\right|_{\mathbf{x}^0, \mathbf{u}^0}$$

(14-57)

The matrices in (14-56) and (14-57) are called *Jacobian matrices*. Note that the linearized model (14-55) is of the standard form of (14-49).

In summary, if the Taylor series expansion of (14-48) is valid about the points $(\mathbf{x}^0, \mathbf{u}^0)$, these equations may be linearized as shown in (14-55), (14-56), and (14-57). The linearized equations will approximate the nonlinear equations in some neighborhood of the points $(\mathbf{x}^0, \mathbf{u}^0)$. An example is given next to illustrate the linearization of nonlinear differential equations.

Example 14.7

Consider the nonlinear equation

$$\ddot{x} + \dot{x} + \dot{x}x = u$$

The state equations have already been derived, and from (14-47),

$$\dot{x}_1 = x_2 = f_1(\mathbf{x}, u)$$
$$\dot{x}_2 = -x_1 x_2 - x_2 + u = f_2(\mathbf{x}, u)$$

Then, from (14-56),

$$\mathbf{A} = \begin{bmatrix} \dfrac{\partial f_1}{\partial x_1} & \dfrac{\partial f_1}{\partial x_2} \\[2ex] \dfrac{\partial f_2}{\partial x_1} & \dfrac{\partial f_2}{\partial x_2} \end{bmatrix}_{\mathbf{x}^0,\, \mathbf{u}^0} = \begin{bmatrix} 0 & 1 \\ -x_2 & -x_1 - 1 \end{bmatrix}_{\mathbf{x}^0,\, \mathbf{u}^0}$$

and, from (14-57),

$$\mathbf{B} = \begin{bmatrix} \dfrac{\partial f_1}{\partial u} \\[2ex] \dfrac{\partial f_2}{\partial u} \end{bmatrix}_{\mathbf{x}^0,\, \mathbf{u}^0} = \begin{bmatrix} 0 \\ 1 \end{bmatrix}$$

With the system at the operating point $\mathbf{x}^0 = [x_1^0 \ \ x_2^0]^T$, the linearized state equations are

$$\delta \dot{x}_1 = \delta x_2$$
$$\delta \dot{x}_2 = (-x_2^0)\delta x_1 + (-x_1^0 - 1)\delta x_2 + \delta u$$

For example, suppose that the operating point is $\mathbf{x}^0 = [1 \ 0]^T$ and $u^0 = 0$. The linearized equations are then

$$\delta \dot{x}_1 = \delta x_2$$
$$\delta \dot{x}_2 = -2\delta x_2 + \delta u$$

Note that linear terms in the nonlinear equations appear with no changes as linear terms in the linearized equations. This result is reasonable, since if we use the preceding method to linearize a set of linear equations, we expect to obtain the very same linear equations (see Problem 14.12).

When a set of nonlinear equations is linearized, the common practice is to replace $\delta\mathbf{x}$ with $\mathbf{x}$ and $\delta\mathbf{u}$ with $\mathbf{u}$ for convenience. The linearized equation of (14-55) may then be expressed as

$$\dot{\mathbf{x}}(t) = \mathbf{A}\mathbf{x}(t) + \mathbf{B}\mathbf{u}(t) \tag{14-58}$$

where $\mathbf{x}(t)$ represents the *variation* of the states about some operating point (operating state) and $\mathbf{u}(t)$ represents the variation in the inputs. In fact, this same statement applies to all models used in this book. The at-rest condition of a physical system is not necessarily the condition in which all states are zero. For example, in the design of autopilots for aircraft, the linear model (14-58) may apply for the case that the aircraft operating point is a certain speed at a certain altitude. If either of these values change by a significant amount, the linear model of (14-58) may change.

14.10 EQUILIBRIUM STATES AND LYAPUNOV STABILITY

In this section the concept of Lyapunov stability is introduced, but first the term *equilibrium states* must be defined. Suppose that we have an autonomous system in which all states have settled to constant values (not necessarily zero values). The system is then said to be in an equilibrium state. Consider, for example, the simple pendulum shown in Figure 14.29. This pendulum obviously has two equilibrium states—one with $\theta = \dot{\theta} = 0$ and the other with $\theta = \pi$, $\dot{\theta} = 0$. Note that in the first case all states are zero, but one state is nonzero in the second case. Note also that in an equilibrium state, all derivative terms must be zero, or else some variables in the system are still changing.

Figure 14.29 Simple pendulum.

If a system is in an equilibrium state, no states are varying with time. Thus we have the following definition of an equilibrium state.

Given the system described by

$$\dot{\mathbf{x}} = \mathbf{f}(\mathbf{x}) \tag{14-59}$$

The state $\mathbf{x}_e$ is an equilibrium state of this system if and only if

$$\dot{\mathbf{x}}_e = \mathbf{f}(\mathbf{x}_e) = \mathbf{0} \tag{14-60}$$

The calculation of equilibrium states is illustrated with an example.

Example 14.8

The equilibrium states of the nonlinear system described by the state equations

$$\dot{x}_1 = x_2$$
$$\dot{x}_2 = -x_1 - x_1^2 - x_2$$

are now found. From (14-60),

$$\dot{x}_{1e} = x_{2e} = 0$$
$$\dot{x}_{2e} = -x_{1e} - x_{1e}^2 - x_{2e} = 0$$

From the first equation, x_{2e} is equal to zero. From the second equation

$$x_{1e}^2 + x_{1e} = x_{1e}(x_{1e} + 1) = 0$$

which has the solution $x_{1e} = 0$ and $x_{1e} = -1$. Thus there are two equilibrium states, given by

$$\mathbf{x}_{e1} = \begin{bmatrix} 0 \\ 0 \end{bmatrix} \qquad \mathbf{x}_{e2} = \begin{bmatrix} -1 \\ 0 \end{bmatrix}$$

Note that (14-60) is satisfied by both $\mathbf{x}_{e1}$ and $\mathbf{x}_{e2}$.

As stated earlier, there are many definitions of stability for nonlinear systems, because no one definition applies to all cases. One of the most commonly used definitions is called Lyapunov stability. We define Lyapunov stability with the help of Figure 14.30. For a second-order system, we define the plot of the state x_2 versus x_1 (with time as a parameter) as the *system trajectory*, as illustrated in Figure 14.30. Such a plot in n-dimensional space, if possible, is the system trajectory for an nth-order system. We cannot plot a system trajectory for orders higher than third, but we still think in terms of a system trajectory for such systems. We now define *Lyapunov stability* [1].

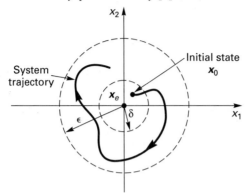

Figure 14.30 Illustration of Lyapunov stability.

Lyapunov stability   An equilibrium state $\mathbf{x}_e$ of an autonomous dynamic system is *stable* in the sense of Lyapunov if for every $\varepsilon > 0$, there exists a $\delta > 0$, where δ depends only on ε, such that $\|\mathbf{x}_0 - \mathbf{x}_e\| < \delta$ results in $\|\mathbf{x}(t;\mathbf{x}_0) - \mathbf{x}_e\| < \varepsilon$ for all $t > t_0$ ($\mathbf{x}_0$ is the initial state at time $t = t_0$).

In this definition we consider the norm of $\mathbf{x}$, denoted by $\|\mathbf{x}\|$, to be the Euclidean norm

$$\|\mathbf{x}\| = [x_1^2 + x_2^2 + \cdots + x_n^2]^{1/2}$$

We are saying in the definition, then, that if an equilibrium state is Lyapunov stable, the trajectory will remain within a given neighborhood of the equilibrium point if the initial state is close to the equilibrium point. In particular, for a Lyapunov stable equilibrium point, if you give us *any* value ε, then we can give you a value of δ such that the definition is satisfied. Figure 14.30 illustrates Lyapunov stability.

Now the definition of unstable in the sense of Lyapunov will be given.

Lyapunov instability   An equilibrium state $\mathbf{x}_e$ of an autonomous dynamic system is *unstable* in the sense of Lyapunov if there exists an ε such that no δ can be found to satisfy the definition of Lyapunov stability.

Note that it is necessary that only one such ε exist such that no δ can be found to satisfy Lyapunov stability. Next asymptotic stability is defined.

Asymptotic stability   An equilibrium state $\mathbf{x}_e$ of an autonomous dynamic system is *asymptotically stable* if

1. It is Lyapunov stable.
2. There is a number $\delta_a > 0$ such that every trajectory starting in the δ_a neighborhood of $\mathbf{x}_e$ converges to $\mathbf{x}_e$ as $t \to \infty$.

Note that a BIBO stable linear system is asymptotically stable, since the initial-condition response dies out with increasing time. Figure 14.30 would illustrate asymptotic stability if the trajectory eventually converged to the origin.

Now we can state a simplified version of Lyapunov's first theorem on stability [1].

Theorem.   For an autonomous system $\dot{\mathbf{x}} = \mathbf{f}(\mathbf{x})$, suppose that $\delta\dot{\mathbf{x}} = \mathbf{A}\delta\mathbf{x}$ is a valid model about the equilibrium point $\mathbf{x}_e$. The characteristic values of the linear model (roots of the characteristic equation) are given by

$$|s\mathbf{I} - \mathbf{A}| = 0 \tag{14-61}$$

Then

a. If the characteristic values all have negative real parts, the equilibrium point is asymptotically stable.

b. If at least one of the characteristic values has a positive real part, the equilibrium point is unstable.

c. If one or more of the characteristic values have a zero real part, with the remaining characteristic values having negative real parts, *no conclusions* can be drawn by the study of the linearized model.

From our studies of linear systems, parts (a) and (b) of this theorem are reasonable. However, part (c) is surprising. An example is given to illustrate Lyapunov's first theorem.

Example 14.9

Consider the nonlinear system from Example 14.8, described by

$$\dot{x}_1 = x_2$$
$$\dot{x}_2 = -x_1 - x_1^2 - x_2$$

As was shown in Example 14.8, this system has two equilibrium states, at the points

$$\mathbf{x}_{e1} = \begin{bmatrix} 0 \\ 0 \end{bmatrix} \qquad \mathbf{x}_{e2} = \begin{bmatrix} -1 \\ 0 \end{bmatrix}$$

The linearized system matrix is given by

$$\mathbf{A} = \begin{bmatrix} \dfrac{\partial f_1}{\partial x_1} & \dfrac{\partial f_1}{\partial x_2} \\ \dfrac{\partial f_2}{\partial x_1} & \dfrac{\partial f_2}{\partial x_2} \end{bmatrix}_{\mathbf{x}_e} = \begin{bmatrix} 0 & 1 \\ -1 - 2x_1 & -1 \end{bmatrix}_{\mathbf{x}_e}$$

Hence, in the neighborhood of $\mathbf{x}_{e1}$,

$$\mathbf{A} = \begin{bmatrix} 0 & 1 \\ -1 & -1 \end{bmatrix}$$

and

$$|s\mathbf{I} - \mathbf{A}| = \begin{vmatrix} s & -1 \\ 1 & s+1 \end{vmatrix} = s^2 + s + 1$$

The characteristic values are the roots of this polynomial, which are equal to $-\frac{1}{2} \pm j\sqrt{3}/2$. Thus $\mathbf{x}_{e1}$ is asymptotically stable.

About $\mathbf{x}_{e2}$,

$$\mathbf{A} = \begin{bmatrix} 0 & 1 \\ 1 & -1 \end{bmatrix}$$

and

$$|s\mathbf{I} - \mathbf{A}| = \begin{vmatrix} s & -1 \\ -1 & s+1 \end{vmatrix} = s^2 + s - 1$$

Hence $\mathbf{x}_{e2}$ is obviously unstable, by the Routh–Hurwitz criterion. A SIMULINK simulation of this system is depicted in Figure 14.31. It is worthwhile for the reader to experiment with this simulation. Initial conditions of $[-0.8\ \ 0]^T$ and $[-1.2\ \ 0]^T$ with a simulation time 5 s illustrate the equilibrium points very well. Choosing different initial conditions illustrates some of the problems for the simulation of nonlinear systems.

Note that Lyapunov's theorem is limited to determining stability in small regions about equilibrium points. The Lyapunov's theorem determines *stability in the small*. Techniques that determine stability in larger regions of state space are said to give *stability in the large*. Techniques that determine stability in all of state space are *global stability* techniques. The Nyquist criterion applied to linear time-invariant systems proves global stability.

14.11 STATE PLANE ANALYSIS

State plane analysis is applicable primarily to second-order systems. This method is presented for two reasons. First, many physical systems can be accurately modeled as second-order systems. Second, application of this method to examples will familiarize the reader with some of the types of responses that appear in nonlinear systems.

State plane (often called phase plane) analysis is simply the plotting of many different system trajectories in the (x_1,x_2)-plane. This plot is called a phase portrait.

From the phase portrait a conception of the system responses for many different sets of initial conditions is obtained. This technique is introduced by examples.

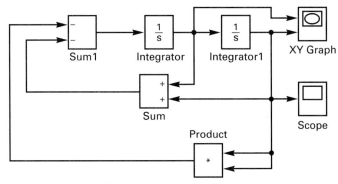

Figure 14.31   State plane example.

Example 14.10

To introduce state plane methods, we use a linear system. Consider the linear model

$$\ddot{x} + 5\dot{x} + 10x = 0$$

which has the state model

$$x_1 = x$$
$$\dot{x}_1 = x_2$$
$$\dot{x}_2 = -10x_1 - 5x_2$$

Since the system characteristic equation is given by

$$s^2 + 5s + 10 = 0$$

the system is underdamped. Hence a typical initial-condition response for $x_1(t)$ is as sketched in Figure 14.32(a). Since the state x_2 is the derivative of x_1, x_2 must be sketched in the figure for the given x_1. From the sketches of x_1 and x_2 we can sketch x_2 versus x_1, with time as a parameter, as shown in Figure 14.32(b). A set of these sketches, for typical initial conditions, form the state plane analysis. Note that the state plane plots can be obtained experimentally on an xy recorder by connecting the physical signal x_1 to the x-input and the physical signal x_2 to the y-input and recording initial-condition responses.

Example 14.11

As a second example, consider the rigid satellite shown in Figure 14.33(a). The model of this satellite was developed in Section 2.6 and has been used in many examples and problems. The purpose of an attitude-control system for this satellite is to maintain the attitude angle θ at some specified value by firing the thrustors shown in the figure. The system can be modeled as

$$\ddot{\theta} = \frac{\tau}{J} = u$$

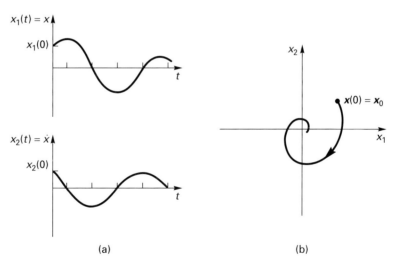

(a)   (b)

Figure 14.32 State plane example.

where τ is the torque generated by the thrustors and J is the moment of inertia of the satellite, and thus u can be considered to be the normalized torque. The state model of the satellite is then

$$x_1 = \theta$$
$$\dot{x}_1 = x_2 = \dot{\theta}$$
$$\dot{x}_2 = u = \ddot{\theta}$$

It is assumed that when the thrustors fire, the thrust is constant, and thus $u(t) = \pm U$. In Figure 14.33(a), since the thrustors that are shown firing tend to drive θ negative, for these thrustors $u(t) = -U$. For the other two thrustors, $u(t) = U$.

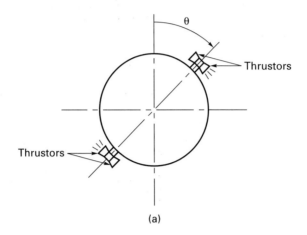

(a)

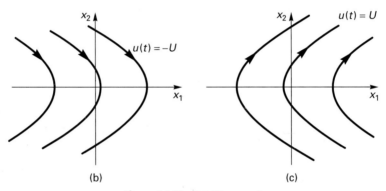

(b)   (c)

Figure 14.33   Satellite example.

Consider first the case that $u(t) = -U$. We can write

$$\frac{dx_2/dt}{dx_1/dt} = \frac{dx_2}{dx_1} = \frac{u(t)}{x_2} = \frac{-U}{x_2}$$

or

$$x_2 dx_2 = -U dx_1$$

We have been able to separate the variables and can integrate this equation to yield

$$\frac{x_2^2}{2} = -U x_1 + C_1$$

where C_1 is a constant of integration and is determined by the initial conditions, that is,

$$C_1 = U x_1(0) + \frac{x_2^2(0)}{2}$$

The response equation describes a family of parabolas in the (x_1, x_2)-plane, and typical system trajectories are plotted in Figure 14.33(b). For $u(t) = U$, solution of the state equations by the preceding technique leads to

$$\frac{x_2^2}{2} = Ux_1 + C_2$$

where C_2 is determined by the initial conditions. Typical system trajectories for this case are shown in Figure 14.33(c).

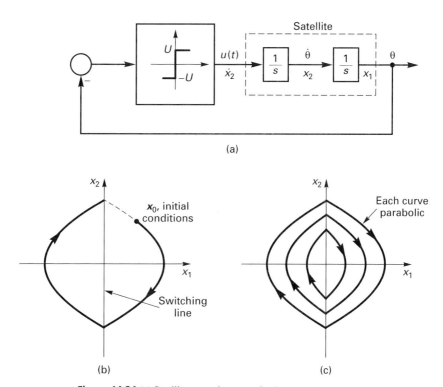

Figure 14.34 (a) Satellite-control system, (b, c) responses.

Example 14.12

The satellite in Example 14.11 is now placed in a feedback configuration in order to maintain the attitude θ [see Figure 14.33(a)] at $0°$. This feedback control system, called an attitude control system, is shown in Figure 14.34(a) . When θ is other than $0°$, the appropriate thrustors will fire to force θ toward $0°$. When $\theta = x_1$ is greater than 0, $u(t) = -U$ and the trajectories of Figure 14.33(b) apply. When x_1 is less than 0, $u(t) = U$ and the trajectories of Figure 14.33(c) apply. Note that switching of $u(t)$ occurs at $x_1 = 0$. Thus the line $x_1 = 0$ (the x_2-axis) is called the *switching line*. From this discussion we see that Figure 14.34(b) illustrates a typical trajectory for the system. The system response is then a limit cycle. A phase portrait is shown in Figure 14.34(c). Hence this control system is not acceptable.

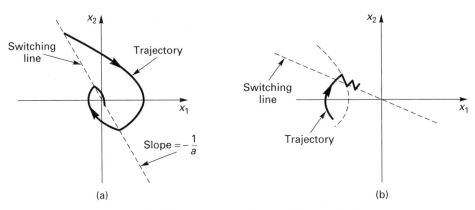

(a)   (b)

Figure 14.36   Satellite-control system with rate feedback.

Example 14.13

The study of the satellite control system is continued in this example. Suppose that we add rate feedback, using a rate gyro with a gain of a as the sensor, to the control system, as shown in Figure 14.35. Note that the input to the satellite is still $\pm\, U$, and thus the response remains the family of parabolas shown in Figures 14.33(b) and 14.33(c). However, the switching of $u(t)$ is now different. Switching occurs when the error signal is zero, or when

$$x_1 + ax_2 = 0$$

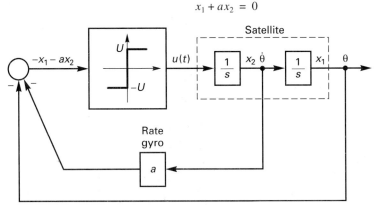

Figure 14.35   Satellite-control system.with rate feedback.

This equation describes the switching line for this case, and hence the switching line is a straight line in the state plane with slope of $dx_2/dx_1 = -1/a$. A typical response for this system is shown in Figure 14.36(a). It is seen that the system response is much improved with the addition of the phase-lead compensation, which is a form of PD compensation as shown in Section 7.6. Furthermore, the origin (an equilibrium point) is now asymptotically stable.

Systems that contain the ideal relay characteristic, as in the system in this example, can develop a response known as chatter. This condition is illustrated in Figure 14.36(b). With the response close to the origin, if the slope of the switching line is small, the condition shown in Figure 14.36(b) develops. The trajectory is always driven to the opposite side of the switching line, independent of which side the trajectory is on. This condition results in the trajectory sliding along the switching line towards the origin.

In the preceding examples, both the analysis and the design of a nonlinear control system by state plane techniques are illustrated. However, note that the technique is limited in general to second-order systems.

14.12 LINEAR-SYSTEM RESPONSE

This section investigates the response of linear systems in the vicinity of equilibrium points. These responses then apply to nonlinear systems that can be linearized by the techniques of Section 14.9.

Consider a second-order linear autonomous system described by

$$\ddot{y} + a\dot{y} + by = 0 \tag{14-62}$$

The characteristic equation for this system is

$$s^2 + as + b = (s - \lambda_1)(s - \lambda_2) = 0 \tag{14-63}$$

The response, for $\lambda_1 \neq \lambda_2$, is of the form

$$y(t) = k_1 e^{\lambda_1 t} + k_2 e^{\lambda_2 t} \tag{14-64}$$

and for $\lambda_1 = \lambda_2$,

$$y(t) = k_1 e^{\lambda_1 t} + k_2 t e^{\lambda_1 t} \tag{14-65}$$

For both responses the constants k_1 and k_2 are determined by the system's initial conditions.

Suppose that we obtain a state model for (14-62). As we usually do, let $x_1 = y$. Then

$$\begin{aligned} \dot{x}_1 &= x_2 = \dot{y} \\ \dot{x}_2 &= -bx_1 - ax_2 \end{aligned} \tag{14-66}$$

Then, from (14-64) and (14-66), for $\lambda_1 \neq \lambda_2$,

$$\begin{aligned} x_1(t) &= y(t) = k_1 e^{\lambda_1 t} + k_2 e^{\lambda_2 t} \\ x_2(t) &= \dot{y}(t) = k_1 \lambda_1 e^{\lambda_1 t} + k_2 \lambda_2 e^{\lambda_2 t} \end{aligned} \tag{14-67}$$

From these equations we can determine the nature of responses in the vicinity of equilibrium points in the (x_1, x_2) plane. Note, however, from (14-66), that the system has only one equilibrium point, and that point occurs at the origin, $\mathbf{x} = \mathbf{0}$. The responses for several cases are now presented.

14.12.1 Case A

For case A, both λ_1 and λ_2 are real and of the same sign. Suppose first that λ_1 and λ_2 are negative. From (14-67), both $x_1(t)$ and $x_2(t)$ approach zero as time becomes large, and each may change sign at most once. This response is illustrated by the phase portrait of Figure 14.37(a). For this case, the equilibrium point is called a *stable node*.

For the case that both λ_1 and λ_2 are both positive and real, the system is unstable and the phase portrait of Figure 14.37(b) applies. For this case, the equilibrium point is called an *unstable node*.

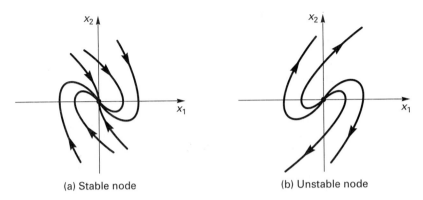

(a) Stable node   (b) Unstable node

Figure 14.37   Phase portrait for λ_1, λ_2 real.

14.12.2 Case B

For case B, λ_1 and λ_2 are complex with nonzero real parts. Letting

$$\lambda_1 = \alpha + j\beta$$
$$\lambda_2 = \alpha - j\beta \tag{14-68}$$

then the states can be expressed as

$$x_1(t) = k_3 e^{\alpha t} \sin(\beta t + \theta_1)$$
$$x_2(t) = k_4 e^{\alpha t} \sin(\beta t + \theta_2) \tag{14-69}$$

For the real parts of λ_1, λ_2 negative, the phase portrait appears as in Figure 14.38(a), and the origin is called a *stable focus*. For the real parts of λ_1, λ_2 positive, the phase portrait appears as in Figure 14.38(b), and the origin is called an *unstable focus*. Example 14.10 presents an example of a stable focus in a linear system.

14.12.3 Case C

For case C, λ_1 and λ_2 are imaginary, and (14-69) becomes

$$x_1(t) = k_3 \sin(\beta t + \theta_1)$$
$$x_2(t) = \beta k_3 \cos(\beta t + \theta_1) \tag{14-70}$$

The phase trajectories are then elliptical, as shown in Figure 14.39, and the origin is called a *center*, or a *vortex*.

14.12.4 Case D

For case D, λ_1 and λ_2 are real, with $\lambda_1 > 0$ and $\lambda_2 < 0$. From (14-67),

$$x_1(t) = k_1 e^{\lambda_1 t} + k_2 e^{\lambda_2 t}$$
$$x_2(t) = k_1 \lambda_1 e^{\lambda_1 t} + k_2 \lambda_2 e^{\lambda_2 t} \tag{14-71}$$

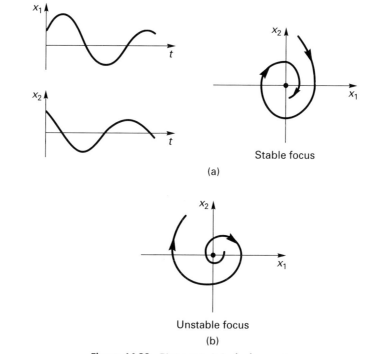

(a)

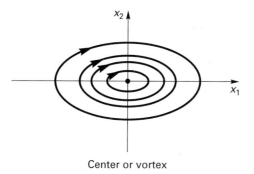

Unstable focus

(b)

Figure 14.38   Phase portrait for λ_1, λ_2 complex.

Center or vortex

Figure 14.39   Phase portrait for λ_1, λ_2 imaginary.

Except for the case that $k_1 = 0$, both x_1 and x_2 become unbounded as time goes to infinity. The phase portrait is as shown in Figure 14.40, and the origin for this case is called a *saddle point*.

14.13 SUMMARY

The describing-function technique for the prediction of limit cycles in nonlinear systems was presented. Since this technique is approximate, certain constraints must be satisfied for the results to be reasonably accurate. First, the nonlinear system can contain only one non-linearity. Second, the linear portion of the system must be sufficiently lowpass to attenuate all harmonics generated by a limit cycle in the system. Last, the technique is applied only to autonomous systems. All these constraints can be relaxed to a degree, but the resulting analysis becomes exceedingly complex and is beyond the scope of this book.

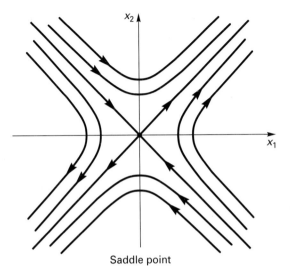

Saddle point

Figure 14.40 Phase portrait for λ_1, λ_2 real and of opposite sign.

Next the nonlinear analysis technique of linearization and the mathematical foundations for linearization were presented. Since all physical systems are inherently nonlinear, all linear system models can be considered to result from the linearization of nonlinear system models.

Lyapunov stability was defined, and Lyapunov's first stability theorem was presented. Lyapunov's first stability theorem is limited to proving stability in small regions about equilibrium points and gives no information concerning stability in other regions.

It should be evident to the reader, after studying the material of this chapter, that no general nonlinear analysis techniques exist. Given a linear autonomous system, we can always determine stability, for example, by using the Nyquist criterion. Given a nonlinear autonomous system, we may or may not be able to determine stability by one of the techniques presented here. Other nonlinear analysis techniques are available, but none is general. Thus for nonlinear systems, simulation becomes a necessity; for high-order systems with more than one nonlinearity (the usual physical system), simulation is the only method available for determining system characteristics from system models.

REFERENCES

1. J. C. Hsu and A. V. Meyer. *Modern Control Principles and Applications.* New York: McGraw-Hill, 1968.

2. D. P. Atherton. *Nonlinear Control Engineering.* London: Van Nostrand Reinhold, 1982.

3. N. Minorsky. *Theory of Nonlinear Control Systems.* New York: McGraw-Hill, 1969.

4. J. D. Irwin. *Basic Engineering Circuit Analysis,* 5th ed. Upper Saddle River, NJ: Prentice Hall, 1996.

5. C. R. Wylie and L. C. Barrett. *Advanced Engineering Mathematics,* 6 th ed. New York: McGraw-Hill, 1995.

PROBLEMS

14.1. Derive the describing function for the nonlinearity shown in Figure P14.1. For this nonlinearity,

$$\text{output} = \begin{cases} A & \text{input} \geq 0 \\ 0 & \text{input} < 0 \end{cases}$$

Ignore the dc component in the output of the nonlinearity.

14.2. Derive the describing function for the nonlinearity shown in Figure P14.2.

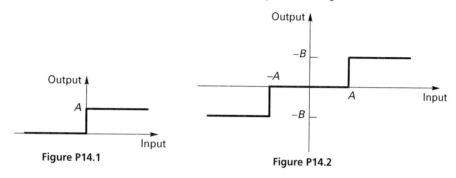

Figure P14.1

Figure P14.2

14.3. (a) Derive the describing function for the nonlinearity shown in Figure P14.3.

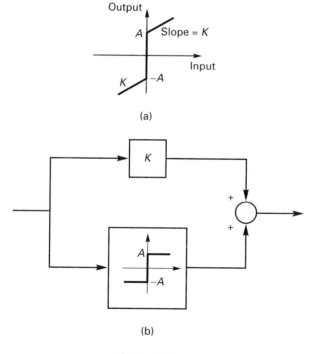

(a)

(b)

Figure P14.3

(b) A realization of this nonlinearity is shown in Figure P14.3(b). Show that the sum of the describing functions of the two blocks, from Table 14.1, is equal to the result in (a).

14.4. Derive the describing function for the nonlinearity shown in Figure P14.4. For this nonlinearity,

$$\text{output} = \begin{cases} 0, & \text{input} < 0 \\ K(\text{input}) & \text{input} \geq 0 \end{cases}$$

Ignore the dc component in the nonlinearity output.

14.5. Consider the system of Figure P14.5, in which the nonlinearity is an ideal relay.

(a) Use the describing-function technique to investigate the possibility of a limit cycle in this system. If a limit cycle is predicted, determine its amplitude and frequency, and investigate its stability.

(b) Verify the results of (a) using SIMULINK.

14.6. Repeat Problem 14.5 for each of the following plant transfer functions.

(a) $G(s) = \dfrac{(s+1)^2}{2s^3}$

(b) $G(s) = \dfrac{2(s+5)^2}{s^2(s+1)}$

(c) Verify all results by simulation.

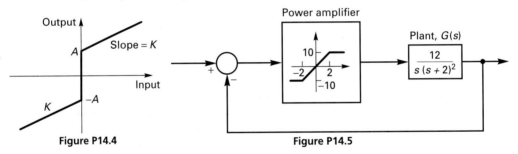

Figure P14.4   Figure P14.5

14.7. Consider the system shown in Figure P14.7, in which the nonlinear element is a power amplifier with gain 5 which saturates for input magnitudes greater than 2. A saturation characteristic such as the one shown is always encountered with physical amplifiers.

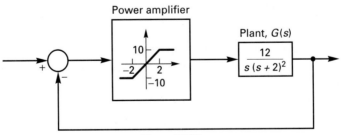

Figure P14.7

(a) Use the describing-function analysis to investigate the possibility of a limit cycle in this system. If a limit cycle is predicted, determine its amplitude and frequency, and investigate its stability.

(b) Verify the results of (a) by simulation.

14.8. Repeat Problem 14.7 for each of the following plant transfer functions.

(a) $G(s) = \dfrac{(s+1)^2}{2s^3}$

(b) $G(s) = \dfrac{2(s+5)^2}{s^2(s+1)}$

(c) Verify all results by simulation.

14.9. It was shown in Problem 14.5 that the input to the nonlinearity in that figure is $m(t) \approx 1.16 \sin 3.16t$.

(a) Accurately sketch $n(t)$, the output of the nonlinearity.

(b) Calculate the fundamental component in the plant output, $c_1(t) = C_1 \sin 3.16t$; that is, find C_1.

(c) Calculate the third harmonic component in the plant output, $c_3(t) = C_3 \sin [(3)3.16t + \theta_3]$; that is, find C_3.

14.10. Consider the system in Figure P14.10. Note that, with $K = 1$ and $A = 1$, this system is the same as that in Example 14.5.

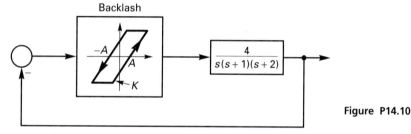

Backlash

Figure P14.10

(a) With $K = 0.75$ and $A = 1$, use the describing-function analysis to investigate the possibility of a limit cycle in this system. If a limit cycle is predicted, determine its amplitude and frequency, and investigate its stability. Figure 14.11 and Table 4.3 are useful in this analysis.

(b) From the results of (a), discuss the general effects on the possibility of a limit cycle of decreasing K on the backlash characteristic.

(c) With $K = 1$ and $A = 1.5$, use the describing-function analysis to investigate the possibility of a limit cycle in this system. If a limit cycle is predicted, determine its amplitude and frequency, and investigate its stability.

(d) In general, what is the effect of more looseness (increasing A) on the possibility of a limit cycle, in gears?

(e) Use a simulation to verify the stable limit cycle in (c).

14.11. Use the simulation of Example 14.5 to verify both the stable limit cycle and the unstable limit cycle.

14.12. Use (14-55) to linearize the linear state equations

$$\dot{\mathbf{x}}(t) = \begin{bmatrix} 0 & 1 \\ -1 & -2 \end{bmatrix} \mathbf{x}(t) + \begin{bmatrix} 0 \\ 3 \end{bmatrix} u(t)$$

Compare the linearized equations with the original equations.

14.13. Given the nonlinear equations

$$\dot{x} = (x+1)y$$
$$\dot{y} = (y+1)x$$

(a) Find all equilibrium points.
(b) Derive the linearized state equations about each equilibrium point.
(c) Determine the stability, in the sense of Lyapunov, of each equilibrium point.
(d) Verify the stability of each equilibrium point using simulation.

14.14. Given the nonlinear equation $\ddot{x} + x^2(\dot{x}-1) + x = 0$.
(a) Find all equilibrium points.
(b) Derive the linearized state equations about each equilibrium point.
(c) Determine the stability, in the sense of Lyapunov, of each equilibrium point.

14.15. Given the nonlinear equation $\ddot{x} + \dot{x} + x^2 - 1 = 0$.
(a) Find all equilibrium points.
(b) Derive the linearized state equations about each equilibrium point.
(c) Determine the stability, in the sense of Lyapunov, of each equilibrium point.

14.16. A nonlinear system is described by the equation

$$\ddot{e} + \dot{e}|e| + e - e^2 = 0 \qquad |e| = \begin{cases} e, & e \geq 0 \\ -e & e < 0 \end{cases}$$

(a) Locate and identify all equilibrium points.
(b) Determine the stability, in the sense of Lyapunov, of each equilibrium point.

14.17. Consider the system of Figure P14.17. Note the *positive* feedback. This system is the model of a certain type of electronic oscillator.

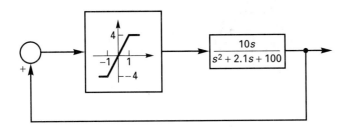

Figure P14.17

(a) Locate and identify all equilibrium points.
(b) Determine the stability of each equilibrium point.
(c) Verify the stability of each equilibrium point using simulation.

A

Matrices

This appendix presents a very brief review of matrices. Those readers requiring more detail are referred to Ref. 1 to 5. MATLAB statements are given for performing the mathematical operations, where appropriate.

Matrices are useful in the study of linear simultaneous equations. For example, consider the equations

$$2x_1 + x_2 + x_3 = 4$$
$$x_1 + x_2 - x_3 = 1 \tag{A-1}$$
$$2x_1 + x_2 + 3x_3 = 6$$

In a *vector-matrix* format we can write these equations as

$$\begin{bmatrix} 2 & 1 & 1 \\ 1 & 1 & -1 \\ 2 & 1 & 3 \end{bmatrix} \begin{bmatrix} x_1 \\ x_2 \\ x_3 \end{bmatrix} = \begin{bmatrix} 4 \\ 1 \\ 6 \end{bmatrix} \tag{A-2}$$

For convenience of notation and for manipulations, we define the following:

$$\mathbf{A} = \begin{bmatrix} 2 & 1 & 1 \\ 1 & 1 & -1 \\ 2 & 1 & 3 \end{bmatrix} \qquad \mathbf{x} = \begin{bmatrix} x_1 \\ x_2 \\ x_3 \end{bmatrix} \qquad \mathbf{u} = \begin{bmatrix} 4 \\ 1 \\ 6 \end{bmatrix} \tag{A-3}$$

Using these definitions, we can write (A-2) in the convenient notation

$$\mathbf{Ax} = \mathbf{u} \tag{A-4}$$

627

In this equation **A** is a 3×3 (3 rows, 3 columns) *matrix,* **x** is a 3×1 matrix, and **u** is a 3×1 matrix. A general matrix is $m \times n$ (*m* rows, *n* columns). If either *m* or *n* is equal to 1, the matrix is often called a *vector;* that is, a vector is a matrix of either one row or one column. If both *m* and *n* are equal to 1, the quantity is a scalar. In the preceding equations, both **x** and **u** are vectors, and, for example, x_1 is a scalar.

One statement for entering the matrix **A** into MATLAB is

$$A = [2\ 1\ 1; 1\ 1\ -1;\ 2\ 1\ 3];$$

The general matrix **A** is written as

$$
\mathbf{A} = \begin{bmatrix} a_{11} & a_{12} & \cdots & a_{1n} \\ a_{21} & a_{22} & \cdots & a_{2n} \\ \vdots & \vdots & & \vdots \\ a_{m1} & a_{m2} & \cdots & a_{mn} \end{bmatrix}
\tag{A-5}
$$

where a_{ij} denotes the element that is common to both the *i*th row and the *j*th column. Generally, for a vector, only one subscript is used to indicate the row of a column vector or the column of a row vector. See **x** in (A-3) for an example of subscripts in a vector.

Some useful definitions are now given.

Diagonal matrix   A *diagonal matrix* is a square matrix $(n \times n)$ in which all elements $a_{ij} = 0$ for $i \neq j$; that is, all elements of the matrix not on the main diagonal are zero. An example of a diagonal matrix is **D**:

$$
\mathbf{D} = \begin{bmatrix} d_{11} & 0 & 0 \\ 0 & d_{22} & 0 \\ 0 & 0 & d_{33} \end{bmatrix}
\tag{A-6}
$$

Identity matrix   The *identity matrix* is a diagonal matrix in which all diagonal elements are equal to unity. The identity matrix is usually denoted by **I**. For example, the (3×3) identity matrix is

$$
\mathbf{I} = \begin{bmatrix} 1 & 0 & 0 \\ 0 & 1 & 0 \\ 0 & 0 & 1 \end{bmatrix}
\tag{A-7}
$$

The 3×3 identity matrix is entered into MATLAB with the statement

$$I = eye(3);$$

The identity matrix has the property that, for **A** square matrix,

$$\mathbf{AI} = \mathbf{IA} = \mathbf{A} \tag{A-8}$$

If **A** is $(m \times n)$, then

$$\mathbf{AI}_n = \mathbf{A} \qquad \mathbf{I}_m\mathbf{A} = \mathbf{A} \tag{A-9}$$

where $\mathbf{I}_r$ denotes the $(r \times r)$ identity matrix. Note that the identity matrix is to matrix algebra what the value of unity is to scalar algebra.

Transpose of a matrix   The *transpose* of a matrix is formed by interchanging the rows and columns. As an example

$$
\mathbf{A} = \begin{bmatrix} 2 & 1 & 1 \\ 1 & 1 & -1 \\ 2 & 1 & 3 \end{bmatrix} \qquad \mathbf{A}^T = \begin{bmatrix} 2 & 1 & 2 \\ 1 & 1 & 1 \\ 1 & -1 & 3 \end{bmatrix} \tag{A-10}
$$

where $\mathbf{A}^T$ denotes the transpose of $\mathbf{A}$. The transpose has the property

$$
(\mathbf{AB})^T = \mathbf{B}^T \mathbf{A}^T \tag{A-11}
$$

The transpose in MATLAB is denoted with the apostrophe; that is, A' is the transpose of the matrix A.

Trace   The *trace* of a square matrix $\mathbf{A}$ denoted by tr $\mathbf{A}$, is the sum of the diagonal elements of that matrix:

$$
\text{tr } \mathbf{A} = a_{11} + a_{22} + \cdots + a_{nn} \tag{A-12}
$$

Eigenvalues   The *eigenvalues* (characteristic values) of a square matrix are the roots of the polynomial equation

$$
|\lambda \mathbf{I} - \mathbf{A}| = 0 \tag{A-13}
$$

where $|\cdot|$ denotes the determinant, as does det $(\cdot)$.

Eigenvectors   The *eigenvectors* (characteristic vectors) of the square matrix $\mathbf{A}$ are the vectors x_i that satisfy the equation.

$$
\lambda_i \mathbf{x}_i = \mathbf{A} \mathbf{x}_i \tag{A-14}
$$

where λ_i are the eigenvalues of $\mathbf{A}$. The eigenvalues and the eigenvectors are calculated with the MATLAB statement

```
[V,D] = eig (A)
```

where D denotes a diagonal matrix with the eigenvalues as the diagonal elements, and the columns of V are the corresponding eigenvectors.

Properties   Two properties of an $(n \times n)$ matrix $\mathbf{A}$ are

$$
|\mathbf{A}| = \prod_{i=1}^{n} \lambda_i \tag{A-15}
$$

$$
\text{tr } \mathbf{A} = \sum_{i=1}^{n} \lambda_i \tag{A-16}
$$

Determinants   With both $\mathbf{A}$ and $\mathbf{B}$ $n \times n$,

$$
|\mathbf{AB}| = |\mathbf{A}||\mathbf{B}| \tag{A-17}
$$

where $|\mathbf{A}|$ denotes the determinant of $\mathbf{A}$. The MATLAB statement for the determinant is

$$d = \text{det(A)}$$

Minor   The *minor* m_{ij} is defined for the element of a_{ij} of a square matrix **A** as the determinant of the matrix that results from deleting the ith row and the jth column of **A**. For example, for the matrix **A** of (A-10),

$$m_{31} = \begin{vmatrix} 1 & 1 \\ 1 & -1 \end{vmatrix} = -1 - 1 = -2$$

Cofactor   The *cofactor* c_{ij} of an element a_{ij} of the square matrix **A** is defined as

$$c_{ij} = (-1)^{i+j} m_{ij} \tag{A-18}$$

In the preceding example,

$$c_{31} = (-1)^{3+1}(-2) = -2$$

Adjoint   The *adjoint* of a square matrix is defined as the transpose matrix of cofactors. For the matrix **A** in (A-10), the adjoint of **A**, denoted as Adj **A**, can be calculated to be

$$\text{Adj } \mathbf{A} = \begin{bmatrix} c_{11} & c_{12} & c_{13} \\ c_{21} & c_{22} & c_{23} \\ c_{31} & c_{32} & c_{33} \end{bmatrix}^T = \begin{bmatrix} 4 & -5 & -1 \\ -2 & 4 & 0 \\ -2 & 3 & 1 \end{bmatrix}^T \tag{A-19}$$

Inverse   The *inverse* of a square matrix **A**, denoted by $\mathbf{A}^{-1}$, is given by

$$\mathbf{A}^{-1} = \frac{\text{Adj } \mathbf{A}}{|\mathbf{A}|} \tag{A-20}$$

The MATLAB statement for the inverse is

```
Ainv = inv(A)
```

For the matrix **A** given earlier, $|\mathbf{A}| = 2$ and

$$\mathbf{A}^{-1} = \begin{bmatrix} 2 & -1 & -1 \\ -\frac{5}{2} & 2 & \frac{3}{2} \\ -\frac{1}{2} & 0 & \frac{1}{2} \end{bmatrix}$$

Two properties of the inverse matrix are

$$\mathbf{A}\mathbf{A}^{-1} = \mathbf{A}^{-1}\mathbf{A} = \mathbf{I} \tag{A-21}$$

$$(\mathbf{A}\mathbf{B})^{-1} = \mathbf{B}^{-1}\mathbf{A}^{-1} \tag{A-22}$$

Note that the matrix inverse is defined only for a square matrix and exists only if the determinant of the matrix is nonzero.

A.1 ALGEBRA OF MATRICES

The algebra of matrices must be defined such that algebraic operations on matrix equations give the same results as equivalent operations on the linear simultaneous equations upon which the matrix equations are based.

Addition   To form the *sum* of two matrices **A** and **B**, the matrices must be of the same order, and we add the corresponding elements a_{ij} and b_{ij} for each *ij*. For example,

$$\begin{bmatrix} 1 & 2 \\ 3 & 4 \end{bmatrix} + \begin{bmatrix} 2 & 4 \\ 6 & 8 \end{bmatrix} = \begin{bmatrix} 3 & 6 \\ 9 & 12 \end{bmatrix}$$

Multiplication by a scalar   The *product of a scalar k and a matrix* **A** is a matrix formed by multiplying each element of the matrix **A** by the scalar k. For example,

$$\mathbf{B} = k\mathbf{A} = k\begin{bmatrix} a_{11} & a_{12} \\ a_{21} & a_{22} \end{bmatrix} = \begin{bmatrix} ka_{11} & ka_{12} \\ ka_{21} & ka_{22} \end{bmatrix}$$

Multiplication of vectors   The *multiplication* of a $(1 \times n)$ vector (row) with an $(n \times 1)$ vector (column) is defined as

$$\begin{bmatrix} x_1 & x_2 & \cdots & x_n \end{bmatrix} \begin{bmatrix} y_1 \\ y_2 \\ \vdots \\ y_n \end{bmatrix} = x_1 y_1 + x_2 y_2 + \cdots + x_n y_n \qquad \text{(A-23)}$$

Note that this vector multiplication results in a scalar.

Multiplication of matrices   An $(m \times p)$ matrix **A** may be multiplied only by a $(p \times n)$ matrix **B**, that is, the number of columns of **A** must equal the number of rows of **B**. The product matrix is then $(m \times n)$. Let

$$\mathbf{C} = \mathbf{AB}$$

The *ij*th element of the product matrix **C** is equal to the multiplication, as vectors, of the *i*th row of **A** and the *j*th column of **B**. Multiplication in MATLAB is performed by the statement

```
C = A*B
```

As an example, consider the multiplication of $\mathbf{AA}^{-1}$ for the (3×3) matrix given earlier:

$$\mathbf{AA}^{-1} = \begin{bmatrix} 2 & 1 & 1 \\ 1 & 1 & -1 \\ 2 & 1 & 3 \end{bmatrix} \begin{bmatrix} 2 & -1 & -1 \\ -\frac{5}{2} & 2 & \frac{3}{2} \\ -\frac{1}{2} & 0 & \frac{1}{2} \end{bmatrix} = \begin{bmatrix} 1 & 0 & 0 \\ 0 & 1 & 0 \\ 0 & 0 & 1 \end{bmatrix} = \mathbf{I}$$

A.2 OTHER RELATIONSHIPS

Some other important relationships are now given. For **A** square and $|\mathbf{A}| \neq 0$,

$$(\mathbf{A}^{-1})^T = (\mathbf{A}^T)^{-1} \tag{A-24}$$

$$|\mathbf{A}^{-1}| = \frac{1}{|\mathbf{A}|} \tag{A-25}$$

The solution of the vector-matrix equation (A-4)

$$\mathbf{A}\mathbf{x} = \mathbf{u}$$

is

$$\mathbf{x} = \mathbf{A}^{-1}\mathbf{u}$$

MATLAB performs this operation with the statement

```
x = inv(A)*u
```

Differentiation   The derivative of a matrix is obtained by differentiating the matrix element by element. For example, let

$$\mathbf{x} = \begin{bmatrix} x_1 \\ x_2 \end{bmatrix}$$

Then

$$\frac{d\mathbf{x}}{dt} = \begin{bmatrix} \dfrac{dx_1}{dt} \\ \dfrac{dx_2}{dt} \end{bmatrix} \tag{A-26}$$

Integration   The *integral* of a matrix is obtained by integrating the matrix element by element. For example, for the vector **x**,

$$\int \mathbf{x}\, dt = \begin{bmatrix} \displaystyle\int x_1\, dt \\ \displaystyle\int x_2\, dt \end{bmatrix} \tag{A-27}$$

A useful determinant   Given the partitioned matrix

$$\mathbf{H} = \begin{bmatrix} \mathbf{D} & \mathbf{E} \\ \mathbf{F} & \mathbf{G} \end{bmatrix}$$

where **D**, **E**, **F**, and **G** are each $(n \times n)$. Then **H** is a $(2n \times 2n)$ matrix. The determinant of **H** is given by [6]

$$|\mathbf{H}| = |\mathbf{G}||\mathbf{D} - \mathbf{E}\mathbf{G}^{-1}\mathbf{F}| = |\mathbf{D}||\mathbf{G} - \mathbf{F}\mathbf{D}^{-1}\mathbf{E}| \tag{A-28}$$

provided that the indicated inverse matrices exist.

REFERENCES

1. F. R. Gantmacher. *Theory of Matrices,* vols. I and II. New York: Chelsea Publishing Company, 1959.

2. P. M. DeRusso, R. J. Roy, and C. M. Close. *State Variables for Engineers.* New York: Wiley, 1965.

3. K. Ogata. *State Space Analysis of Control Systems.* Englewood Cliffs, NJ: Prentice Hall, 1967.

4. G. Strang. *Linear Algebra and Its Applications.* Orlando, FL: Academic Press, 1976.

5. G. H. Golub and C. F. Loan. *Matrix Computations.* Baltimore, MD: Johns Hopkins University Press, 1983.

6. T. E. Fortman. "A Matrix Inversion Identity," *IEEE Trans. Autom. Control,* AC-15 (October 1970): 599.

B

Laplace Transform

A brief review of the Laplace transform is presented in this appendix. We will see that the Laplace transform is useful in the modeling of a linear time-invariant analog system as a transfer function. The Laplace transform may also be used to solve for the response of this type of system; however, we generally use *simulations* (machine solutions of the system equations) for this purpose. For those readers wanting to delve more deeply into the Laplace transform, Refs. 1 to 5 are suggested for supplemental reading.

B.1 INTRODUCTION

By definition, the Laplace transform of a function of time $f(t)$ is [1]

$$F(s) = \mathcal{L}[f(t)] = \int_0^\infty f(t)e^{-st}\, dt \tag{B-1}$$

where L indicates the Laplace transform. Note that the variable time has been integrated out of the equation and that the Laplace transform is a function of the complex variable s. The inverse Laplace transform is given by

$$f(t) = \mathcal{L}^{-1}[F(s)] = \frac{1}{2\pi j}\int_{\sigma-j\infty}^{\sigma+j\infty} F(s)e^{st}\, ds \tag{B-2}$$

where L^{-1} indicates the inverse transform and $j = \sqrt{-1}$.

Equations (B-1) and (B-2) form the Laplace transform pair. Given a function $f(t)$, we integrate (B-1) to find its Laplace transform $F(s)$. Then if this function $F(s)$ is used to evaluate (B-2), the result will be the original value of $f(t)$. The value of σ in (B-2) is determined by the singularities of $F(s)$ [4]. We seldom use (B-2) to evaluate an inverse Laplace transform; instead we use (B-1) to construct a table of transforms for useful time functions. Then, when possible, we use this table to find the inverse transform rather than integrating (B-2).

635

As an example, we will find the Laplace transform of the exponential function e^{-at}. From (B-1),

$$F(s) = \int_0^\infty e^{-at} e^{-st}\, dt = \int_0^\infty e^{-(s+a)t}\, dt = \frac{-e^{-(s+a)t}}{s+a}\bigg|_0^\infty \tag{B-3}$$

$$= \frac{1}{s+a} \qquad \text{Re}(s+a) > 0$$

where $\text{Re}(\cdot)$ indicates the real part of the expression. Of course, Laplace transform tables were derived long ago, and we will not derive any additional transforms. A short table of commonly required transforms is given in Table B.1. Appendix C contains a rather extensive table of both Laplace transforms and z-transforms. The first two columns of this table are a more complete table of Laplace transforms. The remaining column in this table is useful when we consider digital control systems in Chapters 11, 12, and 13.

From the definition of the Laplace transform, (B-1),

$$\mathcal{L}\,[kf(t)] = k\,\mathcal{L}\,[f(t)] = kF(s) \tag{B-4}$$

TABLE B.1 LAPLACE TRANSFORMS

Name	Time function, $f(t)$	Laplace transform, $F(s)$
Unit impulse	$\delta(t)$	1
Unit step	$u(t)$	$\dfrac{1}{s}$
Unit ramp	t	$\dfrac{1}{s^2}$
nth-Order ramp	t^n	$\dfrac{n!}{s^{n+1}}$
Exponential	e^{-at}	$\dfrac{1}{s+a}$
nth-Order exponential	$t^n e^{-at}$	$\dfrac{n!}{(s+a)^{n+1}}$
Sine	$\sin bt$	$\dfrac{b}{s^2+b^2}$
Cosine	$\cos bt$	$\dfrac{s}{s^2+b^2}$
Damped sine	$e^{-at}\sin bt$	$\dfrac{b}{(s+a)^2+b^2}$
Damped cosine	$e^{-at}\cos bt$	$\dfrac{s+a}{(s+a)^2+b^2}$
Diverging sine	$t\sin bt$	$\dfrac{2bs}{(s^2+b^2)^2}$
Diverging cosine	$t\cos bt$	$\dfrac{s^2-b^2}{(s^2+b^2)^2}$

for k constant, and

$$\mathcal{L}\left[f_1(t) + f_2(t)\right] = \mathcal{L}\left[f_1(t)\right] + \mathcal{L}\left[f_2(t)\right] = F_1(s) + F_2(s) \qquad \text{(B-5)}$$

The use of these two relationships greatly extends the application of Table B.1.

We now present some examples of the Laplace transform and of the inverse Laplace transform. First, however, we need to note that using the complex inversion integral (B-2) to evaluate the inverse Laplace transform results in $f(t) = 0$ for $t < 0$ [5]. Hence, to be consistent, we will always assign a value of zero to $f(t)$ for all negative time. Also, to simplify notation, we define the unit step function $u(x)$ to be

$$u(x) = \begin{cases} 0 & x < 0 \\ 1 & x \ge 0 \end{cases} \qquad \text{(B-6)}$$

In Equation (B-3), the Laplace transform of e^{-at} was derived. Note that the Laplace transform of $e^{-at}u(t)$ is the same function. Thus for any function $f(t)$,

$$\mathcal{L}\left[f(t)\right] = \mathcal{L}\left[f(t)u(t)\right] = F(s) \qquad \text{(B-7)}$$

Example B.1

The Laplace transform of the time function

$$f(t) = 5u(t) + 3e^{-2t}$$

will now be found. From Table B.1 and (B-4),

$$\mathcal{L}\left[5u(t)\right] = 5\,\mathcal{L}\left[u(t)\right] = \frac{5}{s}$$

$$\mathcal{L}\left[3e^{-2t}\right] = 3\,\mathcal{L}\left[e^{-2t}\right] = \frac{3}{s+2}$$

Then, from (B-5),

$$F(s) = \mathcal{L}\left[5u(t) + 3e^{-2t}\right] = \frac{5}{s} + \frac{3}{s+2}$$

This Laplace transform can also be expressed as

$$F(s) = \frac{5}{s} + \frac{3}{s+2} = \frac{8s+10}{s(s+2)}$$

The transforms are usually easier to manipulate in the combined form than in the sum-of-terms form. This result is verified with the MATLAB program

```
syms s t
f = 5+3*exp(-2*t);
laplace(f,t,s)
```

This example illustrates an important point. As stated, we usually work with the Laplace transform expressed as a ratio of polynomials in the variable s (we call this ratio of polynomials a *rational function*). However, the tables used to find inverse transforms contain only low-order functions. Hence a method is required for converting from a general rational function to the forms that appear in the tables. This method is called the *partial-fraction expansion* method. A simple example is illustrated in the relationship

$$\frac{c}{(s+a)(s+b)} = \frac{k_1}{s+a} + \frac{k_2}{s+b}$$

Given the constants a, b, and c, the problem is to find the coefficients of the partial-fraction expansion k_1 and k_2. We now derive the general relationships required.

Consider the general rational function

$$F(s) = \frac{b_m s^m + \cdots + b_1 s + b_0}{s^n + a_{n-1} s^{n-1} + \cdots + a_1 s + a_0} = \frac{N(s)}{D(s)} \qquad m < n \qquad \text{(B-8)}$$

where $N(s)$ is the numerator polynomial and $D(s)$ is the denominator polynomial. To perform a partial-fraction expansion, first the roots of the denominator must be found. Then $F(s)$ can be expressed as

$$F(s) = \frac{N(s)}{D(s)} = \frac{N(s)}{\prod_{i=1}^{n}(s - p_i)} = \frac{k_1}{s - p_1} + \frac{k_2}{s - p_2} + \cdots + \frac{k_n}{s - p_n} \qquad \text{(B-9)}$$

where Π indicates the product of terms. Suppose that we wish to calculate the coefficient k_j. We first multiply (B-9) by the term $(s - p_j)$.

$$(s - p_j)F(s) = \frac{k_1(s - p_j)}{s - p_1} + \cdots + k_j + \cdots + \frac{k_n(s - p_j)}{s - p_n} \qquad \text{(B-10)}$$

If this equation is evaluated for $s = p_j$, we see then that all terms on the right side are zero except the jth term, and thus

$$k_j = (s - p_j)F(s)\big|_{s = p_j} \qquad j = 1, 2, \ldots, n \qquad \text{(B-11)}$$

In mathematics, k_j is called the *residue* of $F(s)$ in the pole at $s = p_j$.

If the denominator polynomial of $F(s)$ has repeated roots, $F(s)$ can be expanded as in the example

$$\begin{aligned} F(s) &= \frac{N(s)}{(s - p_1)(s - p_2)^r} \\ &= \frac{k_1}{s - p_1} + \frac{k_{21}}{s - p_2} + \frac{k_{22}}{(s - p_2)^2} + \cdots + \frac{k_{2r}}{(s - p_2)^r} \end{aligned} \qquad \text{(B-12)}$$

where it is seen that a denominator root of multiplicity r yields r terms in the partial-fraction expansion. The coefficients of the repeated-root terms are calculated from the equation

$$k_{2j} = \frac{1}{(r - j)!} \frac{d^{r-j}}{ds^{r-j}} \left[(s - p_2)^r F(s) \right] \Big|_{s = p_2} \qquad \text{(B-13)}$$

This equation is given without proof [2].

The preceding development applies to complex poles as well as real poles. Consider the case that $F(s)$ has a pair of complex poles. If we let $p_1 = a - jb$ and $p_2 = a + jb$, (B-9) can be written as

$$F(s) = \frac{k_1}{s - a + jb} + \frac{k_2}{s - a - jb} + \frac{k_3}{s - p_3} + \cdots + \frac{k_n}{s - p_n} \tag{B-14}$$

The coefficients k_1 and k_2 can be evaluated using (B-11) as before. It will be found, however, that these coefficients are complex valued, and that k_2 is the conjugate of k_1. In order to achieve a convenient form for the inverse transform, we will use the following approach. From (B-11),

$$k_1 = (s - a + jb)F(s)\big|_{s = a - jb} = Re^{j\theta} \tag{B-15}$$
$$k_2 = (s - a - jb)F(s)\big|_{s = a + jb} = Re^{-j\theta} = k_1^*$$

where the asterisk indicates the conjugate of the complex number. Define $f_1(t)$ as the inverse transform of the first two terms of (B-14). Hence,

$$\begin{aligned} f_1(t) &= Re^{j\theta}e^{(a - jb)t} + Re^{-j\theta}e^{(a + jb)t} \\ &= 2Re^{at}\left[\frac{e^{j(bt - \theta)} + e^{-j(bt - \theta)}}{2}\right] \\ &= 2Re^{at}\cos(bt - \theta) \end{aligned} \tag{B-16}$$

by Euler's identity [1]. This approach expresses the inverse transform in a convenient form and the calculations are relatively simple. The damped sinusoid has an amplitude of $2R$ and a phase angle of θ, where R and θ are defined in (B-15). Three examples of finding the inverse Laplace are given next.

Example B.2

In this example the inverse Laplace transform of a rotational function is found.

$$F(s) = \frac{5}{s^2 + 3s + 2} = \frac{5}{(s + 1)(s + 2)}$$

First the partial fractional expansion is derived:

$$F(s) = \frac{5}{(s + 1)(s + 2)} = \frac{k_1}{s + 1} + \frac{k_2}{s + 2}$$

The coefficients in the partial-fraction expansion are calculated from (B-11):

$$k_1 = (s + 1)F(s)\big|_{s = -1} = \frac{5}{s + 2}\bigg|_{s = -1} = 5$$

$$k_2 = (s + 2)F(s)\big|_{s = -2} = \frac{5}{s + 1}\bigg|_{s = -2} = -5$$

Thus the partial-fraction expansion is

$$\frac{5}{(s + 1)(s + 2)} = \frac{5}{s + 1} + \frac{-5}{s + 2}$$

This expansion can be verified by recombining the terms on the right side to yield the left side of the equation. The inverse transform of $F(s)$ is then

$$\mathcal{L}^{-1}[F(s)] = (5e^{-t} - 5e^{-2t})u(t)$$

The function $u(t)$ is often omitted, but we must then understand that the inverse transform can be nonzero only for positive time and must be zero for negative time. The inverse transform is verified with the MATLAB program

```
syms F s t
F = 5/((s+1)*(s+2));
ilaplace(F,s,t)
```

Example B.3

As a second example of finding the inverse Laplace transform, consider the function

$$F(s) = \frac{2s + 3}{s^3 + 2s^2 + s} = \frac{2s + 3}{s(s + 1)^2} = \frac{k_1}{s} + \frac{k_{21}}{s + 1} + \frac{k_{22}}{(s + 1)^2}$$

The coefficients k_1 and k_{22} can easily be evaluated:

$$k_1 = sF(s)\big|_{s = 0} = \frac{2s + 3}{(s + 1)^2}\bigg|_{s = 0} = 3$$

$$k_{22} = (s + 1)^2 F(s)\big|_{s = -1} = \frac{2s + 3}{s}\bigg|_{s = -1} = -1$$

We use (B-13) to find k_{21}:

$$k_{21} = \frac{1}{(2 - 1)!} \frac{d}{ds}[(s + 1)^2 F(s)]\bigg|_{s = -1} = \frac{d}{ds}\left[\frac{2s + 3}{s}\right]\bigg|_{s = -1}$$

$$= \frac{s(2) - (2s + 3)(1)}{s^2}\bigg|_{s = -1} = \frac{-2 - 1}{1} = -3$$

Thus the partial-fraction expansion yields

$$F(s) = \frac{2s + 3}{s(s + 1)^2} = \frac{3}{s} + \frac{-3}{s + 1} + \frac{-1}{(s + 1)^2}$$

Then, from Table B.1,

$$f(t) = 3 - 3e^{-t} - te^{-t}$$

Example B.4

To illustrate the inverse transform of a function having complex poles, consider

$$F(s) = \frac{10}{s^3 + 4s^2 + 9s + 10} = \frac{10}{(s + 2)(s^2 + 2s + 5)} = \frac{10}{(s + 2)[(s + 1)^2 + 2^2]}$$

$$= \frac{k_1}{s + 2} + \frac{k_2}{s + 1 + j2} + \frac{k_2^*}{s + 1 - j2}$$

Evaluating the coefficient k_1 as before,

$$k_1 = (s + 2)F(s)\big|_{s = -2} = \frac{10}{(s + 1)^2 + 4}\bigg|_{s = -2} = \frac{10}{5} = 2$$

Coefficient k_2 is calculated from (B-15).

$$k_2 = (s + 1 + j2)F(s) = \left.\frac{10}{(s+2)(s+1-j2)}\right|_{s=-1-j2}$$

$$= \frac{10}{(-1-j2+2)(-1-j2+1-j2)} = \frac{10}{(1-j2)(-j4)}$$

$$= \frac{10}{(2.236 \;\underline{/-63.4°})(4 \;\underline{/-90°})} = 1.118\underline{/153.4°} = R\underline{/\theta}$$

Therefore, using (B-16),

$$f(t) = 2e^{-2t} + 2.236e^{-t}\cos(2t - 153.4°)$$

MATLAB performs a partial-fraction expansion for this example with the statements

```
Fnum = [0 0 0 10];
Fden = [1 4 9 10];
[r,p,k] = residue(Fnum,Fden)
```

A MATLAB program that finds the inverse transform is

```
SYMS F s t
F = 10/(s^3+4*s^2+9*s+10);
ilaplace (F,s,t)
```

B.2 THEOREMS OF THE LAPLACE TRANSFORM

The Laplace transform was defined in the last section. For the analysis and design of control systems, however, we require several theorems of the Laplace transform. As an example of the theorems, we derive the final-value theorem. We shall see later that this theorem is very useful in control system analysis and design.

Suppose that we wish to calculate the final value of $f(t)$, that is, $\lim_{t\to\infty} f(t)$. However, we wish to calculate this final value directly from the Laplace transform $F(s)$ without finding the inverse Laplace transform. The final-value theorem allows us to do this. To derive this theorem, it is first necessary to find the Laplace transform of the derivative of a general function $f(t)$.

$$\mathcal{L}\left[\frac{df}{dt}\right] = \int_0^\infty e^{-st}\frac{df}{dt}\,dt \qquad (B\text{-}17)$$

This expression can be integrated by parts, with

$$u = e^{-st} \qquad dv = \frac{df}{dt}\,dt$$

Thus

$$\mathcal{L}\left[\frac{df}{dt}\right] = uv\Big|_0^\infty - \int_0^\infty v\,du = f(t)e^{-st}\Big|_0^\infty + s\int_0^\infty e^{-st}f(t)\,dt \tag{B-18}$$

$$= 0 - f(0) + sF(s) = sF(s) - f(0)$$

To be mathematically correct, the initial-condition term must be $f(0^+)$ [4,5] where

$$f(0^+) = \lim_{t\to0} f(t) \qquad t > 0 \tag{B-19}$$

However, for convenience, we use the notation $f(0)$.

Now the final-value theorem can be derived. From (B-17),

$$\lim_{;\to0}\left[\mathcal{L}\left(\frac{df}{dt}\right)\right] = \lim_{s\to0}\int_0^\infty e^{-st}\frac{df}{dt}\,dt \tag{B-20}$$

$$= \int_0^\infty \frac{df}{dt}\,dt = \lim_{t\to\infty} f(t) - f(0)$$

Then, from (B-18) and (B-19),

$$\lim_{t\to\infty} f(t) - f(0) = \lim_{s\to0}\,[sF(s) - f(0)] \tag{B-21}$$

or,

$$\lim_{t\to\infty} f(t) = \lim_{s\to0} sF(s) \tag{B-22}$$

provided that the limit on the left side of this relationship exists. The right-side limit may exist without the existence of the left-side limit. For this case, the right side of (B-22) gives the incorrect value for the final value of $f(t)$.

Table B.2 lists several useful theorems of the Laplace transform. No further proofs of these theorems are given here; interested readers should see Refs. 4 and 5. An example of the use of these theorems is given next.

Example B.5

As an example of applying the theorems, consider the time function $\cos at$.

$$F(s) = \mathcal{L}[f(t)] = \mathcal{L}[\cos at] = \frac{s}{s^2 + a^2}$$

Then, from Table B.2,

$$\mathcal{L}\left[\frac{df}{dt}\right] = \mathcal{L}[-a\sin at] = sF(s) - f(0) = \frac{s^2}{s^2 + a^2} - 1 = \frac{-a^2}{s^2 + a^2}$$

which agrees with the transform from Table B.1. Also,

$$\mathcal{L}\left(\int_0^t f(\tau)\,d\tau\right) = \mathcal{L}\left(\frac{\sin at}{a}\right) = \frac{F(s)}{s} = \frac{1}{s^2 + a^2}$$

which also agrees with Table B.1. The initial value of $f(t)$ is

$$f(0) = \lim_{s\to\infty} sF(s) = \lim_{s\to\infty}\left[\frac{s^2}{s^2 + a^2}\right] = 1$$

TABLE B.2 LAPLACE TRANSFORM THEOREMS

Name	Theorem
Derivative	$\mathcal{L}\left[\dfrac{df}{dt}\right] = sF(s) - f(0^+)$
nth-Order derivative	$\mathcal{L}\left[\dfrac{d^n f}{dt^n}\right] = s^n F(s) - s^{n-1} f(0^+)$
	$\quad - \cdots f^{n-1}(0^+)$
Integral	$\mathcal{L}\left[\displaystyle\int_0^t f(\tau)\, d\tau\right] = \dfrac{F(s)}{s}$
Shifting	$\mathcal{L}\left[f(t - t_0)u(t - t_0)\right] = e^{-t_0 s} F(s)$
Initial value	$\displaystyle\lim_{t \to 0} f(t) = \lim_{s \to \infty} sF(s)$
Final value	$\displaystyle\lim_{t \to \infty} f(t) = \lim_{s \to 0} sF(s)$
Frequency shift	$\mathcal{L}\left[e^{-at} f(t)\right] = F(s + a)$
Convolution integral	$\mathcal{L}^{-1}[F_1(s)F_2(s)] = \displaystyle\int_0^t f_1(t - \tau) f_2(\tau)\, d\tau$
	$\quad = \displaystyle\int_0^t f_1(\tau) f_2(t - \tau)\, d\tau$

which, of course, is correct. If we carelessly apply the final-value theorem, we obtain

$$\lim_{t \to \infty} f(t) = \lim_{s \to 0} sF(s) = \lim_{s \to 0}\left[\frac{s^2}{s^2 + a^2}\right] = 0$$

which is incorrect, since cos at does not have a final value; the function continues to vary between 1 and -1 as time increases without bound. This last exercise emphasizes the point that the final-value theorem does not apply to functions that have no final value.

Example B.6

As a second example, we consider the time function $f(t) = e^{-0.5t}$, which is then delayed by 4 s. Thus the function that we consider is

$$f_1(t) = f(t - 4)u(t - 4) = e^{-0.5(t-4)}u(t - 4)$$

Both $f(t)$ and $f_1(t)$ are shown in Figure B.1. Note that $f(t)$ is delayed by 4 s and that the value of the delayed function is zero for time less than 4 s (the amount of the delay). Both of these conditions are necessary in order to apply the shifting theorem of Table B.2. From this theorem,

$$\mathcal{L}\left[f(t-t_0)u(t-t_0)\right] = e^{-t_0 s}F(s) \qquad F(s) = \mathcal{L}\left[f(t)\right]$$

For this example, the unshifted function is $e^{-0.5t}$, and thus $F(s) = 1/(s+0.5)$. Hence

$$\mathcal{L}\left[e^{-0.5(t-4)}u(t-4)\right] = \frac{e^{-4s}}{s+0.5}$$

Note that for the case that the time function is delayed, the Laplace transform is not a ratio of polynomials in s but contains the exponential function.

B.3 DIFFERENTIAL EQUATIONS AND TRANSFER FUNCTIONS

In control system analysis and design, the Laplace transform is used to transform constant-coefficient linear differential equations into algebraic equations. The algebraic equations are much easier to manipulate and analyze, simplifying the analysis of the differential equations. We generally model analog physical systems with linear differential equations with constant coefficients when possible (when the system can be accurately modeled by these equations). Thus the Laplace transform simplifies the analysis and design of analog linear systems.

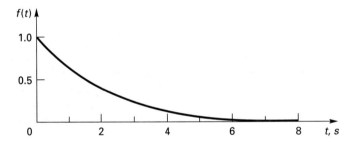

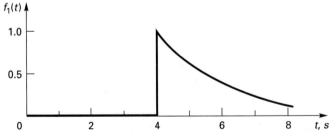

Figure B.1   Delayed time function.

An example of a linear differential equation modeling a physical phenomenon is Newton's law,

$$M\frac{d^2x(t)}{dt^2} = f(t) \tag{B-23}$$

where $f(t)$ is the force applied to a mass M, with the resulting displacement $x(t)$. It is assumed that the units in (B-23) are consistent. Assume that we know the mass M and the applied force $f(t)$. The Laplace transform of (B-23) is, from Table B.2,

$$M[s^2X(s) - sx(0) - \dot{x}(0)] = F(s) \tag{B-24}$$

where $\dot{x}(t)$ denotes the derivative of $x(t)$. Thus to solve for the displacement of the mass, we must know the applied force, the initial displacement, $x(0)$, and the initial velocity, $\dot{x}(0)$. Then we can solve this equation for $X(s)$ and take the inverse Laplace transform to find the displacement $x(t)$. We now solve for $X(s)$:

$$X(s) = \frac{F(s)}{Ms^2} + \frac{x(0)}{s} + \frac{\dot{x}(0)}{s^2} \tag{B-25}$$

For example, suppose that the applied force $f(t)$ is zero. Then the inverse transform of (B-25) is

$$x(t) = x(0) + \dot{x}(0)t \qquad t \geq 0 \tag{B-26}$$

If the initial velocity, $\dot{x}(0)$, is also zero, the mass will remain at its initial position $x(0)$. If the initial velocity is not zero, the displacement of the mass will increase at a constant rate equal to that initial velocity.

Note that if the initial conditions are all zero, (B-25) becomes

$$X(s) = \frac{1}{Ms^2}F(s) \tag{B-27}$$

Consider a physical phenomenon (system) that can be modeled by a linear differential equation with constant coefficients. The Laplace transform of the response (output) of this system can be expressed as the product of the Laplace transform of the forcing function (input) times a function of s (provided all initial conditions are zero), which we call the *transfer function*. We usually denote the transfer function by $G(s)$; for a mass, we see from (B-27) that the transfer function is

$$G(s) = \frac{1}{Ms^2} \tag{B-28}$$

An example is now given.

Example B.7

Suppose that a system is modeled by the differential equation

$$\frac{d^2x(t)}{dt^2} + 3\frac{dx(t)}{dt} + 2x(t) = 2f(t)$$

In this equation, $f(t)$ is the forcing function, or the input, and $x(t)$ is the response function (output). If we take the Laplace transform of this equation, we have

$$s^2X(s) - sx(0) - \dot{x}(0) + 3[sX(s) - x(0)] + 2X(s) = 2F(s)$$

Solving for this equation for the response $X(s)$,

$$X(s) = \frac{2F(s) + (s+3)x(0) + \dot{x}(0)}{s^2 + 3s + 2}$$

The transfer function is obtained by setting the initial conditions to zero.

$$G(s) = \frac{X(s)}{F(s)} = \frac{2}{s^2 + 3s + 2}$$

Suppose that we wish to find the response with no initial conditions and with the system input equal to a unit step function. Then $F(s) = 1/s$, and

$$X(s) = G(s)F(s) = \left[\frac{2}{s^2 + 3s + 2}\right]\left[\frac{1}{s}\right]$$

or

$$X(s) = \frac{2}{s(s+1)(s+2)} = \frac{1}{s} + \frac{-2}{s+1} + \frac{1}{s+2}$$

by partial-fraction expansion. The inverse transform of this expression is then

$$x(t) = 1 - 2e^{-t} + e^{-2t} \qquad t \geq 0$$

Note that $x(0)$ and $\dot{x}(0)$ are zero, as assumed, and that after a very long time, $x(t)$ is approximately equal to unity. This solution is verified with the MATLAB program

```
dsolve('D2x=-3*Dx-2*x+2, x(0)=0,Dx(0)=0')
```

In Example B.7 the response $X(s)$ can be expressed as

$$X(s) = G(s)F(s) + \frac{(s+3)x(0) + \dot{x}(0)}{s^2 + 3s + 2} = X_f(s) + X_{ic}(s) \qquad \text{(B-29)}$$

The term $X_f(s)$ is the *forced* (also called the *zero-state*) *response*, and the term $X_{ic}(s)$ is the *initial-condition* (*zero-input*) *response*. This result is general. We see then that the total response is the sum of two terms. The forcing-function term is independent of the initial conditions, and the initial-condition term is independent of the forcing function. This characteristic is a function of linear equations, and is discussed in Chapter 2 where the term *linear system* is defined.

The concept of a transfer function is basic to the study of feedback control systems. To generalize the results of the preceding paragraphs, let a system having an output $c(t)$ and an input $r(t)$ be described by the nth-order differential equation

$$\frac{d^n c}{dt^n} + a_{n-1}\frac{d^{n-1}c}{dt^{n-1}} + \cdots + a_1\frac{dc}{dt} + a_0 c$$

$$= b_m\frac{d^m r}{dt^m} + b_{m-1}\frac{d^{m-1}r}{dt^{m-1}} + \cdots + b_1\frac{dr}{dt} + b_0 r \qquad \text{(B-30)}$$

If we ignore all initial conditions, the Laplace transform of (B-30) yields

$$(s^n + a_{n-1}s^{n-1} + \cdots + a_1 s + a_0)C(s)$$

$$= (b_m s^m + b_{m-1}s^{m-1} + \cdots + b_1 s + b_0)R(s) \qquad \text{(B-31)}$$

Ignoring the initial conditions allows us to solve for $C(s)/R(s)$ as a rational function of s, namely,

$$\frac{C(s)}{R(s)} = \frac{b_m s^m + b_{m-1} s^{m-1} + \cdots + b_1 s + b_0}{s^n + a_{n-1} s^{n-1} + \cdots + a_1 s + a_0} \tag{B-32}$$

Note that the denominator polynomial of (B-32) is the coefficient of $C(s)$ in (B-31). The reader will recall from studying classical methods for solving linear differential equations that this same polynomial set equal to zero is the *characteristic equation* of the differential equation (B-30).

Since most of the physical systems that we encountered were described by differential equations, we frequently referred to the *characteristic equation* of the *system,* meaning, of course, the characteristic equation of the differential equation that described the system. It was apparent in Chapter 2 that the a_i coefficients in (B-30) were parameters of the physical system described by the differential equation, such as mass, friction coefficient, spring constant, inductance, and resistance. It follows, therefore, that the characteristic equation does indeed *characterize* the system, since its roots are dependent only upon the system parameters; these roots determine that portion of the system's response (solution) whose form does not depend upon the input $r(t)$. This part of the solution is, of course, the complementary solution of the differential equation.

REFERENCES

1. C. L. Phillips and J. M. Parr. *Signals, Systems, and Transforms,* 2nd ed. Upper Saddle River, NJ: Prentice Hall, 1999.

2. J. D. Irwin. *Basic Engineering Circuit Analysis,* 5th ed. Upper Saddle River, NJ: Prentice Hall, 1996.

3. W. A. Blackwell and L. L. Grigsby. *Introductory Network Theory.* Boston: Prindle, Weber & Schmidt, 1985.

4. G. Doetsch. *Guide to the Applications of the Laplace and z-Transforms.* New York: Springer-Verlag, 1970.

5. W. Kaplan. *Operational Methods for Linear Systems.* Reading, MA: Addison-Wesley, 1962.

PROBLEMS

B.1. Using the defining integral of the Laplace transform, (B-1), derive the Laplace tansform of
  (a) $f(t) = u(t - 2.5)$
  (b) $f(t) = e^{-4t}$
  (c) $f(t) = t$
  (d) Verify the results in (b) and (c) with MATLAB.

B.2. Use the Laplace transform tables to find the transforms of each function.
  (a) $f(t) = -3te^{-t}$
  (b) $f(t) = -5 \cos t$
  (c) $f(t) = t \sin 3t$
  (d) $f(t) = 7e^{-0.5t} \cos 3t$
  (e) $f(t) = 5 \cos (4t + 30°)$
  (f) $f(t) = 6e^{-2t} \sin (t - 45°)$
  (g) Verify all the results with MATLAB.

B.3. Find the inverse Laplace transform of each.

(a) $F(s) = \dfrac{5}{s(s+1)(s+2)}$

(b) $F(s) = \dfrac{1}{s^2(s+1)}$

(c) $F(s) = \dfrac{2s+1}{s^2+2s+10}$

(d) $F(s) = \dfrac{s-30}{s(s^2+4s+29)}$

(e) Verify all the results with MATLAB.

B.4. Given the Laplace transform

$$F(s) = \frac{s+5}{s^2+4s+13}$$

(a) Express the inverse transform as the sum of two complex exponentials.

(b) Using Euler's identity, manipulate the result of (a) into the form $f(t) = Be^{-at}\sin(bt+\Phi)$.

(c) Use the procedure of Section B.1 to express the inverse transform as $f(t) = Ae^{-at}\cos(bt+\Phi)$.

(d) Verify all the results with MATLAB.

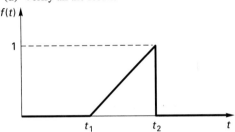

Figure PB.5

B.5. (a) Find and plot $f(t)$ if its Laplace transform is given by

$$F(s) = \frac{e^{-t_1 s} - e^{-t_2 s}}{s} \qquad t_2 > t_1$$

(b) The time function in part (a) is a rectangular pulse. Find the Laplace transform of the triangular pulse shown in Figure PB.5.

B.6. Given that $f(t) = 4e^{-2(t-3)}$

(a) Find $\mathcal{L}\,[df(t)/dt]$ by differentiating $f(t)$ and then using the Laplace transform tables.

(b) Find $\mathcal{L}\,[df(t)/dt]$ using the theorem for differentiation.

(c) Repeat (a) and (b), for the case that $f(t) = 4e^{-2(t-3)}u(t-3)$.

B.7. Find the Laplace transforms of

(a) $f_1(t) = 5e^{-2(t-1)}$

(b) $f_2(t) = 5e^{-2(t-1)}u(t-1)$

(c) Sketch the two time functions.

(d) Why are the two transforms different?

B.8. For the functions of Problem B.3,

(a) Which inverse Laplace transforms do not have final values; that is, for which of the inverse transforms do the $\lim_{t\to\infty} f(t)$ not exist?

(b) Find the final values for those functions that have final values.

B.9. Given the differential equation

$$\frac{d^2x}{dt^2} + 5\frac{dx}{dt} + 4x = 10u(t)$$

(a) Find $x(t)$ for the case that the initial conditions are zero.
(b) Find $x(t)$ for the case that $x(0) = 1$ and $\dot{x}(0) = 1$. Show that your solution yields the correct initial conditions, that is, solve for $x(0)$ and $\dot{x}(0)$ using your solution.
(c) Verify all the results with MATLAB.

B.10. Given the differential equation

$$\frac{d^2x}{dt^2} + 2\frac{dx}{dt} + x = 5 \qquad t \ge 0$$

(a) Find $x(t)$ for the case that the initial conditions are zero.
(b) Find $x(t)$ for the case that $x(0) = 0$ and $\dot{x}(0) = 2$. Show that your solution yields the correct initial conditions; that is, solve for $x(0)$ and $\dot{x}(0)$ using your solution.
(c) Verify all the results with MATLAB.

B.11. For each of the systems, find the system differential equation if $G(s) = C(s)/R(s)$ is given by

(a) $G(s) = \dfrac{60}{s^2 + 10s + 60}$

(b) $G(s) = \dfrac{3s + 20}{s^2 + 4s^2 + 8s + 20}$

(c) $G(s) = \dfrac{s + 1}{s^2}$

(d) $G(s) = \dfrac{7e^{-0.2s}}{s^2 + 5s + 32}$

B.12. Give the characteristic equations for the systems of Problem B.11.

C

Laplace Transform and z-Transform Tables

Time function $e(t)$	Laplace transform $E(s)$	z-Transform $E(z)$
$u(t)$	$\dfrac{1}{s}$	$\dfrac{z}{z-1}$
t	$\dfrac{1}{s^2}$	$\dfrac{Tz}{(z-1)^2}$
$\dfrac{t^2}{2}$	$\dfrac{1}{s^3}$	$\dfrac{T^2 z(z+1)}{2(z-1)^3}$
t^{k-1}	$\dfrac{(k-1)!}{s^k}$	$\displaystyle \lim_{a\to 0}(-1)^{k-1}\dfrac{\partial^{k-1}}{\partial a^{k-1}}\left[\dfrac{z}{z-e^{-aT}}\right]$
e^{-at}	$\dfrac{1}{s+a}$	$\dfrac{z}{z-e^{-aT}}$
te^{-at}	$\dfrac{1}{(s+a)^2}$	$\dfrac{Tze^{-aT}}{(z-e^{-aT})^2}$
$t^{k-1}e^{-at}$	$\dfrac{(k-1)!}{(s+a)^k}$	$(-1)^k\dfrac{\partial^k}{\partial a^k}\left[\dfrac{z}{z-e^{-aT}}\right]$
$1-e^{-at}$	$\dfrac{a}{s(s+a)}$	$\dfrac{z(1-e^{-aT})}{(z-1)(z-e^{-aT})}$
$t-\dfrac{1-e^{-at}}{a}$	$\dfrac{a}{s^2(s+a)}$	$\dfrac{z[(aT-1+e^{-aT})z+(1-e^{-aT}-aTe^{-aT})]}{a(z-1)^2(z-e^{-aT})}$
$1-(1+at)e^{-at}$	$\dfrac{a^2}{s(s+a)^2}$	$\dfrac{z\{[1-e^{-aT}-aTe^{-aT}]z+[e^{-2aT}+(aT-1)e^{-aT}]\}}{(z-1)(z-e^{-aT})^2}$
$e^{-at}-e^{-bt}$	$\dfrac{b-a}{(s+a)(s+b)}$	$\dfrac{(e^{-aT}-e^{-bT})z}{(z-e^{-aT})(z-e^{-bT})^2}$
$\sin bt$	$\dfrac{b}{s^2+b^2}$	$\dfrac{z\sin bT}{z^2-2z\cos bT+1}$
$\cos bt$	$\dfrac{s}{s^2+b^2}$	$\dfrac{z(z-\cos bT)}{z^2-2z\cos bT+1}$
$t\sin bt$	$\dfrac{2bs}{(s^2+b^2)^2}$	
$t\cos bt$	$\dfrac{s^2-b^2}{(s^2+b^2)^2}$	
$e^{-at}\sin bt$	$\dfrac{b}{(s+a)^2+b^2}$	$\dfrac{ze^{-aT}\sin bT}{z^2-2ze^{-aT}\cos bT+e^{-2aT}}$
$e^{-at}\cos bt$	$\dfrac{s+a}{(s+a)^2+b^2}$	$\dfrac{z^2-ze^{-aT}\cos bT}{z^2-2ze^{-aT}\cos bT+e^{-2aT}}$
$1-e^{-at}\left(\cos bt+\dfrac{a}{b}\sin bt\right)$	$\dfrac{a^2+b^2}{s[(s+a)^2+b^2]}$	$\dfrac{z(Az+B)}{(z-1)(z^2-2ze^{-aT}\cos bT+e^{-2aT})}$

$$A = 1-e^{-aT}\left[\cos bT+\left(\dfrac{a}{b}\right)\sin bT\right]$$

$$B = e^{-2aT}+e^{-aT}\left[\left(\dfrac{a}{b}\right)\sin bT-\cos bT\right]$$

Index